Applied and Numerical Harmonic Analysis

Series Editor
John J. Benedetto
University of Maryland

Applied and Numerical Harmonic Analysis

Published titles

J.M. Cooper: *Introduction to Partial Differential Equations with MATLAB*
(ISBN 0-8176-3967-5)

C.E. D'Attellis and E.M. Fernández-Berdaguer: *Wavelet Theory and Harmonic Analysis in Applied Sciences* (ISBN 0-8176-3953-5)

H.G. Feichtinger and T. Strohmer: *Gabor Analysis and Algorithms*
(ISBN 0-8176-3959-4)

T.M. Peters, J.H.T. Bates, G.B. Pike, P. Munger, and J.C. Williams: *Fourier Transforms and Biomedical Engineering* (ISBN 0-8176-3941-1)

A.I. Saichev and W.A. Woyczyński: *Distributions in the Physical and Engineering Sciences*
(ISBN 0-8176-3924-1)

R. Tolimierei and M. An: *Time-Frequency Representations* (ISBN 0-8176-3918-7)

G.T. Herman: *Geometry of Digital Spaces* (ISBN 0-8176-3897-0)

A. Procházka, J. Uhlíř, P.J.W. Rayner, and N.G. Kingsbury: *Signal Analysis and Prediction*
(ISBN 0-8176-4042-8)

J. Ramanathan: *Methods of Applied Fourier Analysis* (ISBN 0-8176-3963-2)

A. Teolis: *Computational Signal Processing with Wavelets* (ISBN 0-8176-3909-8)

W.O. Bray and Č.V. Stanojević: *Analysis of Divergence* (ISBN 0-8176-4058-4)

G.T. Herman and A. Kuba: *Discrete Tomography* (ISBN 0-8176-4101-7)

J.J. Benedetto and P.J.S.G. Ferreira: *Modern Sampling Theory*
(ISBN 0-8176-4023-1)

A. Abbate, C.M. DeCusatis, and P.K. Das: *Wavelets and Subbands*
(ISBN 0-8176-4136-X)

L. Debnath: *Wavelet Transforms and Time-Frequency Signal Analysis*
(ISBN 0-8176-4104-1)

K. Gröchenig: *Foundations of Time-Frequency Analysis* (ISBN 0-8176-4022-3)

D.F. Walnut: *An Introduction to Wavelet Analysis* (ISBN 0-8176-3962-4)

O. Bratelli and P. Jorgensen: *Wavelets through a Looking Glass* (ISBN 0-8176-4280-3)

H. Feichtinger and T. Strohmer: *Advances in Gabor Analysis* (ISBN 0-8176-4239-0)

O. Christensen: *An Introduction to Frames and Riesz Bases* (ISBN 0-8176-4295-1)

L. Debnath: *Wavelets and Signal Processing* (ISBN 0-8176-4235-8)

J. Davis: *Methods of Applied Mathematics with a MATLAB Overview* (ISBN 0-8176-4331-1)

G. Bi and Y. Zeng: *Transforms and Fast Algorithms for Signal Analysis and Representations*
(ISBN 0-8176-4279-X)

(Continued after the Index)

Guoan Bi
Yonghong Zeng

Transforms and Fast Algorithms for Signal Analysis and Representations

Birkhäuser
Boston • Basel • Berlin

Guoan Bi
Nanyang Technical University
School of Electrical and Electronic Engineering
Singapore 63798
Singapore

Yonghong Zeng
The University of Hong Kong
Department of Electrical and Electronic Engineering
Hong Kong

Library of Congress Cataloging-in-Publication Data

Bi, Guoan, 1954-
    Transforms and fast algorithms for signal analysis and representations / Guoan Bi,
Yonghong Zeng.
        p. cm. — (Applied and numerical harmonic analysis)
    Includes bibliographical references and index.
    ISBN 0-8176-4279-X (alk. paper) – ISBN 3-7643-4279-X (alk. paper)
        1. Signal processing–Mathematics. 2. Transformations (Mathematics) 3. Algorithms.
I. Zeng, Yonghong, 1962- II. Title. III. Series.

TK5102.9 .B52 2003
621.382'2'01515723–dc21   2002035634

AMS Subject Classifications: 42B10, 65T50, 42C20, 11T06

Printed on acid-free paper
©2004 Birkhäuser Boston

Birkhäuser ®

ISBN 0-8176-4279-X   SPIN 10851283
ISBN 3-7643-4279-X

Typeset by the authors.
Printed in the United States of America.

9 8 7 6 5 4 3 2 1

Birkhäuser Boston • Basel • Berlin
A member of BertelsmannSpringer Science+Business Media GmbH

Contents

List of Figures

List of Tables

Preface

... that is what learning is. You suddenly understand something you've understood all your life, but in a new way.

Various transforms have been widely used in diverse applications of science, engineering and technology. New transforms are emerging to solve many problems, which may have been left unsolved in the past, or newly created by modern science or technologies. Various methods have been continuously reported to improve the implementation of these transforms. Early developments of fast algorithms for discrete transforms have significantly stimulated the advance of digital signal processing technologies. More than 40 years after fast Fourier transform algorithms became known, several discrete transforms, including the discrete Hartley transform and discrete cosine transform, were proposed and widely used for numerous applications. Although they all are related to the discrete Fourier transform, different fast algorithms and their implementations have to be separately developed to minimize computational complexity and implementation costs. In spite of the tremendous increase in the speed of computers or processors, the demands for higher processing throughout seemingly never ends. Fast algorithms have become more important than ever for modern applications to become a reality. Many new algorithms recently reported in the literature have led to important improvements upon a number of issues, which will be addressed in this book.

Some discrete transforms are not suitable for signals that have time-varying frequency components. Although several approaches are available for such applications, various inherent problems still remain unsolved. Methods of time-frequency signal analysis have been actively researched for many years. The use of existing and new transforms has been studied and reported in the literature. For example, more theoretical results on the fractional Fourier transform and its numerical calculations can be found in many publications. New transforms such as the harmonic transform and tomographic time-frequency transform have also been reported for the time-frequency analysis of particular types of signals. These transforms and their fast algorithms have a high potential for use in many applications.

Several books have specialized in various transforms and fast algorithms. However, books containing recent developments, including advances in transforms and their fast algorithms, have not been available. This book is intended to cover the development of transforms and their fast algorithms based on the research work done by the authors in the School of Electrical and Electronic Engineering, Nanyang Technological University. In many cases, new options or flexibilities provided by some improved or recently developed fast algorithms for the discrete Fourier transform, discrete Hartley transform, and discrete cosine transform have not been well known in the community of digital signal processing. Integer transforms have been recognized to be important for supporting wireless applications because they can well approximate these discrete linear transforms with a much reduced computational complexity. Because integer transforms currently exist only for some special cases, it is important to realize that the concepts of integer transforms can be generalized for all these discrete transforms of various sequence lengths. This book provides the general groundwork to derive integer transforms by using the concept of lifting steps. Newly developed transforms for time-frequency analysis are also introduced in this book. We propose some transforms whose kernel functions coherently include time-varying components so that problems encountered by other approaches can be avoided.

The book is intended for students who have acquired knowledge of advanced digital signal processing, practicing engineers who must access the latest signal processing literature, and researchers who apply the theory of signal analysis and processing in their applications. Therefore, this book can be used as a textbook for senior undergraduates and postgraduates, and as a reference book for engineers and applied mathematicians whose studies or work are related to the fields of electrical engineering, electronics, signal processing, image and speech processing, digital design, and communications.

Acknowledgments. The authors would like to thank the School of Electrical and Electronic Engineering, Nanyang Technological University, Singapore, because most of the materials presented in this book are the research outputs of the authors' work there. Many colleagues, friends and students have been helpful in the preparation of this book. Special thanks to Dr. Feng Zhang, who made a special contribution in his Ph.D research on which the content of Chapter 9 is based.

The authors have been very fortunate to have worked with Birkhäuser on this book.

Transforms and Fast Algorithms
for Signal Analysis
and Representations

1

Introduction

1.1 DISCRETE LINEAR TRANSFORMS

Many fundamental physical phenomena, such as heat, air pressure, temperature, electrical voltage or current and electromagnetic field, change in time. They are generally known as waveforms or signals which are described by mathematical functions of time. Most of these signals can be described as a series that is a linear combination of orthogonal basis functions $\varphi_k(n)$ $(k = \cdots, -2, -1, 0, 1, 2, \cdots)$, where the basis functions are in a discrete format. If these basis functions are in a complete orthogonal set, these signals can be accurately defined. For example, a signal $x(n)$, where n is the data index in the time domain, can be expressed as

$$x(n) = \sum_{k=-\infty}^{\infty} C_k\, \varphi_k(n), \tag{1.1}$$

where C_k $(k = \cdots, -2, -1, 0, 1, 2, \cdots)$ are known as coefficients, which can be calculated using integrals, and $\varphi_k(n)$ is the transform kernel that contains a set of orthogonal basis functions. The generalized discrete Fourier transform, for example, describes signal $x(n)$ in terms of sinusoids of different frequencies. The coefficients of the generalized discrete Fourier transform of input sequence $x(n)$ are computed as

$$X(k) = \frac{1}{N} \sum_{n=0}^{N-1} x(n) e^{\frac{-j\pi(2n+n_0)(2k+k_0)}{2N}}, \tag{1.2}$$

where $e^{\frac{-j\pi(2n+n_0)(2k+k_0)}{2N}}$ is the transform kernel, and the constants n_0 and k_0 are the parameters that specify the shifts in the time and frequency domains. By using a different set of n_0 and k_0 values, a different type of discrete Fourier transform can be defined. Because signal $x(n)$ can be precisely recovered from $X(k)$, we can also say that signal $x(n)$ is completely described by the set of coefficients $X(k)$ in a different domain. In many engineering applications, it is common to describe or solve a problem in a suitable domain because some

1

characteristics of the signal or physical phenomena can only be better revealed in a particular domain. For example, the propagation characteristics of light or magnetic waves can
be conveniently described or observed in the frequency domain. The Fourier transform and
its related transforms are widely used as mathematical tools to solve problems in science
and engineering, including linear systems analysis, antenna studies, optics, random process
modeling, probability theory, quantum physics and boundary-value problems. The Fourier
transform is also used as a physical tool to alter a problem into one that can be more easily
solved. Some scientists understand Fourier theory as a physical phenomenon, not simply
as a mathematical tool. In some branches of science, the Fourier transform of one function
may yield another physical function [2].

Mathematical tools for linear transforms have steadily advanced since Joseph Fourier
developed the Fourier transform for representation of aperiodic signals at the beginning of
the nineteenth century. One direction of the development is to find various transforms, each
of which may be particularly useful for dealing with one or a few types of signals. A few
good textbooks (for example, [2, 5, 6]) provide excellent materials on the topic of generalized
transforms. Here we mention a few examples that are related to the contents of this book.
As seen in (1.1), a transform is defined by suitable selection of the transform kernel. If the
transform kernel $cas(\ ) = \cos(\ ) + \sin(\ )$ is used in (1.1), the generalized discrete Hartley
transform of signal $x(n)$ can be defined as

$$X(k) = \frac{1}{\sqrt{N}} \sum_{n=0}^{N-1} x(n) cas \frac{\pi(2n_1 + n_0)(2k_1 + k_0)}{2N}, \tag{1.3}$$

where n_0 and k_0 are the parameters that define the different types of discrete Hartley
transforms. Although the generalized discrete Hartley transform shares similar properties
with the Fourier transform, the Hartley transform does not use any complex quantities
and is particularly useful for processing real signals. Another difference from the Fourier
transform, as seen in the latter chapters, is that the Hartley transform and its inverse use the
same kernel function so that the same software or hardware implementation of the Hartley
transform can be used for signal analysis or synthesis. The Hartley transform was originally
designed for steady-state and transient analysis of telephone transmission problems and has
been used in diversified areas of applications, including general signal processing, acoustic
and speech signal processing, multidimensional optics and image processing, biomedical
engineering, digital communications, *etc.* [7].

Another example is the discrete cosine transform which has been widely used in many
fields of applications. It is defined by

$$X(k) = \sum_{n=0}^{N-1} x(n) \cos \frac{\pi(2n + n_0)(2k + k_0)}{4N}, \tag{1.4}$$

where the normalization factor in the definition is omitted for simplicity, and again the values
of parameters n_0 and k_0 define the types of discrete cosine transform. The discrete cosine
transform has been found to be superior to other transforms for data compression problems
because it is asymptotically equivalent to the Karhunen–Loeve transform in terms of data
compression performance. In addition, the availability of fast algorithms allows the discrete
cosine transform to readily replace the Karhunen–Loeve transform in nearly all applications
that involve data compression.

1.2 FAST ALGORITHMS

For about the past 40 years, many fast algorithms have been reported for the computation of discrete Fourier transforms, discrete Hartley transforms and discrete cosine transforms. Before the 1960s, it was not possible for many cases to compute the discrete Fourier transform of a reasonable size because the prohibitive computational load could not be supported by the digital computers which were still at their infant stage. This problem has been partially solved by the fast algorithms that were reported by a few pioneers [3, 4, 8]. Based on the symmetry and periodicity properties of the transform kernel, fast Fourier transform algorithms were derived to significantly reduce the computational complexity from the order of $O(N^2)$ needed by a direct computation of a one-dimensional (1D) discrete Fourier transform of N points to the order of $O(N\log N)$ needed by the fast algorithms. With today's computer technologies and fast algorithms, the computational complexity for 1D discrete Fourier transforms required by most applications can be supported at an affordable cost.

Although the discrete Hartley transform and discrete cosine transform are closely related to the discrete Fourier transform, different fast algorithms and their implementations have to be developed to minimize the computational complexity and implementation costs. Early developments of fast algorithms for various transforms have significantly stimulated the advance of digital signal processing for numerous applications. The recent success in communication technologies also creates many multimedia applications that are primarily supported by the advance of digital signal processing technologies. Although the operating speeds of computers or processors have been increased tremendously, the demands for higher processing throughput seemingly never stop. In particular, multidimensional transforms are widely used for real time applications, which requires a formidable computational complexity within a short time interval. Fast algorithms have become more important than ever for modern applications to become a reality.

Some people believe that the most computationally efficient fast algorithms are those designed for the transform size being a power of two because the length 2 and length 4 transforms, which are the basic computational blocks in the fast algorithms, do not need any multiplications. However, R. E. Blahut pointed out in his book (page 19, [1]) that *one unfortunate consequence of the popularity of the Cooley–Tukey FFT is the widespread belief that the discrete Fourier transform is practical only if the block length is a power of two. This belief tends to result in the FFT algorithms dictating the design parameters of an application rather than the applications dictating the choice of FFT algorithm. In fact, there are good FFT algorithms for just about any block length.* Today, we can still observe this tendency of using a power of two in many areas, from the FIR filter length selected by some experienced engineers to the number of channels for future communication systems reported in well-known journal papers. When the required transform size cannot be supported, a zero padding technique, which adds a number of zero valued data, is used to increase the length of the data sequence to match the transform size that is supported by the fast algorithm. Such a practice generally decreases the overall system efficiency because a portion of the available processing power is wasted on zero valued data. By looking into the literature of fast algorithm development for linear transforms, one finds that simple and efficient fast algorithms for transform size rather than a power of two are the minority. These reported fast algorithms for transform sizes rather than a power of two appear to be more complex and require more computational complexity.

One objective of this book is to introduce fast algorithms for the discrete Fourier transform, discrete Hartley transform and discrete cosine transform. Most of the presented algorithms have a relatively simple computational structure, require a computational complexity that is not more than that needed for the sequency length being a power of two and support a wider choice of sequence lengths. These fast algorithms can be generally divided into two categories according to the method of derivation. One category includes 1D and multidimensional fast algorithms that are developed by using the symmetry and periodicity properties of the transform kernel. The fast algorithms in the other category are derived with the help of polynomial transforms for multidimensional linear transforms. Chapters 3 through 7 are devoted to the recently developed fast algorithms in these two categories.

1.3   NEW TRANSFORMS

The other direction of development is to find new transforms that have some special capabilities to meet new requirements for current and future applications. To further reduce the overall computational complexity and implementation cost, integer transforms have been in demand for wireless applications. Because of the limitations of current technologies, it is generally difficult to design power limited portable devices that can provide sufficient and complex signal processing functions. Because integer transforms do not use floating point multiplications, the computational complexity of the transforms can be substantially reduced if fast algorithms are available. By using lifting steps and matrix decomposition techniques, it has been reported that Fourier-related transforms can be converted into integer transforms. However, a few fast algorithms for integer transforms have only been avaliable recently to support limited applications. More general and powerful fast algorithms are definitely needed to support both wireless and general applications. Chapter 8 of this book provides useful materials for the development of integer transforms and their fast algorithms. These fast algorithms are general and support 1D and multidimensional integer Hartley and cosine transforms. Performance comparisons with conventional transforms are also provided.

Different transforms were reported based on various definitions of kernel functions. A good introduction with sufficient depth of development in theory and applications was given in [7]. Each transform has its own set of properties and therefore is suited to a particular category of signals only. It is often found that some of the transforms, when applied to solve complex problems, have their inherent shortcomings in dealing with certain types of signals. For example, the Fourier transform often fails to provide meaningful information for time-varying signals which contain frequency components that change in time. Time-varying (or non-stationary) signals have been extensively used to describe many natural phenomena. It can be easily understood that Fourier-related transforms were not capable of dealing with non-stationary signals because the kernel functions do not have any facility to describe the changes of frequencies in time. A number of time-frequency transforms are reported to analyze and/or synthesize various signals in both the time and frequency domains simultaneously. According to the ways of accommodating the nature of varying frequency, some time-frequency transforms can be loosely classified into the window-based category. Typical examples are the short-time Fourier transform and a few of the modified Wigner distributions, such as the pseudo Wigner distribution and smoothed Wigner distribution. Although the window functions are used under different assumptions and for different purposes, these functions all have the shortcoming of undesirable effects on the useful signal components in the time-frequency distribution. The resolution in time and/or frequency is often reduced because the window unavoidably suppresses a part of the information in the signal. Based on the uncertainty principle, it seems difficult to solve the problem associated with the window-based time-frequency transform.

In the other category, the time-frequency transforms have their kernel functions that inherently describe the time-varying nature of the signal. One example is the fractional Fourier transform whose kernel function has a variable α to rotate the time-frequency distribution in the time-frequency plane. When $\alpha = \pi/2$, the fractional Fourier transform is the Fourier transform. In Chapter 9 of this book, two time-frequency transforms of this type are introduced. The first one is the harmonic transform that is particularly designed for signals containing a fundamental and harmonics, such as music and voiced speech. This transform uses a kernel function $e^{j\omega\phi(t)}$, where $\phi(t)$ is related to the phase of the fundamental. It is noted that $\phi(t)$ is a function of time, which is different from the kernel of Fourier-related transforms which do not change in time. If $\phi(t)$ is known, we can perform the linear integral of the signal along the phase direction to produce the corresponding time-frequency distribution with a better resolution. The other one, known as the tomographic time-frequency transform, is developed based on the fact that the linear integral of the signal in the time-frequency plane along the time axis is the Fourier transform of the signal. Because the fractional Fourier transform can be considered as the linear integral of the signal's time-frequency distributions along all directions, we can use the inverse Radon transform to obtain the time-frequency distribution of the signal. Because no window function is explicitly involved, the resolution in time and frequency depends on the accuracy of the inverse Radon transform. Chapter 9 provides the details of these new time-frequency transforms and examples showing that substantial improvements on the resolution in time and frequency can be made compared to other time-frequency analysis methods.

1.4   ORGANIZATION OF THE BOOK

This book presents recent developments on widely used discrete transforms and their fast algorithms. Most materials of this book are the research results obtained by the authors at the school of Electrical and Electronic Engineering, Nanyang Technological University, Singapore. The book is intended not only for students who are acquiring the knowledge of digital signal processing, but also for practicing engineers who must access the signal processing literature, and for researchers who must apply the theory of signal analysis and processing. Therefore, this book can be used as a textbook for senior undergraduates and postgraduates, and as a reference book for engineers and applied mathematicians whose studies or work are related to the fields of electrical, electronics, signal processing, image and speech processing, digital design and communications.

In particular, the following special features make the book different from others on the market.

- A large number of recently developed fast algorithms are presented for 1D or multidimensional linear discrete transforms. In particular, these fast algorithms support transforms of composite transform sizes without increasing the computational complexity.

- Recent developments on multidimensional integer transforms and their fast algorithms are included to support wireless applications.

- New theoretical developments are used for polynomial transform-based fast algorithms.

- New time-frequency transforms, such as the harmonic transform and tomographic time-frequency transform, are provided. The harmonic transform is particularly designed for signals containing rich harmonics, and the tomographic time-frequency transform avoids the window effect to achieve high resolution in the time and frequency domains.

Because there are too many different fast algorithms in the open literature, it is not possible to cover all aspects in the area of transforms and fast algorithms. This book is based on the research work that further refines or complements the early reported fast algorithms. In many cases, new options or flexibilities are provided by the improved or newly developed fast algorithms, which have not been well known in the community of digital signal processing.

This book contains nine chapters. Each chapter focuses on the fast algorithms of a particular transform, or the development of new transforms. After the introduction, Chapter 2 introduces the concept of polynomial transforms that will be frequently used in the following chapters for the development of polynomial transform-based fast algorithms. After some useful theorems are presented, fast algorithms are presented for 1D and two-dimensional (2D) polynomial transforms.

Chapter 3 focuses on fast algorithms for the discrete Fourier transform (DFT). In Section 3.2, the well-known radix-2 and split-radix algorithms are used as examples to introduce some basic concepts that are essential in the development of fast algorithms for various transforms. To make improvements on the flexibility of supporting various sequence lengths, Section 3.3 presents generalized algorithms for 1D DFTs. It can be shown that the well-known split-radix algorithms reported in the early literature (or discussed in Section 3.1) are a special case of the generalized algorithms. Section 3.4 presents prime factor algorithms that achieve improvements on the regularity of the index mapping process and support in-place computation. The concepts used in Section 3.3 are extended for fast algorithms of 2D DFTs in Section 3.5. Section 3.6 presents the fast algorithms of generalized DFTs whose kernel functions are shifted in time and frequency. Finally, polynomial transform-based fast algorithms for 2D DFTs are provided in Section 3.7.

Chapter 4 deals with fast algorithms for the 1D discrete Hartley transform (DHT). After the introduction in Section 4.1, split radix 2/4 and 3/9 algorithms are described in Section 4.2. It is shown in Section 4.3 that the split-radix algorithms can be generalized to deal with different sequence lengths. Section 4.4 discusses radix-2 algorithms for type-II, -III and -IV DHTs. Section 4.5 describes prime factor algorithms which provide improvements on the complexity of index mapping and in-place computation. Section 4.6 presents radix-q algorithms to decompose the entire computation into smaller ones based on odd factors of the sequence lengths. Finally, Section 4.7 shows that the generalized DHT can be efficiently computed by using the type-I DHTs.

Chapter 5 is devoted to fast algorithms for multidimensional (MD) DHTs. By extending some of the concepts used for 1D cases considered in Chapter 4, we first derive the generalized split-radix algorithms for 2D type-I DHTs in Section 5.2. By decomposing the computation task into subtasks, Sections 5.3 and 5.4 present two different decomposition methods to derive fast algorithms for the type-II, -III and -IV 2D DHTs. Based on polynomial transforms, Sections 5.5 and 5.6 present fast algorithms for multidimensional type-II and type-III DHTs, respectively. Finally, the polynomial transform-based algorithms are described in Sections 5.7 and 5.8 for multidimensional type-I and -II DHTs, respectively.

Fast algorithms for the 1D discrete cosine transform (DCT) are presented in Chapter 6. Section 6.2 considers the radix-2 algorithms for type-II, -III and -IV DCTs. Section 6.3 presents prime factor algorithms that achieve desirable improvements on the index mapping and in-place computation. In Section 6.4, radix-q algorithms are derived to decompose the entire computation task when the transform sizes contain odd factors. Based on the odd factors in the transform sizes, Section 6.5 presents other decomposition methods that allow us to compute the type-II and -III DCTs in terms of the type-I DCT and the discrete sine transform (DST). A few appendices are also included for implementation of these fast algorithms.

Fast algorithms for MD DCTs are presented in Chapter 7. The algorithms based on odd factors in the transform sizes are derived for 2D type-I, -II and -III DCTs in Section 7.2. The concept of prime factor algorithms is extended for MD type-II and -III DCTs in Section 7.3. Then polynomial transform-based algorithms for MD type-II and -III DCTs with sizes 2^m are developed in Sections 7.4 and 7.5, respectively. The polynomial transform-based algorithms for MD type-II and -III DCTs with sizes q^m are presented in Sections 7.6 and 7.7, respectively.

Integer transforms generally exist only for some special cases in the literature. Chapter 8 generalizes the concept of integer transforms for the DCT and DHT. The integer transforms have a good potential to be used for applications in which reduction of computational complexity and power consumption is critical. Section 8.2 briefs the basic concepts of lifting steps that are used to derive the integer transforms. Section 8.3 provides the definitions and derivation of the integer DCTs. Fast algorithms are also detailed this sections. Similarly, Section 8.4 is devoted to integer DHTs and their fast algorithms. Sections 8.5 and 8.6 derive the MD integer DCT and DHT, respectively. Analysis and comparison on computational complexity and approximation performance are also given.

Chapter 9 presents two recently developed transforms for time-frequency signal analysis and representation. Some basic concepts of time-frequency transforms are quickly reviewed in Section 9.2. The main properties and the drawbacks of the existing time frequency transforms are also discussed. Section 9.3 introduces the harmonic transform that is particularly designed for signals containing rich time-varying harmonics. By adding time-varying components into the kernel definition, the harmonic transform avoids a few drawbacks of other time-frequency transforms and increases the analysis resolution. Based on the marginal condition of the general time-frequency distribution and the physical interpretation of the fractional Fourier transform, the tomographic time-frequency transform is described in Section 9.4 Both theoretical development and discrete calculation by using the fractional Fourier transform are given. Examples are also provided to show the improvements in the resolution in the time and frequency domains compared with other reported time-frequency transforms.

REFERENCES

1. R. E. Blahut, *Fast Algorithms for Digital Signal Processing*, Addison-Wesley Publishing Company, Reading, MA, 1985.

2. R. N. Bracewell, *The Fourier Transform and its Applications*, McGraw-Hill Higher Education, New York, 2000.

3. J. W. Cooley and J. W. Tukey, An algorithm for the machine calculation of complex Fourier series, *Math. Comp.*, vol. 9, no. 90, 297 – 301, 1965.

4. I. J. Good, The interaction algorithm and practical Fourier analysis, *J. Roy. Statist. Society*, S. V, vol. 20, 361 – 375, 1958 and vol. 22, 372 – 375, 1960.

5. D. F. Elliott and K. R. Rao, *Fast Transforms – Algorithms, Analyses, Applications*, Academic Press Inc., New York, 1982.

6. O. Ersoy, *Fourier-related Transforms, Fast Algorithms and Applications*, Prentice-hall International, Inc., Upper Saddle River, NJ, 1997.

7. A. D. Poularikas, *The Transforms and Applications Handbook, IEEE Press*, Washington, D. C., 1996.

8. C. M. Rader, Discrete Fourier transform when the number of data samples is prime, *Proc. IEEE*, vol. 56, 1107 – 1108, 1968.

2

Polynomial Transforms and Their Fast Algorithms

This chapter presents some essential and important concepts that are frequently used in the theory of polynomial transforms and fast algorithms.

- Section 2.1 presents basic definitions and theorems of the number theory. These are essentially important for understanding the polynomial transforms and fast algorithms.

- Section 2.2 extends the definitions and concepts given in Section 2.1 into the polynomial theory. In particular, the concepts of congruence and the Chinese remainder theorem for polynomials are introduced.

- Section 2.3 presents the one-dimensional (1D) polynomial transform, the formation of the transform and its associated properties.

- Section 2.4 discusses the fast algorithms for the 1D polynomial transform. In particular, the radix-2 and radix-p algorithms are provided.

- Section 2.5 considers the multidimensional (MD) polynomial transform and its associated fast algorithms.

It is hoped that readers who are not familiar with the number theory, polynomial theory and polynomial transform may quickly understand these essential concepts that are frequently used in the following chapters.

The polynomial transform (PT) was first proposed by H. J. Nussbaumer in 1978 [10, 11] as a tool for computing the two-dimensional (2D) discrete Fourier transform (DFT) and convolution. Since then, extensive research on its theory and applications has been done [2, 4, 5]. By using the PT, for example, many new fast algorithms for various types of discrete Hartley transforms (DHTs) and discrete cosine transforms (DCTs) have been reported in the literature [3, 12, 16, 17, 18, 19, 20, 21] to minimize the required number of arithmetic operations. Theoretically, the PT can be viewed as a generalized DFT defined on polynomial modulo rings. Therefore, the well-known fast Fourier transform (FFT) algorithm can be easily generalized to the fast polynomial transform (FPT) algorithm which will be discussed

9

in this chapter. Since we will use the PT in the following chapters, this chapter provides the definition, construction, property and fast algorithms of the PTs. Some basic knowledge on the number theory and polynomial theory, which are essential for understanding the PT and other chapters of this book, is also introduced.

2.1 BASIC NUMBER THEORY

2.1.1 Divisibility, greatest common divisor (GCD) and Euclid's algorithm

Definition 1 *An integer a is said to be divisible by an integer b ($b \neq 0$) if and only if there exists an integer q such that $a = bq$.*

If a is divisible by b, we write $b|a$. The notation $b \nmid a$ means that a is not divisible by b. Many other expressions have the same meaning as a *is divisible by* b, such as, b *divides* a, a *is a multiple of* b, b *is a divisor of* a and b *is a factor of* a.

Every integer is divisible by 1 and itself, that is, $1|a$, $a|a$. Also -1 and $-a$ are divisors of a if $a \neq 0$. Some integers may have other factors and some do not. We list some properties of divisibility as follows.

Properties of divisibility

1. $b|a$ if and only if $\pm b| \pm a$.

2. If $c|b$ and $b|a$, then $c|a$ (the transitive property).

3. If $b|a$ and a is positive, then $b \leq a$.

4. If $c|a$ and $c|b$, then $c|(ax + by)$, where x and y are any integers (the linear combination property).

Theorem 1 *If a and b are integers and $b > 0$, then there exist uniquely integers q (quotient) and r (remainder) such that*

$$a = qb + r, \ 0 \leq r < b. \tag{2.1}$$

For example,

$$25 = 3 \cdot 7 + 4, \ -25 = (-4) \cdot 7 + 3.$$

Sometimes, the notation $\langle a \rangle_b$ is used to represent the remainder r. If the remainder $r = 0$, b divides a.

Definition 2 *An integer greater than 1 is a prime (or a prime number) if and only if it has no positive divisors other than 1 and itself.*

For example, 2, 3, 5, 7, 11, 13, 17, 19, 23 are primes. In general, 2 is the only even prime and 1 is not considered to be a prime. Other positive integers greater than 1 which are not primes are called *composites* or *composite numbers*. For example, 4, 6, 8, 9, 10, 12, 14, 15, 18, 20 are composites. It is known that there are infinitely many primes. Every integer can be expressed as the products of some primes.

Theorem 2 (Fundamental Theorem of Arithmetic) *Every integer $n > 1$ can be written uniquely in the form*

$$n = p_1^{l_1} p_2^{l_2} \cdots p_k^{l_k}, \tag{2.2}$$

where p_i $(1 \leq i \leq k)$ are primes, $p_1 < p_2 < \cdots < p_k$ and l_i $(1 \leq i \leq k)$ are positive integers. (2.2) is called the prime factorization of n.

For example,

$$126 = 2 \cdot 3^2 \cdot 7$$

$$125 = 5^3, \ 3072 = 2^{10} \cdot 3.$$

If $d|a$, $d|b$ and $a \neq b$, we call d a *common divisor* of a and b. For example, 5 and 35 are common divisors of 70 and 385.

Definition 3 *A positive integer d is said to be the greatest common divisor (GCD) of two positive integers a and b if and only if d is the largest common divisor of a and b.*

We use the notation $\mathrm{GCD}(a, b)$ to represent the GCD of a and b. For example, it is obvious that 35 is the largest positive integer among the common factors of 70 and 385. So, $\mathrm{GCD}(70, 385) = 35$. The definition can be extended to many integers. We use $\mathrm{GCD}(a_1, a_2, \ldots, a_k)$ to denote the largest common factors of k integers a_i $(i = 1, 2, \ldots, k)$. For example, $\mathrm{GCD}(70, 210, 385) = 35$. The GCD can be extended to negative integers by defining

$$\mathrm{GCD}(-a, b) = \mathrm{GCD}(a, b)$$
$$\mathrm{GCD}(a, -b) = \mathrm{GCD}(a, b)$$
$$\mathrm{GCD}(-a, -b) = \mathrm{GCD}(a, b)$$

for any positive integers a and b. Some properties of the GCD are listed below.

Properties of GCD

1. $\mathrm{GCD}(a, b) = \mathrm{GCD}(b, a)$.

2. $\mathrm{GCD}(na, nb) = n\mathrm{GCD}(a, b)$ for all positive integers a, b and n.

3. If c is a common factor of a and b, then $c|\mathrm{GCD}(a, b)$ and $\mathrm{GCD}(a/c, b/c) = \mathrm{GCD}(a, b)/c$.

4. If $\mathrm{GCD}(a, b) = 1$, then $\mathrm{GCD}(c, ab) = \mathrm{GCD}(c, a)\mathrm{GCD}(c, b)$.

5. If $\mathrm{GCD}(c, a) = \mathrm{GCD}(c, b) = 1$, then $\mathrm{GCD}(c, ab) = 1$.

6. If $ax + by = c$ for integers x and y, then $\mathrm{GCD}(a, b)|c$.

7. $\mathrm{GCD}(a_1, a_2, \ldots, a_k) = \mathrm{GCD}(\mathrm{GCD}(a_1, a_2, \ldots, a_{k-1}), a_k)$.

If we know the prime factorizations of n and m, that is,

$$n = p_1^{n_1} p_2^{n_2} \cdots p_k^{n_k}, \quad m = p_1^{m_1} p_2^{m_2} \cdots p_k^{m_k},$$

where n_i and m_i are non-negative integers and p_i are primes, then we can easily get the GCD of n and m as

$$\mathrm{GCD}(n, m) = p_1^{l_1} p_2^{l_2} \cdots p_k^{l_k},$$

where $l_i = \min(n_i, m_i)$ $(i = 1, 2, \ldots, k)$. For example, since

$$60 = 2^2 \cdot 3^1 \cdot 5^1 \cdot 7^0, \ 210 = 2^1 \cdot 3^1 \cdot 5^1 \cdot 7^1$$

we get

$$GCD(60, 210) = 2^1 \cdot 3^1 \cdot 5^1 \cdot 7^0 = 30.$$

This can also be extended to many integers. However, it is generally not an easy task to find the prime factorization of an integer, especially when the integer is very large, because an efficient algorithm is not available yet. Some crypto systems such as the RSA are based on the intractability of the factorization problem. So, using the prime factorization of integers to compute their GCD is not a practical method. Fortunately, Euclid knew a fast method to find the GCD which does not use the factorization. This method, called *Euclid's algorithm*, is very important in number theory and digital signal processing. To find the GCD of two positive integers a and b, Euclid used the following algorithm.

Euclid's algorithm: If

$$a = q_1 b + r_1, \; 0 < r_1 < b$$
$$b = q_2 r_1 + r_2, \; 0 < r_2 < r_1$$
$$r_1 = q_3 r_2 + r_3, \; 0 < r_3 < r_2$$
$$\vdots$$
$$r_{k-2} = q_k r_{k-1} + r_k, \; 0 < r_k < r_{k-1}$$
$$r_{k-1} = q_{k+1} r_k + 0$$

then $r_k = GCD(a, b)$.

Example 1 *Find the GCD of 385 and 210, and the GCD of 134 and 785.*

Solution. Since

$$385 = 1 \cdot 210 + 175$$
$$210 = 1 \cdot 175 + 35$$
$$175 = 5 \cdot 35 + 0$$

we know that $GCD(385, 210) = 35$.

$$785 = 5 \cdot 134 + 115$$
$$134 = 1 \cdot 115 + 19$$
$$115 = 6 \cdot 19 + 1$$
$$19 = 19 \cdot 1 + 0$$

So, $GCD(134, 785) = 1$.

If the GCD of a and b is 1, we call a and b relatively prime. This is a very special and useful property. For example, we can prove that if $c|ab$ and $GCD(c, a) = 1$, then $c|b$.

An integer d is said to be a *common multiple* of two integers a and b if d is the multiple of a and b, that is, $a|d$ and $b|d$. For example, ab is a common multiple of a and b, and so is any multiple of ab. Therefore, there are infinitely many common multiples of a and b.

Definition 4 *A positive integer d is said to be the least common multiple (LCM) of two positive integers a and b if and only if d is the smallest common multiple of a and b.*

We use $LCM(a, b)$ to denote the LCM of a and b. For example, $LCM(70, 385) = 770$. This concept can be extended to the LCM of many positive integers. Some properties of the LCM are listed below.

Properties of LCM

1. If d is a common multiple of a and b, then $\text{LCM}(a,b)|d$.

2. $\text{LCM}(na, nb) = n\text{LCM}(a,b)$.

3. $\text{GCD}(a,b) \cdot \text{LCM}(a,b) = ab$.

4. $\text{LCM}(a_1, a_2, \ldots, a_k) = \text{LCM}(\text{LCM}(a_1, a_2, \ldots, a_{k-1}), a_k)$.

If the prime factorizations of n and m are known, that is,

$$n = p_1^{n_1} p_2^{n_2} \cdots p_k^{n_k}, \ m = p_1^{m_1} p_2^{m_2} \cdots p_k^{m_k},$$

where n_i and m_i are non-negative integers and p_i are primes, then the LCM of n and m is

$$\text{LCM}(n,m) = p_1^{e_1} p_2^{e_2} \cdots p_k^{e_k},$$

where $e_i = \max(n_i, m_i)$ $(i = 1, 2, \ldots, k)$. Based on this observation, property 3 is easily proved since $\max(n_i, m_i) + \min(n_i, m_i) = n_i + m_i$. For example, since

$$60 = 2^2 \cdot 3^1 \cdot 5^1 \cdot 7^0, \quad 210 = 2^1 \cdot 3^1 \cdot 5^1 \cdot 7^1$$

we get

$$\text{LCM}(60, 210) = 2^2 \cdot 3^1 \cdot 5^1 \cdot 7^1 = 420.$$

We see that $\text{GCD}(60, 210) \cdot \text{LCM}(60, 210) = 30 \cdot 420 = 60 \cdot 210$. Property 3 is very important in theory and practice. Since we have the Euclid's algorithm for finding the GCD, it can be used to get the LCM by first finding the GCD and then using property 3.

2.1.2   Congruence and Chinese remainder theorem

Definition 5 *Let a, b and m be integers with $m > 1$. We say that a is congruent to b modulo m and write $a \equiv b \bmod m$ if and only if $m|(a-b)$. If $m \nmid (a-b)$, we say that a is not congruent (or incongruent) to b modulo m and write $a \not\equiv b \bmod m$.*

For example, $35 \equiv 14 \bmod 7$, $35 \not\equiv 14 \bmod 5$, $120 \equiv 65 \bmod 11$ and $120 \not\equiv 65 \bmod 3$.

From the definition, if $a \equiv b \bmod m$, then $a - b = cm$ or $a = cm + b$ for an integer c, that is, the difference of a and b is a multiple of m. Sometimes, if $a \equiv b \bmod m$, we also call a a *residue* of b modulo m. The residue of b modulo m may be different from the remainder of b divided by m $(\langle b \rangle_m)$ defined in (2.1). The reason is that the remainder is assumed to be between 0 and $m - 1$, while the residue has no limitation. The remainder is also a residue. For example, 1 is the remainder of 7 divided by 3. However, for any integer c, $1 + 3c$ is a residue of 7 modulo 3.

The symbol "$\equiv$" is used to indicate the congruence partly because the *congruence* and the *equality*, represented by the symbol "$=$", do share many properties. We list some properties of congruence below.

Properties of congruence

1. $a \equiv a \bmod m$ for every integer a (reflexive property).

2. If $a \equiv b \bmod m$, then $b \equiv a \bmod m$ (symmetric property).

3. If $a \equiv b \bmod m$ and $b \equiv c \bmod m$, then $a \equiv c \bmod m$ (transitivity property).

4. If $a \equiv b \bmod m$ and $c \equiv d \bmod m$, then $a \pm c \equiv b \pm d \bmod m$.

5. If $a \equiv b \bmod m$ and $c \equiv d \bmod m$, then $ac \equiv bd \bmod m$.

6. If $ka \equiv kb \bmod m$ $(k \neq 0)$, then $a \equiv b \bmod m/\text{GCD}(k, m)$.

Some properties of equality cannot be used for congruence. For example, if $ka = kb$ and $k \neq 0$, then $a = b$. This is not correct for $\equiv$. Even if $ka \equiv kb \bmod m$, sometimes we do not have $a \equiv b \bmod m$. For example, $3 \cdot 7 \equiv 3 \cdot 14 \bmod 21$, but obviously $7 \not\equiv 14 \bmod 21$. Property 6 tells us that $7 \equiv 14 \bmod (21/3 = 7)$.

Definition 6 *Assume that m is a positive integer. A set C of integers is called a complete residue system (CRS) modulo m if and only if*

- *for every pair of distinct elements x and y of C, $x \not\equiv y \bmod m$,*

- *every integer is congruent modulo m to some elements of C.*

For example, for module $m = 4$, $C = \{-1, 2, 0, 5\}$ is a CRS, but $\{-1, 2, 0, 3\}$ is not because $-1 \equiv 3 \bmod 4$. For every positive integer m, $\{0, 1, \ldots, m-1\}$ and $\{1, 2, \ldots, m\}$ are two CRSs modulo m. These two CRSs are the most commonly used. If C is a CRS modulo m, it should satisfy

- the number of elements in C is m, and

- every integer is congruent to only one element of C.

Every set C with m elements and satisfying the first condition in Definition 6 is a CRS modulo m.

Definition 7 *Assume that m is a positive integer. A set R of integers is called a reduced residue system (RRS) modulo m if and only if*

- *for every pair of distinct elements x and y of C, $x \not\equiv y \bmod m$;*

- *every element of C is relatively prime to m; and*

- *every integer relatively prime to m is congruent modulo m to some elements of C.*

For example, for module $m = 8$, $C = \{-1, 3, 5, 1\}$ is a RRS. Neither $\{-1, 7, 5, 3\}$ nor $\{1, 3, 5, 7, 4\}$ is a RRS modulo 8 because $-1 \equiv 7 \bmod 8$ and 4 is not relatively prime to 8. If p is a prime, then $\{1, 2, \ldots, p-1\}$ is a RRS modulo p. A RRS R is a subset of a CRS C. In fact, if C is a CRS modulo m, by deleting the elements in C which are not relatively prime to m we get a RRS modulo m. For example, $\{0, 1, 2, 3, 4, 5\}$ is a CRS of 6, where 0, 2, 3 and 4 are not relatively prime to 6. Deleting these elements we get a set $\{1, 5\}$ which is a RRS modulo 6. We know that every CRS modulo m contains m elements, but what is the number of elements in a RRS modulo m? To answer this question, we should know the number of positive integers that are smaller than m and relatively prime to m.

Definition 8 (Euler's totient function) *Assume that m is a positive integer. The number of positive integers r satisfying $1 \leq r \leq m$ and $\text{GCD}(r, m) = 1$ is denoted by $\phi(m)$, the Euler's totient function.*

From the definition, $\phi(1)=1$, $\phi(6) = 2$ and $\phi(8) = 4$. The number of elements in every RRS modulo m $(m > 1)$ is $\phi(m)$. Any set R which is composed of $\phi(m)$ integers and satisfies the first two conditions in Definition 7 is a RRS modulo m. Some properties of the Euler's totient function are listed below.

Properties of the Euler's totient function

1. If p is a prime, then $\phi(p) = p - 1$.

2. If p is a prime and l is a positive integer, then $\phi(p^l) = p^{l-1}(p - 1) = p^l(1 - 1/p)$.

3. If $\mathrm{GCD}(a, b) = 1$, then $\phi(ab) = \phi(a)\phi(b)$.

From properties 2 and 3, we have

$$\phi(p_1^{l_1}p_2^{l_2}\cdots p_k^{l_k}) = p_1^{l_1}p_2^{l_2}\cdots p_k^{l_k}(1 - \frac{1}{p_1})\cdots(1 - \frac{1}{p_k}),$$

where for $i = 1, 2, \ldots, k$, every p_i is a prime and every l_i is a positive integer. Therefore, if we know the prime factorization of an integer n, it is easy to find $\phi(n)$.

A congruence involving unknown variables is called a *Diophantine equation*. For example, $5x \equiv 7 \bmod 12$ and $31x^2 \equiv 6 \bmod 125$ are two Diophantine equations, where x is an unknown variable. A *linear Diophantine equation* is anything of the form

$$ax \equiv b \bmod m,$$

where a, b and m are known integers and x is an unknown variable. The theory of solving various Diophantine equations is a major part of number theory. There are still a lot of unsolved problems in this area. It is easy to solve a *linear Diophantine equation*. We have the following theorem.

Theorem 3 *A linear Diophantine equation $ax \equiv b \bmod m$ is solvable, that is, there exists an integer x_0 such that $ax_0 \equiv b \bmod m$, if and only if $\mathrm{GCD}(a, m)|b$.*

For example, $5x \equiv 7 \bmod 12$ is solvable because $\mathrm{GCD}(5, 12) = 1|7$. $x_0 = -1$ is a solution. $x_1 = -1 + 12c$ is also a solution for any integer c. In general, if a Diophantine equation is solvable, it will have infinitely many solutions. In fact, if x_0 is a solution, then $x_0 + cm$ is also a solution for every integer c. However, since every solution $x_0 + cm$ is congruent to x_0 modulo m, they are essentially the same and we view them all as *one* solution. For a Diophantine equation, we define a set of solutions to be *different* if they are pairwise incongruent modulo m. The number of solutions of a Diophantine equation is defined as the number of essentially different solutions (pairwise incongruent modulo m).

Theorem 4 *If a linear Diophantine equation $ax \equiv b \bmod m$ is solvable, then the number of its solutions is $\mathrm{GCD}(a, m)$.*

A special linear Diophantine equation is $ax \equiv 1 \bmod m$ where $\mathrm{GCD}(a, m) = 1$. Such an equation has only one solution. We use the notation $a^{-1} \bmod m$ to denote the solution and call it the *inverse* of a modulo m. For simplification sometimes we use a^{-1} to represent it if we know what the module is. It should not be confused with the real number a^{-1}. For example, $5^{-1} \bmod 7 = 3$ because $5 \cdot 3 \equiv 1 \bmod 7$. Of course, any integer of the form $3 + 7c$, where c is an integer, is also an inverse of 5 modulo 7. Usually we choose $5^{-1} \bmod 7$ to be the smallest positive integer among them.

Now we consider a set of linear Diophantine equations:

$$x \equiv b_1 \bmod m_1$$
$$x \equiv b_2 \bmod m_2$$
$$\vdots$$
$$x \equiv b_n \bmod m_n, \tag{2.3}$$

where for $i = 1, 2, \ldots, n$, m_i are positive integers and b_i are integers. We call it a set of *simultaneous linear Diophantine equations*.

Are these simultaneous linear Diophantine equations solvable and how can we solve them? Now let us consider a special case when m_i $(i = 1, 2, \ldots, n)$ are pairwise relatively prime.

Theorem 5 (Chinese Remainder Theorem (CRT)) *Assume that m_i $(i = 1, 2, \ldots, n)$ are positive integers and pairwise relatively prime. Then (2.3) is solvable and its solution is*

$$x_0 \equiv N_1 M_1 b_1 + N_2 M_2 b_2 + \cdots + N_n M_n b_n \bmod M, \tag{2.4}$$

where $M = m_1 m_2 \cdots m_n$, $M_i = M/m_i$ and N_i is the inverse of M_i modulo m_i $(i = 1, 2, \ldots, n)$.

Proof. Since $N_i M_i \equiv 1 \bmod m_i$ and $m_j | M_i$ for $j \neq i$, we have

$$N_i M_i b_i \equiv \begin{cases} b_i \bmod m_i \\ 0 \bmod m_j \end{cases}.$$

For $i = 1, 2, \ldots, n$,

$$N_1 M_1 b_1 + N_2 M_2 b_2 + \cdots + N_n M_n b_n \equiv b_i \bmod m_i$$

that is, (2.4) is a solution of (2.3). If x_1 is also a solution of (2.3), then $x_1 \equiv x_0 \bmod m_i$, that is, $m_i | (x_1 - x_0)$ $(i = 1, 2, \ldots, n)$. Since m_i are pairwise relatively prime, we know that $M | (x_1 - x_0)$, that is, $x_1 \equiv x_0 \bmod M$. So, every solution of (2.3) can be expressed by (2.4). There is essentially only one solution.

Example 2 *Find the solution of the equations*

$$x \equiv 1 \bmod 7$$
$$x \equiv 2 \bmod 8$$
$$x \equiv 3 \bmod 9$$
$$x \equiv 4 \bmod 11$$

Solution. Since $m_1 = 7$, $m_2 = 8$, $m_3 = 9$ and $m_4 = 11$, we have

$$M = 5544, \ M_1 = 792, \ M_2 = 693, \ M_3 = 616, \ M_4 = 504$$

and

$$N_1 = 1, \ N_2 = 5, \ N_3 = 7, \ N_4 = 5.$$

The solution is

$$x_0 \equiv (1 \cdot 792) \cdot 1 + (5 \cdot 693) \cdot 2 + (7 \cdot 616) \cdot 3 + (5 \cdot 504) \cdot 4 \bmod 5544$$
$$\equiv 3018 \bmod 5544.$$

The CRT is widely used in signal processing and computer science [8]. One of the popular applications is the *residue number system* (RNS). Choose n positive integers m_i ($i = 1, 2, \ldots, n$) such that they are pairwise relatively prime. From the CRT, any integer x between 0 and $M - 1$, where $M = m_1 m_2 \cdots m_n$, can be uniquely expressed as an n-dimensional vector $(b_n, b_{n-1}, \ldots, b_1)$, where b_i is the least non-negative residue of x modulo m_i, that is, $b_i \equiv x \bmod m_i$ and $0 \leq b_i < m_i$. The CRT also tells us how to recover x from the vector. Parallel computation can be easily supported by the RNS to achieve high signal processing throughput [14, 15].

2.1.3 Euler's theorem and primitive roots

Theorem 6 (Euler's Theorem) *If* $\mathrm{GCD}(a, m) = 1$, *then* $a^{\phi(m)} \equiv 1 \bmod m$.

A special case of the theorem is the Fermat's theorem.

Fermat's theorem: If p is a prime and $p \nmid a$, then $a^{p-1} \equiv 1 \bmod p$.

From Definition 8, for example, we have $\phi(15) = 8$. Based on the Euler's theorem, we have $\mathrm{GCD}(2, 15) = 1$, $2^8 = 256 = 17 \cdot 15 + 1 \equiv 1 \bmod 15$. Another example of the Fermat's theorem is $3^{7-1} \equiv 1 \bmod 7$.

To compute $a^n \bmod m$ (the residue of a^n modulo m), where n is a positive integer, we do not need to compute a^n and then divide it by m. For example, to compute $3^6 \bmod 7$, we can use the following algorithm: $3^2 = 9 \equiv 2 \bmod 7$, $3^6 = (3^2)^3 \equiv 2^3 \equiv 1 \bmod 7$. The Euler's theorem can be used to facilitate the computation. If $\mathrm{GCD}(a, m) = 1$, then $a^{\phi(m)} \equiv 1 \bmod m$. Divide n by $\phi(m)$ and let $n = q\phi(m) + r$, where $0 \leq r < \phi(m)$. Then $a^n \equiv a^r \bmod m$.

Is there any positive integer t that is smaller than $\phi(m)$ and satisfies $a^t \equiv 1 \bmod m$ for any integer a relatively prime to m? The answer is yes for some m, but not for all positive integers. For example, if $m = 24$, then $\phi(24) = 8$. For every integer a such that $\mathrm{GCD}(a, 24) = 1$, we have $a^4 \equiv 1 \bmod 24$. The reason is that $a^4 \equiv 1 \bmod 8$ and $a^4 \equiv 1 \bmod 3$ (from Euler's theorem). If $m = 7$, however, there is no positive integer t smaller than 6 such that $a^t \equiv 1 \bmod 7$ for any integer a relatively prime to 7. In fact, if $0 < t < 6$, then $3^t \not\equiv 1 \bmod 7$.

We know that if $\mathrm{GCD}(a, m) = 1$, then a has an inverse d modulo m, that is, $ad \equiv 1 \bmod m$, where d is usually denoted by a^{-1}. If there is a positive integer such that $a^t \equiv 1 \bmod m$, then we will show that $d^t \equiv 1 \bmod m$. Since $ad \equiv 1 \bmod m$, we have

$$(ad)^t \equiv 1 \bmod m$$

$$(ad)^t = a^t d^t \equiv d^t \equiv 1 \bmod m.$$

Therefore, $d^t \equiv 1 \bmod m$. For simplification, we usually use a^{-t} for $(a^{-1})^t$. So, if $a^t \equiv 1 \bmod m$, then $a^{-t} \equiv 1 \bmod m$.

Is there any integer a with $\mathrm{GCD}(a, m) = 1$ such that $(a^0, a^1, \ldots, a^{\phi(m)-1})$ is an RRS modulo m? If so, the construction of RRS modulo m becomes very simple. Such an integer a is therefore of special importance.

Definition 9 *An integer g is said to be a primitive root of positive integer m if and only if*

$$g^0, g^1, \ldots, g^{\phi(m)-1}$$

constitute a RRS modulo m.

For example, 2 is a primitive root of 5 because $(2^0, 2^1, 2^2, 2^3) \bmod 5 = (1, 2, 4, 3)$ is a RRS modulo 5. However, 2 is not a primitive root of 7 because $(2^0, 2^1, 2^2, 2^3, 2^4, 2^5) \bmod 7 = (1, 2, 4, 1, 2, 4)$ is obviously not a RRS modulo 7. Very few positive numbers have primitive roots.

Theorem 7 (Primitive Root Theorem) *Only positive integers*

$$1, \ 2, \ 2^2, \ p^l, \ 2p^l,$$

where p is any odd prime and l is any positive integer, have primitive roots. Furthermore, there is at least one odd primitive root for p^l.

Primitive roots have found many applications in fast algorithms. Sometimes we need to reorder the index to eliminate the computation redundancy. For example, if the indices are $1, 2, \ldots, p-1$ (an RRS modulo p) with p being a prime, and g is a primitive root of p, then $1, \langle g^1 \rangle_p, \ldots, \langle g^{p-2} \rangle_p$ are the reordered indices, where and henceforth $\langle x \rangle_p$ is used to represent the least non-negative residue of integer x modulo p. The reason is that based on the definition of a primitive root, $(1, \langle g^1 \rangle_p, \ldots, \langle g^{p-2} \rangle_p)$ is an RRS modulo p and each of them is between 1 and $p-1$. This property is the theoretical basis of Winograd FFT algorithms [2, 9]. Table 2.1 lists the smallest positive primitive root of the primes between 1 and 500.

Table 2.1 Smallest primitive roots of primes below 500.

p	g	p	g	p	g	p	g	p	g
2	1	3	2	5	2	7	3	11	2
13	2	17	3	19	2	23	5	29	2
31	3	37	2	41	6	43	3	47	5
53	2	59	2	61	2	67	2	71	7
73	5	79	3	83	2	89	3	97	5
101	2	103	5	107	2	109	6	113	3
127	3	131	2	137	3	139	2	149	2
151	6	157	5	163	2	167	5	173	2
179	2	181	2	191	19	193	5	197	2
199	3	211	2	223	3	227	2	229	6
233	3	239	7	241	7	251	6	257	3
263	5	269	2	271	6	277	5	281	3
283	3	293	2	307	5	311	17	313	10
317	2	331	3	337	10	347	2	349	2
353	3	359	7	367	6	373	2	379	2
383	5	389	2	397	5	401	3	409	21
419	2	421	2	431	7	433	5	439	15
443	2	449	3	457	13	461	2	463	2
467	2	479	13	487	3	491	2	499	7

Although 2^t $(t \geq 3)$ has no primitive root, we can construct its RRS by a simple method stated in the following theorem.

Theorem 8 $(-1)^i 3^j$ $(i = 0, 1, \; j = 0, 1, \ldots, 2^{t-2} - 1)$ *is a RRS modulo* 2^t $(t \geq 3)$.

For $m = 8$, for example, $(1, 3, -1, -3)$ is a RRS. For $m = 16$,

$$(1, 3, 9, 27, -1, -3, -9, -27) \bmod 16 = (1, 3, 9, 11, 15, 13, 7, 5) \bmod 16$$

is a RRS. In general, the set

$$(3^0, 3^1, \ldots, 3^{2^{t-2}-1}, -3^0, -3^1, \ldots, -3^{2^{t-2}-1}) \bmod 2^t$$

is a reordering of the set of consecutive odd integers

$$(1, 3, 5, \ldots, 2^t - 1).$$

This property can be used to express the DFT of length 2^t into convolutions [4].

2.2   BASIC POLYNOMIAL THEORY

Definition 10 *A polynomial of degree n over a number field F is an expression of the form*

$$p(z) = \sum_{i=0}^{n} a_i z^i, \tag{2.5}$$

where $a_i \in F$ are known as the coefficients of the polynomial and $a_n \neq 0$, n is called the degree of the polynomial and is denoted by $\deg[p(z)]$ or $\deg[p]$ and z is an indeterminate variable which can be replaced by other variables such as x or y. If $a_n = 1$, the polynomial is monic.

For example,

$$p(z) = 2z^3 + 3z + 1$$

is a polynomial of degree 3, while

$$q(z) = z^4 + 5z^2 + 2z + 1$$

is a monic polynomial of degree 4. We can also use $p(y) = 2y^3 + 3y + 1$ or $p(x) = 2x^3 + 3x + 1$ to represent the polynomial.

Two or more polynomials can be added or multiplied and the result is still a polynomial. For example, adding the two polynomials defined above, we get

$$p(z) + q(z) = z^4 + 2z^3 + 5z^2 + 5z + 2$$

while their product is

$$p(z)q(z) = 2z^7 + 13z^5 + 5z^4 + 17z^3 + 11z^2 + 5z + 1.$$

Regarding the degree of the polynomials, we have

$$\deg[p(z) + q(z)] \leq \max(\deg[p(z)], \deg[q(z)])$$

$$\deg[p(z)q(z)] = \deg[p(z)] + \deg[q(z)].$$

2.2.1   Divisibility, GCD and Euclid's algorithm

Theorem 9 *Given two polynomials $f(z)$ and $g(z)$ with $g(z) \neq 0$, there exist uniquely two polynomials $q(z)$ and $r(z)$ such that*

$$f(z) = q(z)g(z) + r(z), \ r(z) = 0 \text{ or } \deg[r(z)] < \deg[g(z)],$$

where $q(z)$ is known as the quotient of the polynomial $f(z)$ divided by $g(z)$ and $r(z)$ is the remainder.

Sometimes we use $\langle f(z) \rangle_{g(z)}$ to denote the remainder $r(z)$. For example,

$$z^4 + 5z^2 + 2z + 1 = \frac{1}{2}z(2z^3 + 3z + 1) + \frac{7}{2}z^2 + \frac{3}{2}z + 1$$

that is, the quotient of $z^4 + 5z^2 + 2z + 1$ divided by $2z^3 + 3z + 1$ is $\frac{1}{2}z$ and the remainder is $\frac{7}{2}z^2 + \frac{3}{2}z + 1$ or $\langle z^4 + 5z^2 + 2z + 1 \rangle_{2z^3+3z+1} = \frac{7}{2}z^2 + \frac{3}{2}z + 1$. In general, finding the remainder requires long division of polynomials. In some special cases, the operation can be simplified. For example, $\langle f(z) \rangle_{z-a} = f(a)$.

Example 3 *Let $f(z) = \sum_{i=0}^{M} a_i z^i$. Find the remainder of $f(z)$ divided by $z^N + 1$.*

Solution. We write $M = KN + L$ with $0 \leq L < N$. Since $z^N = (z^N + 1) - 1$, we know that the remainder of z^N divided by $z^N + 1$ is -1. Similarly, it is easy to verify that the remainder of z^{kN} is $(-1)^k$, that is, there exists $q(z)$ such that

$$z^{kN} = q(z)(z^N + 1) + (-1)^k.$$

For any integer i, $0 \leq i < N$, we get

$$z^{kN+i} = z^i q(z)(z^N + 1) + (-1)^k z^i$$

that is, the remainder of z^{kN+i} is $(-1)^k z^i$. Therefore we have

$$
\begin{aligned}
\langle f(z) \rangle_{z^N+1} &= \sum_{i=0}^{N-1} \left[\sum_{k=0}^{K-1} (-1)^k a_{kN+i} \right] z^i + \sum_{l=0}^{L} (-1)^K a_{KN+l} z^l \\
&= \sum_{l=0}^{L} \left[\sum_{k=0}^{K} (-1)^k a_{kN+l} \right] z^l + \sum_{l=L+1}^{N-1} \left[\sum_{k=0}^{K-1} (-1)^k a_{kN+l} \right] z^l.
\end{aligned}
$$

Definition 11 *Given two polynomials $f(z)$ and $g(z)$ with $g(z) \neq 0$, if there exists a polynomial $q(z)$ such that*

$$f(z) = q(z)g(z)$$

then we say that $f(z)$ is divisible by $g(z)$, which is represented by the notation $g(z)|f(z)$. Therefore $g(z)$ is a factor or divisor of $f(z)$ and $f(z)$ is a multiple of $g(z)$. The notation $g(z) \nmid f(z)$ means that $f(z)$ is not divisible by $g(z)$.

Obviously, we have the following properties.

Properties of polynomial divisibility

   1. $f(z)|f(z)$, if $f(z) \neq 0$.

2. If $f(z)|g(z)$ and $g(z)|f(z)$, then $f(z)$ and $g(z)$ can only differ in a factor of constant number, that is, $f(z) = cg(z)$ for some nonzero numbers in F.

3. If $h(z)|g(z)$ and $g(z)|f(z)$, then $h(z)|f(z)$.

4. If $g(z)|f_i(z)$ $(i = 1, 2, \ldots, n)$, then

$$g(z)| \sum_{i=1}^{n} h_i(z) f_i(z)$$

for any polynomials $h_i(z)$.

5. If $g(z)|f(z)$, then $\deg[g(z)] \leq \deg[f(z)]$.

6. If $g(z)|f(z)$ and c_1 and c_2 are nonzero numbers from field F, then $c_1 g(z)|c_2 f(z)$.

Definition 12 *If polynomial $g(z)$ is a factor (divisor) of each polynomial $f_i(z)$ $(i = 1, 2, \ldots)$, then $g(z)$ is called a common factor (divisor) of $f_i(z)$ $(i = 1, 2, \ldots, n)$. If $D(z)$ is a common factor of $f_i(z)$ $(i = 1, 2, \ldots, n)$ and its degree is greater than or equal to that of any other common factor, then $D(z)$ is called a greatest common divisor (GCD) of $f_i(z)$ and is denoted by $D(z) = \mathrm{GCD}(f_1(z), f_2(z), \ldots, f_n(z))$.*

If $d(z)$ is a factor of a polynomial $f(z)$, then $cd(z)$ is also a factor of $f(z)$, where c is a nonzero constant number in F. So if $D(z)$ is a GCD of some polynomials, then for any nonzero constant number c in F, $cD(z)$ is also a GCD. The GCD is not unique and can differ in a nonzero constant. We view the GCDs as the same if they only differ in nonzero constant numbers. Some properties of the GCD are as follows.

Properties of GCD

1. $\mathrm{GCD}(f(z), g(z)) = \mathrm{GCD}(g(z), f(z))$.

2. $\mathrm{GCD}(h(z)f(z), h(z)g(z)) = h(z)\mathrm{GCD}(f(z), g(z))$ for all nonzero polynomials $f(z)$, $g(z)$ and $h(z)$.

3. If $h(z)$ is a common factor of $f(z)$ and $g(z)$, then $h(z)|\mathrm{GCD}(f(z), g(z))$ and $\mathrm{GCD}(f(z)/h(z), g(z)/h(z)) = \mathrm{GCD}(f(z), g(z))/h(z)$.

4. If $\mathrm{GCD}(f(z), g(z)) = 1$, then $\mathrm{GCD}(h(z), f(z)g(z)) = \mathrm{GCD}(h(z), f(z)) \cdot \mathrm{GCD}(h(z), g(z))$.

5. If $\mathrm{GCD}(h(z), f(z)) = \mathrm{GCD}(h(z), g(z)) = 1$, then $\mathrm{GCD}(h(z), f(z)g(z)) = 1$.

6. $\mathrm{GCD}(f_1(z), f_2(z), \ldots, f_k(z)) = \mathrm{GCD}(\mathrm{GCD}(f_1(z), f_2(z), \ldots, f_{k-1}(z)), f_k(z))$.

Euclid's algorithm is an efficient method for finding the GCD of two polynomials. Let $f(z)$ and $g(z)$ be two polynomials and $g(z) \neq 0$. Euclid's algorithm uses the following steps to find the GCD.

Euclid's algorithm: If

$$
\begin{aligned}
f(z) &= q_1(z)g(z) + r_1(z), &\quad r_1(z) \neq 0, \ \deg[r_1(z)] < \deg[g(z)] \\
g(z) &= q_2(z)r_1(z) + r_2(z), &\quad r_2(z) \neq 0, \ \deg[r_2(z)] < \deg[r_1(z)] \\
r_1(z) &= q_3(z)r_2(z) + r_3(z), &\quad r_3(z) \neq 0, \ \deg[r_3(z)] < \deg[r_2(z)] \\
&\ \ \vdots \\
r_{k-2}(z) &= q_k(z)r_{k-1}(z) + r_k(z), &\quad r_k(z) \neq 0, \ \deg[r_k(z)] < \deg[r_{k-1}(z)] \\
r_{k-1}(z) &= q_{k+1}(z)r_k(z) + 0
\end{aligned}
$$

then $r_k(z) = \text{GCD}(f(z), g(z))$.

Example 4 *Find the GCD of $z^4 + 2z^3 + 1$ and $z^3 + 1$.*

Solution. Using the Euclid's algorithm, we have

$$z^4 + 2z^3 + 1 = (z + 2)(z^3 + 1) - z - 1$$
$$z^3 + 1 = (-z^2 + z - 1)(-z - 1) + 0.$$

So, $\text{GCD}(z^4 + 2z^3 + 1, z^3 + 1) = -z - 1$.

Theorem 10 *Assume that at least one of the polynomials $f_i(z)$ $(i = 1, 2, \ldots, n)$ is nonzero and $D(z)$ is their GCD. Then there exist polynomials $h_i(z)$ $(i = 1, 2, \ldots, n)$ such that*

$$D(z) = h_1(z)f_1(z) + h_2(z)f_2(z) + \cdots + h_n(z)f_n(z).$$

Definition 13 *If $\text{GCD}(f_1(z), f_2(z), \ldots, f_n(z)) = 1$, then $f_1(z), f_2(z), \ldots, f_n(z)$ are called relatively prime.*

Definition 14 *Assume that $f(z)$ is a polynomial and its degree is greater than 0. If it has no factors with degree greater than 0 except itself, then $f(z)$ is called irreducible. Otherwise, it is reducible.*

A reducible polynomial can be factored into products of polynomials with a lower degree, while an irreducible polynomial cannot. For example, $z^4 - 1$ is reducible because it can be factored into

$$z^4 - 1 = (z^2 + 1)(z + 1)(z - 1).$$

Polynomial $z^2 + 1$ is irreducible if the field F is the real number field or the rational number field. However, if F is the complex number field, then it is reducible because

$$z^2 + 1 = (z + j)(z - j),$$

where $j = \sqrt{-1}$. Therefore, the reducibility of a polynomial depends upon the related number field.

Theorem 11 *Every polynomial $f(z)$ with degree greater than 0 can be written uniquely in the form*

$$f(z) = ap_1(z)^{l_1}p_2(z)^{l_2}\cdots p_k(z)^{l_k},$$

where a is a number in F, $p_i(z)$ $(i = 1, 2, \ldots, k)$ are monic irreducible polynomials, $\deg[p_1(z)] < \deg[p_2(z)] < \cdots < \deg[p_k(z)]$ and l_i $(1 \le i \le k)$ are positive integers.

2.2.2   Congruence and CRT

Definition 15 *Let $f(z)$, $g(z)$ and $m(z)$ be polynomials with $m(z) \ne 0$. If*

$$m(z)|(f(z) - g(z))$$

then we say that $f(z)$ is congruent to $g(z)$ modulo $m(z)$ and write

$$f(z) \equiv g(z) \bmod m(z).$$

If

$$m(z) \nmid (f(z) - g(z))$$

then we say that $f(z)$ is not congruent (or incongruent) to $g(z)$ modulo $m(z)$ and write

$$f(z) \not\equiv g(z) \bmod m(z).$$

For example,

$$z^4 \equiv 1 \bmod z^2 + 1$$

$$z^{81} + z^{49} + z^{25} + z^9 + z \equiv 5z \bmod z^8 - z.$$

Some useful properties are listed as follows.

Properties of polynomial congruence

1. $f(z) \equiv f(z) \bmod m(z)$.

2. If $f(z) \equiv g(z) \bmod m(z)$, then $g(z) \equiv f(z) \bmod m(z)$.

3. If $f(z) \equiv g(z) \bmod m(z)$ and $g(z) \equiv h(z) \bmod m(z)$, then $f(z) \equiv h(z) \bmod m(z)$.

4. If $f_1(z) \equiv g_1(z) \bmod m(z)$ and $f_2(z) \equiv g_2(z) \bmod m(z)$, then

$$f_1(z) \pm f_2(z) \equiv g_1(z) \pm g_2(z) \bmod m(z)$$

$$f_1(z)f_2(z) \equiv g_1(z)g_2(z) \bmod m(z).$$

5. If $f(z)h(z) \equiv g(z)h(z) \bmod m(z)$ and $h(z)$ is relatively prime to $m(z)$, then $f(z) \equiv g(z) \bmod m(z)$.

Let $r(z)$ be the remainder of $f(z)$ divided by $m(z)$. Then from the definition of the remainder we know that $f(z)$ is congruent to $r(z)$ modulo $m(z)$, and $r(z)$ either equals 0 or has a lower degree than $m(z)$. Let us assume $n = \deg[m(z)] > 0$. We use the notation $F(z)/(m(z))$ to denote the set of all polynomials over field F with a degree lower than n, that is,

$$F(z)/(m(z)) = \{a_{n-1}z^{n-1} + \cdots + a_1 z + a_0 | a_i \in F, \ i = 0, 1, \ldots, n-1\}.$$

The sum of any two polynomials in the set is still in the set. The product of two polynomials in the set may be out of the set. However, the remainder of the result divided by $m(z)$ is in the set. So we define the multiplication in set $F(z)/(m(z))$ by

$$h_1(z) * h_2(z) = \langle h_1(z)h_2(z) \rangle_{m(z)}$$

such that $h_1(z) * h_2(z) \in F(z)/(m(z))$ for any polynomials $h_1(z)$ and $h_2(z)$ in $F(z)/(m(z))$. According to the operations of addition and multiplication defined above, the set $F(z)/(m(z))$ is a ring. If $m(z)$ is irreducible, then $F(z)/(m(z))$ is a field.

If $f(z)$ is relatively prime to $m(z)$, then from Theorem 10 we can find two polynomials $h_1(z)$ and $h_2(z)$ such that

$$h_1(z)f(z) + h_2(z)m(z) = 1$$

that is,

$$m(z)|(h_1(z)f(z) - 1) \text{ or } h_1(z)f(z) \equiv 1 \bmod m(z).$$

So $h_1(z)$ is the *reciprocal* or *inverse* of $f(z)$ modulo $m(z)$. We use the notation $f^{-1}(z) \bmod m(z)$ to denote $h_1(z)$. Sometimes when $m(z)$ is known, we simply use $f^{-1}(z)$ to represent it.

For example, for any positive integer N, z is relatively prime to $z^N + 1$. Since $(-z^{N-1})z - 1 = -(z^N + 1)$, we see that $(-z^{N-1})z \equiv 1 \bmod z^N + 1$. So $z^{-1} \bmod z^N + 1 = -z^{N-1}$. We can also treat $f(z)$ and $h_1(z)$ as two elements in the ring $F(z)/(m(z))$. Then $h_1(z) * f(z) = 1$ in the ring. $f(z)$ is invertible in the ring and its inverse is $h_1(z)$. We use $f^{-n}(z)$ to represent $[f^{-1}(z)]^n$, where n is a positive integer.

Example 5 *Let R be the real number field. The polynomial $z^2 + 1$ is irreducible over R and*

$$R(z)/(z^2 + 1) = \{az + b | a, b \in R\}.$$

The $+$ and $*$ operations in this ring are defined as

$$(a_1 z + b_1) + (a_2 z + b_2) = (a_1 + a_2)z + (b_1 + b_2).$$

Since

$$
\begin{aligned}
(a_1 z + b_1)(a_2 z + b_2) &= a_1 a_2 z^2 + (a_1 b_2 + a_2 b_1)z + b_1 b_2 \\
&\equiv (a_1 b_2 + a_2 b_1)z + (b_1 b_2 - a_1 a_2) \bmod z^2 + 1
\end{aligned}
$$

we have

$$(a_1 z + b_1) * (a_2 z + b_2) = (a_1 b_2 + a_2 b_1)z + (b_1 b_2 - a_1 a_2).$$

It is the same operation of multiplying two complex numbers if we replace the indeterminate z by $j = \sqrt{-1}$. In algebra, $R(z)/(z^2 + 1)$ is treated as identical to the complex number field C. Replacing the complex number field C by $R(z)/(z^2 + 1)$ can reduce the computational complexity for some problems [2, 4]. If $a^2 + b^2 \neq 0$, then $az + b$ has an inverse $\frac{-az + b}{a^2 + b^2}$ because

$$(az + b)\frac{-az + b}{a^2 + b^2} \equiv 1 \bmod z^2 + 1.$$

Theorem 12 (Chinese Remainder Theorem) *Assume that $m_1(z), m_2(z), \ldots, m_n(z)$ are nonzero polynomials and pairwise relatively prime. Then the simultaneous linear congruence*

$$f(z) \equiv f_1(z) \bmod m_1(z)$$
$$f(z) \equiv f_2(z) \bmod m_2(z)$$
$$\vdots$$
$$f(z) \equiv f_n(z) \bmod m_n(z)$$

is solvable and its solution is

$$f(z) \equiv \sum_{i=1}^{n} N_i(z)M_i(z)f_i(z) \bmod M(z),$$

where $f_i(z)$ $(i = 1, 2, \ldots, n)$ are arbitrary polynomials and

$$
\begin{aligned}
M(z) &= m_1(z)m_2(z)\cdots m_n(z) \\
M_i(z) &= M(z)/m_i(z) \\
N_i(z)M_i(z) &\equiv 1 \bmod m_i(z)
\end{aligned}
$$

for $i = 1, 2, \ldots, n$.

Example 6 *Solve the simultaneous linear congruences*

$$f(z) \equiv f_0 \bmod z - 1$$
$$f(z) \equiv f_1(z) \bmod z + 1$$
$$f(z) \equiv f_2(z) \bmod z^2 + 1$$
$$\vdots$$
$$f(z) \equiv f_n(z) \bmod z^{2^{n-1}} + 1,$$

where $f_i(z)$ $(i = 1, 2, \ldots, n)$ are known polynomials and f_0 is a constant.

Solution. The modules are

$$m_0(z) = z - 1, \ m_i(z) = z^{2^{i-1}} + 1$$

$$M(z) = \prod_{i=0}^{n} m_i(z) = z^{2^n} - 1$$

$$M_0(z) = \frac{M(z)}{m_0(z)} = \frac{z^{2^n} - 1}{z - 1}$$

$$M_i(z) = \frac{z^{2^n} - 1}{z^{2^{i-1}} + 1}.$$

Since $z - 1, z + 1, z^2 + 1, \ldots, z^{2^{n-1}} + 1$ are pairwise relatively prime, from the CRT we know that the simultaneous linear congruences are solvable and the solution is

$$f(z) = s_0 \frac{z^{2^n} - 1}{z - 1} f_0 + \sum_{i=1}^{n} s_i(z) \frac{z^{2^n} - 1}{z^{2^{i-1}} + 1} f_i(z) \bmod z^{2^n} - 1,$$

where

$$s_0 \frac{z^{2^n} - 1}{z - 1} \equiv 1 \bmod z - 1$$

$$s_i(z) \frac{z^{2^n} - 1}{z^{2^{i-1}} + 1} \equiv 1 \bmod z^{2^{i-1}} + 1.$$

We can simply choose

$$s_0 = \frac{1}{2^n}, \ s_i(z) = -\frac{1}{2^{n+1-i}}.$$

Therefore, the solution is

$$f(z) \equiv \frac{1}{2^n} \frac{z^{2^n} - 1}{z - 1} f_0 - \sum_{i=1}^{n} \frac{1}{2^{n+1-i}} \frac{z^{2^n} - 1}{z^{2^{i-1}} + 1} f_i(z) \bmod z^{2^n} - 1.$$

This example is useful for the fast computation of MD convolutions [2, 4, 9, 11].

Example 7 *Use the CRT to compute polynomial multiplication*

$$h(z) \equiv f(z)g(z) \bmod z^{2^n} - 1,$$

where $f(z)$ and $g(z)$ are known polynomials.

Solution. We know that $z^{2^n} - 1$ can be factored into

$$z^{2^n} - 1 = (z - 1) \prod_{i=1}^{n} \left(z^{2^{i-1}} + 1 \right).$$

Let

$$f_0 \equiv f(z) \bmod z - 1, \qquad g_0 \equiv g(z) \bmod z - 1,$$
$$f_i(z) \equiv f(z) \bmod z^{2^{i-1}} + 1, \quad g_i(z) \equiv g(z) \bmod z^{2^{i-1}} + 1,$$

where $i = 1, 2, \ldots, n$, and

$$h_0 = f_0 g_0$$
$$h_i(z) \equiv f_i(z) g_i(z) \bmod z^{2^{i-1}} + 1. \tag{2.6}$$

Then

$$h(z) \equiv h_0 \bmod z - 1$$
$$h(z) \equiv h_1(z) \bmod z + 1$$
$$h(z) \equiv h_2(z) \bmod z^2 + 1$$
$$\vdots$$
$$h(z) \equiv h_n(z) \bmod z^{2^{n-1}} + 1.$$

Example 6 tells us that

$$h(z) = \frac{1}{2^n} \frac{z^{2^n} - 1}{z - 1} h_0 - \sum_{i=1}^{n} \frac{1}{2^{n+1-i}} \frac{z^{2^n} - 1}{z^{2^{i-1}} + 1} h_i(z) \bmod z^{2^n} - 1.$$

It shows that the problem of polynomial multiplication modulo $z^{2^n} - 1$ becomes a series of polynomial multiplications modulo $z - 1$ or modulo $z^{2^{i-1}} + 1$ $(i = 1, 2, \ldots, n)$ in (2.6). A complex calculation is converted into simple ones, which may lead to more efficient algorithms.

2.3   1D POLYNOMIAL TRANSFORM

2.3.1   Definition

Let R denote the real number field, that is, the whole set of real numbers. All polynomials to be discussed are assumed to have coefficients of real numbers.

Definition 16 *Let $M(z)$ and $G(z)$ be two nonzero polynomials, and $H_n(z)$ $(n = 0, 1, \ldots, N-1)$ be a polynomial sequence. We define*

$$\bar{H}_k(z) \equiv \sum_{n=0}^{N-1} H_n(z) G^{nk}(z) \bmod M(z), \quad k = 0, 1, \ldots, N - 1 \tag{2.7}$$

to be a PT if it has an inverse operation

$$H_n(z) \equiv \frac{1}{N} \sum_{k=0}^{N-1} \bar{H}_k(z) G^{-nk}(z) \bmod M(z), \quad n = 0, 1, \ldots, N - 1 \tag{2.8}$$

which is called an inverse polynomial transform (IPT).

The PT in (2.7) transforms the polynomial sequence $H_n(z)$ to another polynomial sequence $\bar{H}_k(z)$, while the IPT in (2.8) restores $H_n(z)$ from $\bar{H}_k(z)$. A PT is composed of three elements: $M(z)$, $G(z)$ and N, which are known as *module*, *factor* (or *root*) and *length*, respectively. For convenience, we use $(M(z), G(z), N)$ to represent the transform.

2.3.2 Formation of polynomial transform

In general, not all choices of the three elements can produce a PT. For example, if $M(z) = z^2 + 1$, $G(z) = 1$ and $N = 2$, (2.7) is not invertible. This can be seen from the fact that a polynomial sequence $H_n(z)$ $(n = 0, 1)$ is transformed into a single polynomial $H_0(z) + H_1(z)$. It is impossible to recover the two polynomials from their sum. Let us consider how to use the three elements to generate a PT.

Theorem 13 $(M(z), G(z), N)$ *can generate a PT if and only if*

$$\frac{1}{N} \sum_{n=0}^{N-1} G^{nk}(z) \equiv \begin{cases} 1 \mod M(z), & k \equiv 0 \mod N \\ 0 \mod M(z), & k \not\equiv 0 \mod N \end{cases}. \tag{2.9}$$

Proof. We first consider the sufficiency of condition (2.9). If (2.9) holds, we have

$$G^N(z) - 1 \equiv (G(z) - 1) \sum_{n=0}^{N-1} G^n(z) \equiv 0 \mod M(z).$$

Then we have

$$G^N(z) \equiv 1 \mod M(z)$$

and for $n = 0, 1, \ldots, N - 1$,

$$\frac{1}{N} \sum_{k=0}^{N-1} \bar{H}_k(z) G^{-nk}(z) \equiv \frac{1}{N} \sum_{k=0}^{N-1} \sum_{m=0}^{N-1} H_m(z) G^{mk}(z) G^{-nk}(z)$$

$$\equiv \sum_{m=0}^{N-1} H_m(z) \left[\frac{1}{N} \sum_{k=0}^{N-1} G^{(m-n)k}(z) \right]$$

$$\equiv H_n(z) \mod M(z),$$

which means that the transform in (2.7) is invertible and the inversion can be done by (2.8).

Now we prove the necessity of condition (2.9). If (2.7) is a PT, for any polynomial sequence: $H_n(z)$ we have

$$\frac{1}{N} \sum_{k=0}^{N-1} \bar{H}_k(z) G^{-nk}(z) \equiv \sum_{m=0}^{N-1} H_m(z) \left[\frac{1}{N} \sum_{k=0}^{N-1} G^{(m-n)k}(z) \right]$$

$$\equiv H_n(z) \mod M(z). \tag{2.10}$$

Since (2.10) is correct for any polynomial sequence $H_n(z)$ $(n = 0, 1, \ldots, N-1)$, it is easy to show that (2.9) holds.

Theorem 13 is essential, but it is difficult to use it for constructing a PT. Now we consider the next theorem.

Theorem 14 $(M(z), G(z), N)$ *can generate a PT if and only if*

$$G^N(z) \equiv 1 \bmod M(z) \tag{2.11}$$

and for any positive integer n which is a factor of N such that the quotient N/n is a prime number,

$$\mathrm{GCD}(M(z), G^n(z) - 1) = 1, \tag{2.12}$$

where GCD *means the greatest common divisor.*

With some knowledge of the number theory, this theorem can be proven (see [4]). Using this theorem, we can derive two types of PTs, which are particularly useful for constructing fast algorithms.

Theorem 15 $(M(z), G(z), p^l)$ *is a PT if p is a prime number, l is a positive integer, $G(z)$ is a nonzero polynomial and*

$$M(z) = \frac{G^{p^l}(z) - 1}{G^{p^{l-1}}(z) - 1}. \tag{2.13}$$

Proof. It is obvious that $G^{p^l}(z) \equiv 1 \bmod M(z)$ and factors of p^l are p^i $(i = 0, 1, \ldots, l)$. Among these factors, there exists only one factor, p^{l-1}, such that the quotient is a prime number. Based on Theorem 14 we only need to prove that

$$\mathrm{GCD}(M(z), G^{p^{l-1}}(z) - 1) = 1.$$

Since

$$
\begin{aligned}
M(z) &= \frac{G^{p^l}(z) - 1}{G^{p^{l-1}}(z) - 1} \\
&= G^{p^{l-1}(p-1)}(z) + G^{p^{l-1}(p-2)}(z) + \cdots + G^{p^{l-1}}(z) + 1,
\end{aligned}
$$

we have

$$M(z) \equiv p \bmod G^{p^{l-1}}(z) - 1.$$

Therefore,

$$\mathrm{GCD}(M(z), G^{p^{l-1}}(z) - 1) = \mathrm{GCD}(p, G^{p^{l-1}}(z) - 1) = 1.$$

By properly choosing $G(z)$, we can generate many useful PTs, as shown in the following example.

Example 8 $G(z) = z^q$, $M(z) = (z^{qp^l} - 1)/(z^{qp^{l-1}} - 1)$, *where q is a positive integer and p is a prime. Then $(M(z), z^q, p^l)$ is a PT. Especially, by choosing q and l to be 1 or 2, respectively, the following PTs can be achieved.*

- $((z^p - 1)/(z - 1), z, p)$
- $((z^{p^2} - 1)/(z^p - 1), z, p^2)$
- $((z^{2p} - 1)/(z^2 - 1), z^2, p)$
- $((z^{2p^2} - 1)/(z^{2p} - 1), z^2, p^2)$

If $p = 2$ and $q = 2^j$, we know that $(z^{2^{j+l-1}} + 1, z^{2^j}, 2^l)$ is a PT. If $j = 0$ and 1, respectively, we obtain the following two most useful PTs.

- $(z^{2^{l-1}} + 1, z, 2^l)$

- $(z^{2^l} + 1, z^2, 2^l)$

These two PTs were used for computing MD DFTs and MD convolutions [9, 11]. They are the simplest and have FFT-like fast algorithms.

Theorem 16 *$(M(z), G(z), 2p^l)$ is a PT if p is an odd prime number, l is a positive integer, $G(z)$ is a nonzero polynomial and*

$$M(z) = \frac{G^{p^l}(z) + 1}{G^{p^{l-1}}(z) + 1}. \tag{2.14}$$

The proof can be easily performed in a similar way to that for Theorem 15. Based on Theorem 16 and by properly choosing $G(z)$, we can generate a few useful PTs in the following example.

Example 9 *$G(z) = z^q$, where q is a positive integer, $M(z) = (z^{qp^l} + 1)/(z^{qp^{l-1}} + 1)$. Then $(M(z), z^q, 2p^l)$ is a PT, where p is an odd prime number. Especially, by choosing q to be 1 or p, we have*

- $((z^p + 1)/(z + 1), z, 2p)$

- $((z^{p^l} + 1)/(z^{p^{l-1}} + 1), z, 2p^l)$

- $((z^{p^{l+1}} + 1)/(z^{p^l} + 1), z^p, 2p^l)$

Table 2.2 lists a number of useful PTs.

Table 2.2   Commonly used polynomial transforms.

Module	Factor	Length	Remarks
$z^{2^{l-1}} + 1$	z	2^l	l: positive integer
$z^{2^{j+l-1}} + 1$	z^{2^j}	2^l	l: positive integer
$(z^p - 1)/(z - 1)$	z	p	p: prime
$(z^p - 1)/(z - 1)$	$-z$	$2p$	p: odd prime
$(z^{p^2} - 1)/(z^p - 1)$	z	p^2	p: prime
$(z^{p^2} - 1)/(z^p - 1)$	$-z$	$2p^2$	p: odd prime
$(z^{qp^l} - 1)/(z^{qp^{l-1}} - 1)$	z^q	p^l	p: prime; q, l: integers
$(z^{p^2} + 1)/(z^p + 1)$	z^p	$2p$	p: odd prime
$(z^{qp^l} + 1)/(z^{qp^{l-1}} + 1)$	z^q	$2p^l$	p: odd prime; q, l: integers
$(z^{qp^l} + 1)/(z^{qp^{l-1}} + 1)$	z^{2q}	p^l	p: odd prime; q, l: integers

2.3.3   Properties of polynomial transform

Let $(M(z), G(z), N)$ be a PT. We use $H_n(z)$ and $\text{PT}(H_n(z))$ $(n = 0, 1, \ldots, N-1)$ to denote the input and the output of the polynomial transform. Similar to the DFT, the PT has many useful properties. Some of them are described as follows.

- **Linearity**
$$\text{PT}(c_1 H_n(z) + c_2 X_n(z)) = c_1 \text{PT}(H_n(z)) + c_2 \text{PT}(X_n(z)),$$

where c_1 and c_2 are real numbers.

- **Symmetries**
  If $H_n(z)$ $(n = 0, 1, \ldots, N-1)$ is symmetric, that is,
$$H_n(z) \equiv H_{N-n}(z) \bmod M(z),$$

then its PT output $\bar{H}_k(z)$ is also symmetric, that is,
$$\bar{H}_k(z) \equiv \bar{H}_{N-k}(z) \bmod M(z), \ k = 0, 1, \ldots, N-1.$$

If $H_n(z)$ $(n = 0, 1, \ldots, N-1)$ is anti-symmetric, that is,
$$H_n(z) \equiv -H_{N-n}(z) \bmod M(z),$$

then its PT output $\bar{H}_k(z)$ is also anti-symmetric, that is,
$$\bar{H}_k(z) \equiv -\bar{H}_{N-k}(z) \bmod M(z), \ k = 0, 1, \ldots, N-1.$$

- **Shift property**
  Let m be a fixed integer and $\bar{H}_k(z) = \text{PT}(H_n(z))$. Then
$$\text{PT}(H_{\langle n+m \rangle}(z)) = \bar{H}_k(z) G(z)^{-mk} \bmod M(z)$$

and
$$\text{PT}(G(z)^{mn} H_n(z) \bmod M(z)) = \bar{H}_{\langle k+m \rangle}(z),$$

where $\langle l \rangle$ means the least non-negative residue of integer l modulo N. For example, if $N = 16$, $l = -3$, then $\langle -3 \rangle = 13$.

- **Convolutional property**
  Let $H_n(z)$ and $X_n(z)$ $(n = 0, 1, \ldots, N-1)$ be two polynomial sequences and let $Y_n(z)$ be their cyclic convolution, that is,
$$Y_n(z) \equiv \sum_{m=0}^{N-1} X_m(z) H_{\langle n-m \rangle}(z) \bmod M(z), \ \ n = 0, 1, \ldots, N-1. \qquad (2.15)$$

Then,
$$\text{PT}(Y_n(z)) \equiv \text{PT}(X_n(z)) \text{PT}(H_n(z)) \bmod M(z). \qquad (2.16)$$

Proof. Let
$$\bar{H}_k(z) = \text{PT}(H_n(z)), \ \bar{X}_k(z) = \text{PT}(X_n(z)), \ \bar{Y}_k(z) = \text{PT}(Y_n(z)).$$

We only need to prove that the IPT of $\bar{X}_k(z)\bar{H}_k(z)$ is $Y_n(z)$. The IPT of $\bar{X}_k(z)\bar{H}_k(z)$ is

$$\frac{1}{N}\sum_{k=0}^{N-1}\bar{X}_k(z)\bar{H}_k(z)G^{-nk}(z) \bmod M(z)$$

$$= \frac{1}{N}\sum_{k=0}^{N-1}\left[\sum_{l=0}^{N-1}X_l(z)G^{lk}(z)\right]\left[\sum_{m=0}^{N-1}H_m(z)G^{mk}(z)\right]G^{-nk}(z) \bmod M(z)$$

$$= \sum_{l=0}^{N-1}X_l(z)\left\{\sum_{m=0}^{N-1}H_m(z)\left[\frac{1}{N}\sum_{k=0}^{N-1}G^{(m+l-n)k}(z)\right]\right\} \bmod M(z). \tag{2.17}$$

Since $(M(z), G(z), N)$ is a PT, we have

$$\frac{1}{N}\sum_{n=0}^{N-1}G^{(m+l-n)k}(z) \equiv \begin{cases} 1 \bmod M(z), & m \equiv n-l \bmod N \\ 0 \bmod M(z), & m \not\equiv n-l \bmod N \end{cases}.$$

Therefore, (2.17) becomes

$$\sum_{l=0}^{N-1}X_l(z)H_{\langle n-l\rangle}(z) \bmod M(z) = Y_n(z), \quad n = 0, 1, \ldots, N-1.$$

This property is similar to the cyclic convolutional property of the DFT. It is an important property for using the PT to compute convolutions [9, 11].

- **Parseval property**
  Let $\bar{H}_k(z) = \mathrm{PT}(H_n(z))$ and $\bar{X}_k(z) = \mathrm{PT}(X_n(z))$. Then

$$N\sum_{n=0}^{N-1}X_n(z)H_n(z) \equiv \sum_{k=0}^{N-1}\bar{X}_k(z)\bar{H}_{\langle N-k\rangle}(z) \bmod M(z)$$

and

$$N\sum_{n=0}^{N-1}X_n(z)H_{\langle N-n\rangle}(z) \equiv \sum_{k=0}^{N-1}\bar{X}_k(z)\bar{H}_k(z) \bmod M(z).$$

2.4   FAST POLYNOMIAL TRANSFORM

2.4.1   Radix-2 fast polynomial transform

It is easy to derive a fast algorithm for a PT whose length is a power of two. Let N and M be positive integers and
$$N = 2^t, \ M = 2^{r-1}N = 2^{t+r-1}.$$

It is proved that $(z^M + 1, z^{2^r}, N)$ is a PT. For simplicity, we represent z^{2^r} by $\tilde{z}$. When the input is $H_n(z)$ $(n = 0, 1, \ldots, N-1)$, the output of the PT is

$$\bar{H}_k(z) \equiv \sum_{n=0}^{N-1}H_n(z)\tilde{z}^{nk} \bmod z^M + 1, \quad k = 0, 1, \ldots, N-1. \tag{2.18}$$

For simplicity, we assume that the degree of the input polynomials $H_n(z)$ is smaller than M. Therefore, we have

$$H_n(z) = \sum_{m=0}^{M-1} h(n,m) z^m, \tag{2.19}$$

where $h(n,m)$ are the coefficients of $H_n(z)$.

Both polynomial addition and multiplication (modulo $z^M + 1$) are needed in (2.18). The polynomial addition uses M real additions by simply adding the corresponding coefficients, respectively. Multiplication of two polynomials is generally more difficult, and is actually a skew-cyclic convolution [2, 4, 5]. For our computation, fortunately, we do not need general polynomial multiplication. We only need to multiply general polynomials $H_n(z)$ with special polynomials $\tilde{z}^{nk} = z^{2^r nk}$. Since

$$z^M \equiv -1 \bmod z^M + 1,$$

we can replace z^{lM} by $(-1)^l$ in our computation. In general, let $2^r nk = lM + q$, where $0 \leq q \leq M - 1$. Then

$$H_n(z) z^{2^r nk} \equiv H_n(z)(-1)^l z^q \bmod z^M + 1,$$

which can be expressed into the computation of

$$B(z) \equiv A(z) z^q \bmod z^M + 1, \tag{2.20}$$

where $A(z)$ is a polynomial with a degree smaller than M. Let the coefficients of $A(z)$ and $B(z)$ be a_m and b_m, respectively,

$$A(z) = \sum_{m=0}^{M-1} a_m z^m, \quad B(z) = \sum_{m=0}^{M-1} b_m z^m.$$

It is easily seen that b_m is the reordering of a_m with possible sign changes. In fact, we have

$$b_m = \begin{cases} -a_{m+M-q}, & m = 0, 1, \ldots, q-1 \\ a_{m-q}, & m = q, \ldots, M-1 \end{cases}.$$

It shows that (2.20) needs no additions and multiplications at all but only some simple index mappings. Therefore, we call the operation in (2.20) a *polynomial rotation*.

If we compute (2.18) directly, the numbers of real additions and polynomial rotations needed are

$$A_d = NM(N-1)$$

and

$$R_t = N^2.$$

respectively. By using an algorithm known as the *fast polynomial transform* (FPT), we can dramatically reduce the number of additions and polynomial rotations. The FPT is described as follows.

Most fast algorithms are derived by using the *divide and conquer* method. By properly *dividing* a large computational task into a number of smaller ones which can be much more easily *conquered*, we can significantly reduce the total computational complexity. For example, the entire computation task can be divided into two smaller ones, each of which is again further divided into halves. This decomposition technique is used recursively until the size of the divided tasks are small enough to be easily solved. Such a technique can be generally applied by *decimation-in-time* and *decimation-in-frequency*, as will be seen in the following chapters.

To start with the *decimation-in-time* decomposition, we divide the length N $H_n(z)$ into $H_{2n}(z)$ and $H_{2n+1}(z)$ which are length $N/2$. The PT can be expressed by

$$\bar{H}_k(z) \equiv \sum_{n=0}^{N/2-1} H_{2n}(z)\tilde{z}^{2nk} + \sum_{n=0}^{N/2-1} H_{2n+1}(z)\tilde{z}^{(2n+1)k} \mod z^M + 1$$

$$\equiv \sum_{n=0}^{N/2-1} H_{2n}(z)(\tilde{z}^2)^{nk} + \tilde{z}^k \sum_{n=0}^{N/2-1} H_{2n+1}(z)(\tilde{z}^2)^{nk} \mod z^M + 1.$$

Let

$$\bar{H}_k^1(z) \equiv \sum_{n=0}^{N/2-1} H_{2n}(z)(\tilde{z}^2)^{nk} \mod z^M + 1 \tag{2.21}$$

$$\bar{H}_k^2(z) \equiv \sum_{n=0}^{N/2-1} H_{2n+1}(z)(\tilde{z}^2)^{nk} \mod z^M + 1, \tag{2.22}$$

where $k = 0, 1, \ldots, N/2 - 1$. Because

$$\tilde{z}^{N/2} = z^M \equiv -1 \mod z^M + 1,$$

we have

$$\bar{H}_k(z) \equiv \bar{H}_k^1(z) + \tilde{z}^k \bar{H}_k^2(z) \mod z^M + 1 \tag{2.23}$$

$$\bar{H}_{k+N/2}(z) \equiv \bar{H}_k^1(z) - \tilde{z}^k \bar{H}_k^2(z) \mod z^M + 1, \tag{2.24}$$

where $k = 0, 1, \ldots, N/2 - 1$. The computation of PT in (2.18) is now changed into the computation of (2.21) and (2.22). Because $(z^M + 1, \tilde{z}^2, N/2)$ is also a PT, a length N PT can be computed with two length $N/2$ PTs and a post-processing stage defined by (2.23) and (2.24). Similarly, every length $N/2$ PT can be further divided into two length $N/4$ PTs and a post-processing stage. Finally, after $t = \log_2 N$ stages, the length N PT is decomposed into N length 1 PTs. Obviously, no operation is needed for a length 1 PT. Let $A_d^I(N)$ and $R_t^I(N)$ be the number of additions and polynomial rotations, respectively, for the computation of a length N PT. We have

$$A_d^I(N) = 2A_d^I\left(\frac{N}{2}\right) + NM \tag{2.25}$$

$$R_t^I(N) = 2R_t^I\left(\frac{N}{2}\right) + N - 2. \tag{2.26}$$

By recursively using (2.25) and (2.26), we finally get

$$A_d^I(N) = NM \log_2 N \tag{2.27}$$

$$R_t^I(N) = (N - 2) \log_2 N. \tag{2.28}$$

Compared with direct computation, the above FPT algorithm significantly reduces the computational complexity.

A more general method can be used to derive the fast algorithm described above to show the computational steps more clearly. Let the binary expressions of k and n be $(k_{t-1}k_{t-2}\cdots k_0)$ and $(n_{t-1}n_{t-2}\cdots n_0)$, respectively, that is,

$$n = n_{t-1}2^{t-1} + n_{t-2}2^{t-2} + \cdots + n_0$$

$$k = k_{t-1}2^{t-1} + k_{t-2}2^{t-2} + \cdots + k_0,$$

where $0 \le k, \, n \le 2^t - 1$, and $k_i, \, n_i \in \{0,1\}$. We convert the polynomial transform (2.18) into

$$\bar{H}_k(z) = \bar{H}_{k_{t-1}k_{t-2}\cdots k_0}(z)$$
$$\equiv \sum_{n_0=0}^{1}\sum_{n_1=0}^{1}\cdots\sum_{n_{t-1}=0}^{1} H_{n_{t-1}n_{t-2}\cdots n_0}(z)\tilde{z}^{nk} \bmod z^M + 1,$$

where $\tilde{z}^{nk}$ can be expressed as

$$\tilde{z}^{nk} \equiv \tilde{z}^{2^{t-1}n_{t-1}k_0}\tilde{z}^{2^{t-2}n_{t-2}(2k_1+k_0)}\cdots \tilde{z}^{n_0(2^{t-1}k_{t-1}+\cdots+k_0)} \bmod z^{2M} + 1.$$

We further define

$$\bar{H}^0_{n_0n_1\cdots n_{t-1}}(z) = H_{n_{t-1}n_{t-2}\cdots n_0}(z),$$

$$\bar{H}^j_{n_0\cdots n_{t-j-1}k_{j-1}\cdots k_0}(z) \equiv \sum_{n_{t-j}=0}^{1}\cdots\sum_{n_{t-1}=0}^{1} H_{n_{t-1}n_{t-2}\cdots n_0}(z)$$
$$\cdot\tilde{z}^{2^{t-1}n_{t-1}k_0+2^{t-2}n_{t-2}(2k_1+k_0)+\cdots+2^{t-j}n_{t-j}(2^{j-1}k_{j-1}+\cdots+k_0)} \bmod z^M + 1, \quad (2.29)$$

where $j = 1, 2, \ldots, t$ and sequence $\{\bar{H}^0_n(z)\}$ is the binary inverse of sequence $\{H_n(z)\}$. Because

$$\bar{H}^t_{k_{t-1}k_{t-2}\cdots k_0}(z) = \bar{H}_{k_{t-1}k_{t-2}\cdots k_0}(z)$$

or

$$\bar{H}^t_k(z) = \bar{H}_k(z), \quad k = 0, 1, \ldots, 2^t - 1,$$

we need only to compute $\bar{H}^t_k(z)$. Let

$$\begin{aligned}\hat{n} &= (n_0\cdots n_{t-j-1}), \quad \bar{k} = (k_{j-1}\cdots k_0)\\ \bar{n} &= (n_{t-1}\cdots n_{t-j}), \quad n' = (n_{t-j-1}\cdots n_0),\end{aligned} \quad (2.30)$$

where n' is the binary inverse of $\hat{n}$. We can rewrite (2.29) as

$$\bar{H}^j_{\hat{n}2^j+\bar{k}}(z) \equiv \sum_{\bar{n}=0}^{2^j-1} H_{\bar{n}2^{t-j}+n'}(z)\tilde{z}^{2^{t-j}\bar{n}\bar{k}} \bmod z^M + 1 \quad (2.31)$$

$$\bar{k} = 0, 1, \ldots, 2^j - 1, \, \hat{n} = 0, 1, \ldots, 2^{t-j} - 1,$$

where n' is the binary inverse of $\hat{n}$. Equation (2.31) tells us that $\bar{H}^j_{\hat{n}2^j+\bar{k}}(z)$ is the PT (length 2^j, module $z^M + 1$ and root $\tilde{z}^{2^{t-j}}$) of $H_{\bar{n}2^{t-j}+n'}(z)$. This equation will be used in the following chapters for deriving PT-based algorithms for the MD DCT and DHT. From (2.29) and (2.31), we can prove that

$$\bar{H}^j_{\hat{n}2^j+\bar{k}}(z) \equiv \bar{H}^{j-1}_{\hat{n}2^j+\bar{k}}(z) + \bar{H}^{j-1}_{\hat{n}2^j+(\bar{k}+2^{j-1})}(z)\tilde{z}^{2^{t-j}\bar{k}} \bmod z^M + 1 \quad (2.32)$$

$$\bar{H}^j_{\hat{n}2^j+(\bar{k}+2^{j-1})}(z) \equiv \bar{H}^{j-1}_{\hat{n}2^j+\bar{k}}(z) - \bar{H}^{j-1}_{\hat{n}2^j+(\bar{k}+2^{j-1})}(z)\tilde{z}^{2^{t-j}\bar{k}} \bmod z^M + 1, \quad (2.33)$$

where $\hat{n} = 0, 1, \ldots, 2^{t-j}-1$, $\bar{k} = 0, 1, \ldots, 2^{j-1}-1$, and $j = 1, 2, \ldots, t$. These are the t stages of the decimation-in-time radix-2 FPT. The last stage produces the outputs $\bar{H}_k^t(z) = \bar{H}_k(z)$.

Algorithm 1 *Decimation-in-time radix-2 FPT*

Step 1. *Reorder the input polynomial sequence $\{H_n(z)\}$ in binary inverse order to get $\{\bar{H}_n^0(z)\}$, that is, $\{\bar{H}_n^0(z)\} = \{H_{n'}(z)\}$, where n' is the binary inverse of n, $0 \leq n$, $n' \leq 2^t - 1$.*

Step 2. *For j from 1 to t, compute the t stages defined by (2.32) and (2.33). The last stage produces the $\bar{H}_k(z)$, that is, the PT of $H_n(z)$.*

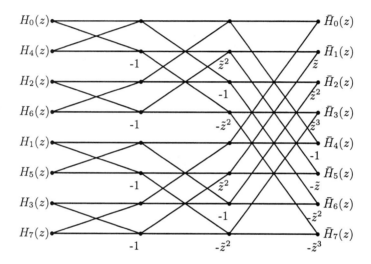

Figure 2.1   Signal flow graph of 8-point FPT (decimation-in-time).

Figure 2.1 shows the signal flow graph of the FPT for $N = 8$. It is easy to obtain the signal flow graph of the FPT for any length N. In general, the input polynomial sequence must be binary inversely ordered. For example, for $N = 8$, $3 = 0 \cdot 2^2 + 1 \cdot 2 + 1$, the binary inverse of 3 is $1 \cdot 2^2 + 1 \cdot 2 + 0 = 6$, which means that $H_3(z)$ must be at the 6th place. The binary inverses of 0, 1, 2, 3, 4, 5, 6 and 7 are, respectively, 0, 4, 2, 6, 1, 5, 3 and 7. The FPT also supports in-place computation which minimizes the consumption of memory space.

Now, let us consider the *decimation-in-frequency* decomposition.

$$\bar{H}_{2k}(z) \equiv \sum_{n=0}^{N/2-1} H_n(z)\tilde{z}^{2nk} + \sum_{n=0}^{N/2-1} H_{n+N/2}(z)\tilde{z}^{2nk} \bmod z^M + 1$$

$$\equiv \sum_{n=0}^{N/2-1} (H_n(z) + H_{n+N/2}(z))(\tilde{z}^2)^{nk} \bmod z^M + 1$$

$$\bar{H}_{2k+1}(z) \equiv \sum_{n=0}^{N/2-1} H_n(z)\tilde{z}^n\tilde{z}^{2nk} - \sum_{n=0}^{N/2-1} H_{n+N/2}(z)\tilde{z}^n\tilde{z}^{2nk} \bmod z^M + 1$$

$$\equiv \sum_{n=0}^{N/2-1} (H_n(z) - H_{n+N/2}(z))\tilde{z}^n(\tilde{z}^2)^{nk} \bmod z^M + 1,$$

where the properties

$$\tilde{z}^N \equiv 1 \bmod z^M + 1, \ \tilde{z}^{N/2} \equiv -1 \bmod z^M + 1$$

are used. Let

$$H_n^1(z) \equiv H_n(z) + H_{n+N/2}(z) \bmod z^M + 1 \qquad (2.34)$$

$$H_n^2(z) \equiv (H_n(z) - H_{n+N/2}(z))\tilde{z}^n \bmod z^M + 1, \qquad (2.35)$$

where $n = 0, 1, \ldots, N/2 - 1$. Then we have

$$H_{2k}(z) \equiv \sum_{n=0}^{N/2-1} H_n^1(z)(\tilde{z}^2)^{nk} \bmod z^M + 1 \qquad (2.36)$$

$$H_{2k+1}(z) \equiv \sum_{n=0}^{N/2-1} H_n^2(z)(\tilde{z}^2)^{nk} \bmod z^M + 1, \qquad (2.37)$$

where $k = 0, 1, \ldots, N/2 - 1$. Similar to the decimation-in-time FPT, we also convert the length N PT in (2.18) into two length $N/2$ PTs in (2.36) and (2.37). A *pre-processing* stage defined in (2.34) and (2.35) is needed. Every length $N/2$ PT can be further decomposed into two length $N/4$ PTs with a pre-processing stage. Finally, after $t = \log_2 N$ steps, the length N PT is converted into N length 1 PTs. The required numbers of additions and polynomial rotations are the same as those for the decimation-in-time FPT. The signal flow graph of the FPT for $N = 8$ is shown in Figure 2.2.

Fast algorithms for the IPT can be designed in a similar way, which is not repeated here.

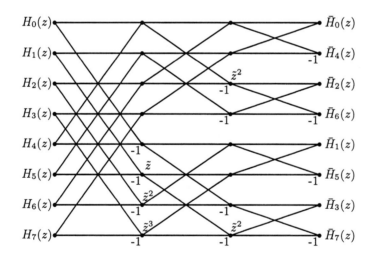

Figure 2.2   Signal flow graph of 8-point FPT (decimation-in-frequency).

2.4.2   Radix-p fast polynomial transform

Fast algorithms for other types of PTs can be similarly designed. This subsection presents the computation of such PTs as $((z^{qp^t} + 1)/(z^{qp^{t-1}} + 1), z^{2q}, p^t)$, where p is an odd prime number, t is a positive integer and q is an integer. For convenience, let $N = p^t$, $M = qp^t$ and $\hat{z} = z^{2q}$. The output of the transform is

$$\bar{H}_k(z) \equiv \sum_{n=0}^{N-1} H_n(z)\hat{z}^{nk}(z) \bmod (z^M + 1)/(z^{M/p} + 1), \qquad (2.38)$$

where $k = 0, 1, \ldots, N-1$ and $H_n(z)$ $(n = 0, 1, \ldots, N-1)$ is the input. Since the degree of the module polynomial is $M(1 - 1/p)$, the degree of the input polynomials is assumed to be smaller than $M(1 - 1/p)$. Let

$$\hat{H}_k(z) \equiv \sum_{n=0}^{N-1} H_n(z)\hat{z}^{nk} \bmod (z^M + 1). \qquad (2.39)$$

It is noted that (2.39) is not a PT because $(z^M + 1, \hat{z}, N)$ does not satisfy the requirements in Theorem 13 or Theorem 14. However, we need to further investigate (2.39) because it will be used in the fast algorithms for MD transforms in the following chapters of this book. It is obvious that

$$\bar{H}_k(z) \equiv \hat{H}_k(z) \bmod (z^M + 1)/(z^{M/p} + 1). \qquad (2.40)$$

We also use the *divide and conquer* technique to design an FPT algorithm for the computation of (2.39). By using an index mapping:

$$k = pk_1 + k_0, \quad n = \frac{N}{p}n_0 + n_1, \qquad (2.41)$$

where

$$k_0 = 0, 1, \ldots, p-1, \ k_1 = 0, 1, \ldots, \frac{N}{p} - 1$$

$$n_0 = 0, 1, \ldots, p-1, \ n_1 = 0, 1, \ldots, \frac{N}{p} - 1,$$

we get

$$\begin{aligned}
\hat{H}_k(z) &= \hat{H}_{pk_1+k_0}(z) \\
&\equiv \sum_{n_1=0}^{N/p-1} \sum_{n_0=0}^{p-1} H_{\frac{N}{p}n_0+n_1}(z)\hat{z}^{(\frac{N}{p}n_0+n_1)(pk_1+k_0)} \bmod (z^M + 1) \\
&\equiv \sum_{n_1=0}^{N/p-1} \left[\sum_{n_0=0}^{p-1} H_{\frac{N}{p}n_0+n_1}(z)\hat{z}^{(\frac{N}{p}n_0+n_1)k_0} \right] \hat{z}^{pn_1k_1} \bmod (z^M + 1).
\end{aligned}$$

Let

$$H'_{n_1,k_0}(z) \equiv \sum_{n_0=0}^{p-1} H_{\frac{N}{p}n_0+n_1}(z)\hat{z}^{(\frac{N}{p}n_0+n_1)k_0} \bmod (z^M + 1). \qquad (2.42)$$

Then

$$\hat{H}_{pk_1+k_0}(z) \equiv \sum_{n_1=0}^{N/p-1} H'_{n_1,k_0}(z)(\hat{z}^p)^{n_1k_1} \bmod (z^M + 1) \qquad (2.43)$$

$$k_0 = 0, 1, \ldots, p-1, \ k_1 = 0, 1, \ldots, \frac{N}{p} - 1.$$

For the derivation of the above equations, we have used the property

$$\hat{z}^N \equiv 1 \bmod (z^M + 1).$$

For fixed k_0, (2.43) is the same as (2.39) except that the length is now N/p. This means that (2.39) is decomposed into p smaller transforms (for $k_0 = 0, 1, \ldots, p-1$) of length N/p at the cost of a pre-processing stage (2.42). Since

$$(\hat{z}^p)^{N/p} = \hat{z}^N \equiv 1 \bmod (z^M + 1),$$

each transform in (2.43) can be decomposed further in the same way until the size is p. Finally, the length p transform can be directly computed. The computation of the pre-processing stage (2.42) needs $(p-1)NM$ additions and $(p-1)N$ polynomial rotations (including some trivial rotations). Let $A_d^{II}(N)$ and $R_t^{II}(N)$ be the number of additions and polynomial rotations, respectively, for the computation of (2.39). We have

$$A_d^{II}(N) = pA_d^{II}(\frac{N}{p}) + (p-1)NM \tag{2.44}$$

$$R_t^{II}(N) = pR_t^{II}(\frac{N}{p}) + (p-1)N \tag{2.45}$$

or in a closed form

$$A_d^{II}(N) = (p-1)NM \log_p N \tag{2.46}$$

$$R_t^{II}(N) = (p-1)N \log_p N. \tag{2.47}$$

After obtaining (2.39), we compute (2.40) to get (2.38). In general, the *Euclid's algorithm* can be used for such a computation. A more efficient algorithm, however, is available if the polynomial $(z^M + 1)/(z^{M/p} + 1)$ has some special properties.

When $N = 3^2 = 9$, the FPT algorithm for (2.39) is shown below in Example 10.

Example 10 *Decimation-in-frequency radix-3 FPT for (2.39) ($N = 9$)*

$$H'_{0,0}(z) = H_0(z) + H_3(z) + H_6(z)$$
$$H'_{1,0}(z) = H_1(z) + H_4(z) + H_7(z)$$
$$H'_{2,0}(z) = H_2(z) + H_5(z) + H_8(z)$$
$$H'_{0,1}(z) \equiv H_0(z) + H_3(z)\hat{z}^3 + H_6(z)\hat{z}^6 \bmod (z^M + 1)$$
$$H'_{1,1}(z) \equiv [H_1(z) + H_4(z)\hat{z}^3 + H_7(z)\hat{z}^6]\hat{z} \bmod (z^M + 1)$$
$$H'_{2,1}(z) \equiv [H_2(z) + H_5(z)\hat{z}^3 + H_8(z)\hat{z}^6]\hat{z}^2 \bmod (z^M + 1)$$
$$H'_{0,2}(z) \equiv H_0(z) + H_3(z)\hat{z}^6 + H_6(z)\hat{z}^3 \bmod (z^M + 1)$$
$$H'_{1,2}(z) \equiv [H_1(z) + H_4(z)\hat{z}^6 + H_7(z)\hat{z}^3]\hat{z}^2 \bmod (z^M + 1)$$
$$H'_{2,2}(z) \equiv [H_2(z) + H_5(z)\hat{z}^6 + H_8(z)\hat{z}^3]\hat{z}^4 \bmod (z^M + 1)$$
$$H'_{1,2}(z) \equiv [H_1(z) + H_4(z)\hat{z}^6 + H_7(z)\hat{z}^3]\hat{z}^2 \bmod (z^M + 1)$$

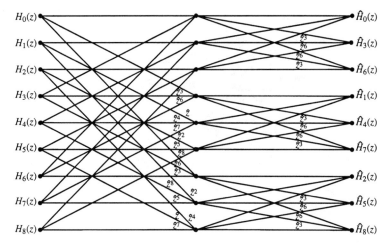

Figure 2.3   Signal flow graph of radix-3 FPT.

$$H'_{2,2}(z) \equiv [H_2(z) + H_5(z)\hat{z}^6 + H_8(z)\hat{z}^3]\hat{z}^4 \bmod (z^M + 1)$$

$$\hat{H}_0(z) = H'_{0,0}(z) + H'_{1,0}(z) + H'_{2,0}(z)$$

$$\hat{H}_3(z) = H'_{0,0}(z) + H'_{1,0}(z)\hat{z}^3 + H'_{2,0}(z)\hat{z}^6 \bmod (z^M + 1)$$

$$\hat{H}_6(z) = H'_{0,0}(z) + H'_{1,0}(z)\hat{z}^6 + H'_{2,0}(z)\hat{z}^3 \bmod (z^M + 1)$$

$$\hat{H}_1(z) = H'_{0,1}(z) + H'_{1,1}(z) + H'_{2,1}(z)$$

$$\hat{H}_4(z) = H'_{0,1}(z) + H'_{1,1}(z)\hat{z}^3 + H'_{2,1}(z)\hat{z}^6 \bmod (z^M + 1)$$

$$\hat{H}_7(z) = H'_{0,1}(z) + H'_{1,1}(z)\hat{z}^6 + H'_{2,1}(z)\hat{z}^3 \bmod (z^M + 1)$$

$$\hat{H}_2(z) = H'_{0,2}(z) + H'_{1,2}(z) + H'_{2,2}(z)$$

$$\hat{H}_5(z) = H'_{0,2}(z) + H'_{1,2}(z)\hat{z}^3 + H'_{2,2}(z)\hat{z}^6 \bmod (z^M + 1)$$

$$\hat{H}_8(z) = H'_{0,2}(z) + H'_{1,2}(z)\hat{z}^6 + H'_{2,2}(z)\hat{z}^3 \bmod (z^M + 1),$$

where $\hat{z}^9 \equiv 1 \bmod (z^M + 1)$. The above computation needs 36 polynomial additions, i.e. 36M real number additions, and 30 polynomial rotations. The signal flow graph is shown in Figure 2.3.

In general, a decimation-in-time algorithm can be derived in a similar way as described above. Instead of doing so, we derive the decimation-in-time algorithm based on a different approach. If we express n and k by $(n_{t-1}n_{t-2}\cdots n_0)$ and $(k_{t-1}k_{t-2}\cdots k_0)$ into

$$n = n_{t-1}p^{t-1} + n_{t-2}p^{t-2} + \cdots + n_0$$

$$k = k_{t-1}p^{t-1} + k_{t-2}p^{t-2} + \cdots + k_0,$$

respectively, where $0 \leq n_i,\ k_i \leq p - 1$, then (2.39) can be written as

$$\hat{H}_k(z) = \hat{H}_{k_{t-1}k_{t-2}\cdots k_0}(z)$$
$$\equiv \sum_{n_0=0}^{p-1} \sum_{n_1=0}^{p-1} \cdots \sum_{n_{t-1}=0}^{p-1} H_{n_{t-1}n_{t-2}\cdots n_0}(z)\hat{z}^{nk} \bmod z^M + 1,$$

where $\hat{z}^{nk}$ can be expressed as

$$\hat{z}^{nk} \equiv \hat{z}^{p^{t-1}n_{t-1}k_0} \hat{z}^{p^{t-2}n_{t-2}(pk_1+k_0)} \ldots \hat{z}^{n_0(p^{t-1}k_{t-1}+\cdots+k_0)} \mod z^M + 1.$$

We further define

$$\hat{H}^0_{n_0 n_1 \cdots n_{t-1}}(z) = H_{n_{t-1} n_{t-2} \cdots n_0}(z)$$

$$\hat{H}^j_{n_0 \cdots n_{t-j-1} k_{j-1} \cdots k_0}(z) \equiv \sum_{n_{t-j}=0}^{p-1} \cdots \sum_{n_{t-1}=0}^{p-1} H_{n_{t-1} n_{t-2} \cdots n_0}(z)$$

$$\cdot \hat{z}^{p^{t-1}n_{t-1}k_0+p^{t-2}n_{t-2}(pk_1+k_0)+\cdots+p^{t-j}n_{t-j}(p^{j-1}k_{j-1}+\cdots+k_0)} \mod z^M + 1, \quad (2.48)$$

where $j = 1, 2, \ldots, t$. It is clear that

$$\hat{H}^t_{k_{t-1} \cdots k_0}(z) = \hat{H}_{k_{t-1} \cdots k_0}(z)$$

or

$$\hat{H}^t_k(z) = \hat{H}_k(z), \quad k = 0, 1, \ldots, p^t - 1.$$

From (2.48) we can deduce that

$$\hat{H}^j_{n_0 \cdots n_{t-j-1} k_{j-1} \cdots k_0}(z) \equiv \sum_{n_{t-j}=0}^{p-1} \hat{H}^{j-1}_{n_0 \cdots n_{t-j-1} n_{t-j} k_{j-2} \cdots k_0}(z)$$

$$\cdot \hat{z}^{p^{t-j}n_{t-j}(p^{j-1}k_{j-1}+\cdots+k_0)} \mod z^M + 1. \quad (2.49)$$

Let

$$\hat{n} = (n_0 \cdots n_{t-j-1}), \quad \bar{n} = (n_{t-1} \cdots n_{t-j}), \quad n' = (n_{t-j-1} \cdots n_0)$$

$$\bar{k} = (k_{j-1} \cdots k_0), \quad \hat{k} = (k_{j-2} \cdots k_0),$$

where n' is the p-ary inverse of $\hat{n}$. Then we can prove that

$$\hat{H}^j_{\hat{n}p^j + \bar{k}}(z) \equiv \sum_{\bar{n}=0}^{p^j - 1} H_{\bar{n}p^{t-j}+n'}(z) \hat{z}^{p^{t-j}\bar{n}\bar{k}} \mod z^M + 1, \quad (2.50)$$

where $\bar{k} = 0, 1, \ldots, p^j - 1$, $\hat{n} = 0, 1, \ldots, p^{t-j} - 1$ and

$$\hat{H}^j_{\hat{n}p^j + \hat{k} + k_{j-1}p^{j-1}}(z) \equiv \sum_{n_{t-j}=0}^{p-1} \left[\hat{H}^{j-1}_{\hat{n}p^j + \hat{k} + n_{t-j}p^{j-1}}(z) \hat{z}^{p^{t-j}n_{t-j}\hat{k}} \right]$$

$$\cdot \hat{z}^{p^{t-1}n_{t-j}k_{j-1}} \mod z^M + 1, \quad (2.51)$$

where $j = 1, 2, \ldots, t$. Equation (2.50) will be used in the following chapters. Equation (2.51) is the jth stage of the decimation-in-time radix-p FPT algorithm for $j = 1, 2, \ldots, t$. The last stage $\hat{H}^t_k(z)$ produces the outputs $\hat{H}_k(z)$.

Algorithm 2 *Decimation-in-time radix-p FPT*

Step 1.   *Reorder the input polynomial sequence $\{H_n(z)\}$ in p-ary inverse order to get $\{\hat{H}^0_n(z)\}$, that is, $\{\hat{H}^0_n(z)\} = \{H_{n'}(z)\}$, where n' is the p-ary inverse of n, $0 \leq n$, $n' \leq p^t - 1$.*

Step 2.   *For j from 1 to t, compute the t stages defined by (2.51). The last stage produces $\hat{H}_k(z)$, the transform of $H_n(z)$.*

The algorithm uses the same number of operations as the decimation-in-frequency algorithm, which is shown in (2.46) and (2.47).

2.5 MD POLYNOMIAL TRANSFORM AND FAST ALGORITHM

2.5.1 MD polynomial transform

The MD PT is generated through the row-column (tensor-product) method based on 1D PTs.

Definition 17 *Let $(M(z), G_i(z), N_i)$ $(i = 1, 2, \ldots, r)$ be 1D PTs and $H_{n_1, n_2, \ldots, n_r}(z)$ $(n_i = 0, 1, \ldots, N_i - 1; i = 1, 2, \ldots, r)$ be polynomials. The rD PT is defined as*

$$\bar{H}_{k_1, k_2, \ldots, k_r}(z) \equiv \sum_{n_1=0}^{N_1-1} \cdots \sum_{n_r=0}^{N_r-1} H_{n_1, n_2, \ldots, n_r}(z) G_1^{n_1 k_1}(z) \cdots G_r^{n_r k_r}(z) \bmod M(z), \quad (2.52)$$

where $k_i = 0, 1, \ldots, N_i - 1$; $i = 1, 2, \ldots, r$. We denote such an rD PT by $(M(z), G_1(z), N_1) \times \cdots \times (M(z), G_r(z), N_r)$.

From the definition, we know that an rD PT transforms an rD polynomial sequence into another rD polynomial sequence. An rD PT is constructed by r 1D PTs using the tensor-product method. Since each 1D PT is invertible, the rD PT is also invertible and its inverse is

$$H_{n_1, n_2, \ldots, n_r}(z) \equiv \frac{1}{N_1 \cdots N_r} \sum_{k_1=0}^{N_1-1} \cdots \sum_{k_r=0}^{N_r-1} \bar{H}_{k_1, k_2, \ldots, k_r}(z)$$
$$\cdot G_1^{-n_1 k_1}(z) \cdots G_r^{-n_r k_r}(z) \bmod M(z), \quad (2.53)$$

where $n_i = 0, 1, \ldots, N_i - 1$ and $i = 1, 2, \ldots, r$. Equation (2.53) is also known as the rD inverse polynomial transform (rD IPT).

Example 11 *Let $N_i = 2^{l_i}$ $(i = 1, 2, \ldots, r)$ where l_i are positive integers, and $M = 2^t$, $t \geq l_i - 1$. Then $(z^M + 1, z^{2^{t-l_i+1}}, N_i)$ are 1D PTs. Therefore, their tensor product $(z^M + 1, z^{2^{t-l_1+1}}, N_1) \times \cdots \times (z^M + 1, z^{2^{t-l_r+1}}, N_r)$ is an rD PT. This rD PT will be used in the following chapters.*

We now consider the 2D PT. From (2.52), the 2D PT is defined by

$$\bar{H}_{k_1, k_2}(z) \equiv \sum_{n_1=0}^{N_1-1} \sum_{n_2=0}^{N_2-1} H_{n_1, n_2}(z) G_1^{n_1 k_1}(z) G_2^{n_2 k_2}(z) \bmod M(z), \quad (2.54)$$

where $k_1 = 0, 1, \ldots, N_1 - 1$ and $k_2 = 0, 1, \ldots, N_2 - 1$. Its inverse is

$$H_{n_1, n_2}(z) \equiv \frac{1}{N_1 N_2} \sum_{k_1=0}^{N_1-1} \sum_{k_2=0}^{N_2-1} \bar{H}_{k_1, k_2}(z) G_1^{-n_1 k_1}(z) G_2^{-n_2 k_2}(z) \bmod M(z), \quad (2.55)$$

where $n_1 = 0, 1, \ldots, N_1 - 1$ and $n_2 = 0, 1, \ldots, N_2 - 1$. Table 2.3 lists some useful 2D PTs. In the table, all numbers are assumed to be integers.

Table 2.3   Commonly used 2D polynomial transforms.

Module	$G_1(z)$	$G_2(z)$	N_1	N_2	Remarks
$z^{2^{l-1}}+1$	z	z	2^l	2^l	l: positive
$z^{2^{j+l-1}}+1$	z^{2^j}	z^{2^l}	2^l	2^j	l: positive
$(z^p-1)/(z-1)$	z	z	p	p	p: prime
$(z^p-1)/(z-1)$	$-z$	$-z$	$2p$	$2p$	p: odd prime
$(z^{p^2}-1)/(z^p-1)$	z	z	p^2	p^2	p: prime
$(z^{p^2}-1)/(z^p-1)$	$-z$	$-z$	$2p^2$	$2p^2$	p: odd prime
$(z^{qp^l}-1)/(z^{qp^{l-1}}-1)$	z^q	z^q	p^l	p^l	p: prime; $q>0$
$(z^{p^2}+1)/(z^p+1)$	z^p	z^p	$2p$	$2p$	p: odd prime
$(z^{qp^l}+1)/(z^{qp^{l-1}}+1)$	z^q	z^q	$2p^l$	$2p^l$	p: odd prime; $q>0$
$(z^{qp^l}+1)/(z^{qp^{l-1}}+1)$	z^{2q}	z^{2q}	p^l	p^l	p: odd prime; $q>0$

2.5.2   Fast algorithm for MD polynomial transform

We can use the FPT for a 1D PT to compute the rD PT. A simple method known as the row-column algorithm is to be presented. Let

$$H^{(0)}_{n_1,n_2,\ldots,n_r}(z) = H_{n_1,n_2,\ldots,n_r}(z) \tag{2.56}$$

and

$$H^{(j)}_{k_1,\ldots,k_j,n_{j+1},\ldots,n_r}(z) \equiv \sum_{n_j=0}^{N_j-1} H^{(j-1)}_{k_1,\ldots,k_{j-1},n_j,\ldots,n_r}(z)G_j^{n_jk_j}(z) \bmod M(z), \tag{2.57}$$

where $k_i, n_i = 0,1,\ldots,N_i-1$, $i=1,2,\ldots,r$ and $j=1,2,\ldots,r$. From the definition of rD PT, we know that

$$\bar{H}_{k_1,k_2,\ldots,k_r}(z) = H^{(r)}_{k_1,k_2,\ldots,k_r}(z).$$

For any fixed $(k_1,\ldots,k_{j-1},n_{j+1},\ldots,n_r)$, (2.57) is a PT. Therefore, an rD PT can be computed by a number of 1D PTs. More accurately, (2.57) defines $\frac{\hat{N}}{N_j}$ 1D PTs of length N_j, where $\hat{N} = N_1N_2\cdots N_r$. The computation of an rD PT in (2.52) requires us to compute $\frac{\hat{N}}{N_1}$ 1D PTs of length N_1, $\frac{\hat{N}}{N_2}$ 1D PTs of length N_2, ..., and $\frac{\hat{N}}{N_r}$ 1D PTs of length N_r.

Example 12 (*The row-column algorithm for the 2D PT*) *When $r=2$, the row-column algorithm discussed above becomes*

$$\tilde{H}_{k_1,n_2}(z) \equiv \sum_{n_1=0}^{N_1-1} H_{n_1,n_2}(z)G_1^{n_1k_1}(z) \bmod M(z) \tag{2.58}$$

$$\bar{H}_{k_1,k_2}(z) \equiv \sum_{n_2=0}^{N_2-1} \tilde{H}_{k_1,n_2}(z)G_2^{n_2k_2}(z) \bmod M(z), \tag{2.59}$$

where for (2.58) and (2.59), $k_1 = 0,1,\ldots,N_1-1$; $k_2 = 0,1,\ldots,N_2-1$ and $n_2 = 0,1,\ldots,N_2-1$. Equation (2.58) implements a 1D PT along each column of the 2D input $H_{n_1,n_2}(z)$ and obtains a 2D output $\tilde{H}_{k_1,n_2}(z)$. Then (2.59) implements a 1D PT along each row of the 2D input $\tilde{H}_{k_1,n_2}(z)$ and finally obtains the 2D output $\bar{H}_{k_1,k_2}(z)$. The algorithm needs to compute N_2 1D PTs of length N_1 and N_1 1D PTs of length N_2.

Let $A_d(N_1, N_2, \ldots, N_r)$ and $R_t(N_1, N_2, \ldots, N_r)$ be the number of additions and polynomial rotations of the row-column algorithm for the rD PT. Then we have

$$A_d(N_1, N_2, \ldots, N_r) = \frac{\hat{N}}{N_1} A_d^I(N_1) + \cdots + \frac{\hat{N}}{N_r} A_d^I(N_r) \tag{2.60}$$

$$R_t(N_1, N_2, \ldots, N_r) = \frac{\hat{N}}{N_1} R_t^I(N_1) + \cdots + \frac{\hat{N}}{N_r} R_t^I(N_r), \tag{2.61}$$

where $A_d^I(N_i)$ and $R_t^I(N_i)$ are the numbers of real additions and polynomial rotations of the FPT for a 1D PT of length N_i. If the length of the 1D PTs is powers of 2 as shown in Example 11, we have

$$A_d^I(N_i) = N_i M \log_2 N_i, \ R_t^I(N_i) = N_i \log_2 N_i - 2 \log_2 N_i.$$

Then

$$A_d(N_1, N_2, \ldots, N_r) = \hat{N} M \log_2 \hat{N} \tag{2.62}$$

$$R_t(N_1, N_2, \ldots, N_r) = \hat{N} \log_2 \hat{N} - \frac{2\hat{N}}{N_1} \log_2 N_1 - \cdots - \frac{2\hat{N}}{N_r} \log_2 N_r, \tag{2.63}$$

where $\hat{N} = N_1 N_2 \cdots N_r$.

2.6   CHAPTER SUMMARY

Number theory and polynomial theory are fundamental for understanding fast algorithms as well as signal processing and computer science. For the self-sufficiency of this book, this chapter attempts to provide basic concepts for the readers to understand the fast algorithms to be presented in the following chapters. More information on number theory and polynomial theory can be found in [13, 14].

The definition, construction, properties and fast algorithms of the PT are introduced in this chapter. Unlike many discrete transforms such as the DFT and DCT, the PT itself is seldom used directly for processing digital signals. The output of the PT seems to have little meaning in physics. However, PTs have found many applications in the area of fast algorithms, especially for MD discrete transform and convolution algorithms. Some of these fast algorithms will be discussed in the following chapters.

Auslander, Feig and Winograd have found the relationship between the PT-based algorithm for DFTs and the Abelian semi-simple algebra [1]. We think that the PT has theoretic merit in algebra and computational complexity. Also, it is found in [6, 7] that the PT is related to the discrete Radon transform. Other possible applications of the PT may be in coding theory and cryptography since polynomial theory is an essential tool in these fields.

REFERENCES

1. L. Auslander, E. Feig and S. Winograd, Abelian semi-simple algebra and algorithms for the discrete Fourier transform, *Adv. Appl. Math.*, vol. 5, 31–55, 1984.

2. R. E. Blahut, *Fast Algorithms for Digital Signal Processing*, Addison-Wesley, Reading, MA, 1984.

3. P. Duhamel and C. Guillemot, Polynomial transform computation of the 2-D DCT, *Proc. ICASSP*, 1515–1518, 1990.

4. Z. R. Jiang and Y. H. Zeng, *Polynomial Transform and its Applications*, National University of Defense Technology Press, Changsha, P. R. China, 1989 (in Chinese).

5. Z. R. Jiang, Y. H. Zeng and P. N. Yu, *Fast Algorithms*, National University of Defense Technology Press, Changsha, P. R. China, 1994 (in Chinese).

6. E. V. Labunets, V. G. Labunets, K. Egiazarian and J. Astola, New fast algorithms of multidimensional Fourier and Radon discrete transforms, *Proc. ICASSP*, 3193–3196, 1999.

7. W. Ma, Algorithms for computing two-dimensional discrete Hartley transform of size $p^n \times p^n$, *Electron. Lett.*, vol. 26, no. 11, 1795–1797, 1990.

8. J. H. McClellan and C. M. Radar, *Number Theory in Digital Signal Processing*, Prentice-Hall, Inc., Englewood Cliffs, NJ, 1979.

9. H. J. Nussbaumer, *Fast Fourier Transform and Convolutional Algorithms*, Springer-Verlag, New York, 1981.

10. H. J. Nussbaumer and P. Quandalle, Computation of convolutions and discrete Fourier transforms by polynomial transform, *IBM J. Res. Develop.*, vol. 22, no. 2, 134–144, 1978.

11. H. J. Nussbaumer, New polynomial transform algorithms for multi-dimensional DFT's and convolutions, *IEEE Trans. ASSP*, vol. 29, no. 1, 74–84, 1981.

12. J. Prado and P. Duhamel, A polynomial transform based computation of the 2-D DCT with minimum multiplicative complexity, *Proc. ICASSP*, vol. 3, 1347–1350, 1996.

13. K. H. Rosen, *Elementary Number Theory*, Addison-Wesley, Reading MA, fourth edition, 2000.

14. M. R. Schroeder, *Number Theory in Science and Communication*, Springer-Verlag, New York, 1986.

15. M. Soderstrand, W. K. Jenkins, G. A. Jullien and F. J. Taylor, *Residue Number System Arithmetic: Modern Applications in Digital Signal Processing*, IEEE Press, Washington, 1986.

16. Y. H. Zeng, G. Bi and A. R. Leyman, New polynomial transform algorithm for multidimensional DCT, *IEEE Trans. Signal Process.*, vol. 48, no. 10, 2814–2821, 2000.

17. Y. H. Zeng, G. Bi and A. C. Kot, New algorithm for multidimensional type-III DCT, *IEEE Trans. Circuits Syst. II*, vol. 47, no. 12, 1523–1529, 2000.

18. Y. H. Zeng, G. Bi and A. R. Leyman, New algorithm for r-dimensional DCT-II with size $q^{l_1} \times q^{l_2} \times \cdots \times q^{l_r}$, *IEE Proc., Vision, Image Signal Process.*, vol. 148, no. 1, 1–8, 2001.

19. Y. H. Zeng, L. Z. Cheng and M. Zhou, *Parallel algorithms for digital signal processing*, National University of Defense Technology Press, Changsha, P. R. China, 1998 (in Chinese).

20. Y. H. Zeng and X. M. Li, Multidimensional polynomial transform algorithms for multidimensional discrete W transform, *IEEE Trans. Signal Process.*, vol. 47, no. 7, 2050–2053, 1999.

21. Y. H. Zeng, G. Bi and A. C. Kot, Combined polynomial transform and radix-q algorithm for MD discrete W transform, *IEEE Trans. Signal Process.*, vol. 49, no. 3, 634–641, 2001.

3

Fast Fourier Transform
Algorithms

This chapter is devoted to the fast algorithms for various types of discrete Fourier transforms (DFTs).

- Section 3.2 illustrates, by using radix-2 and split-radix algorithms as examples, some basic concepts and techniques that are frequently used in the development of fast algorithms for discrete linear transforms.

- Section 3.3 presents a generalized split-radix algorithm that makes a few improvements on the previous split-radix algorithms.

- Section 3.4 derives a prime factor algorithm to simplify the index mapping procedures and support in-place computation.

- Section 3.5 extends the concepts of the generalized split-radix algorithms for two-dimensional (2D) DFTs.

- Section 3.6 considers the fast algorithms for one-dimensional (1D) generalized DFTs.

- Section 3.7 presents polynomial transform (PT)-based algorithms for multidimensional (MD) DFTs.

Details of the mathematical derivations and comparisons in terms of the computational complexity and structures are also provided for these algorithms. The readers will obtain understanding of the main issues and the techniques that are frequently used to evaluate the merits of various fast algorithms.

3.1 INTRODUCTION

The DFT has a wide range of applications in digital signal processing. Many fast algorithms ([1, 2, 7, 8, 9, 10, 12, 13, 14, 21, 22, 23, 24, 25], for example) were reported in the literature. These algorithms can be generally classified into two broad categories: the radix type and the non-radix type. The radix type category includes the well-known radix-2, radix-4 and split-radix algorithms and others that decompose the entire DFT computation based on a particular radix. The split-radix algorithms [7, 21] for a DFT of length 2^m, where m is a positive integer, generally require a smaller number of arithmetic operations and a regular computational structure compared to other radix type fast algorithms. In general, the radix type algorithms require a relatively simple computational structure and a reasonable amount of computational complexity. The algorithms are recursive and can be used for many different transform sizes. The concept of decomposition process can be easily understood; therefore, implementation of these algorithms becomes easy. The radix type fast algorithms can be generalized to deal with DFTs whose sizes are a power of an integer, for example, 3^m and 5^m. However, fast algorithms for such sequence lengths generally require complex computational structures and more computational complexity than that for length 2^m [22, 23, 25].

The non-radix type fast algorithms can be used for DFTs whose sequence length contains many different factors. For example, the Winograd Fourier transform algorithm [26], Rader's fast algorithm [20], the prime factor algorithm [15] and the Good–Thomas fast algorithm [12] were also reported. Although these fast algorithms provide the possibilities of efficiently dealing with DFTs whose transform length is not a power of an integer, there exist a few limitations that prevent them from being widely used in practice. In general, the non-radix type algorithms do not have a good regularity of the computational structure. The irregularity can seen from the following two aspects. One is the computational structure and the other is that the algorithm used for a particular transform size cannot be easily extended for other transform sizes. The fast algorithms generally require an index mapping to rearrange the sequence of the input data. Such a mapping generally requires complex arithmetic operations, such as modulo operations that consume substantial computational overheads. Another possible problem is that the in-place computation may not be possible, which requires more memory space for the computation.

Although mature and cost-effective algorithms for length 2^m DFT work quite well, computational overhead usually occurs when the input sequence length is not matched to the available algorithm. To use the fast algorithm for a particular size of a DFT, the zero-padding technique is often used to augment the input sequence length, which generally results in unnecessary computation. Therefore, fast algorithms supporting arbitrarily composite transform sizes are desired. This chapter presents two fast algorithms for 1D and 2D DFTs of composite sequence lengths. Special techniques are used to avoid the requirement of data swapping operations so that in-place computation can be achieved. Compared to other algorithms, substantial reduction of arithmetic operations can be achieved.

The first algorithm for 1D DFTs, presented in Section 3.3, is a generalized split-radix algorithm that supports transform sizes of $q \cdot 2^m$, where q is an arbitrarily odd integer [1]. By setting different q values, many transform sizes can be supported by the fast algorithm. It can be shown that when $q = 1$, the algorithm is similar to the split-radix algorithm reported in [21]. Section 3.4 describes the second algorithm, which is based on a prime factor decomposition to support DFT computation of composite sequence length [3]. The decomposition method is based upon the Good–Thomas algorithm [12] and is combined with the split-radix algorithm to jointly utilize their desirable features. These two algorithms have regular computational structures and use a smaller number of arithmetic operations than those reported in the literature. Fast algorithms for 2D DFTs are considered in Section 3.5, which extends the concept used in Section 3.3 to MD cases [2]. Section 3.6 presents fast algorithms for 1D generalized DFTs [4]. Finally, the PT-based algorithms for MD DFTs are presented in Section 3.7. Detailed comparisons with other reported algorithms are provided in terms of the computational complexity and the regularity of the computational structure.

3.2   RADIX-2 AND SPLIT-RADIX ALGORITHMS

The strategy of "divide and conquer" has always been used by all beings for their activities. In engineering applications, it means dividing a large problem into smaller ones that can be easily tackled. The solution to the big problem can be obtained by collecting the solutions to these smaller problems in an organized manner according to the dividing procedures. Often such a derived solution to the big problem is modular and systematic, which is advantageous for many issues in implementation of the solution. One excellent example of using this divide and conquer strategy in signal processing is the development of fast algorithms for discrete linear transforms.

It seems that the complexity of signal processing problems has increased at a faster speed than that supported by new technologies. We can be see that the advance of signal processing power in speed and intelligence is far behind that needed for today's ambitions. Therefore, the divide and conquer strategy becomes even more fundamental in solving our problems. Without exception, the development of all algorithms presented in this book is based on this strategy. By using the radix-2 and split-radix algorithms as examples, this section shows the main steps and concepts used in the development of fast algorithms. It is hoped that the experience obtained in this section will be useful in understanding other fast algorithms to be presented in this book.

Consider a finite duration discrete-time signal $x(n)$ $(n = 0, 1, \ldots, N-1)$. The DFT of $x(n)$ is defined as

$$X(k) = \sum_{n=0}^{N-1} x(n) W_N^{nk} \tag{3.1}$$

for $0 \le k < N$, where $W_N = e^{-j2\pi/N}$. The original signal $x(n)$ can be exactly recovered from $X(k)$ by the inverse DFT (IDFT)

$$x(n) = \frac{1}{N} \sum_{k=0}^{N-1} X(k) W_N^{-nk} \tag{3.2}$$

for $0 \le n \le N-1$.

Since the DFT and IDFT basically require the same type of computation, we will mainly focus on the fast algorithms for the computation of the DFT. The same algorithm can be easily applied to the computation of the IDFT. A direct computation of (3.1) requires N^2 complex multiplications and $N(N-1)$ complex additions. When N is large, the computational complexity becomes prohibitive. The computational complexity can be significantly reduced if the symmetry property

$$W_N^{k+N/2} = -W_N^k$$

and the periodicity property

$$W_N^{k+N} = W_N^k$$

are utilized in the DFT computation. Numerous efficient algorithms based on these properties have been reported in the literature and they are generally known as the fast Fourier transform (FFT) algorithms. In the rest of this subsection, the radix-2 and the split-radix 2/4 algorithms are used to demonstrate the use of the symmetric and periodic properties in deriving fast algorithms. Some basic concepts and terminologies are particularly essential because similar concepts are also applied to derive fast algorithms for other discrete transforms.

3.2.1   Radix-2 algorithm

Let us assume that the sequence length N is even. By using the relationship $W_N^2 = W_{N/2}$, the DFT defined in (3.1) can be expressed as

$$
\begin{aligned}
X(k) &= \sum_{n=0}^{N-1} x(n) W_N^{nk} \\
&= \sum_{n=0}^{N/2-1} x(2n) W_N^{2nk} + \sum_{n=0}^{N/2-1} x(2n+1) W_N^{(2n+1)k} \\
&= \sum_{n=0}^{N/2-1} x(2n) W_{N/2}^{nk} + W_N^k \sum_{n=0}^{N/2-1} x(2n+1) W_{N/2}^{nk} \\
&= F(k) + W_N^k G(k),
\end{aligned}
\tag{3.3}
$$

where

$$
F(k) = \sum_{n=0}^{N/2-1} x(2n) W_{N/2}^{nk}, \qquad k = 0, 1, \ldots, \frac{N}{2} - 1
\tag{3.4}
$$

$$
G(k) = \sum_{n=0}^{N/2-1} x(2n+1) W_{N/2}^{nk}, \qquad k = 0, 1, \ldots, \frac{N}{2} - 1.
\tag{3.5}
$$

Both $F(k)$ and $G(k)$ are the length $N/2$ DFT of subsequences $x(2n)$ and $x(2n+1)$, respectively. When the periodicity property is applied, it can easily verify that $F(k+N/2) = F(k)$, $G(k+N/2) = G(k)$ and $W_N^{k+N/2} = -W_N^k$, which leads to

$$
X(k + \frac{N}{2}) = F(k) - W_N^k G(k).
\tag{3.6}
$$

Therefore, the length N DFT can be computed by

$$
X(k) = F(k) + W_N^k G(k), \qquad k = 0, 1, \ldots, \frac{N}{2} - 1
\tag{3.7}
$$

$$
X(k + \frac{N}{2}) = F(k) - W_N^k G(k), \qquad k = 0, 1, \ldots, \frac{N}{2} - 1.
\tag{3.8}
$$

Figure 3.1 shows the signal flow graph for a length 8 DFT. The length $N/2$ DFT given in (3.4) and (3.5) can be decomposed further in the same way as long as the length of subsequences is even. The decomposition technique is performed by decimating the input sequence $x(n)$ into two subsequences $x(2n)$ and $x(2n+1)$. Because the input sequence of the DFT is traditionally considered to be given in the time domain, the computation technique derived in such a way is generally known as the *decimation-in-time* fast algorithm.

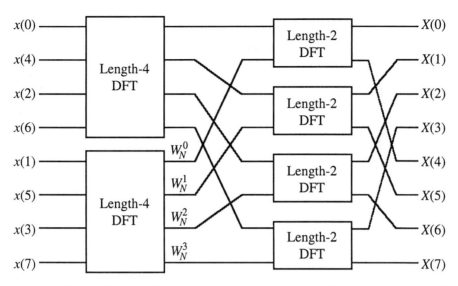

Figure 3.1   Signal flow graph of the decimation-in-time radix-2 algorithm for $N = 8$.

The divide and conquer technique can also be performed by splitting (or decimating) the transformed outputs into even and odd subsequences. For example, the DFT computation can be divided into

$$X(2k) = \sum_{n=0}^{N/2-1} \left[x(n) + x(n + \frac{N}{2}) \right] W_{N/2}^{nk} \qquad (3.9)$$

$$X(2k+1) = \sum_{n=0}^{N/2-1} \left[x(n) - x(n + \frac{N}{2}) \right] W_N^{n(2k+1)}$$

$$= \sum_{n=0}^{N/2-1} \left\{ \left[x(n) - x(n + \frac{N}{2}) \right] W_N^n \right\} W_{N/2}^{nk}, \qquad (3.10)$$

where $k = 0, 1, \ldots, N/2 - 1$. If $x(n) + x(n + \frac{N}{2})$ is considered as the input sequence, (3.9) becomes a length $N/2$ DFT. Similarly, (3.10) is also a length $N/2$ DFT if $[x(n) - x(n + \frac{N}{2})]W_N^n$ is considered as the input sequence. According to the definition of the DFT, it can be easily proven that the computation $x(n) + x(n + N/2)$ and $x(n) - x(n + N/2)$ can be considered to the operation of a length 2 DFT. The computation derived in such a way is known as the *decimation-in-frequency* algorithm. Figure 3.2 shows the signal flow graph for a length 8 DFT. It can be verified that Figure 3.2 is the mirror image of Figure 3.1.

Now let us consider a few basic concepts that are commonly used for developing fast algorithms of discrete transforms.

Index reordering

One observation is that the input data indices in Figure 3.1 (or the output indices in Figure 3.2) are not in a natural order. Therefore, index reordering is needed for the input sequence or the output sequence or both. If the computational structure of the fast algorithm is regular, new data indices can be easily generated. When the sequence length $N = 2^m$, for example, a simple approach known as the *bit-reversal scheme* has been widely

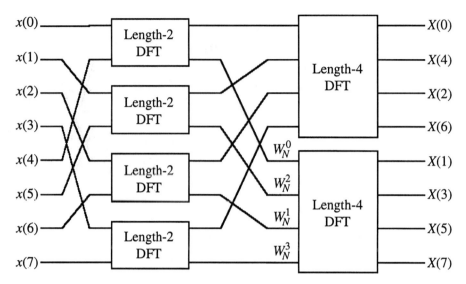

Figure 3.2   Signal flow graph of the decimation-in-frequency radix-2 algorithm for $N = 8$.

used for the radix-2 algorithm [19], as shown in Table 3.1. However, some algorithms have a complex computational structure so that the generation of new indices is time consuming and becomes a substantial computational overhead. One example is that the index mapping process needed by the prime factor algorithm has to use arithmetic operations to calculate the reordered indices.

Table 3.1   Bit-reversal representation for $N = 8$.

Index	Binary representation	Bit reversed representation	Bit reversed index
0	000	000	0
1	001	100	4
2	010	010	2
3	011	110	6
4	100	001	1
5	101	101	5
6	110	011	3
7	111	111	7

Butterfly operation

A *butterfly operation* is a basic computational block that is performed on a pair of complex (or real) numbers. According to the types of decomposition methods used in deriving the fast algorithms, the butterfly operation can be performed in two different ways, as shown in Figure 3.3. The left side is the butterfly operation used in the decimation-in-time fast algorithm, and the right side is the one employed in the decimation-in-frequency algorithm. Each butterfly operation generally needs one complex multiplication and two complex additions. When the multiplication factor becomes 1, the butterfly computation becomes a length 2 DFT which is the basic computational element used in the fast algorithm. Figure 3.4 shows a complete signal flow graph for $N = 16$ that uses the length 2 DFT as the based computational element.

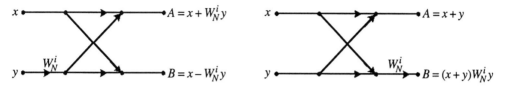

Figure 3.3   Signal flow graph of the butterfly operation.

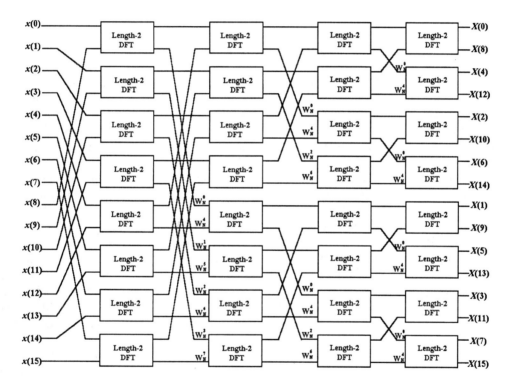

Figure 3.4   Signal flow graph of the decimation-in-frequency algorithm for a length 16 DFT.

In-place computation   For any fast algorithm, memory spaces are needed to hold the input sequence, output sequence and the partially processed data. When the sequence length is large, the requirement of memory space becomes important and critical. Therefore, a good fast algorithm should use a minimum amount of memory space. One technique known as *in-place computation* is particularly useful to minimize the required amount of memory space. It can be explained by using the concept of the *butterfly operation*, as just discussed.

Once a butterfly operation is performed on a pair of complex numbers (x, y) to produce the desired outputs (A, B), the data (x, y) are no longer useful and their memory space can be reused by the newly generated data (A, B). If in-place computation can be achieved, N complex data stores are enough for the entire computation of a length N DFT.

Twiddle factor

The multiplication factor W_N^i is known as a *twiddle factor* which can be expressed by

$$W_N^i = \cos\frac{2\pi i}{N} - j\sin\frac{2\pi i}{N}.$$

Multiplying a complex number with a twiddle factor is equivalent to rotating a complex number with a given angle in the complex plane. For computation of discrete transforms, the rotation operation is frequently used and can be generalized into the process described by

$$A = x\cos(\alpha) + y\sin(\alpha)$$
$$B = x\sin(\alpha) + y\cos(\alpha),$$

where x, y, A and B are real numbers. The rotation operation can be realized with three multiplications and three additions or four multiplications and two additions on real numbers. In general, the total number of arithmetic operations (multiplication plus addition) used by a complex multiplication (or rotation operation) is always six regardless of the implementation scheme used. It is possible that the twiddle factor becomes trivial, i.e., $W_N^i = \pm 1$ or $\pm j$, in which case we do not need multiplication and addition in the rotation operation. When the twiddle factor becomes $(1 \pm j)/\sqrt{2}$, multiplications can also be saved. The savings in computational complexity due to the trivial twiddle factor can be substantial for some fast algorithms.

We now consider the computational complexity required by the decimation-in-time radix-2 algorithm. It can be easily proven that the decimation-in-time and the decimation-in-frequency algorithms need the same computational complexity. In general, the radix-2 algorithm for a length 2^m DFT needs m stages of butterfly computation. Each stage uses $N/2$ butterfly operations. It can be easily seen that (3.7) and (3.8) need N complex additions and $N/2$ complex multiplications. In addition, two length $N/2$ DFTs $F(k)$ and $G(k)$ have to be calculated. It we assume that the rotation operation is implemented with 4 real multiplications and 2 real additions,[1] we have

$$M(N) = 2M(\frac{N}{2}) + 2N - M_t \tag{3.11}$$

$$A(N) = 2A(\frac{N}{2}) + 3N - A_t, \tag{3.12}$$

where $M(N)$ and $A(N)$ are the number of real multiplications and additions needed by the radix-2 algorithm for a length N DFT, and M_t and A_t are the number of real multiplications and additions saved from trivial twiddle factors. When $N = 0$ and $N/2$, the twiddle factor becomes ± 1 so that four multiplications and two additions can be saved. When $n = N/8$ and $3N/8$, the twiddle factor is in the form of $\pm(1 \pm j)/\sqrt{2}$ to save two multiplications. If

---

[1] In this book, the computational complexity is expressed by the number of operations on real numbers. Therefore, multiplication and addition mean the operations for real numbers.

all trivial twiddle factors are considered, we have $M_t = 12$ and $A_t = 4$. The computational complexities in (3.11) and (3.12) are recursive. To obtain a closed form expression, the initial values of these equations are needed. In this case, the initial values can be the number of operations that are needed by the length 4 DFT, i.e., $M(4) = 0$ and $A(4) = 16$. Therefore, (3.11) and (3.12) become

$$M(N) = 2N \log_2 N - 7N + 12 \tag{3.13}$$
$$A(N) = 3N \log_2 N - 3N + 4, \tag{3.14}$$

where $N > 2$.

If the rotation operation is implemented with 3 multiplications and 3 additions, we have

$$M(N) = 2M(\frac{N}{2}) + \frac{3N}{2} - M_t \tag{3.15}$$
$$A(N) = 2A(\frac{N}{2}) + \frac{7N}{2} - A_t, \tag{3.16}$$

where $M_t = A_t = 8$. With $M(4) = 0$ and $A(4) = 16$, the closed form expressions are

$$M(N) = \frac{3N}{2} \log_2 N - 5N + 8 \tag{3.17}$$
$$A(N) = \frac{7N}{2} \log_2 N - 5N + 8, \tag{3.18}$$

Because multiplication was generally more expensive than addition in the past, it was often desirable to implement the rotation operation with 3 multiplications and 3 additions so that the number of multiplications could be minimized. With the advance of computing technologies, minimizing the number of multiplications becomes much less important because the multiplication time needed by many processors is comparable with the addition time. We often use the total number of arithmetic operations as a criterion to compare the computational complexity of a fast algorithm. It can be easily verified that the total number of arithmetic operations needed by a fast algorithm is alway the same regardless of the method of implementing the rotation operation. Table 3.2 lists the number of operations needed by the radix-2 algorithm.

Table 3.2   The number of operations needed by the radix-2 and split-radix algorithms.

	Radix-2 algorithm				Split-radix 2/4 algorithm				
N	4-2 scheme		3-3 scheme		Total	4-2 scheme		3-3 scheme		Total
4	0	16	0	16	16	0	16	0	16	16
8	4	52	4	52	56	4	52	4	52	56
16	28	148	24	152	176	24	144	20	148	168
32	108	388	88	408	496	84	372	68	388	456
64	332	964	264	1032	1296	248	912	196	388	456
128	908	2308	712	2504	3216	660	2164	516	2308	2824
256	2316	5380	1800	5896	7696	1656	5008	1284	5380	6664

3.2.2   Split-radix algorithm

The radix-2 algorithm recursively decomposes the length N DFT into two length $N/2$ DFTs. In contrast, the split-radix algorithm decomposes the length N DFT using different radices.

When $N = 2^m$, for example, the length N DFT can be decomposed into a length $N/2$ DFT computing the even indexed transform outputs $X(2k)$, which can be considered to use radix-2 decomposition, and two length $N/4$ DFTs calculating the odd indexed transform outputs $X(4k+1)$ and $X(4k+3)$, which can be considered to use the radix-4 algorithm. Similar to the radix-2 algorithm, both decimation-in-time and decimation-in-frequency decompositions can be used to derive the split-radix algorithm. In this subsection, the decimation-in-frequency approach is used for illustration as follows. The even indexed transform outputs $X(2k)$ are computed by (3.9) and the odd indexed transform outputs $X(2k+1)$ defined in (3.10) are further decomposed into

$$
\begin{aligned}
X(4k+1) &= \sum_{n=0}^{N/2-1} \left[x(n) - x(n+\frac{N}{2}) \right] W_N^{n(4k+1)} \\
&= \sum_{n=0}^{N/4-1} \left\{ \left[x(n) - x(n+\frac{N}{2}) \right] - j \left[x(n+\frac{N}{4}) - x(n+\frac{3N}{4}) \right] \right\} W_N^n W_{N/4}^{kn} \quad (3.19)
\end{aligned}
$$

$$
\begin{aligned}
X(4k+3) &= \sum_{n=0}^{N/2-1} \left[x(n) - x(n+\frac{N}{2}) \right] W_N^{n(4k+3)} \\
&= \sum_{n=0}^{N/4-1} \left\{ \left[x(n) - x(n+\frac{N}{2}) \right] + j \left[x(n+\frac{N}{4}) - x(n+\frac{3N}{4}) \right] \right\} W_N^{3n} W_{N/4}^{kn}, \quad (3.20)
\end{aligned}
$$

where $k = 0, 1, \ldots, N/4 - 1$. Now the length N DFT is decomposed into a length $N/2$ DFT in (3.9) and two length $N/4$ DFTs in (3.19) and (3.20). Each of them can be further decomposed in the same way. The basic computational block is shown in Figure 3.5. Based

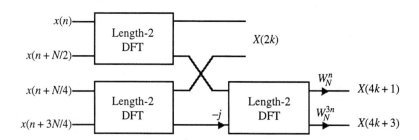

Figure 3.5   The basic computational block of the split-radix algorithm.

on (3.9), (3.19) and (3.20) and with the rotation operation implemented by 4 multiplications and 2 additions, the computational complexity of the split-radix algorithm can be expressed by

$$
M(N) = M(\frac{N}{2}) + 2M(\frac{N}{4}) + 2N - 12 \quad (3.21)
$$

$$
A(N) = A(\frac{N}{2}) + 2A(\frac{N}{4}) + 4N - 4 \quad (3.22)
$$

with initial values $M(2) = 0$, $A(2) = 4$ and $M(4) = 0$ and $A(4) = 16$. If the rotation is realized by using 3 multiplications and 3 additions, the computational complexity becomes

$$M(N) = M(\frac{N}{2}) + 2M(\frac{N}{4}) + \frac{3N}{2} - 8 \tag{3.23}$$

$$A(N) = A(\frac{N}{2}) + 2A(\frac{N}{4}) + \frac{9N}{2} - 8 \tag{3.24}$$

with the same initial values as mentioned before. Table 3.2 lists the number of operations needed by the split-radix algorithm for various sequence lengths.

3.2.3   Discussions

In general, there exist numerous fast algorithms for the DFT and other discrete transforms. There are a few issues to be considered for selecting a suitable algorithm for a particular application. Based on these issues, let us compare the two well-known and widely used fast algorithms described in the previous subsection. These issues are generally valid for comparing fast algorithms of any discrete linear transform.

Recursion and regularity - It is important to understand that recursion provides regularity. With recursive decomposition, the smaller problems after decomposition have the same nature and properties as the big problem before the decomposition, which directly leads to the regularity of the solutions to these small problems. Regularity is very desirable for implementation, testing, maintenance and components reuse. The regularity of a fast algorithm can be considered for the following three issues. The first one is the recursion in the decomposition process. As demonstrated, recursive decomposition can be achieved for different transform sizes such as 2^m. The second issue is the possibility that the same recursive decomposition can be used for different numbers of dimensions.

The last issue involves in the implementation of the particular fast algorithm. We have seen that input/output indices may be regenerated by a fast algorithm. The generation of the indices can be extremely simple or extremely difficult. If the new indices cannot be simply expressed by some formula, the fast algorithm tends to have a complex indexing structure. The consequences are that the computational overhead for indexing is substantially increased and in-place computation may not be possible. Often the fast algorithm cannot be used for computation of different transform sizes or different numbers of dimensions.

Computational complexity - One of the main objectives for developing fast algorithms is to significantly reduce the computational complexity and/or implementation complexity. It is hoped that the fast algorithm needs a minimum number of arithmetic operations. Although new technologies have increased significantly the processing throughput, this issue is still dominant for present and future applications. In addition to other factors, the minimization of computational complexity becomes a critical factor in selecting the available fast algorithms. It can be seen in Table 3.2 that the split-radix algorithm uses substantially fewer arithmetic operations than those needed by the radix-2 algorithm. When $N = 256$, for example, the split-radix algorithm uses 6664 operations compared to 7696 operations needed by the radix-2 algorithm.

Computational structure - The computational structure is closely related to the implementation complexity. A regular computational structure allows the same algorithm to support many different sequence lengths and provides convenience for either software or hardware implementation. The radix-2 algorithm has an extremely regular computational structure, as seen in Figure 3.4. It supports DFTs whose sequence length is 2^m and the entire computation can be recursively achieved using only a basic computation module – the length 2 DFT. In comparison, the split-radix algorithm has a more complex computational structure than that of the radix-2 algorithm because both radix-2 and radix-4 decompositions are used in the split-radix algorithm. The difference in computational structure between the radix and the split-radix algorithms is not substantial. However, it is possible that some fast algorithms have an irregular computational structure. For example, the prime factor algorithms generally require an irregular index mapping process so that in-place computation is not possible. A particular prime factor decomposition supports only one sequence length and cannot be extended for other sequence lengths.

Memory consumption - The amount of memory space needed by a particular fast algorithm should also be estimated. It is very much desired that the fast algorithms support in-place computation to minimize the amount of memory space. Very often the fast algorithms that do not have a regular computational structure need more memory due to the difficulties in supporting in-place computation. Because both the radix-2 and the split-radix algorithms have a regular computational structure, their requirements of memory space are about the same.

Sequence lengths - A fast algorithm can support DFTs of one or a few sequence lengths. In practice, these sequence lengths do not have any relation to the specifications of the signal processing system. For example, the number of filter coefficients can be determined only from the filter specifications and has nothing to do with the transform length. When the filter operation is performed in the frequency domain, the number of filter coefficients may not match the sequence length supported by the fast algorithm. To solve the problem, zero padding techniques are often used, which often reduces the efficiency of the signal processing system. A fast algorithm should ideally support composite sequence length without sacrificing other desirable properties. Both the radix-2 and the split-radix algorithms support the sequence length $N = 2^m$ only.

3.3   GENERALIZED SPLIT-RADIX ALGORITHM

We assume that the 1D DFT of a sequence $x(n)$ ($n = 0, \ldots, N - 1$) defined in (3.1) has a sequence length $N = q \cdot 2^m$, where m is a positive integer and q is an odd integer. Various sequence lengths can be represented by setting different values of m and q. In this section, the fast algorithms based on both decimation-in-time and decimation-in-frequency decompositions are described. It can be seen that the split-radix algorithm described in Section 3.2 is only a special case of the generalized split-radix algorithm to be presented in this section. The main achievements are that the generalized split-radix algorithms can provide substantial savings in the number of arithmetic operations and support a wider range of choices on sequence lengths.

3.3.1   Decimation-in-frequency algorithm

The DFT defined in (3.1) can be decomposed into even and odd indexed transform outputs. It is straightforward to compute the even indexed transform outputs by

$$X(2k) = \sum_{n=0}^{N/2-1} u(n)W_{N/2}^{nk}, \qquad 0 \le k < \frac{N}{2} - 1, \tag{3.25}$$

where $u(n) = x(n) + x(n + N/2)$. The even indexed outputs can be obtained by a length $N/2$ DFT.

Now let us consider the computation of odd indexed outputs. If $m = 1$ or $N = 2q$, the odd indexed transform outputs can be computed by

$$X[(2k + q) \bmod N] = \sum_{n=0}^{N-1} x(n)W_N^{n(2k+q)} = \sum_{n=0}^{N-1} x(n)(-1)^n W_{N/2}^{nk}$$

$$= \sum_{n=0}^{q-1} (-1)^n \left[x(n) - x(n + q)\right] W_q^{nk}, \qquad 0 \le k < q, \tag{3.26}$$

which is a length q DFT. It is noted that in (3.26) $X(2k+q)$ is used instead of $X(2k+1)$, which is different from the split-radix algorithms discussed in section 3.2. The advantage of such an arrangement is that multiplications for non-trivial twiddle factors, which are needed by other algorithms, are not required since (3.26) has only the trivial twiddle factor $(-1)^n$.

If $m > 1$, the odd indexed transform outputs can be further decomposed. For example, we have

$$X(4k \pm q) = \sum_{n=0}^{N/2-1} v(n)W_N^{n(4k\pm q)}$$

$$= \sum_{n=0}^{N/2-1} v(n)\left[\cos\frac{2\pi n}{N/q} \mp j\sin\frac{2\pi n}{N/q}\right] W_{N/4}^{nk}, \tag{3.27}$$

where $v(n) = x(n) - x(n + N/2)$. Based on (3.27), we form new sequences

$$F(k) = \frac{X[(N + 4k - q) \bmod N] + X[(4k + q) \bmod N]}{2}$$

$$= \sum_{n=0}^{N/2-1} v(n)\cos\frac{2\pi n}{N/q} \, W_{N/4}^{nk}$$

$$= \sum_{n=0}^{N/4-1} \left[v(n)\cos\frac{2\pi n}{N/q} + (-1)^{(q+1)/2}v(n + \frac{N}{4})\sin\frac{2\pi n}{N/q}\right] W_{N/4}^{nk} \tag{3.28}$$

and

$$G(k) = \frac{X[(N + 4k - q) \bmod N] - X[(4k + q) \bmod N]}{2}$$

$$= \sum_{n=0}^{N/2-1} v(n)\sin\frac{2\pi n}{N/q} \, W_{N/4}^{nk}$$

$$= \sum_{n=0}^{N/4-1} \left[v(n)\sin\frac{2\pi n}{N/q} + (-1)^{(q+1)/2}v(n + \frac{N}{4})\cos\frac{2\pi n}{N/q}\right] W_{N/4}^{nk}, \tag{3.29}$$

where both (3.28) and (3.29), $k = 0, 1, \ldots, N/4 - 1$, are length $N/4$ DFTs whose inputs are rotated by twiddle factors $\cos(2\pi nq/N)$ and $\sin(2\pi nq/N)$, respectively. The final odd indexed transform outputs are obtained by

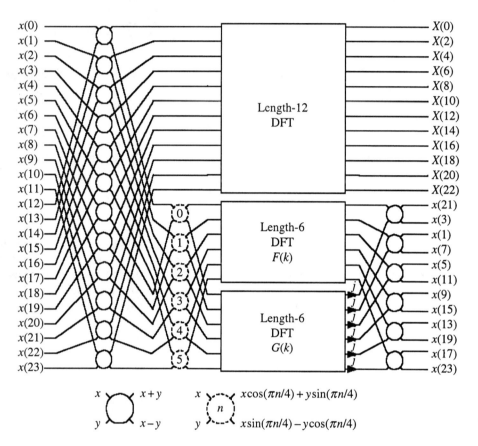

Figure 3.6 Signal flow graph of the decimation-in-frequency decomposition of the generalized split-radix algorithm ($N = 24$).

$$X[(N + 4k - q) \bmod N] = F(k) + jG(k) \tag{3.30}$$
$$X[(4k + q) \bmod N] = F(k) - jG(k), \tag{3.31}$$

where $0 \leq k < N/4$. Equations (3.25), (3.28) and (3.29) can be further decomposed until length q and length $2q$ DFTs are reached. The first decomposition step is illustrated in Figure 3.6 for $N = 24$. The dashed circle represents the operations needed by the twiddle factors.

Now let us consider the role of the parameter q in (3.28) and (3.29) by examples. Table 3.3 shows the values of the twiddle factors for $N = 24$. By comparing with the number of trivial twiddle factors for $q = 1$, it can be seen that with $q = 3$, more trivial twiddle factors are generated, which can further reduce the number of additions and multiplications used by the twiddle factors. Arithmetic operations can be saved if twiddle factors $\cos(2\pi nq/N)$ or $\sin(2\pi nq/N)$ in (3.28) and (3.29) become 0, 1 or $\sqrt{2}/2$. When $q = 3$, the total number of trivial twiddle factors is three times that when $q = 1$, which is equivalent to the split-radix algorithm reported in [7]. It can be easily generalized that the total number of trivial twiddle factors is proportional to the value of q.

Table 3.3   Generation of twiddle factors.

n	$q = 1, N = 24$		$q = 3, N = 24$	
	$\cos(\frac{2\pi n}{N/q})$	$\sin(\frac{2\pi n}{N/q})$	$\cos(\frac{2\pi n}{N/q})$	$\sin(\frac{2\pi n}{N/q})$
0	1.000	0.000	1.000	0.000
1	0.965	0.258	0.707	0.707
2	0.866	0.500	0.000	1.000
3	0.707	0.707	-0.707	0.707
4	0.500	0.866	-1.000	0.000
5	0.258	0.965	-0.707	-0.707
6	0.000	1.000	0.000	-1.000
7	-0.258	0.965	0.707	-0.707
8	-0.500	0.866	1.000	0.000
9	-0.707	0.707	0.707	0.707
10	-0.866	0.500	0.000	1.000
11	-0.965	0.258	-0.707	0.707
12	-1.000	0.000	-1.000	0.000
13	-0.965	-0.258	-0.707	-0.707
14	-0.866	-0.500	0.000	-1.000
15	-0.707	-0.707	0.707	-0.707
16	-0.500	-0.866	1.000	0.000
17	-0.258	0.965	0.707	0.707
18	0.000	-1.000	0.000	1.000
19	0.258	-0.965	-0.707	0.707
20	0.500	-0.866	-1.000	0.000
21	0.707	-0.707	-0.707	-0.707
22	0.866	-0.500	0.000	-1.000
23	0.965	-0.258	0.707	-0.707

3.3.2   Decimation-in-time algorithm

Now let us consider the fast algorithm achieved by the decimation-in-time decomposition. The 1D DFT defined in (3.1) can be expressed by

$$X(k) = \sum_{n=0}^{N/2-1} x(2n)W_{N/2}^{nk} + \sum_{n=0}^{N/2-1} x(2n+q)W_N^{qk}\, W_{N/2}^{nk} \tag{3.32}$$

and

$$X(k+N/2) = \sum_{n=0}^{N/2-1} x(2n)W_{N/2}^{nk} - \sum_{n=0}^{N/2-1} x(2n+q)W_N^{qk}\, W_{N/2}^{nk}, \tag{3.33}$$

where for (3.32) and (3.33), $0 \le k < N/2$. If $m = 1$ or $N = 2q$, the twiddle factor W_N^{qk} in the second term of (3.32) and (3.33) becomes $(-1)^k$, which does not need any multiplication.

The second term in (3.32) and (3.33) can be further decomposed into two length $N/4$ DFTs. For example, we can express it as a summation of $B(k)$ and $C(k)$ defined by

$$B(k) = W_N^{qk} \sum_{n=0}^{N/4-1} x\left[(4n+q) \bmod N\right] W_{N/4}^{nk} \tag{3.34}$$

$$C(k) = W_N^{-qk} \sum_{n=0}^{N/4-1} x\left[N + (4n-q) \bmod N\right] W_{N/4}^{nk}. \tag{3.35}$$

Therefore, the length N DFT is decomposed into one length $N/2$ and two length $N/4$ DFTs. If we define

$$A(k) = \sum_{n=0}^{N/2-1} x(2n) W_{N/2}^{nk}, \quad 0 \le k < \frac{N}{2}, \tag{3.36}$$

the length N DFT can be achieved by

$$X(k) = A(k) + [B(k) + C(k)] \tag{3.37}$$

$$X(k + \frac{N}{4}) = A(k + \frac{N}{4}) - j(-1)^{(q-1)/2}[B(k) + C(k)] \tag{3.38}$$

$$X(k + \frac{N}{4}) = A(k) - [B(k) + C(k)] \tag{3.39}$$

$$X(k + \frac{3N}{4}) = A(k + \frac{N}{4}) + j(-1)^{(q-1)/2}[B(k) + C(k)], \tag{3.40}$$

where $0 \le k < N/4$. When $N = 4q$, the twiddle factors in (3.34) and (3.35) are trivial so that no multiplications are needed. When $N = 8q$, the non-trivial twiddle factors associated with $B(k)$ and $C(k)$ are 0.707 so that some multiplications can be saved. Similar to the algorithm based on the decimation-in-frequency decomposition, the number of trivial twiddle factors is also proportional to the value of q.

3.3.3   Computational complexity

Let us now consider the computational complexity in terms of the number of additions and multiplications required by the described algorithms. It is assumed that a complex multiplication needs four real multiplications and two real additions.

Complex input sequence

Based on (3.25) and (3.26), the number of multiplications and additions for $N = 2q$ is

$$M(N) = 2M(q) \tag{3.41}$$
$$A(N) = 2A(q) + 4q. \tag{3.42}$$

When $N = 4q$, both (3.28) and (3.29) are length q DFTs whose twiddle factors become trivial, i.e., $\sin(n\pi/2)$ and $\cos(n\pi/2)$. The required numbers of multiplications and additions are

$$M(N) = M(2q) + 2M(q) \tag{3.43}$$
$$A(N) = A(2q) + 2A(q) + 12q. \tag{3.44}$$

It is observed that when $N = 2q$ and $4q$, all twiddle factors become trivial and do not need any multiplications. This is a general property for split-radix algorithms. For example, it is well known that the length 2 and 4 DFTs do not need any multiplications, which corresponds to the case when $q = 1$.

When $N = 8q$, savings on multiplications can also be achieved by further decomposing (3.28) or (3.29) into a length q DFT and a scaled length q DFT. For example, we write (3.28) as

$$
\begin{aligned}
F(k) &= \sum_{n=0}^{q-1} \left[v(2n) \cos \frac{\pi n}{2} + (-1)^{\frac{q+1}{2}} v(2n + \frac{N}{4}) \sin \frac{\pi n}{2} \right] W_q^{nk} \\
&+ (-1)^k \sum_{n=0}^{q-1} \left\{ v[(2n+q) \bmod (2q)] \cos \frac{\pi[(2n+q) \bmod (2q)]}{4} \right. \\
&+ \left. (-1)^{q+1/2} v \left[(2n+q) \bmod (2q) + \frac{N}{4} \right] \sin \frac{\pi[(2n+q) \bmod (2q)]}{4} \right\} W_q^{nk}. (3.45)
\end{aligned}
$$

The addition related to the first sum of (3.45) is redundant because either $\cos(\pi n/2)$ or $\sin(\pi n/2)$ is zero. The multiplication of the length q DFT by a constant factor $\pm\sqrt{2}/2$ (i.e., the twiddle factors in the second and third lines) can be achieved by a scaled length q DFT. Compared to the standard DFTs, for example, 6 and 12 extra multiplications are needed by the scaled length 9 and 15 DFTs while the same number of additions is needed. Therefore, the required number of multiplications is

$$
M(N) = M(4q) + 2M(q) + 2M_s(q), \quad N = 8q, \tag{3.46}
$$

where $M_s(N)$ is the number of multiplications needed by a scaled DFT. The number of additions is

$$
A(N) = A(4q) + 4A(q) + 36q, \quad N = 8q. \tag{3.47}
$$

According to (3.25), (3.28) and (3.29), the total number of multiplications required by a length N DFT is

$$
M(N) = M(\frac{N}{2}) + 2M(\frac{N}{4}) + 2N - M_t, \quad N > 8q \tag{3.48}
$$

and the total number of additions is

$$
A(N) = A(\frac{N}{2}) + 2A(\frac{N}{4}) + 4N - A_t, \quad N > 8q, \tag{3.49}
$$

where M_t and A_t are the number of multiplications and additions saved from trivial twiddle factors. In this algorithm, $12q$ multiplications and $4q$ additions can be saved from the trivial twiddle factors compared to 12 multiplications and 4 additions saved in other reported split-radix algorithms [7, 21].

The computation of small DFTs, such as length 5 and 9, can be easily optimized. Appendix A shows that a length 9 DFT requires 16 multiplications and 84 additions for a complex sequence. Other reported realizations need 32 multiplications and 80 additions [23] and 16 multiplications and 90 additions [25]. By using optimized length 5 DFTs which need 8 multiplications and 36 additions, Appendix B shows that a length 15 DFT requires 30 multiplications and 168 additions. A subroutine for a length 5 DFT is given in Appendix

Table 3.4   Computational complexity of the FFT algorithm for complex input sequence.

| $N = q \cdot 2^m$ | | | | | |
| $q = 1$ | | | $q = 3$ | | |
N	$M(N)$	$A(N)$	N	$M(N)$	$A(N)$
16	24	144	12	8	96
32	84	372	24	24	252
64	248	912	48	100	624
128	660	2164	96	304	1500
256	1656	5008	192	852	3504
512	3988	11380	384	2192	8028
1024	9336	25488	768	5396	18096
2048	21396	56436	1536	12816	40284
$q = 9$			$q = 15$		
N	$M(N)$	$A(N)$	N	$M(N)$	$A(N)$
18	32	204	15	30	168
36	64	480	30	60	396
72	140	1140	60	120	264
144	448	2640	120	264	2124
288	1196	6036	240	804	4848
576	3136	13584	480	2112	10956
1152	7724	30228	960	5460	24432
2304	18496	66576	1920	13344	53964

C. Based on the computational complexity of the small DFTs, the number of arithmetic operations required by the described algorithm for various sequence lengths can be calculated as shown in Table 3.4 for $q = 1, 3, 9$ and 15.

Real input sequence

The DFT of a real input sequence has a symmetric property described by

$$X(k) = X^*(N - k), \tag{3.50}$$

where $X^*(\ )$ is a conjugate of $X(\ )$. Based on this property, it is known that (3.30) can be obtained from (3.31) or vice versa. Furthermore, two additions can be saved from (3.30) because $F(0)$ is real and $G(0)$ is imaginary. Therefore, the number of additions required in this case is

$$\begin{aligned} A_r(N) &= 2A_r(q) + 2q, & N &= 2q \\ A_r(N) &= A_r(2q) + 2A_r(q) + 6q - 2, & N &= 4q \end{aligned} \tag{3.51}$$

$$\begin{aligned} A_r(N) &= A_r(4q) + 4A_r(q) + 18q - 4, & N &= 8q \\ A_r(N) &= A_r(\tfrac{N}{2}) + 2A_r(\tfrac{N}{4}) + 2N - 2q - 4, & N &> 8q, \end{aligned} \tag{3.52}$$

where the subscript r indicates the additive complexity for a real input sequence. It can be shown that for a real input sequence, the length 3, 9 and 15 DFTs need 4, 34 and 70 additions, respectively. The number of multiplications needed for a real input sequence is

Table 3.5   Computational complexity of the FFT algorithm for real input sequence.

$N = q \cdot 2^m$							
$q = 1$		$q = 3$		$q = 9$		$q = 15$	
N	$A(N)$	N	$A(N)$	N	$A(N)$	N	$A(N)$
16	58	12	38	18	86	15	70
32	156	24	104	36	208	30	170
64	394	48	266	72	500	60	398
128	956	96	656	144	1178	120	944
256	2250	192	1562	288	2732	240	2186
512	5180	384	3632	576	6218	480	5000
1024	11722	768	8282	1152	13964	960	11258
2048	26172	1536	18604	2304	30986	1920	25064

half of that for a complex input sequence. Table 3.5 shows the number of multiplications and additions required for a real input sequence. When a complex multiplication uses three real multiplications and three additions, the number of operations required by the FFT computation is different from that listed in Tables 3.4 and 3.5. However, the total number of arithmetic operations needed by the algorithms is always the same.

Comparison

Let us compare the computational complexity needed by various split-radix fast algorithms in terms of the required number of additions and multiplications. When $q = 1$ in (3.28) to (3.31), the described algorithm becomes the split-radix algorithm reported in [7]. Both algorithms need about the same number of arithmetic operations (additions plus multiplications). For $q > 1$, the number of multiplications and additions saved from trivial twiddle factors is proportionally increased with the value of q. Compared to that for $q = 1$, Figure 3.7 (a) shows that our algorithm needs fewer additions for $q = 3$ and more additions for $q = 9$ and 15. Figure 3.7 (b) shows that a significant reduction of the number of multiplications can be made. It is important to compare the total number of arithmetic operations when the computation time for a multiplication is comparable to that for an addition. Figure 3.7 (c) compares the total number of arithmetic operations needed for $q = 1, 3, 9$ and 15. When $N > 100$, the generalized algorithm needs a smaller number of arithmetic operations for $q = 3, 9$ and 15 than for $q = 1$.

3.4   PRIME FACTOR ALGORITHMS

Although mature and cost-effective algorithms in the radix category work quite well, computational overhead usually occurs when the input sequence length is not matched to the available algorithm. Therefore, there exist fast algorithms in the non-radix category that can be used for various transform sequence lengths. For example, the prime factor algorithm (PFA) [13] and Winograd Fourier transform algorithm [26] were often used for sequence lengths other than a power of two. However, these algorithms generally require complex computational structures. For example, their common drawback is the requirement of an irregular index mapping process to reorder the original input data sequence during computation. A substantial portion of computation time and additional memory space are

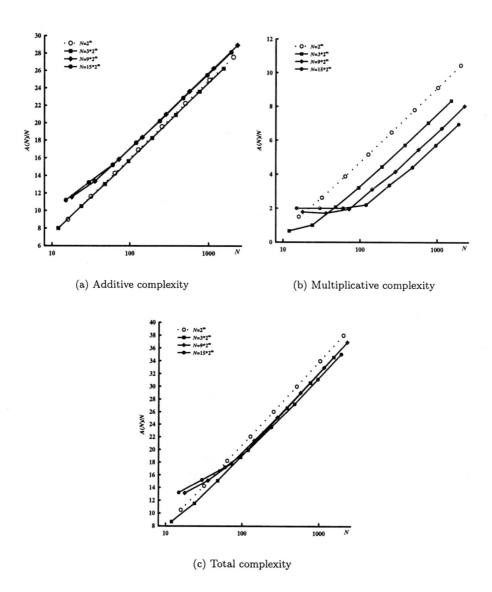

(a) Additive complexity

(b) Multiplicative complexity

(c) Total complexity

Figure 3.7   Computational complexity for complex input sequence.

generally needed. The desirable property of in-place computation, which is easily achieved by the algorithms in the radix category, is difficult to achieve. In this section, we describe a prime factor algorithm for 1D DFTs which can achieve a regular index mapping process and support in-place computation.

3.4.1   Algorithm

Let us assume that the length N of the DFT is $p \cdot q$, where q is an odd integer, and p and q are relatively prime. There is a one-to-one mapping between k $(k = 0, 1, \ldots, N-1)$, and a pair of integers k_1 $(k_1 = 0, 1, \ldots, p-1)$ and m $(m = 0, 1, \ldots, q-1)$ so that (3.1) can be expressed by

$$
\begin{aligned}
X[(qk_1 + mp) \bmod M] &= \sum_{n=0}^{N-1} x(n) W_q^{nm} W_p^{nk_1} \\
&= \sum_{n=0}^{p-1} \left[\sum_{n=0}^{q-1} x(n+ip) W_q^{(n+ip)m} \right] W_p^{nk_1} \\
&= \sum_{n=0}^{p-1} U(n,m) W_p^{nk_1},
\end{aligned}
\tag{3.53}
$$

where for $0 \le n < p$ and $0 \le m < q$,

$$
U(n,m) = \sum_{i=0}^{q-1} x(n+ip) W_q^{(n+ip)m}.
\tag{3.54}
$$

Equation (3.53) becomes a length p DFT of sequence $U(n,m)$. When q and p are mutually prime, it can be proven that for each i $(i = 0, 1, \ldots, q-1)$ there is a one-to-one mapping between i and l where

$$
d_{n,l} = (n + ip) \bmod q.
\tag{3.55}
$$

Therefore, (3.54) can be computed by

$$
U(n,m) = U_r(n,m) - jU_j(n,m) = \sum_{d_{n,l}=0}^{q-1} x(d_{n,l}) W_q^{d_{n,l}m},
\tag{3.56}
$$

where $0 \le n < p$ and $0 \le m < q$. Thus, (3.56) becomes a length q DFT whose input sequence is rearranged. When the input sequence is real, we note that $U(n,m)$ has a symmetric property. In general, we have

$$
X(k) = \begin{cases} \begin{bmatrix} X^*(N-k) \\ IM\{X(0)\} = IM\{X(\frac{N}{2})\} = 0 \end{bmatrix}, & N \bmod 2 = 0 \\ X^*(N-k), IM\{X(0)\} = 0, & N \bmod 2 = 1 \end{cases}
\tag{3.57}
$$

where $0 \le k < N/2$ and $X^*(\ )$ is the complex conjugate of $X(\ )$. Therefore, (3.53) becomes a length p real-valued DFT for $m = 0$ and complex-valued DFTs for $1 \le m \le (q-1)/2$, which can be achieved by separately computing two real-valued DFTs. Figure 3.8 shows an example of the algorithm for $N = 3 \cdot 5$ in which the complex-valued DFT is achieved by combining real-valued DFTs of sequence $U_r(n,m)$ and $U_j(n,m)$.

One undesirable requirement of the algorithm is that a process of regrouping the input sequence at each computation stage is needed as indicated in (3.55). Because the issue on input data reordering or index mapping was not properly addressed, it may be a complicated and time-consuming process as seen in a few early reported algorithms. However, we can show that such a requirement can be eliminated by carefully generating data indices. Let us consider an example for $N = 3 \cdot 5$. The input sequence is divided into 5 groups and each contains 3 elements whose original input index is represented by $d_{n,l}$. Observing Figure 3.9, we can conclude that the input index can be obtained by

$$
\begin{aligned}
d_{n,l} &= d_{n-1,l-1} + 1, \quad 1 \le n < p, \ 1 \le l < q \\
d_{n,0} &= d_{n-1,q-1} + 1, \quad 1 \le n < p,
\end{aligned}
\tag{3.58}
$$

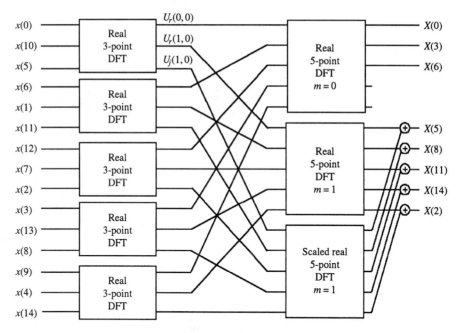

Figure 3.8   Signal flow graph of the PFA ($N = 3 \cdot 5$).

where $d_{0,l}, 0 \leq l < q$, can be pre-calculated. The calculation of $d_{n,l}$ is general as long as p and q are mutually prime. The modulo operation used in (3.55) is replaced by increment operation. Once $d_{n,l}$ are available for a particular value of n, the length 3 DFT can be computed. In this way, both data swapping and modulo operations are avoided. In-place computation can also be obtained by writing the outputs of the length 3 DFT into the data locations (represented by $f_{n,l}$ in Figure 3.9) which are no longer used by the input data. One important property is that the data for the next stage computation are in a natural order so that the same data indexing scheme can be used for further decomposition.

	$l = 0$	1	2			
$n = 0$	0	10	5	0	5	10
1	6	1	11	1	6	11
2	12	7	2	2	7	12
3	3	13	8	3	8	13
4	9	4	14	4	9	14
		$d_{n,l}$			$f_{n,l}$	

Figure 3.9   Input/output indices for the first stage.

The above decomposition is based upon the Good–Thomas algorithm. The steps listed below summarize the algorithm:

Step 1.   calculate the array indices according to (3.58),

Step 2. calculate $U(n, m)$ according to (3.56), and

Step 3. calculate the final transform outputs according to (3.53).

The method for index mapping can be extended to sequence lengths containing more than two factors that are mutually prime.

3.4.2 Computational complexity

For real input sequences, the number of additions required is

$$A_r(N) = A_r(q) + qA_r(p) + \frac{(p-1)}{2}A_c(q) \tag{3.59}$$

and the number of multiplications is

$$M_r(N) = M_r(q) + qM_r(p) + \frac{(p-1)}{2}M_c(q), \tag{3.60}$$

where the subscripts c and r indicate complex and real, respectively. When q is small, the multiplicative complexity expressed by (3.59) can be further reduced. When $q = 3$, for example, the imaginary part of $U(m)$ becomes

$$U_j(1) = -U_j(2) = \frac{\sqrt{3}}{2}[x(1) - x(2)]. \tag{3.61}$$

The multiplication required by (3.61) can be saved if the second term of (3.56) becomes a scaled length q DFT by a factor of $\sqrt{3}/2$. Such an arrangement is also possible for other q values. For example, the imaginary parts of a length 5 real-valued DFT can be expressed by

$$U_j(1) = -U_j(4) = \sin\frac{2\pi}{5}\left\{[x(1) - x(4)] + 2\cos\frac{2\pi}{5}[x(2) - x(3)]\right\} \tag{3.62}$$

$$U_j(2) = -U_j(3) = \sin\frac{2\pi}{5}\left\{2\cos\frac{2\pi}{5}[x(1) - x(4)] - [x(2) - x(3)]\right\}. \tag{3.63}$$

When the scaling factors outside the braces in (3.62) and (3.63) are not counted, the length 5 real-valued DFT uses 3 multiplications and 13 additions. For a length 9 real valued DFT, the scaling factor $\sqrt{3}/2$ can be separated and combined with the next stage computation. Therefore, the real-valued scaled length 9 DFT used by the algorithm requires 6 multiplications and 34 additions. Similarly, 2 scaling factors can be separated from a standard length 15 real-valued DFT that needs 18 multiplications and 67 additions.

In general, the reduction of multiplications obtained from rearranging the scaling factors is Kp, where K is the number of scaling factors that can be separated from the imaginary parts of a standard length q DFT and p is the number of length q real-valued DFTs needed by (3.56). All these scaling factors are absorbed by a length q DFT of input sequence $U_j(n, m)$. For $N = 2^m$, the scaled DFT for a real input sequence can be decomposed into a scaled length $N/2$ DFT and two standard length $N/4$ DFTs. To minimize the computational complexity for a scaled DFT, the scaling factor for the odd indexed transform outputs can be combined with the twiddle factors, or equivalently, the DFT computation uses scaled kernel values. The number of additions needed by a scaled length N DFT is the same as that needed by a standard length N DFT, and the required number of multiplications is

$$M_r^s(N) = M_r^s(\frac{N}{2}) + 2M_r(\frac{N}{4}) + N - 2, \quad N > 16, \tag{3.64}$$

Table 3.6   Computational complexity for length 2^m DFT.

N	Algorithm in [7] $M(N)$	$A(N)$	Scaled by $\sqrt{3}/2$ $M(N)$	$A(N)$
8	2	20	8	20
16	12	58	22	58
32	42	156	56	156
64	124	394	142	394
128	330	956	352	956
256	828	2250	854	2250
512	1994	5180	2024	5180
1024	4668	11722	4702	11722

where M_r^s is for a scaled DFT of real input sequence. The number of extra multiplications for a scaled DFT is small. For example, Table 3.6 shows that for $N = 1024$, 34 extra multiplications are needed compared to a standard length 1024 DFT. Based on (3.59) and (3.60) and the above discussion, Table 3.7 shows the number of additions and multiplications needed by a length $p \cdot q$ DFT for $q = 3, 5, 9$ and 15. The described algorithm can easily be used for DFTs of complex data. The transform outputs of the complex sequence can be obtained by combining the DFTs of real and imaginary parts of the input sequence. Therefore, the number of multiplications required by a DFT of complex data is two times that for real input data and the number of additions is given by

$$A_c(N) = \begin{cases} 2A_r(\frac{N}{2}) + 2N - 4, & \frac{N}{2} \text{ is even} \\ 2A_r(\frac{N}{2}) + 2N - 2, & \frac{N}{2} \text{ is odd} \end{cases} . \tag{3.65}$$

Table 3.7   The computational complexity for DFT of real input.

	$M(N)$	$A(N)$
$q = 3$	$2M_r(q) + M_r^s(p)$	$3A_r(p) + 6p - 4$
5	$3M_r(p) + 2M_r^s(p) + 3p$	$5A_r(p) + 17p - 8$
9	$5M_r(p) + 4M_r^s(p) + 7p$	$9A_r(p) + 42p - 16$
15	$9M_r(p) + 6M_r^s(p) + 12p$	$15A_r(p) + 71p - 28$

3.5   GENERALIZED 2D SPLIT-RADIX ALGORITHMS

The generalized 1D split-radix algorithm in Section 3.3 and the PFA in Section 3.4 can be easily extended for MD DFTs. A 2D complex sequence $x(m_1, m_2), 0 \leq m_1, m_2 < N$, has an $N \times N$ DFT defined by

$$X(k_1, k_2) = \sum_{m_1=0}^{N-1} \sum_{m_2=0}^{N-1} x(m_1, m_2) W_N^{m_1 k_1 + m_2 k_2} \tag{3.66}$$

$$0 \le k_1, k_2 < N - 1.$$

The sequence length N is assumed to be $q \cdot 2^m$, where q is an odd integer. Equation (3.66) can be easily rewritten as

$$X(k_1, k_2) = \sum_{m_1=0}^{N-1} \left[\sum_{m_2=0}^{N-1} x(m_1, m_2) W_N^{m_1 k_1} \right] W_N^{m_2 k_2}, \qquad (3.67)$$

which has two summations. Because each summation can be considered to be a 1D DFT along each row (for $k_1 = 0$ to $N-1$) or column (for $k_2 = 0$ to $N-1$), $2N$ 1D DFTs are needed. Since the computation is performed according to the row and column indices, such an approach is known as the *row-column algorithm*. The row-column concept is very simple and can be easily implemented by using fast algorithms for 1D DFTs. Row-column algorithms are also available for computation of MD discrete cosine transforms (DCTs) and discrete Hartley transforms (DHTs). However, this algorithm is not computationally efficient. This subsection describes the fast algorithms for 2D DFTs by extending the concepts used in the previous subsections to support the computation of a $q \cdot 2^m \times q \cdot 2^m$ DFT. It can be shown that the existing algorithms reported in [6, 16] can be derived from the presented one by setting $q = 1$.

3.5.1   Algorithm

The 2D DFT defined in (3.66) can be decomposed by using the split-radix concept. The even-even indexed transform outputs can be calculated by

$$X(2k_1, 2k_2) = \sum_{m_1=0}^{N/2-1} \sum_{m_2=0}^{N/2-1} z_{0,0}(m_1, m_2) W_{N/2}^{m_1 k_1 + m_2 k_2} \qquad (3.68)$$

$$0 \le k_1, k_2 < \frac{N}{2} - 1.$$

When $N = 2q$, we have

$$X(2k_1 + r_1 q, 2k_2 + r_2 q) = \sum_{m_1=0}^{N/2-1} \sum_{m_2=0}^{N/2-1} (-1)^{(r_1 m_1 + r_2 m_2)} z_{r_1, r_2}(m_1, m_2) W_q^{m_1 k_1 + m_2 k_2}, \quad (3.69)$$

where $0 \le k_1, k_2 < q - 1$ and for both (3.68) and (3.69)

$$\begin{aligned} z_{r_1, r_2}(m_1, m_2) &= [x(m_1, m_2) + (-1)^{r_1} x(m_1 + q, m_2)] \\ &\quad + (-1)^{r_2} [x(m_1, m_2 + q) + (-1)^{r_1} x(m_1 + q, m_2 + q)], \end{aligned} \qquad (3.70)$$

where the parameters $(r_1, r_2) = (0, 0)$, $(0, 1)$, $(1, 0)$ and $(1, 1)$ indicate even-even, even-odd, odd-even and odd-odd decomposition. Both (3.68) and (3.69) are $q \times q$ DFTs which do not need any non-trivial multiplication for twiddle factors. When $N = 4q$, the entire computation can be decomposed into one $N/2 \times N/2$ DFT, as defined in (3.68), and twelve

$N/4 \times N/4$ DFTs which are defined by

$$X[(4k_1 + r_1 q + N) \bmod N, (4k_2 + r_2 q + N) \bmod N]$$

$$= \sum_{m_1=0}^{N-1} \sum_{m_2=0}^{N-1} x(m_1, m_2) W_{N/q}^{m_1 r_1 + m_2 r_2} \, W_{N/4}^{m_1 k_1 + m_2 k_2}$$

$$= \sum_{m_1=0}^{N/4-1} \sum_{m_2=0}^{N/4-1} w(r_1, r_2) W_{N/q}^{m_1 r_1 + m_2 r_2} \, W_{N/4}^{m_1 k_1 + m_2 k_2}, \tag{3.71}$$

where $(r_1, r_2) = (0, 1), (0, -1), (2, 1), (2, -1), (1, 0), (-1, 0), (1, 2), (-1, 2), (1, 1), (-1, -1),$ $(-1, 1), (1, -),$ and

$$w(r_1, r_2) = \left[z_{r_1, r_2}(m_1, m_2) + e^{-j\pi q r_1/2} z_{r_1, r_2}(m_1 + \frac{N}{4}, m_2) \right]$$

$$+ e^{-j\pi q r_2/2} \left[z_{r_1, r_2}(m_1, m_2 + \frac{N}{4}) \right.$$

$$\left. + e^{-j\pi q r_1/2} z_{r_1, r_2}(m_1 + \frac{N}{4}, m_2 + \frac{N}{4}) \right]. \tag{3.72}$$

The modulo operation in (3.71) implements the property $X(-k) = X(N - k), k > 0$. Furthermore, it can be proven that in (3.72),

$$z_{0,1}(m_1, m_2) = z_{0,-1}(m_1, m_2) = z_{2,1}(m_1, m_2) = z_{2,-1}(m_1, m_2)$$
$$z_{1,0}(m_1, m_2) = z_{-1,0}(m_1, m_2) = z_{1,2}(m_1, m_2) = z_{-1,2}(m_1, m_2)$$
$$z_{1,1}(m_1, m_2) = z_{-1,-1}(m_1, m_2) = z_{1,-1}(m_1, m_2) = z_{1,-1}(m_1, m_2).$$

The decomposition can be continued until the $q \times q$ DFT is reached.

3.5.2 Computational complexity

It is assumed that four real multiplications and two real additions are used for a complex multiplication. The computational complexities for various sequence lengths are

$$M_c(2q \times 2q) = 4M_c(q \times q)$$
$$A_c(2q \times 2q) = 4A_c(q \times q) + 16q^2$$
$$M_c(4q \times 4q) = M_c(2q \times 2q) + 12M_c(q \times q)$$
$$A_c(4q \times 4q) = A_c(2q \times 2q) + 12A_c(q \times q) + 112q^2$$
$$M_c(N \times N) = M_c(\frac{N}{2} \times \frac{N}{2}) + 12M_c(\frac{N}{4} \times \frac{N}{4}) + 3N^2 - 18qN$$
$$A_c(N \times N) = A_c(\frac{N}{2} \times \frac{N}{2}) + 12A_c(\frac{N}{4} \times \frac{N}{4}) + \frac{17}{2}N^2 - 6qN,$$

where $N > 4q$. Because the parameter q in (3.71) is associated with the twiddle factors, multiplication is not needed by twiddle factors when $N = 2q$ and $4q$. Furthermore, the number of operations saved from trivial twiddle factors can be increased by q times. For real input data, the entire DFT computation can be decomposed into one real $N/2 \times N/2$ DFT and six complex $N/4 \times N/4$ DFTs. The number of additions needed by the algorithms is

$$
\begin{aligned}
A_r(2q \times 2q) &= 4A_r(q \times q) + 8q^2 \\
A_r(4q \times 4q) &= A_r(2q \times 2q) + 6A_c(q \times q) + 44q^2 \\
A_r(N \times N) &= A_r(\frac{N}{2} \times \frac{N}{2}) + 6A_c(\frac{N}{4} \times \frac{N}{4}) + \frac{7}{2}N^2 - 3qN,
\end{aligned}
$$

where $N > 4q$. The number of multiplications needed for real data is half of that for complex data.

Table 3.8   Number of operations for complex data.

N	Algorithm in [6] $M_c(N \times N)$	$A_c(N \times N)$
8	48	816
16	720	4496
32	4175	22928
64	24720	111568
128	123216	525712
256	614928	2421072
	Presented algorithm ($q = 1$)	
8	48	816
16	528	4432
32	3600	22736
64	21072	110352
128	111120	521680
256	555984	2401424
	Presented algorithm ($q = 3$)	
6	32	464
12	128	2432
24	944	12464
48	6800	60368
96	40592	286544
192	222416	1320848

Table 3.8 lists the number of additions and multiplications needed by Chan's algorithm [6] and the described one for $q = 1$, and 3. For real input, our algorithm achieves exactly the same number of additions as that in [16] and a smaller number of additions than that in [6]. Because the algorithm has the same computational structure as other split-radix ones, it preserves all the desirable properties, such as in-place computation, regular permutation and simple memory requirements.

3.6   FAST ALGORITHMS FOR GENERALIZED DFT

The generalized DFT (GDFT) has many applications in digital signal processing such as filter banks, convolution and signal representation [10, 11]. Fast algorithms ([5, 11], for example) were reported to reduce the required computational complexity. In particular, a split-radix algorithm [18] decomposes the entire computation of a length N GDFT into the computation of a length $N/2$ and two length $N/4$ GDFTs. Based on this algorithm, efficient computation of the odd-time and odd-frequency DFT can be achieved. This subsection attempts to make improvements on the split-radix GDFT algorithm. Fast algorithms using the split-radix FFT algorithm to compute the GDFT for arbitrary time and frequency origins are presented. The fast algorithms for the odd-time and odd-frequency GDFTs are derived in terms of the DCT. The computational complexity required by the presented algorithms for complex and real data is analyzed.

3.6.1 Algorithm

The GDFT of a sequence $x(n)$ $(n = 0, 1, \ldots, N-1)$ is defined by

$$
\begin{aligned}
X(k) &= \sum_{n=0}^{N-1} x(n) W_N^{(n+n_0)(k+k_0)} & (3.73) \\
&= W_N^{n_0(k+k_0)} \sum_{n=0}^{N-1} x(n) W_N^{nk_0} W_N^{nk}, \quad 0 \le k < N,
\end{aligned}
$$

where n_0 and k_0 are the time and frequency origins or shifts in time and frequency, respectively. When $n_0 = k_0 = 1/2$, the GDFT is known as the odd-squared DFT. When $n_0 = 1/2$ and $k_0 = 0$ (or $n_0 = 0$ and $k_0 = 1/2$, the GDFT is often termed as the odd-time (or odd-frequency) DFT.

Generalized DFT (arbitrary n_0 and k_0)
Equation (3.73) is basically a length N DFT whose input sequence $x(n)$ and output sequence $X(k)$ are rotated by twiddle factors. We can rearrange the twiddle factors so that the split-radix decomposition presented earlier can be applied. The even indexed GDFT outputs $X(2k)$ (k = 0, 1, $\ldots$, N/2) can be computed by

$$
\begin{aligned}
X(2k) &= W_{N/2}^{n_0 k} \sum_{n=0}^{N-1} x(n) W_N^{nk_0 + n_0 k_0} W_{N/2}^{nk} \\
&= W_{N/2}^{n_0 k} \sum_{n=0}^{N/2-1} \left[x(n) + x(n + \frac{N}{2}) e^{-j\pi k_0} \right] W_N^{nk_0 + n_0 k_0} W_{N/2}^{nk}, & (3.74)
\end{aligned}
$$

which is a length $N/2$ DFT whose input and output sequences are rotated by twiddle factors, respectively. The odd indexed outputs can be computed by

$$
\begin{aligned}
X(4k+1) &= W_{N/4}^{n_0(4k+k_0+1)} \sum_{n=0}^{N-1} x(n) W_N^{nk_0} W_N^{n(4k+1)} \\
&= W_N^{n_0(4k+k_0+1)} \sum_{n=0}^{N/4-1} \left\{ \left[x(n) - x(n + \frac{N}{2}) e^{-j\pi k_0} \right] \right. \\
&\quad \left. - j \left[x(n + \frac{N}{4}) - x(n + \frac{3N}{4}) e^{-j\pi k_0} \right] e^{-j\pi k_0/2} \right\} W_N^{n(k_0+1)} W_{N/4}^{nk} \quad (3.75)
\end{aligned}
$$

and

$$X(4k+3) = W_{N/4}^{n_0(4k+k_0+3)} \sum_{n=0}^{N/4-1} \left\{ \left[x(n) - x(n+\frac{N}{2})e^{-j\pi k_0} \right] \right.$$

$$\left. + j \left[x(n+\frac{N}{4}) - x(n+\frac{3N}{4})e^{-j\pi k_0} \right] e^{-j\pi k_0/2} \right\} W_N^{n(k_0+3)} W_{N/4}^{nk}, \quad (3.76)$$

where $k = 0, 1, \ldots, N/4\text{-}1$ in (3.75) and (3.76). In summary, the length N GDFT for arbitrary n_0 and k_0 can be decomposed into one length $N/2$ and two length $N/4$ DFTs whose input and output sequences are rotated by twiddle factors. Other methods to decompose the length N GDFT into one length $N/2$ and two length $N/4$ GDFTs have been reported [18]. These algorithms have no fundamental difference from the presented algorithm since they can be easily converted into each other by manipulating twiddle factors in (3.75) and (3.76). However, the described algorithm allows trivial rotation factors to be exploited to reduce the number of arithmetic operations. For computation of the odd-time, odd-frequency and odd-squared DFTs, it may be necessary to combine some of the twiddle factors in (3.74) – (3.76) to minimize the overall computational complexity.

Odd-squared DFT $(n_0 = k_0 = 0.5)$
When $n_0 = k_0 = 0.5$, (3.73) can be expressed by

$$X(k) = \sum_{n=0}^{N-1} x(n) W_N^{(n+0.5)(k+0.5)}$$

$$= W_N^{0.5(k+0.5)} \sum_{n=0}^{N-1} x(n) W_N^{0.5n} W_N^{nk}. \quad 0 \le k < N$$

It can be converted into

$$X(k) = A(k) - jB(k) \qquad (3.77)$$

$$X(N-k-1) = -[A(k) + jB(k)], \qquad (3.78)$$

where $k = 0, \ldots, N/2\text{-}1$, and

$$A(k) = \sum_{n=0}^{N-1} x(n) \cos \frac{\pi(2n+1)(2k+1)}{2N} \qquad (3.79)$$

$$B(k) = \sum_{n=0}^{N-1} x(n) \sin \frac{\pi(2n+1)(2k+1)}{2N} \qquad (3.80)$$

or

$$B(\frac{N}{2}-1-k) = \sum_{n=0}^{N-1} x(n)(-1)^n \cos \frac{\pi(2n+1)(2k+1)}{2N}. \qquad (3.81)$$

Equation (3.78) can be easily derived by using $A(k) = -A(N-1-k)$ and $B(k) = B(N-1-k)$. According to (3.79), we define an auxiliary sequence

$$
\begin{aligned}
F(k) &= \frac{A(k) + A(k-1)}{2} \\
&= \sum_{n=0}^{N-1} x(n) \cos\frac{\pi(2n+1)}{2N} \cos\frac{\pi(2n+1)k}{2(N/2)} \\
&= \sum_{n=0}^{N/2-1} [x(n) - x(N-1-n)] \cos\frac{\pi(2n+1)}{2N} \cos\frac{\pi(2n+1)k}{2(N/2)},
\end{aligned}
\tag{3.82}
$$

where $k = 0, 1, \ldots, N/2 - 1$. Equation (3.82) is a length $N/2$ DCT of complex data. The sequence $A(k)$ can be achieved by

$$
A(0) = F(0), \quad A(k) = 2F(k) - A(k-1), \quad k = 1, 2, \ldots, \frac{N}{2} - 1.
\tag{3.83}
$$

In the same way, $B(N/2 - 1 - k)$ defined in (3.81) can be obtained by a length $N/2$ DCT computation. Therefore, the computation of the length N odd-squared DFT can be achieved by the following steps:

- compute $A(k)$ and $B(k)$ according to (3.82) and (3.83).

- compute the final transform outputs by (3.77) and (3.78).

The algorithm for the odd-squared DFT is shown in Figure 3.10 in which all operations are complex.

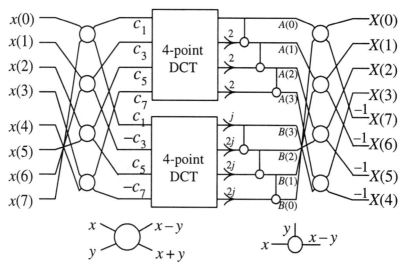

Figure 3.10   The algorithm for odd-squared DFT ($N = 8$).

Odd-time DFT ($n_0 = 0.5, k_0 = 0$)
When $n_0 = 0.5$ and $k_0 = 0$, (3.73) can be decomposed into even and odd indexed transform

outputs. For example,

$$X(2k) = \sum_{n=0}^{N-1} x(n) W_{N/2}^{(n+1/2)k} = \sum_{n=0}^{N/2-1} \left[x(n) + x(n + \frac{N}{2}) \right] W_{N/2}^{(n+1/2)k}, \tag{3.84}$$

which is a length $N/2$ odd-time DFT, and

$$\begin{aligned} X(2k+1) &= \sum_{n=0}^{N-1} x(n) W_{N/2}^{(n+1/2)(k+1/2)} \\ &= \sum_{n=0}^{N/2-1} \left[x(n) - x(n + \frac{N}{2}) \right] W_{N/2}^{(n+1/2)(k+1/2)}, \end{aligned} \tag{3.85}$$

which is a length $N/2$ odd-squared DFT and can be computed by the algorithm shown in the last subsection. The length $N/2$ odd-time DFT can be further decomposed in the same way.

Odd-frequency DFT $(n_0 = 0, k_0 = 0.5)$
When $n_0 = 0$, and $k_0 = 0.5$, (3.73) can be decomposed into

$$\begin{aligned} X(k) &= C(k) + D(k) \\ X(k + N/2) &= C(k) - D(k), \end{aligned} \tag{3.86}$$

where for $0 \leq k < N/2$

$$C(k) = \sum_{n=0}^{N/2-1} x(2n) W_{N/2}^{n(k+1/2)} \tag{3.87}$$

is a length $N/2$ odd-frequency DFT, and

$$D(k) = \sum_{n=0}^{N/2-1} x(2n+1) W_{N/2}^{(n+1/2)(k+1/2)} \tag{3.88}$$

is a length $N/2$ odd-squared DFT. Similarly, (3.87) can be further decomposed and the odd-squared DFT in (3.88) is computed by the previously described algorithm.

Real-valued GDFT
The algorithm for a real-valued odd-squared GDFT can be easily derived from (3.77) – (3.83). $A(k)$ in (3.79) and $B(k)$ in (3.80) directly deal with real data and become the real and the imaginary parts of the final transform outputs, respectively. The additions and subtractions in (3.77) and (3.78) can be eliminated. Equation (3.82) becomes a length $N/2$ DCT of real data. For the odd-time and odd-frequency DFTs, the computation for real data can also be easily obtained from their complex counterparts. In this case, the length $N/2$ odd-time (or odd-frequency) and odd-squared DFTs directly deal with real data without any modification.

3.6.2   Computational complexity

We now consider the computational complexity in terms of the number of real additions and multiplications required by the presented algorithms.

Generalized DFT (arbitrary n_0 and k_0)
The decomposition cost required by (3.74) – (3.76) is:

- $6N$ multiplications and $3N$ additions for twiddle factors, and N additions for the input sequence in (3.74)

- $3N$ multiplications and $3N/2$ additions for twiddle factors, and $3N/2$ additions for the input sequence in (3.75)

- $2N$ multiplications and N additions for twiddle factors, and $N/2$ additions for the input sequence in (3.76)

The twiddle factor $\exp(\text{-j}\pi n_0)$ in (3.74) – (3.76) and $\exp(\text{-j}\pi k_0/2)$ in (3.75) and (3.76) are computed once only, respectively. The number of real multiplications required by the algorithm is

$$M_{GDFT}(N) = M_{DFT}(\frac{N}{2}) + 2M_{DFT}(\frac{N}{4}) + 11N - 12 \qquad (3.89)$$

and the number of real additions is

$$A_{GDFT}(N) = A_{DFT}(\frac{N}{2}) + 2A_{DFT}(\frac{N}{4}) + \frac{17N}{2} - 6, \qquad (3.90)$$

where in both (3.89) and (3.90) the subscript GDFT means the computation for GDFT with arbitrary n_0 and k_0, and the subscript DFT indicates the computation for DFT. The last term in both (3.89) and (3.90) is the number of multiplications or additions saved from the trivial twiddle factors.

Odd-squared DFT $(n_0 = k_0 = 0.5)$
As shown in Figure 3.10, the decomposition cost of the odd-squared DFT is:

- $2N$ additions for summation of input sequence in (3.82)

- $2N$ multiplications for twiddle factors in (3.82)

- $2N$ additions for (3.77) and (3.78), and $2N - 4$ additions for (3.83)

- four length $N/2$ DCTs of real sequences.

Therefore, the number of multiplications needed by a length N odd-squared DFT of complex data is

$$M^{OS}_{CGDFT}(N) = 4M_{DCT}(\frac{N}{2}) + 2N = N(\log_2 N + 1), \quad N > 2 \qquad (3.91)$$

and the number of additions is

$$A^{OS}_{CGDFT}(N) = 4A_{DCT}(\frac{N}{2}) + 6N - 4 = N(3\log_2 N + 1), \quad N > 2, \qquad (3.92)$$

where the superscript OS indicates odd-squared and subscript CGDFT means the complex GDFT. The DCT computation can be performed by fast DCT algorithms to be presented in a latter chapter. When N is a power of two, the number of multiplications needed by a length N DCT is

$$M_{DCT}(N) = \frac{N}{2} \log_2 N \qquad (3.93)$$

and the number of additions is

$$A_{DCT}(N) = \frac{3N}{2} \log_2 N - N + 1. \tag{3.94}$$

For real data, the decomposition cost is halved and the additions/subtraction for (3.77) and (3.78) are redundant. Therefore, the number of multiplications for a length N odd-squared DFT of real data is exactly half of that for complex data, and the number of additions is

$$A^{OS}_{RGDFT}(N) = 2A_{DCT}(\frac{N}{2}) + 2N - 2 = \frac{N}{2}(3\log_2 N - 1), \tag{3.95}$$

where RGDFT indicates the real-valued GDFT.

Odd-time and odd-frequency DFT
It can be easily seen that the computational complexity of old-time or odd-frequency DFTs is

- $2N$ additions

- one length $N/2$ odd-time (or odd-frequency) DFT and one length $N/2$ odd-squared DFT

Therefore, the computational complexity for both odd-time and odd-frequency DFTs is the same. For example, the number of multiplications needed by the complex odd-time DFT is

$$\begin{aligned}
M^{OT}_{CGDFT}(N) &= M^{OT}_{CGDFT}(\frac{N}{2}) + M^{OS}_{CGDFT}(\frac{N}{2}) \\
&= M^{OT}_{CGDFT}(\frac{N}{2}) + \frac{N}{2}\log_2 N, \quad N > 4
\end{aligned} \tag{3.96}$$

and the number of additions is

$$\begin{aligned}
A^{OT}_{CGDFT}(N) &= A^{OT}_{CGDFT}(\frac{N}{2}) + A^{OS}_{CGDFT}(\frac{N}{2}) + 2N \\
&= A^{OT}_{CGDFT}(\frac{N}{2}) + \frac{3N}{2}\log_2 N + N, \quad N > 4.
\end{aligned} \tag{3.97}$$

For real data, the number of multiplications for the odd-time or odd-frequency DFT is half of that for complex data. The number of additions needed by the real-valued odd-time DFT is

$$\begin{aligned}
A^{OT}_{RGDFT}(N) &= A^{OT}_{RGDFT}(\frac{N}{2}) + A^{OS}_{RGDFT}(\frac{N}{2}) + N \\
&= A^{OT}_{RGDFT}(\frac{N}{2}) + \frac{3N}{4}\log_2 N, \quad N > 4.
\end{aligned} \tag{3.98}$$

The same additive complexity as that given in (3.98) is needed by the odd-frequency DFT. It can be proved if we consider the property of the real-valued odd-frequency DFT

$$X(N-1-k) = X^*(k), \quad k = 0, \ldots, \frac{N}{2} - 1 \tag{3.99}$$

and that of the odd-squared DFT

$$X(N-1-k) = -X^*(k), \quad k = 0, \ldots, \frac{N}{2} - 1. \tag{3.100}$$

Table **3.9**   Computational complexity for arbitrary n_0 and k_0.

N	DFT		Pei's algorithm		Presented	
	Mul	Add	Mul	Add	Mul	Add
8	4	52	128	112	76	86
16	24	144	336	296	168	214
32	84	372	816	728	372	514
64	248	912	1936	1736	824	1198
128	660	2164	4464	4024	1812	2738
256	1656	5008	10128	9160	3960	6158

Table **3.10**   Computational complexity for odd-squared DFT.

N	Complex data				Real data			
	Pei's algor.		Presented		Pei's algor.		Presented	
	Mul	Add	Mul	Add	Mul	Add	Mul	Add
4	18	34	12	28	9	11	6	10
8	54	102	32	80	27	35	16	32
16	146	274	80	208	73	97	40	88
32	366	686	192	512	183	247	96	224
64	882	1650	448	1216	441	601	224	544
128	2062	3854	1024	2816	1031	1415	512	1280
256	4722	8818	2304	6400	2361	3257	1152	2944

By using (3.99) and (3.100), N additions are needed by (3.86).

Table 3.9 lists the number of additions and multiplications required by the discussed algorithms and Pei's algorithm [18] for arbitrary k_0 and n_0. The presented algorithm reduces the total computational complexity by exploiting many trivial twiddle factors. Similar savings in the computational complexity can also be achieved by our algorithm for the odd-squared DFTs, as shown in Table 3.10. For the odd-time and odd-frequency DFTs, however, Pei's algorithm and our DCT-based algorithm require the same number of additions and multiplications.

3.7   POLYNOMIAL TRANSFORM ALGORITHMS FOR MD DFT

The polynomial transform (PT)-based algorithm for the 2D DFT was first reported in [17]. Compared with other types of fast algorithms, the PT-based algorithms use a smaller number of multiplications without increasing the number of additions. In this section, we first discuss the PT algorithm for the 2D DFT to show some basic techniques for developing fast algorithms. Then we derive solutions to the MD DFT computation.

3.7.1   Polynomial transform algorithm for 2D DFT

Let us consider an $N \times M$ DFT, where M and N are powers of 2 and $M \geq N$. We can write $M = 2^t$ and $M/N = 2^L$, where $t > 0$ and $L \geq 0$, respectively. When $M < N$, we can simply swap M and N to satisfy the above assumption. The 2D DFT $X(k, l)$ defined in (3.66) can be decomposed into

$$X(k, l) = A(k, l) + B(k, l), \tag{3.101}$$

where

$$A(k, l) = (-1)^u A(k, l + \frac{uM}{2}) = \sum_{n=0}^{N-1} \sum_{m=0}^{M/2-1} x(n, 2m+1) W_N^{nk} W_M^{(2m+1)l} \tag{3.102}$$

$$B(k, l) = B(k, l + \frac{uM}{2}) = \sum_{n=0}^{N-1} \sum_{m=0}^{M/2-1} x(n, 2m) W_N^{nk} W_{M/2}^{ml}, \tag{3.103}$$

where $k = 0, 1, \ldots, N - 1$, $l = 0, 1, \ldots, M/2 - 1$ and u is an integer. Equation (3.103) is a 2D DFT with size $N \times M/2$ and can be further decomposed in the same way.

Let us study the computation of (3.102). Since $A(k, l) = (-1)^u A(k, l + \frac{uM}{2})$ for any integer u, we can consider indices l ($l = 0, 1, \ldots, M/2 - 1$) only. If we use $< (2m+1)p >$ to denote the least non-negative residue of $(2m+1)p$ modulo N, then (3.102) becomes

$$A(k, l) = \sum_{p=0}^{N-1} \sum_{m=0}^{M/2-1} x(< (2m+1)p >, 2m+1) W_N^{<(2m+1)p>k} W_M^{(2m+1)l}$$

$$= \sum_{p=0}^{N-1} \sum_{m=0}^{M/2-1} x(< (2m+1)p >, 2m+1) W_M^{(2m+1)(pk2^L+l)}$$

$$= \sum_{p=0}^{N-1} V_p(pk2^L + l), \tag{3.104}$$

where $k = 0, 1, \ldots, N - 1$, $l = 0, 1, \ldots, M/2 - 1$ and

$$V_p(i) = \sum_{m=0}^{M/2-1} x(< (2m+1)p >, 2m+1) W_M^{(2m+1)i}$$

$$= W_M^i \sum_{m=0}^{M/2-1} x(< (2m+1)p >, 2m+1) W_{M/2}^{mi} \tag{3.105}$$

for $i = 0, 1, \ldots, M/2-1$, and $p = 0, 1, \ldots, N-1$. Equation (3.105) can be computed by N 1D DFTs of length $M/2$ plus some complex multiplications. Since $V_p(i + uM/2) = (-1)^u V_p(i)$ for any integer u, (3.104) can be computed if $V_p(i)$ is known for $i = 0, 1, \ldots, M/2 - 1$.

In general, direct computation of (3.104) requires $\frac{1}{2} NM(N - 1)$ additions. This additive complexity can be substantially reduced by using the following lemma to convert (3.104) into a PT.

Lemma 1 Let $A_k(z)$ be the generating polynomial of $A(k, l)$,

$$A_k(z) = \sum_{l=0}^{M/2-1} A(k, l) z^l$$

and

$$U_p(z) \equiv \sum_{l=0}^{M/2-1} V_p(l)z^l. \tag{3.106}$$

Then $A_k(z)$ is the PT of $U_p(z)$, that is,

$$A_k(z) \equiv \sum_{p=0}^{N-1} U_p(z)\hat{z}^{pk} \bmod (z^{M/2}+1), \quad k = 0, 1, \ldots, N-1, \tag{3.107}$$

where

$$\hat{z} \equiv z^{-2^L} \bmod (z^{M/2}+1).$$

Proof. Based on the property $V_p(i + uM/2) = (-1)^u V_p(i)$ and (3.104), we have

$$
\begin{aligned}
A_k(z) &\equiv \sum_{l=0}^{M/2-1} \left[\sum_{p=0}^{N-1} V_p(pk2^L + l)z^l \right] \bmod (z^{M/2}+1) \\
&\equiv \sum_{p=0}^{N-1} \sum_{l=0}^{M/2-1} V_p(pk2^L + l)z^l \bmod (z^{M/2}+1) \\
&\equiv \sum_{p=0}^{N-1} \sum_{l=0}^{M/2-1} V_p(l)z^{l-pk2^L} \bmod (z^{M/2}+1) \\
&\equiv \sum_{p=0}^{N-1} U_p(z)\hat{z}^{pk} \bmod (z^{M/2}+1).
\end{aligned}
$$

A very simple FFT-like algorithm known as the fast polynomial transform (FPT) discussed in Chapter 2 can be used to compute (3.107), which needs $\frac{1}{2}NM \log_2 N$ complex additions. In general, the PT algorithm for a 2D DFT can be summarized as follows.

Algorithm 3 *PT algorithm for 2D DFT*

Step 1. *Compute the N 1D DFTs and some multiplications in (3.105).*

Step 2. *Compute a PT in (3.107).*

Step 3. *Compute the $N \times M/2$ DFT in (3.103) by recursively using Steps 1 and 2.*

Step 4. *Compute $X(k,l)$ according to (3.101).*

This algorithm can be easily extended to compute an rD DFT, where $r > 2$. The computational complexity will be discussed for a general rD DFT ($r \geq 2$) in the next subsection.

3.7.2   Polynomial transform algorithm for MD DFT

This subsection extends the concepts used in the previous subsection to the rD DFTs whose dimensional sizes are $N_1 \times N_2 \times \cdots \times N_r$, where N_i ($i = 1, 2, \ldots, r$) are powers of 2. Using an index mapping and the MD polynomial transform, we can convert the rD DFT into a series of 1D DFTs, $(r-1)$D PTs and some simple complex multiplications.

The rD DFT with size $N_1 \times N_2 \times \cdots \times N_r$ is defined by

$$X(k_1, k_2, \ldots, k_r) = \sum_{n_1=0}^{N_1-1} \cdots \sum_{n_r=0}^{N_r-1} x(n_1, n_2, \ldots, n_r) W_{N_1}^{n_1 k_1} \cdots W_{N_r}^{n_r k_r}, \qquad (3.108)$$

where $k_i = 0, 1, \ldots, N_i - 1$ and $i = 1, 2, \ldots, r$. We assume that N_i $(i = 1, 2, \ldots, r)$ are powers of 2. Without loss of generality, we also assume that $N_r \geq N_i$ $(i = 1, 2, \ldots, r-1)$. We can write $N_r = 2^t$, $N_r/N_i = 2^{l_i}$ and $l_i \geq 0$.

By dividing the inner sum into two parts, we have

$$\begin{aligned}
X(k_1, k_2, \ldots, k_r) &= A(k_1, k_2, \ldots, k_r) + B(k_1, k_2, \ldots, k_r) \\
&= \sum_{n_1=0}^{N_1-1} \cdots \sum_{n_{r-1}=0}^{N_{r-1}-1} \sum_{n_r=0}^{N_r/2-1} x(n_1, \ldots, n_{r-1}, 2n_r + 1) \\
&\quad \cdot W_{N_1}^{n_1 k_1} \cdots W_{N_{r-1}}^{n_{r-1} k_{r-1}} W_{N_r}^{(2n_r+1)k_r} \\
&\quad + \sum_{n_1=0}^{N_1-1} \cdots \sum_{n_{r-1}=0}^{N_{r-1}-1} \sum_{n_r=0}^{N_r/2-1} x(n_1, \ldots, n_{r-1}, 2n_r) \\
&\quad \cdot W_{N_1}^{n_1 k_1} \cdots W_{N_{r-1}}^{n_{r-1} k_{r-1}} W_{N_r/2}^{n_r k_r}, \qquad (3.109)
\end{aligned}$$

where for $k_r = 0, 1, \ldots, N_r/2 - 1$,

$$A(k_1, k_2, \ldots, k_r + \frac{N_r}{2}) = -A(k_1, k_2, \ldots, k_r) \qquad (3.110)$$

$$B(k_1, k_2, \ldots, k_r + \frac{N_r}{2}) = B(k_1, k_2, \ldots, k_r). \qquad (3.111)$$

Since $B(k_1, k_2, \ldots, k_r)$ is an rD DFT with size $N_1 \times \cdots \times N_{r-1} \times (N_r/2)$, it can be further decomposed in the same way.

Let us consider the computation of $A(k_1, k_2, \ldots, k_r)$ by using the following lemma to convert $A(k_1, k_2, \ldots, k_r)$ into an $(r-1)$D PT and 1D DFTs.

Lemma 2 Let $p_i(n_r)$ be the least non-negative residue of $(2n_r + 1)p_i$ modulo N_i, and $S_1 = \{(n_1, \ldots, n_{r-1}, 2n_r + 1) \mid 0 \leq n_i \leq N_i - 1, 1 \leq i \leq r-1, 0 \leq n_r \leq N_r/2 - 1\}$, $S_2 = \{(p_1(n_r), p_2(n_r), \ldots, p_{r-1}(n_r), 2n_r + 1) \mid 0 \leq p_i \leq N_i - 1, 1 \leq i \leq r-1, 0 \leq n_r \leq N_r/2 - 1\}$. Then $S_1 = S_2$.

Proof. It is obvious that the number of elements in S_1 is the same as that in S_2 and $S_2 \subseteq S_1$. It is sufficient to prove that the elements in S_2 are different from each other. Let $(p_1(n_r), p_2(n_r), \ldots, p_{r-1}(n_r), 2n_r + 1)$ and $(p_1'(n_r'), \ldots, p_{r-1}'(n_r'), 2n_r' + 1)$ be two elements in S_2. If they are equal, then

$$p_i(n_r) = p_i'(n_r'), \quad i = 1, 2, \ldots, r-1, \quad n_r = n_r'.$$

From the definition of $p_i(n_r)$ we see that

$$(2n_r + 1)p_i \equiv (2n_r + 1)p_i' \bmod N_i.$$

Hence

$$(2n_r + 1)(p_i - p_i') \equiv 0 \bmod N_i, \quad i = 1, 2, \ldots, r-1.$$

Since $2n_r + 1$ is relatively prime with N_i, we get $p_i \equiv p_i' \bmod N_i$, that is $p_i = p_i'$

The first term in (3.109) is a multiple sum in which the indices belong to S_1. Since $S_1 = S_2$, the data indices also belong to S_2. Therefore we can express this term as

$$
A(k_1, k_2, \ldots, k_r) = \sum_{p_1=0}^{N_1-1} \cdots \sum_{p_{r-1}=0}^{N_{r-1}-1} \sum_{n_r=0}^{N_r/2-1} x(p_1(n_r), \ldots, p_{r-1}(n_r), 2n_r+1)
$$
$$
\cdot W_{N_1}^{p_1(n_r)k_1} \cdots W_{N_{r-1}}^{p_{r-1}(n_r)k_{r-1}} W_{N_r}^{(2n_r+1)k_r}. \tag{3.112}
$$

Because $p_i(n_r) \equiv p_i(2n_r+1) \bmod N_i$, (3.112) becomes

$$
A(k_1, k_2, \ldots, k_r) = \sum_{p_1=0}^{N_1-1} \cdots \sum_{p_{r-1}=0}^{N_{r-1}-1} \sum_{n_r=0}^{N_r/2-1} x(p_1(n_r), \ldots, p_{r-1}(n_r), 2n_r+1)
$$
$$
\cdot W_{N_r}^{(2n_r+1)(p_1 k_1 2^{l_1} + \cdots + p_{r-1}k_{r-1}2^{l_{r-1}} + k_r)}. \tag{3.113}
$$

Let

$$
V_{p_1,\ldots,p_{r-1}}(m) = \sum_{n_r=0}^{N_r/2-1} x(p_1(n_r), \ldots, p_{r-1}(n_r), 2n_r+1) W_{N_r}^{(2n_r+1)m}
$$
$$
= W_{N_r}^m \sum_{n_r=0}^{N_r/2-1} x(p_1(n_r), \ldots, p_{r-1}(n_r), 2n_r+1) W_{N_r/2}^{n_r m}, \tag{3.114}
$$

where $p_i = 0, 1, \ldots, N_i - 1$, $i = 1, 2, \ldots, r-1$ and $m = 0, 1, \ldots, \frac{N_r}{2} - 1$. Equation (3.114) is $N_1 N_2 \cdots N_{r-1}$ 1D DFTs whose length is $N_r/2$ plus some multiplications. Therefore, (3.113) becomes

$$
A(k_1, k_2, \ldots, k_r) = \sum_{p_1=0}^{N_1-1} \cdots \sum_{p_{r-1}=0}^{N_{r-1}-1} V_{p_1,\ldots,p_{r-1}}(p_1 k_1 2^{l_1} + \cdots + p_{r-1}k_{r-1}2^{l_{r-1}} + k_r).
$$
$$
\tag{3.115}
$$

Since $V_{p_1,\ldots,p_{r-1}}(m)$ has the property of

$$
V_{p_1,\ldots,p_{r-1}}\left(m + \frac{uN_r}{2}\right) = (-1)^u V_{p_1,\ldots,p_{r-1}}(m) \tag{3.116}
$$

for any integer u, (3.115) can be computed if $V_{p_1,\ldots,p_{r-1}}(m)$ $(m = 0, 1, \ldots, N_r/2 - 1)$ is known. The direct computation of (3.115) requires many additions. The required number of additions can be reduced by computing (3.115) with the PT.

Lemma 3 *Let the generating polynomial of $A(k_1, \ldots, k_r)$ and $V_{p_1,\ldots,p_{r-1}}(m)$ be*

$$
A_{k_1,\ldots,k_{r-1}}(z) = \sum_{k_r=0}^{N_r/2-1} A(k_1, \ldots, k_{r-1}, k_r) z^{k_r} \tag{3.117}
$$

and

$$
U_{p_1,\ldots,p_{r-1}}(z) = \sum_{m=0}^{N_r/2-1} V_{p_1,\ldots,p_{r-1}}(m) z^m, \tag{3.118}
$$

respectively. Then

$$A_{k_1,\ldots,k_{r-1}}(z) \equiv \sum_{p_1=0}^{N_1-1} \cdots \sum_{p_{r-1}=0}^{N_{r-1}-1} U_{p_1,\ldots,p_{r-1}}(z) \; z_1^{p_1 k_1} \cdots z_{r-1}^{p_{r-1} k_{r-1}} \bmod (z^{N_r/2}+1), \quad (3.119)$$

where $k_i = 0, 1, \ldots, N_i - 1$, $i = 1, 2, \ldots, r-1$ *and* $z_i \equiv z^{-2^{l_i}} \bmod (z^{N_r/2}+1)$.

Proof. From (3.115) we have

$$A_{k_1,\ldots,k_{r-1}}(z) \equiv \sum_{p_1=0}^{N_1-1} \cdots \sum_{p_{r-1}=0}^{N_{r-1}-1} \sum_{k_r=0}^{N_r/2-1} V_{p_1,\ldots,p_{r-1}}(p_1 k_1 2^{l_1} + \cdots$$

$$+ p_{r-1}k_{r-1}2^{l_{r-1}} + k_r)z^{k_r} \bmod (z^{N_r/2}+1). \quad (3.120)$$

Because $V_{p_1,\ldots,p_{r-1}}(\lambda + k_r)z^{k_r} \bmod (z^{N_r/2}+1)$ is periodic with period $N_r/2$ for variable k_r (for any integer λ), (3.120) becomes

$$A_{k_1,\ldots,k_{r-1}}(z) \equiv \sum_{p_1=0}^{N_1-1} \cdots \sum_{p_{r-1}=0}^{N_{r-1}-1} \sum_{k_r=0}^{N_r/2-1} V_{p_1,\ldots,p_{r-1}}(k_r)$$

$$\cdot z^{k_r - p_1 k_1 2^{l_1} - \cdots - p_{r-1}k_{r-1}2^{l_{r-1}}} \bmod (z^{N_r/2}+1)$$

$$\equiv \sum_{p_1=0}^{N_1-1} \cdots \sum_{p_{r-1}=0}^{N_{r-1}-1} U_{p_1,\ldots,p_{r-1}}(z)z_1^{p_1 k_1} \cdots z_{r-1}^{p_{r-1}k_{r-1}} \bmod (z^{N_r/2}+1).$$

Equation (3.119) is an $(r-1)$D PT. Once $A_{k_1,\ldots,k_{r-1}}(z)$ is computed according to (3.119), $A(k_1,\ldots,k_r)$ is obtained from the coefficients of $A_{k_1,\ldots,k_{r-1}}(z)$. The main steps of the algorithms are summarized below, and Figure 3.11 shows the sequence of these steps.

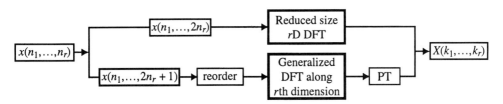

Figure 3.11   Flow chart of the PT algorithm for rD DFT.

Algorithm 4 *PT algorithm for rD DFT*

Step 1.   *Compute $N_1 N_2 \cdots N_{r-1}$ 1D DFTs of length $N_r/2$ and some multiplications according to (3.114) and generate $U_{p_1,\ldots,p_{r-1}}(z)$ according to (3.118).*

Step 2.   *Compute the $(r-1)$D polynomial transform of the polynomial sequence $U_{p_1,\ldots,p_{r-1}}(z)$ according to (3.119) to get $A_{k_1,\ldots,k_{r-1}}(z)$ whose coefficients are $A(k_1,\ldots,k_r)$.*

Step 3.   *Compute $B(k_1,\ldots,k_r)$, that is, the second term in (3.109), which is an rD DFT with size $N_1 \times \cdots \times N_{r-1} \times N_r/2$. The same method as in Steps 1 and 2 can be used for further decomposition.*

Step 4.   *Compute $X(k_1,\ldots,k_r)$ from $A(k_1,\ldots,k_r)$ and $B(k_1,\ldots,k_r)$ as described in (3.109).*

3.7.3   Computational complexity

Now let us consider the number of arithmetic operations needed by Algorithm 4. We compute $N_1 N_2 \cdots N_{r-1}$ 1D DFTs with length $N_r/2$ and some multiplications in Step 1, and an $(r-1)$D PT with size $N_1 \times \cdots \times N_{r-1}$ in Step 2. Using the split-radix algorithm discussed earlier in this chapter to compute the 1D DFT of length $N_r/2$, $\frac{1}{2}N_r \log_2 N_r - 2N_r + 4$ real multiplications and $\frac{3}{2}N_r \log_2 N_r - 3N_r + 4$ additions are needed (with the assumption that a complex multiplication is realized by 3 real multiplications and 3 real additions). The $(r-1)$D PT in (3.119) needs $N \log_2(N_1 \cdots N_{r-1})$ real additions (see Chapter 2), where $N = N_1 N_2 \cdots N_r$. Step 4 also needs $2N$ real additions. Therefore, the total number of multiplications and additions can be expressed, respectively, by

$$M_u(N_1, \ldots, N_{r-1}, N_r) = M_u(N_1, \ldots, N_{r-1}, \frac{N_r}{2}) + \frac{N}{2}\log_2 N_r - \frac{N}{2} - \frac{2N}{N_r} \quad (3.121)$$

and

$$A_d(N_1, \ldots, N_{r-1}, N_r) = A_d(N_1, \ldots, N_{r-1}, \frac{N_r}{2})$$
$$+ N \log_2 N + \frac{N}{2}\log_2 N_r + \frac{N}{2} - \frac{2N}{N_r}. \quad (3.122)$$

Based on (3.121) and (3.122), it is possible to derive a complex formula for the required computational complexity. For simplicity, let us consider the transform whose dimensional sizes are identical, that is, $N_1 = N_2 = \cdots = N_r = 2^t$, and use $M_u(2^t)$ and $A_d(2^t)$ to represent $M_u(2^t, \ldots, 2^t)$ and $A_d(2^t, \ldots, 2^t)$, respectively. By recursively using (3.121) and (3.122), we obtain

$$M_u(2^t) = (1 - \frac{1}{2^r})2^{rt}(t-1) - 4(1 - \frac{1}{2^r})2^{(r-1)t} + M_u(2^{t-1}) \quad (3.123)$$

$$A_d(2^t) = (2r+1)(1 - \frac{1}{2^r})2^{rt}t - (1 - \frac{2r+1}{2^r})2^{rt}$$
$$- 4(1 - \frac{1}{2^r})2^{(r-1)t} + A_d(2^{t-1}). \quad (3.124)$$

We can repeatedly use (3.123) and (3.124) until the dimensional size becomes 2 or the rD DFT whose size is $2 \times \cdots \times 2$. With $M_u(2) = 0$ and $A_d(2) = 2r2^r$, (3.123) and (3.124) become

$$M_u(2^t) = (1 - \frac{1}{2^r})\sum_{i=2}^{t} 2^{ri}(i-1) - 4(1 - \frac{1}{2^r})\sum_{i=2}^{t} 2^{(r-1)i} + M_u(2)$$

$$= Nt - \frac{2^r}{2^r - 1}(N-1) - \frac{2(2^r - 1)}{2^{r-1} - 1}(2^{(r-1)t} - 1) + 2(2^r - 1). \quad (3.125)$$

$$A_d(2^t) = (2r+1)(1 - \frac{1}{2^r})\sum_{i=2}^{t} 2^{ri}i - (1 - \frac{2r+1}{2^r})\sum_{i=2}^{t} 2^{ri} - 4(1 - \frac{1}{2^r})\sum_{i=2}^{t} 2^{(r-1)i} + A_d(2)$$

$$= (2r+1)Nt - \frac{2^r(N-1)}{2^r - 1} - \frac{2(2^r - 1)(2^{(r-1)t} - 1)}{2^{r-1} - 1} + 2(2^r - 1). \quad (3.126)$$

Let us consider the computational complexity of the row-column algorithm that is widely used for DFTs of more than one dimension. For an rD DFT, we need rN/N_1 1D DFTs of length N_1. With the split-radix algorithm, the 1D length N_1 DFT needs $N_1 \log_2 N_1 - 3N_1 + 4$ real multiplications and $3N_1 \log_2 N_1 - 3N_1 + 4$ real additions. Therefore, the total multiplicative and additive complexities of the row-column method can be expressed, respectively, by

$$\overline{M}_u(2^t) = rNt - 3rN + 4r\frac{N}{N_1} = rN(t - 3 + \frac{4}{N_r}) \tag{3.127}$$

$$\overline{A}_d(2^t) = 3rNt - 3rN + 4r\frac{N}{N_1} = rN(3t - 3 + \frac{4}{N_r}). \tag{3.128}$$

In comparison, the PT-based fast algorithm substantially reduces the number of arithmetic

Table 3.11   Comparison of number of operations for 2D DFT.

	PT-based		Row-column	
N_1	M_u	A_d	$\overline{M}_u$	$\overline{A}_d$
2	0	16	0	16
4	0	128	0	128
8	72	840	64	832
16	600	4696	640	4736
32	3576	24056	4352	24832
64	18744	117048	25088	123392
128	92088	550840	132096	590848
256	435384	2532536	657408	2754560
512	2006712	11443896	3149824	12587008
1024	9081528	51024568	14688256	56631296
2048	40532664	225082040	67125248	251674624

operations. The relationship between the computational complexities needed by the PT-based and the row-column algorithms can be approximated to be

$$M_u \approx \frac{1}{r}\overline{M}_u, \ A_d \approx \frac{2r+1}{3r}\overline{A}_d$$

when the dimensional size N is large. The reduction on the computational complexity increases with the number of dimensions.

Table 3.11 lists the number of operations for an $N_1 \times N_1$ 2D DFT, where M_u, $\overline{M}_u$ and A_d, $\overline{A}_d$ represent the number of multiplications and additions for the PT-based and the row-column algorithms, respectively. Table 3.12 shows the comparison for a 3D DFT of size $N_1 \times N_1 \times N_1$.

The structure of the algorithm is relatively simple. The main part of the computation is in Step 1 , which includes the computation of a number of 1D DFTs, and Step 2, which is an $(r - 1)$D PT. In Step 1, an index mapping or sequence reordering for the input array is also needed. For a fixed size problem, the index mapping can be pre-calculated and stored in an array of $N/2$ integers. If the indices are not computed in advance, some integer operations are needed for the index mapping. For example, if we use an array $x_1(n_1, \ldots, n_r)$ with size

Table **3.12**   Comparison of number of operations for 3D DFT.

N_1	PT-based		Row-column	
	M_u	A_d	$\overline{M}_u$	$\overline{A}_d$
2	0	48	0	48
4	0	768	0	768
8	672	9888	768	9984
16	10528	108832	15360	113664
32	121632	1104672	208896	1191936
64	1254176	10691360	2408448	11845632
128	12206880	100287264	25362432	113442816
256	114737952	920044320	252444672	1057751040
512	1053344544	8301101856	2419064832	9666822144

$N/2$ to store the new data indices, the index mapping is realized according to

$$x_1(p_1,\ldots,p_{r-1},n_r) = x(p_1(n_r),\ldots,p_{r-1}(n_r),2n_r+1), \tag{3.129}$$

where indices $p_i(n_r)$ $(i = 1, 2, \ldots, r-1)$ is computed according to $(2n_r+1)p_i \bmod N_i$. The computation of the indices requires $\frac{N}{2} - \frac{N}{N_r}$ additions, $\frac{N}{2} - \frac{N}{N_r}$ multiplications and $\frac{N}{2} - \frac{N}{N_r}$ modulo operations (shift operations are not included). This computational overhead is trivial compared to the overall computational complexity. The presented algorithm has a regular computational structure to support various transform sizes for different numbers of dimensions.

3.8   CHAPTER SUMMARY

This chapter is devoted to the development of fast algorithms for various DFT transforms. Although numerous fast algorithms are available in the literature, many of them can support only a limited range of transform sizes. The fast algorithms described in this chapter attempt to provide more choices of transform sizes without increasing the required computational complexity and structural complexity. The generalized split-radix algorithms naturally utilize a parameter q to support a wider range of transform sizes. The PFA provides the solution to simple index mapping and in-place computation, which substantially simplifies the structural complexity of the algorithm. Fast algorithms for a generalized DFT are also presented to achieve a better computational efficiency. Finally, the PT-based algorithms for an MD DFT are provided. These algorithms can be used for any number of dimensions, provide reduction of computational complexity and have a relatively simple computational complexity.

Appendix A. length 9 DFT computation

It is assumed that the input data are given in arrays of xx and yy. The transformed outputs are given in xx and yy. The transform needs 84 additions and 16 multiplications.

u1= yy1-yy8;	u2= yy2-yy7;
u3= yy3-yy6;	u4= yy4-yy5;
t1=0.98480775(u2+u4);	t2=0.342020143(u1+u2);
t3=0.64278761(u1-u4);	t4=0.866025404u3;
yi3=0.866025404(u1-u2+u4);	t1=t1+t3;
t2=t2+t3;	t2=t2+t3;
t3=t2-t1;	yi1=t1+t4;
yi2=t2-t4;	yi4=t3+t4;
yi5= -yi4;	yi6= -yi3;
yi7= -yi2;	yi8= -yi1;
u1=xx1-xx8;	u2=xx2-xx7;
u3=xx3-xx6;	u4=xx4-xx5;
t1 =0.984807753(u2+u4);	t2 =0.342020143(u1+u2);
t3 =0.64278761(u1-u4);	t4 =0.866025404u3;
yr3 = 0.866025404(u1-u2+u4);	t1 = t1+t3;
t2 = t2+t3;	t3 = t2-t1;
yr1 = t1+t4;	yr2 = t2-t4;
yr4 = t3+t4;	yr5 = -yr4;
yr6 = -yr3;	yr7 = -yr2;
yr8 = -yr1;	u0 = xx0;
u1 = xx1+xx8;	u2 = xx2+xx7;
u3 = xx3+xx6;	u4 = xx4+xx5;
t1 = 0.766044443(u1-u4);	t2 = 0.173648178(u2-u4);
t3 = 0.939692621(u1-u2);	t2 = t1+t2;
t3 = t3-t1;	t1 = -t2-t3;
t4 = xx0 + u3;	xx3=u1+u2+u4;
xx0 = t4+xx3;	xx3=t4-0.5xx3;
t4=u0-0.5u3;	xx1=t2+t4;
xx2 = t3+t4;	xx4=t1+t4;
xx5 = xx4;	xx6=xx3;
xx7 = xx2;	xx8=xx1;
u0 = yy0;	u1 = yy1+yy8;
u2 = yy2+yy7;	u3 = yy3+yy6;
u4 = yy4+yy5;	t1 = 0.766044443(u1-u4);
t2 = 0.173648178(u2-u4);	t3 = 0.939692621(u1-u2);
t2 = t1+t2;	t3 = t3-t1;
t1 = -t2-t3;	t4 = yy0 + u3;
yy3=u1+u2+u4;	yy0 = t4+yy3;
yy3=t4-0.5yy3;	t4 = u0-0.5u3;
yy1=t2+t4;	yy2 = t3+t4;
yy4=t1+t4;	yy5 = yy4;
yy6=yy3;	yy7 = yy2;
yy8=yy1;	
for n=1 to 8 do	begin
xx(n) = xx(n)+yi(n);	yy(n) = yy(n)-yr(n); end;

Appendix B. length 15 DFT computation

The length 15 DFT can be calculated by the following equations.

$$X(3k) = \sum_{n=0}^{4}[x(n) + x(n+5) + x(n+10)]W_5^{nk}, \quad 0 \leq k < 4 \tag{B.1}$$

$$F(k) = \frac{X[(3k+5) \bmod 15] + X[(3k+10) \bmod 15]}{2}$$

$$= \sum_{n=0}^{4}\left[x(n)\cos\frac{2\pi n}{3} + x(n+5)\cos\frac{2\pi(n+5)}{3}\right.$$

$$\left. + x(n+10)\cos\frac{2\pi(n+10)}{3}\right]W_5^{nk} \tag{B.2}$$

$$G(k) = \frac{X[(3k+5) \bmod 15] - X[(3k+10) \bmod 15]}{2j}$$

$$= \sum_{n=0}^{4}\left[x(n)\sin\frac{2\pi n}{3} + x(n+5)\sin\frac{2\pi(n+5)}{3}\right.$$

$$\left. + x(n+10)\sin\frac{2\pi(n+10)}{3}\right]W_5^{nk}, \tag{B.3}$$

where (B.1) – (B.3) are length 5 DFTs. The transformed outputs are obtained by

$$x[(3k+5) \bmod 15] = F(k) + jG(k), \quad x[(3k+10) \bmod 15] = F(k) - jG(k), \tag{B.4}$$

Appendix C. length 5 DFT computation

```
for n=1 to 2 do
begin vr(n)=xx(n)+xx(5-n);    vi(n)=yy(n)+yy(5-n); end;
vr0=xx0+vr1+vr2;              t = 0.309016994(vr1-vr2);
t1=t+0.5vr1;                  vr1=xx0+t-0.5vr2;
vr2=xx0-t1;                   vi0=yy0+vi1+vi2;
t = 0.309016994(vi1-vi2);    t1=t+0.5vi1;
vi1=yy0+t-0.5vi2;            vi2=yy0-t1;
for n=1 to 2 do
begin xx(n)=xx(n)-xx(5-n);    yy(n)=yy(n)-yy(5-n); end;
t=0.587785252(xx1+xx2);      t1=0.587785252(yy1+yy2);
vr3=0.363271264xx1;          vi3=0.363271264yy1;
vr3= vr3+t; vi3= vi3+t1;     vr4=1.538841768xx2;
vi4=1.538841768yy2;          vr4= t-vr4;
vi4= t1-vi4;
xx0=vr0;                     yy0=vi0;
xx1=vr1+vi3;                 yy1=vi1-vr3;
xx2=vr2+vi4;                 yy2=vi2-vr4;
xx3=vr2-vi4;                 yy3=vi2+vr4;
xx4=vr1-vi3;                 yy4=vi1+vr3;
```

The computation needs 36 additions and 8 multiplications.

REFERENCES

1. G. Bi and Y. Q. Chen, Fast DFT algorithms for length $N = q \times 2^m$, *IEEE Trans. Circuits Systems II*, vol. 45, no. 6, 685–690, 1998.

2. G. Bi and Y. Q. Chen, Split-radix algorithm for the 2D DFT, *Electron. Lett.*, vol. 33, no. 3, 203–305, 1997.

3. G. Bi, Fast algorithms for DFT of composite sequence lengths, *Signal Processing,* Elsevier, 70, 139–145, 1998.

4. G. Bi and Y. Q. Chen, Fast generalized DFT and DHT algorithms, *Signal Processing,* Elsevier, 65, 383–390, 1997.

5. G. Bonnerot and M. Bellanger, Odd-time odd-frequency discrete Fourier transform for symmetric real valued series, *Proc. IEEE,* vol. 64, no. 3, 392–393, 1976.

6. S. C. Chan and K. L. Ho, Split vector radix fast Fourier transform, *IEEE Trans. Signal Process.*, vol. 40, no. 8, 2029–2039, 1992.

7. P. Duhamel, Implementation of split-radix FFT algorithms for complex, real, and real-symmetric data, *IEEE Trans. Acoustics, Speech, Signal Process.*, vol. 34, no. 4, 285–295, 1986.

8. P. Duhamel and M. Vetterli, Improved Fourier and Hartley transform algorithm: application to cyclic convolution of real data, *IEEE Trans. Acoustics, Speech, Signal Process.*, vol. 35, no. 6, 818–824, 1987.

9. D. F. Elliott and K. R. Rao, *Fast Transforms : Algorithms, Analyses, Applications,* Academic Press, Inc., New York, 1982.

10. O. K. Ersoy, *Fourier-Related Transforms, Fast Algorithms and Applications,* Prentice-Hall International, Inc., Upper Saddle River, NJ, 1997.

11. O. K. Ersoy, Semisystolic array implementation of circular, skew-circular, and linear convolutions, *IEEE Trans. Comput.*, 34, no. 2, 190–194, 1985.

12. I. J. Good, The relationship between two fast Fourier transforms, *IEEE Trans. Comput.*, vol. 20, 310–317, 1977.

13. M. T. Heideman, C. S. Burrus and H. W Johnson, Prime-factor FFT algorithm for real-valued series, *Int. Conf. ASSP,* 28A.7.1–28A.7.4, 1984.

14. N. Hu and O. K. Ersoy, Fast computation of real discrete Fourier transform for any number of data points, *IEEE Trans. Circuits Syst. II*, vol. 38, no. 11, 1280–1292, 1991.

15. D. P. Kolba and T. W. Parks, A prime factor FFT algorithm using high-speed convolution, *IEEE Trans.*, vol. 25, 90–103, 1977.

16. Z. J. Mou and P. Duhamel, In-place butterfly-style FFT of 2-D real sequences, *IEEE Trans. Acoustics, Speech, Signal Process.*, vol. 36, no. 10, 1642–1650, 1988.

17. H. J. Nussbaumer and P. Quandalle, Computation of convolutions and discrete Fourier transforms by polynomial transform, *IBM J. Res. Develop.*, vol. 22, no. 2, 1978.

18. S. C. Pei and Tzyy-Liang Luo, Split-radix generalized fast Fourier transform, *Signal Processing,* vol. 54, 137–151, 1996.

19. L. R. Rabiner and B. Gold, *Theory and Application of Digital Aignal Processing,* Prentice-Hall, Englewood Cliffs, NJ, 1975.

20. C. M. Rader, Discrete Fourier transforms when the number of data sample is prime, *Proc. IEEE.,* vol. 56, 1107–1108, 1968.

21. V. Sorensen, M. T. Heideman, and C. S. Burreu, On computing the split-radix FFT, *IEEE Trans. Acoustics, Speech, Signal Process.,* vol. 34, no. 2, 152–156, 1986.

22. R. Stasinski, Radix-K FFT's using K-point convolutions, *IEEE Trans. Signal Process.,* vol. 42, no. 4, 743–750, 1994.

23. Y. Suzuki, T. Sone and K. Kido, A new algorithm of radix 3, 6, and 12, *IEEE Trans. Acoustics, Speech, Signal Process.,* vol. 34, no. 2, 389–383, 1986.

24. R. Tolimieri, M. An and C. Liu, *Algorithms for Discrete Fourier Transform and Convolution,* Springer-Verlag, New York, 1989.

25. M. Vetterli and P. Duhamel, Split-radix algorithms for length p^m DFT's, *IEEE Trans. Acoustics, Speech, Signal Process.,* vol. 37, no. 1, 57–64, 1989.

26. S. Winograd, On computing the discrete Fourier transform, *Math. Comput.,* 32, 175–199, 1978.

27. H. R. Wu and F. J. Paoloni, On the two-dimensional vector-radix FFT algorithm, *IEEE Trans. Acoustics, Speech, Signal Process.,* vol. 37, no. 8, 1302–1304, 1989.

4

Fast Algorithms for 1D Discrete Hartley Transform

This chapter is devoted to new fast algorithms for computation of the four types of one-dimensional (1D) discrete Hartley transforms (DHTs).

- Section 4.2 presents split-radix 2/4 and 3/9 algorithms for the type-I DHT. In addition to providing some basic concepts of the split-radix algorithms for the 1D DHT computation, optimization of the computational complexity of the small length DHTs is provided.

- Section 4.3 describes a generalized split-radix algorithm for the type-I DHT to support sequence lengths of $q \cdot 2^m$ without increasing the total computational complexity.

- Section 4.4 presents radix-2 algorithms for type-II, -III and -IV DHTs.

- Section 4.5 describes prime factor algorithms (PFAs) for type-I, type-II and -III DHTs. The PFAs allow both computational complexity and implementation complexity to be minimized.

- Section 4.6 presents a radix-q algorithm to decompose the entire computation into smaller ones based on odd factors. It differs from the prime factor algorithm in that the factors of the sequence length are not necessarily mutually prime.

- Section 4.7 discusses the relationships between all these different types of DHTs. Based on the relationships, it shows that the type-II and -III DHTs can be expressed as type-I DHTs of smaller lengths, and the type-IV DHT can be expressed in terms of type-II or -III DHTs.

4.1 INTRODUCTION

The DHT has gained popularity in the field of digital signal processing over recent years. It has many desirable properties, as detailed in [26, 31], used for many applications ([19, 20, 34,

40], for example). The special issue of IEEE on the Hartley transform published in March 1994 also provided an excellent review on the theoretical developments and its applications [26]. Comparisons with other transforms on the computational complexity have also been made in [9, 30, 33].

In general, there exist four types of DHTs defined as

Type-I

$$X(k) = \sum_{n=0}^{N-1} x(n) \mathrm{cas} \frac{2\pi nk}{N} \tag{4.1}$$

Type-II

$$X(k) = \sum_{n=0}^{N-1} x(n) \mathrm{cas} \frac{\pi(2n+1)k}{N} \tag{4.2}$$

Type-III

$$X(k) = \sum_{n=0}^{N-1} x(n) \mathrm{cas} \frac{\pi n(2k+1)}{N} \tag{4.3}$$

Type-IV

$$X(k) = \sum_{n=0}^{N-1} x(n) \mathrm{cas} \frac{\pi(2n+1)(2k+1)}{2N}, \tag{4.4}$$

where $\mathrm{cas}(\alpha) = \cos(\alpha) + \sin(\alpha)$, $x(n)$ $(n = 0, 1, \ldots, N-1)$ is the input sequence and $X(k)$ $(k = 0, 1, \ldots, N-1)$ is the output sequence. The constant normalization factors are ignored in the above definition for simplicity. In the literature, the type-I DHT is widely known as the discrete Hartley transform and the other types defined in (4.2) – (4.4) are called the generalized discrete Hartley transforms because their definitions contain shifts in the time and/or frequency domains. In the literature, the four types of DHTs are also known as the W transforms which were defined by Wang [36]. The difference between the DHT and the W transform is that they use different scaling factors in their definitions. If these scaling factors are ignored, the same fast algorithms can be used for both the DHT and the W transform. In this book, we denote these transforms the discrete Hartley transform or DHT. For example, (4.1) – (4.4) are known as the type-I, -II, -III, and -IV DHTs, respectively. This chapter discusses the fast algorithms for the 1D DHT and Chapter 5 considers the fast algorithms for the multidimensional (MD) DHT.

There exist many fast algorithms in the literature to reduce the computational complexity and implementation cost. Similar to the discrete Fourier transform (DFT), we have the radix-type fast algorithms [1, 2, 11, 16, 18, 21, 29, 32, 41], and the non-radix-type algorithms [10, 22, 25, 38]. In particular, Duhamel's algorithm [12] decomposes a length N DHT computation into a length $N/2$ DHT and two length $N/4$ DFTs, which uses the lowest additive complexity without increasing the multiplicative complexity. Other fast algorithms [3, 12] using the relation between the DFT and DHT are also available. For the length 2^m type-I DHTs, Hou reported a decomposition method of the DHT computation [16]. Sorensen et al. [32] proposed split-radix algorithms to provide a better computational efficiency. Chan described a different algorithm by using a symmetric cosine structure [11]. These algorithms basically require the same number of multiplications although the rotation operation may be implemented by using 3 multiplications and 3 additions or 4 multiplications and 2 additions. Other methods using fast algorithms of cosine or Fourier transforms also exist [12, 23]. All these mentioned algorithms were designed for the computation of length 2^m DHTs although then can be modified for DHTs of other sequence lengths. The fast algorithms for the generalized DHT or the type-II, -III and -IV DHTs are more complex in terms of computational and structural complexity compared to those for the type-I DHT. Various fast algorithms for type-II, -III and -IV DHTs were reported in the literature ([3, 4, 5, 6, 7, 8, 17, 37, 38, 39], for example).

4.2   SPLIT-RADIX ALGORITHMS

4.2.1   Split-radix 2/4 algorithm

The split-radix 2/4 algorithm is designed for a type-I DHT whose sequence length N is $q \cdot 2^m$, where q is a small positive integer. The even indexed transform outputs can be computed by

$$X(2k) = \sum_{n=0}^{N/2-1} x(n)\text{cas}\,\frac{2\pi nk}{N/2} + x(n+\frac{N}{2})\text{cas}\,\frac{2\pi(n+N/2)k}{N/2}$$

$$= \sum_{n=0}^{N/2-1} u(n)\text{cas}\,\frac{2\pi nk}{N/2}, \tag{4.5}$$

where $u(n) = x(n) + x(n + N/2)$. The periodic property $\text{cas}[2\pi(n + N/2)k]/(N/2) = \text{cas}(2\pi nk)/(N/2)$ is used in the above derivation. It can be seen that (4.5) is a length $N/2$ type-I DHT.

Now let us consider the computation of the odd indexed outputs $X(2k + 1)$. By using the periodic properties of the cas ( ) function, $X(2k + 1)$ can be expressed into

$$X(4k \pm 1) = \sum_{n=0}^{N/2-1} v(n)\text{cas}\,\frac{2\pi n(4k \mp 1)}{N} \tag{4.6}$$

$$= \sum_{n=0}^{N/2-1} v(n)\left[\text{cas}\,\frac{2\pi nk}{N/4}\cos\frac{2\pi n}{N} \pm \text{cas}\,\frac{-2\pi nk}{N/4}\sin\frac{2\pi n}{N}\right],$$

where $k = 0, 1, \ldots, N/4 - 1$ and $v(n) = x(n) - x(n + N/2)$. From (4.6), we form new sequences

$$F(k) = \frac{X(4k+1) + X(4k-1)}{2}$$

$$= \sum_{n=0}^{N/2-1} v(n)\cos\frac{2\pi n}{N}\text{cas}\,\frac{2\pi nk}{N/4}$$

$$= \sum_{n=0}^{N/4-1}\left[v(n)\cos\frac{2\pi n}{N} - v(n+\frac{N}{4})\sin\frac{2\pi n}{N}\right]\text{cas}\,\frac{2\pi nk}{N/4} \tag{4.7}$$

$$G(k) = \frac{X(4k+1) - X(4k-1)}{2}$$

$$= \sum_{n=0}^{N/2-1} v(n) \sin\frac{2\pi n}{N} \, \text{cas}\frac{2\pi n(N/4-k)}{N/4}$$

$$= \sum_{n=0}^{N/4-1} \left[v(n) \sin\frac{2\pi n}{N} + v(n+\frac{N}{4}) \cos\frac{2\pi n}{N} \right] \text{cas}\frac{2\pi n(N/4-k)}{N/4}, \qquad (4.8)$$

where $0 \le k < N/4$. Both (4.7) and (4.8) are type-I length $N/4$ DHTs whose inputs are modified by the twiddle factors $\cos(2\pi n/N)$ and $\sin(2\pi n/N)$. It is noted that in (4.8), $(N/4 - k)$ is used to calculate the transform. The odd indexed transform outputs are finally calculated by

$$X(4k+1) = F(k) + G(k), \qquad X(4k-1) = F(k) - G(k), \qquad (4.9)$$

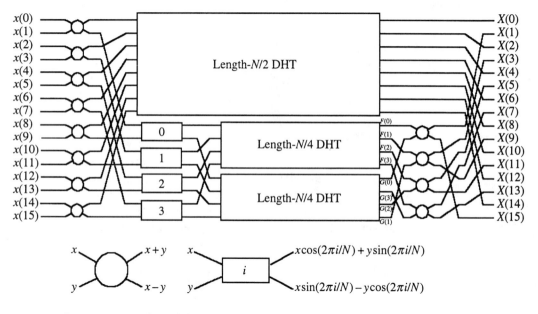

Figure 4.1   Signal flow graph of the split-radix algorithm for $N = 16$.

where $0 \le k < N/4$ and $X(-1) = X(N-1)$ for $k = 0$. The same decomposition process can be continued until the length q and length $2q$ DHTs are reached. Figure 4.1 shows the signal flow graph after the first decomposition step for a length 16 DHT.

The costs of the decomposition process are:

- N additions for $u(n)$ and $v(n)$ in (4.5) and (4.6), respectively.

- $N/2$ additions for (4.9).

- The twiddle factors in (4.7) and (4.8) use $3N/4$ multiplications and $3N/4$ additions if the rotation operation uses 3 multiplications and 3 additions, or N multiplications and $N/2$ additions if the rotation operation uses 4 multiplications and 2 additions.

For the former case, the number of required multiplications for a length N DHT is

$$M(N) = M(\frac{N}{2}) + 2M(\frac{N}{4}) + \frac{3N}{4} - M_t \qquad (4.10)$$

and the number of additions is

$$A(N) = A(\frac{N}{2}) + 2A(\frac{N}{4}) + \frac{9N}{4} - A_t, \qquad (4.11)$$

where M_t and A_t are the number of multiplications and additions saved from the trivial twiddle factors. When $N = 2^m$, the twiddle factors are trivial for $n = 0$ and $N/8$. When $N = 3 \cdot 2^m$, however, the twiddle factors are trivial for $n = 0, N/8, N/6$ and $N/12$. Therefore, $M_t = A_t = 4$ for $N = 2^m$ and $M^t = A^t = 12$ for $N = 3 \cdot 2^m$. The total computational complexity is closely related to their initial value, i.e., the computational complexity of small DHTs. length 2 and length 4 DHTs are widely used as basic computation blocks since they need no multiplication and a minimum number of additions. It is also known that a length 8 DHT uses 2 multiplications and 22 additions. To further minimize the computational complexity, Appendix A presents a procedure for the odd indexed outputs of the length 16 DHT, which requires the same number of multiplications as that in [11, 24] and a smaller number of additions, i.e., 4 additions saved compared to the number needed in [11, 24].

It seems that the most efficient DHT computation is achieved when N is a power of two. However, efficient computations can also be obtained by the presented algorithm for other sequence lengths. For example, we can compute length $3 \cdot 2^m$ DHTs by using the same computational complexity as that for length 2^m DHTs. When N is divisible by 3, such as 6, 12, 24, efficient computations can be made by using a rarely utilized property of the DHT coefficients. For example, a length 6 DHT can be computed by

$$
\begin{bmatrix}
X(0) \\
X(1) \\
X(2) \\
X(3) \\
X(4) \\
X(5)
\end{bmatrix}
=
\begin{bmatrix}
1.0 & 1.0 & 1.0 & 1.0 & 1.0 & 1.0 \\
1.0 & -1.0 & 0.36 & -0.36 & 1.36 & -1.36 \\
1.0 & 1.0 & -1.36 & -1.36 & 0.36 & 0.36 \\
1.0 & -1.0 & 1.0 & -1.0 & -1.0 & 1.0 \\
1.0 & 1.0 & 0.36 & 0.36 & -1.36 & -1.36 \\
1.0 & -1.0 & -1.36 & 1.36 & -0.36 & 0.36
\end{bmatrix}
=
\begin{bmatrix}
x(0) \\
x(3) \\
x(2) \\
x(5) \\
x(1) \\
x(4)
\end{bmatrix}. \qquad (4.12)
$$

In the above equation, the two non-trivial coefficients 0.36 and 1.36 are needed. We can use the relation $1.36x = x + 0.36x$ to eliminate one coefficient. It can be easily proven that the length 6 DHT can be achieved by using 2 multiplications and 20 additions, rather than 8 multiplications and 18 additions needed by a direct computation. The same technique can also be applied to the length 12 DHT. Appendix B presents the subroutines, which need 52 additions and 4 multiplications for a length 12 DHT and 120 additions, 12 multiplications and 2 shift operations for a length 24 DHT.

4.2.2   Split-radix 3/9 algorithm

When $N = 3^m$, the length N type-I DHT can be decomposed into

$$X(3k) = \sum_{n=0}^{N/3-1} \left[x(n) + x(n + \frac{N}{3}) + x(N + \frac{2N}{3}) \right] \text{cas} \frac{2\pi nk}{N/3}, \qquad (4.13)$$

where $0 \leq k < N/3$, and

$$X(9k_1 + k_2) = \sum_{n=0}^{N/9-1} \sum_{i=0}^{8} x(n + \frac{iN}{9}) \text{cas} \frac{2\pi(n + iN/9)(9k_1 + k_2)}{N} \tag{4.14}$$

for $0 \leq k_1 < N/3$ and $k_2 = 1, 2, 4, 5, 7, 8$. Using the property $\text{cas}(\alpha + \beta) = \text{cas}(\alpha)\cos(\beta) + \text{cas}(-\alpha)\sin(\beta)$, (4.14) becomes

$$
\begin{aligned}
X(9k_1 \pm k_2) = & \sum_{n=0}^{N/9-1} \sum_{i=0}^{8} x(n + \frac{iN}{9}) \cos \frac{2\pi[k_2 n + iN(9k_1 + k_2)/9]}{N} \text{cas} \frac{2\pi n k_1}{N/9} \\
& \pm \sum_{n=0}^{N/9-1} \sum_{i=0}^{8} x(n + \frac{iN}{9}) \sin \frac{2\pi[k_2 n + iN(9k_1 + k_2)/9]}{N} \text{cas} \frac{-2\pi n k_1}{N/9}. \tag{4.15}
\end{aligned}
$$

The inner summation in the first term is defined by

$$\sum_{i=0}^{8} x(n + \frac{iN}{9}) \cos \frac{2\pi[k_2 n + iN(9k_1 + k_2)/9]}{N} = \cos \frac{2\pi n k_2}{N} C(n, k_2) - \sin \frac{2\pi n k_2}{N} S(n, k_2)$$

$$\tag{4.16}$$

and similarly, the inner summation in the second term of (4.15) becomes

$$\sum_{i=0}^{8} x(n + \frac{iN}{9}) \sin \frac{2\pi[k_2 n + iN(9k_1 + k_2)/9]}{N} = \sin \frac{2\pi n k_2}{N} C(n, k_2) + \cos \frac{2\pi n k_2}{N} S(n, k_2),$$

$$\tag{4.17}$$

where

$$C(n, k) = C(n, -k) = \sum_{i=0}^{8} x(n + \frac{iN}{9}) \cos \frac{2\pi i k}{9} \tag{4.18}$$

$$S(n, k) = -S(n, -k) = \sum_{i=0}^{8} x(n + \frac{iN}{9}) \sin \frac{2\pi i k}{9}. \tag{4.19}$$

From (4.15), we can derive

$$
\begin{aligned}
F(k_1, k_2) &= \frac{X(9k_1 + k_2) - X(9k_1 - k_2)}{2} \tag{4.20} \\
&= \sum_{n=0}^{N/9-1} \left[C(n, k_2) \cos \frac{2\pi n k_2}{N} - S(n, k_2) \sin \frac{2\pi n k_2}{N} \right] \text{cas} \frac{2\pi n k_1}{N/9}
\end{aligned}
$$

$$
\begin{aligned}
G(k_1, k_2) &= \frac{X(9k_1 + k_2) + X(9k_1 - k_2)}{2} \tag{4.21} \\
&= \sum_{n=0}^{N/9-1} \left[C(n, k_2) \sin \frac{2\pi n k_2}{N} + S(n, k_2) \cos \frac{2\pi n k_2}{N} \right] \text{cas} \frac{-2\pi n k_1}{N/9}.
\end{aligned}
$$

Therefore, the entire DHT is decomposed into one length $N/3$ DHT in (4.13) and six length $N/9$ DHTs defined in (4.20) and (4.21). The final transform outputs are obtained from (4.13) and

$$
\begin{aligned}
X(9k_1 + k_2) &= F(k_1, k_2) + G(k_1, k_2) \\
X(9k_1 - k_2) &= F(k_1, k_2) - G(k_1, k_2).
\end{aligned}
\tag{4.22}
$$

For (4.20), (4.21) and (4.22), $k_1 = 0, 1, \ldots, N/9 - 1$ and $k_2 = 1, 2, 4$. The costs of decomposition are:

- $2N/3$ additions are needed by (4.13).

- $N - 9$ multiplications and additions for the computation defined in (4.16) and (4.17).

- $2N/3$ additions to form $X(9k_1 - k_2)$ and $X(9k_1 + k_2)$ by calculating the sum and difference given in (4.22). When $X(9k_1 - k2), 9k_1 < k_2$, is computed, the property of $X(N - k) = X(-k)$ is used.

Furthermore, $C(n, k)$ and $S(n, k)$ have to be computed according to (4.18) and (4.19). Appendix C shows that each $C(n, k)$ requires 3 multiplications and 14 additions. Similarly, 4 multiplications and 13 additions are needed to compute each $S(n, k)$. The total number of multiplications required by a type-I length N DHT is

$$
M(N) = M(\frac{N}{3}) + 6M(\frac{N}{9}) + \frac{16N}{9} - 9
\tag{4.23}
$$

and the total number of additions is

$$
A(N) = A(\frac{N}{3}) + 6A(\frac{N}{9}) + \frac{16N}{3} - 9.
\tag{4.24}
$$

Table 4.1 lists the number of multiplications and additions needed by the split-radix 3/9 algorithm.

Table 4.1 Computational complexity for split-radix 3/9 algorithm.

N	$M(N)$	$A(N)$
3	1	7
9	8	40
27	53	217
81	236	880
243	977	3469
729	3680	12628
2187	13421	45097

4.3 GENERALIZED SPLIT-RADIX ALGORITHMS

It has been recognized that the split-radix algorithms for both the type-I DHT and the DFT are more computationally efficient than other radix algorithms. These algorithms are particularly useful for sequence lengths of 2^m or 3^m, as shown in Section 4.2. To increase the choices of sequence lengths, the split-radix 2/4 algorithm can be extended for the sequence lengths of $q \cdot 2^m$, where q is an odd integer. To retain the computational efficiency of the split-radix algorithm, the value of q has to be small so that the computational complexity of the length q and the length $2q$ DHTs are used as the initial values of (4.10) and (4.11). When q is large, the computational efficiency of the split-radix 2/4 algorithm is more related to that of the length q and $2q$ DHTs.

Fast DHT algorithms for length 2^m are popular because they generally require a smaller number of arithmetic operations and regular computational structures. In particular, the split-radix algorithm [24, 32] uses length 2 and 4 DHTs as the basic elements, which do not need multiplications for twiddle factors. In fact, this property is also valid for other cases. Similar to the generalized split-radix algorithm for the DFT, this section considers a generalized split-radix algorithm that uses the factor q of the sequence length in the decomposition process. More trivial twiddle factors can be exploited to further reduce the computational complexity. Therefore, this split-radix algorithm naturally supports the sequence length $N = q \cdot 2^m$ and achieves a reduction of computational complexity. It can be shown that the algorithms in [24, 32] are equivalent to the special case of $q = 1$ of the generalized split-radix algorithm.

4.3.1 Algorithm

If the sequence length $N = q \cdot 2^m$, where q is an odd integer, the type-I DHT can be decomposed into even indexed transform outputs

$$X(2k) = \sum_{n=0}^{N/2-1} u(n)\mathrm{cas}\,\frac{2\pi nk}{N/2}, \qquad 0 \le k < \frac{N}{2}, \tag{4.25}$$

where $u(n) = x(n) + x(n + N/2)$. If $m = 1$ or $N = 2q$, the odd transformed outputs can be computed by

$$\begin{aligned} X[(2k + q) \bmod N] &= \sum_{n=0}^{N-1} x(n)\mathrm{cas}\,\frac{2\pi n(2k + q)}{N} \\ &= \sum_{n=0}^{N-1} x(n)(-1)^n\mathrm{cas}\,\frac{2\pi nk}{q} \\ &= \sum_{n=0}^{q-1}(-1)^n\,[x(n) - x(n + q)]\,\mathrm{cas}\,\frac{2\pi nk}{q}. \end{aligned} \tag{4.26}$$

Both (4.25) and (4.26) are length q DHTs. Because $X(2k+q)$, instead of $X(2k+1)$, is used, multiplications are not needed for (4.26). If $m > 1$, the odd indexed transform outputs can be obtained by

$$\begin{aligned} X(4k \pm q) &= \sum_{n=0}^{N/2-1} v(n)\mathrm{cas}\,\frac{2\pi n(4k \pm q)}{N} \\ &= \sum_{n=0}^{N/2-1} v(n)\left[\mathrm{cas}\,\frac{2\pi nk}{N/4}\cos\frac{2\pi n}{N/q} \pm \mathrm{cas}\,\frac{-2\pi nk}{N/4}\sin\frac{2\pi n}{N/q}\right], \end{aligned} \tag{4.27}$$

where $v(n) = x(n) - x(n + N/2)$. From (4.27), we form new sequences

$$
\begin{aligned}
F(k) &= \frac{X(4k+q) + X(4k-q)}{2} = \sum_{n=0}^{N/2-1} v(n) \cos\frac{2\pi n}{N/q} \operatorname{cas}\frac{2\pi n k}{N/4} \\
&= \sum_{n=0}^{N/4-1} \left[v(n) \cos\frac{2\pi n}{N/q} + (-1)^{(q+1)/2} v(n + \frac{N}{4}) \sin\frac{2\pi n}{N/q} \right] \operatorname{cas}\frac{2\pi n k}{N/4} \quad (4.28)
\end{aligned}
$$

and

$$
\begin{aligned}
G(k) &= \frac{X(4k+q) - X(4k-q)}{2} \\
&= \sum_{n=0}^{N/2-1} v(n) \sin\frac{2\pi n}{N/q} \operatorname{cas}\frac{2\pi n(N/4-k)}{N/4} \\
&= \sum_{n=0}^{N/4-1} \left[v(n) \sin\frac{2\pi n}{N/q} + (-1)^{(q-1)/2} v(n + \frac{N}{4}) \cos\frac{2\pi n}{N/q} \right] \operatorname{cas}\frac{2\pi n(N/4-k)}{N/4}. (4.29)
\end{aligned}
$$

Equations (4.28) and (4.29) are length $N/4$ DHTs whose inputs are rotated by the twiddle factors $\cos(2\pi nq/N)$ and $\sin(2\pi nq/N)$. Compared to other split-radix algorithms, the twiddle factors have an additional parameter q. When $q > 1$, it provides substantial savings in the number of multiplications used for data rotations. The final transform outputs are achieved by

$$
\begin{aligned}
X[(4k+q) \bmod N] &= F(k) + G(k) \\
X[(N+4k-q) \bmod N] &= F(k) - G(k), \quad (4.30)
\end{aligned}
$$

where $k = 0, 1, \ldots, N/4 - 1$. Equations (4.28) and (4.29) can be further decomposed in the same way until length $4q$ and length $8q$ DHTs are reached.

4.3.2   Computational complexity

If 3 multiplications and 3 additions are used for a rotation operation, the number of multiplications required by the generalized split-radix algorithm for a length N DHT is

$$
M(N) = M(\frac{N}{2}) + 2M(\frac{N}{4}) + \frac{3N}{4} - M_t, \quad N > 8q \quad (4.31)
$$

and the number of additions is

$$
A(N) = A(\frac{N}{2}) + A(\frac{N}{4}) + \frac{9N}{4} - A_t, \quad N > 8q, \quad (4.32)
$$

where M_t and A_t are the numbers of multiplications and additions saved from trivial twiddle factors. Arithmetic operations can be saved if the factors $\cos(2\pi nq/N)$ or $\sin(2\pi nq/N)$ become 0, 0.5 or 1. In (4.28) and (4.29), $4q$ multiplications and $4q$ additions can be saved from the trivial twiddle factors. When $q = 1$, however, the number of additions and multiplications saved from trivial twiddle factors is the same as that in [24, 32]. The algorithm needs no multiplication for twiddle factors when $N = 2q$ and $4q$, which is similar to the case that the length 2 and 4 DHTs do not need multiplication for twiddle factors. From (4.25) and (4.26), the number of multiplications needed for $N = 2q$ is

$$
M(2q) = 2M(q) \quad (4.33)
$$

and the number of additions is

$$A(2q) = 2A(p) + 2q. \tag{4.34}$$

When $N = 4q$, both (4.28) and (4.29) become length q DHTs whose twiddle factors become trivial, i.e., $\cos(n\pi/2)$ and $\sin(n\pi/2)$. Therefore, the required number of multiplications is

$$M(4q) = M(2q) + 2M(q) \tag{4.35}$$

and the number of additions is

$$A(4q) = A(2q) + 2A(q) + 6q. \tag{4.36}$$

When $N = 8q$, (4.28) and (4.29) can be further decomposed. For example,

$$\begin{aligned}
F(k) &= \sum_{n=0}^{q-1} \left[v(2n) \cos\frac{\pi n}{2} + (-1)^{\frac{q+1}{2}} v(2n + \frac{N}{4}) \sin\frac{\pi n}{2} \right] \cas\frac{2\pi nk}{q} \\
&\quad + (-1)^k \sum_{n=0}^{q-1} \left\{ v[(2n+q) \bmod 2q] \cos\frac{\pi[(2n+q) \bmod 2q]}{4} \right. \\
&\quad \left. + (-1)^{\frac{q+1}{2}} v[(2n+q) \bmod 2q + \frac{N}{4}] \sin\frac{\pi[(2n+q) \bmod 2p]}{4} \right\} \cas\frac{2\pi nk}{q}. \tag{4.37}
\end{aligned}$$

The first sum of (4.37) is a length q DHT. The additions in the square brackets are redundant because either $\cos(\pi n/2)$ or $\sin(\pi n/2)$ is zero. The second sum is a scaled length q DHT since the twiddle factors are $\pm\sqrt{2}/2$. Compared to the standard DHTs, for example, 3 and 6 extra multiplications are needed by scaled length 9 and 15 DHTs (see Appendices D and E). Therefore, the required number of multiplications is

$$M(8q) = M(4q) + 2M(q) + 2M_s(q) \qquad q > 1, \tag{4.38}$$

where M_s is the number of multiplications needed by a scaled DHT. The number of additions can be calculated by (4.32).

4.3.3 Comparison

Based on (4.31) – (4.38) and the number of additions and multiplications needed by a length q DHT, Table 4.2 lists the number of additions and multiplications required by the generalized split-radix algorithm. For $q = 1$, it requires about the same number of arithmetic operations (additions plus multiplications) as that needed by the algorithms in [24, 32]. For $q > 1$, the number of multiplications and additions saved from the trivial twiddle factors proportionally increases with the value of q. This is the main difference from other split-radix algorithms. Savings on both additions and multiplications are achieved compared to that needed by the algorithm reported in [12].

The presented algorithm generally needs more additions compared to the algorithms reported in [24, 32] (see the curve corresponding to $N = 2^m$ in Figure 4.2 (a)). However, Figure 4.2 (b) shows that a significant reduction in the number of multiplications can be made. Figure 4.3 (a) shows the total number of arithmetic operations needed by various algorithms. In general, this algorithm needs a smaller number of arithmetic operations for $q = 3$, 9 and 15 when $N > 100$. The algorithm also consistently requires a smaller number

Table 4.2 Computational complexity for $q = 1, 3, 9$ and 15.

$N = 2^m$ [24]			$N = 3 \cdot 2^m$		
N	$M(N)$	$A(N)$	N	$M(N)$	$A(N)$
16	10	66	12	4	52
32	32	174	24	12	120
64	98	442	48	44	320
128	258	1070	96	128	764
256	642	2522	192	348	1824
512	1538	5806	384	880	4204
1024	3586	13146	768	2140	9568
$N = 9 \cdot 2^m$			$N = 15 \cdot 2^m$		
18	20	98	15	18	81
36	40	232	30	36	192
72	86	554	60	72	444
144	238	1306	120	156	1038
288	590	3026	240	420	2406
576	1462	6898	480	1032	5502
1152	3470	15506	960	2532	12414

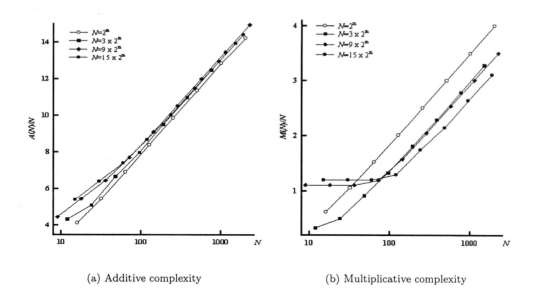

(a) Additive complexity   (b) Multiplicative complexity

Figure 4.2 The numbers of multiplications and additions needed by the algorithm for various sequence lengths.

of arithmetic operations, as shown in Figure 4.3 (a), although fewer operations are used by Meher's algorithm in [25] for certain sequence lengths. Because the computational structure is similar to that of other split-radix algorithms [24, 32], the presented algorithm has the same merits on the issues such as memory requirement and implementation complexity. Extra attention should be paid to the generation of output indices so that modulo operation in (4.30) can be eliminated. One drawback of this algorithm is that bit-reversal method for data swapping [13] cannot be used since the sequence length is not a power of two. Figure 4.3 (b) shows a comparison in terms of computation time with the subroutine reported in [24]. Because the twiddle factors are pre-calculated and stored in a table, Figure 4.3 (b) shows that our subroutine for $q = 1$ uses less computation time than that required by the algorithm in [24] and further reduction of computation time can be made for $q > 1$.

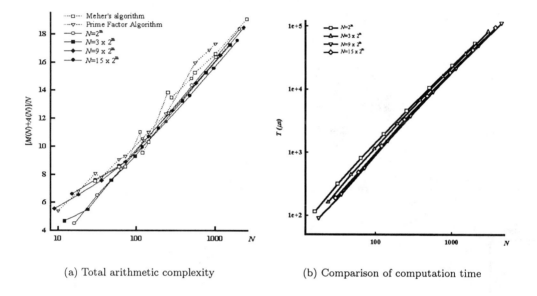

(a) Total arithmetic complexity   (b) Comparison of computation time

Figure 4.3   Computational complexity for complex input sequence.

4.4   RADIX-2 ALGORITHMS FOR TYPE-II, -III AND -IV DHTS

This section considers the fast algorithms for various types of DHTs. By following similar procedures to these presented in Section 3.2, it is straightforward to derive the radix-2 algorithm for a type-I DHT, which is not repeated here. Details of the fast algorithms for type-II, -III and -IV DHTs are provided in this section. It is generally expected that more computation is needed for type-II, -III and -IV DHTs because their definitions contain shifts in the time and/or frequency domain. Some of the decomposition techniques previously used can be applied directly to these DHTs. Of course, special attention is often needed to take care of the difficulties that arise from the variations of the definitions.

4.4.1   Algorithm for type-II DHT

The type-II DHT defined in (4.2) can be decomposed into

$$X(2k) = \sum_{n=0}^{N/2-1} \left[x(n) + x(n + \frac{N}{2}) \right] \operatorname{cas} \frac{\pi(2n + 1)k}{N/2}, \quad 0 \le k < \frac{N}{2}, \qquad (4.39)$$

which is a length $N/2$ type-II DHT, and

$$X(2k+1) = \sum_{n=0}^{N/2-1} \left[x(n) - x(n + \frac{N}{2}) \right] \text{cas} \frac{\pi(2n+1)(2k+1)}{2(N/2)}, \quad 0 \le k < \frac{N}{2}, \quad (4.40)$$

which is a length $N/2$ type-IV DHT. The additive and multiplicative complexities needed by the above decomposition are

$$A_H^{II}(N) = A_H^{II}(\frac{N}{2}) + A_H^{IV}(\frac{N}{2}) + N \quad (4.41)$$

$$M_H^{II}(N) = M_H^{II}(\frac{N}{2}) + M_H^{IV}(\frac{N}{2}), \quad (4.42)$$

where the joint use of the superscript and the subscript specifies the particular type of DHT computation.

4.4.2   Algorithm for type-III DHT

The type-III DHT defined in (4.3) can be decomposed into

$$x(n) = A(n) + B(n), \quad x(n + \frac{N}{2}) = A(n) - B(n), \quad 0 \le n < \frac{N}{2}, \quad (4.43)$$

where

$$A(n) = A(n + \frac{N}{2}) = \sum_{k=0}^{N/2-1} X(2k)\text{cas}\frac{\pi(2n+1)k}{N/2} \quad (4.44)$$

$$B(n) = -B(n + \frac{N}{2}) = \sum_{k=0}^{N/2-1} X(2k+1)\text{cas}\frac{\pi(2n+1)(2k+1)}{2(N/2)}. \quad (4.45)$$

Therefore, the length N type-III DHT is decomposed into a length $N/2$ type-III DHT and a length $N/2$ type-IV DHT. The additive and multiplicative complexities needed by the type-III DHT are

$$M_H^{III}(N) = M_H^{III}(\frac{N}{2}) + M_H^{IV}(\frac{N}{2}) \quad (4.46)$$

$$A_H^{III}(N) = A_H^{III}(\frac{N}{2}) + A_H^{IV}(\frac{N}{2}) + N. \quad (4.47)$$

4.4.3   Algorithm for type-IV DHT

The length N type-IV DHT defined in (4.4) can be computed by

$$\begin{aligned}
F(k) &= \frac{X(2k) + X(2k-1)}{2} \\
&= \sum_{n=0}^{N-1} x(n) \cos \frac{\pi(2n+1)}{2N} \text{cas} \frac{\pi(2n+1)k}{N/2} \\
&= \sum_{n=0}^{N/2-1} [x(n) - x(N-n-1)] \cos \frac{\pi(2n+1)}{2N} \text{cas} \frac{\pi(2n+1)k}{N/2} \quad (4.48)
\end{aligned}$$

and

$$G(k) = \frac{X(2k) - X(2k-1)}{2}$$

$$= \sum_{n=0}^{N/2-1} [x(n) + x(N-n-1)] \sin\frac{\pi(2n+1)}{2N} \text{cas}\frac{-\pi(2n+1)k}{N/2}, \qquad (4.49)$$

where for both (4.48) and (4.49), $0 \le k \le N/2-1$. They are both length $N/2$ type-II DHTs. The final transform outputs are given by

$$X(0) = F(0) + G(0), \quad X(N-1) = F(0) - G(0)$$
$$X(2k) = F(k) + G(k), \quad X(2k-1) = F(k) - G(k) \qquad (4.50)$$
$$1 \le k \le \frac{N}{2} - 1.$$

The computational complexities for the length N type-IV DHT are

$$M_H^{IV}(N) = 2M_H^{II}(\frac{N}{2}) + N \qquad (4.51)$$

$$A_H^{IV}(N) = 2A_H^{II}(\frac{N}{2}) + 2N. \qquad (4.52)$$

A different decomposition is also possible. For example, the type-IV DHT can be computed by

$$X(k) = \sum_{n=0}^{N/2-1} x(2n)\text{cas}\frac{\pi(4n+1)(2k+1)}{2N}$$

$$+ \sum_{n=0}^{N/2-1} x'(2n-1)\text{cas}\frac{\pi(4n-1)(2k+1)}{2N} \qquad (4.53)$$

$$= \cos\frac{\pi(2k+1)}{2N}A(k) + \sin\frac{\pi(2k+1)}{2N}B(k), \qquad 0 \le k < \frac{N}{2}$$

and

$$X(k + \frac{N}{2}) = -\sin\frac{\pi(2k+1)}{2N}A(k) + \cos\frac{\pi(2k+1)}{2N}B(k), \qquad 0 \le k < \frac{N}{2}, \qquad (4.54)$$

where

$$A(k) = \sum_{n=0}^{N/2-1} [x(2n) + x'(2n-1)]\text{cas}\frac{\pi n(2k+1)}{N/2} \qquad (4.55)$$

$$B(k) = \sum_{n=0}^{N/2-1} [x(2n) - x'(2n-1)]\text{cas}\frac{-\pi n(2k+1)}{N/2}, \qquad (4.56)$$

where $x'(-1) = -x(N-1)$ and $x'(2n-1) = x(2n-1)$ $(n = 1, \ldots, N/2-1)$. Therefore, the type-IV DHT is decomposed into two length $N/2$ type-III DHTs, which corresponds to a decimation-in-frequency radix-2 algorithm. The same decomposition can be further applied if the subsequence length is even, which requires a decomposition of a type-III DHT. If the subsequence length contains factors that are mutually prime, however, further decomposition can be achieved by using the prime factor decomposition procedures.

4.5   PRIME FACTOR ALGORITHMS

In the last chapter, the prime factor algorithm (PFA) was discussed for the DFT computation. Using a similar procedure, this section develops the PFAs for type-I, -II and -III
DHTs. Once the readers understand the principle, it is easy to derive the PFA for a type-IV
DHT. In general, the fast algorithms in the non-radix category reported in [10, 15, 25] can be
used to provide more choices of transform sequence lengths. The algorithm in [25] appears
to be the most computationally efficient compared to other algorithms [10, 15, 21] in the
non-radix category. As in the DFT case, a common drawback of these algorithms is the
irregularity of the index mapping. The desirable property of in-place computation, which is
easily achieved by the radix-type algorithms, is difficult to implement.

This section shows that the 1D DHT can be decomposed into p length q DHTs and q length
p DFTs. Attempts were made to minimize the required number of arithmetic operations
and eliminate the irregular index mapping process. The decomposition method is combined
with the split-radix algorithm to utilize their desirable features. It can be shown that we can
achieve a regular computational structure, provide flexibility for composite sequence lengths
and make substantial savings on the computational complexity.

4.5.1   Algorithm for type-I DHT

When $N = p \cdot q$, where p and q are relatively prime, the length N DHT defined in (4.1) can
be decomposed into

$$X(pk) = \sum_{n=0}^{N-1} x(n)\text{cas}\frac{2\pi nk}{N/p} = \sum_{n=0}^{q-1} u(n,0)\text{cas}\frac{2\pi nk}{q}, \qquad (4.57)$$

where $k = 0, 1, \ldots, q-1$, and

$$u(n,0) = \sum_{i=0}^{p-1} x(n+iq). \qquad (4.58)$$

By using the property $\text{cas}(\alpha + \beta) = \text{cas}(\alpha)\cos(\beta) + \text{cas}(-\alpha)\sin(\beta)$, (4.1) can be rewritten
as

$$
\begin{aligned}
X[(pk+mq) \bmod N] &= \sum_{n=0}^{N-1} x(n)\text{cas}\frac{2\pi n(pk+mq)}{N} \\
&= \sum_{n=0}^{N-1} x(n)\cos\frac{2\pi nm}{p}\text{cas}\frac{2\pi nk}{q} + \sum_{n=0}^{N-1} x(n)\sin\frac{2\pi nm}{p}\text{cas}\frac{-2\pi nk}{q} \\
&= \sum_{n=0}^{q-1}\sum_{i=0}^{p-1} x(n+iq)\cos\frac{2\pi(n+iq)m}{p}\text{cas}\frac{2\pi nk}{q} \\
&\quad + \sum_{n=0}^{q-1}\sum_{i=0}^{p-1} x(n+iq)\sin\frac{2\pi(n+iq)m}{p}\text{cas}\frac{-2\pi nk}{q} \\
&= \sum_{n=0}^{q-1} u(n,m)\text{cas}\frac{2\pi nk}{q} + \sum_{n=0}^{q-1} v(n,m)\text{cas}\frac{-2\pi nk}{q}, \qquad (4.59)
\end{aligned}
$$

where $k = 0, 1, \ldots, q - 1$, and $m = 0, 1, \ldots, p - 1$. Both terms in (4.59) are length q DHTs of $u(n, m)$ and $v(n, m)$ which can be obtained by

$$F(n, m) = u(n, m) - jv(n, m) = \sum_{i=0}^{p-1} x(n + iq)e^{-j2\pi(n+iq)m/p} \qquad (4.60)$$

for $n = 0, 1, \ldots, q - 1$, and $m = 0, 1, \ldots, p - 1$. When $m = 0$, $v(n, m) = 0$ and $F(n, m)$ is the same as (4.58). According to the Chinese remainder theorem, it can be proven that if p and q are mutually prime, there exists a one-to-one mapping between $i, 0 \le i < p$ and $l, 0 \le l < p$ such that

$$l = (n + iq) \bmod p \qquad (4.61)$$

for any n, $0 \le n < q$. Therefore, (4.60) can be expressed by

$$F(n, m) = \sum_{l=0}^{p-1} x'(n, l)e^{-j2\pi lm/p}, \qquad m = 0, 1, \ldots, p - 1, \qquad (4.62)$$

where $x'(n, l) = x(n + iq)$, and l and i are related by (4.61). Therefore, (4.62) is exactly a length p DFT of real data. The computation defined in (4.57) and (4.59) is basically a DHT version of the Good–Thomas FFT algorithm [14]. It divides the entire computation into a stage of q length p real-valued DFTs and a stage of p length q DHTs. Figure 4.4 shows the signal flow graph of a length 12 DHT.

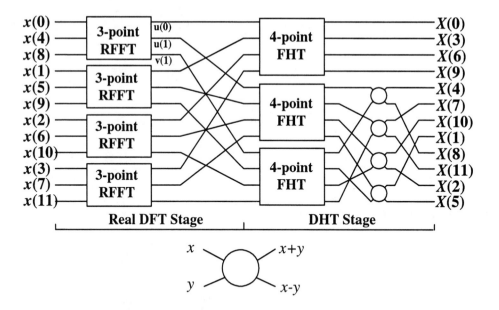

Figure 4.4   Signal flow graph of the prime factor DHT algorithm ($N = 12$).

The decomposition procedures in (4.57) – (4.62) can be extended for composite sequence lengths. Let the sequence length N be a composite number expressed by

$$N = 2^m p_1 p_2 \cdots p_{M-1} p_M, \qquad (4.63)$$

where $p_i(i = 1, \ldots, M)$ are different odd integers and relatively prime, and m is a positive integer. We further define

$$N_i = N_{i-1}/p_i, \tag{4.64}$$

where $N_0 = N$ and N_i $(i = 1, 2, \ldots, M)$ are the subsequence lengths. The length N DHT can be decomposed into $M + 1$ stages. The first M stages consist of real-valued DFTs and the last stage computes length 2^m DHTs, as shown in Figure 4.5. In general, p_i can be a prime number or in the form of k^j where k is an odd integer and j is a positive integer. An efficient computation of a short DFT can be found in [27] and the algorithms for length k^j are given in [35].

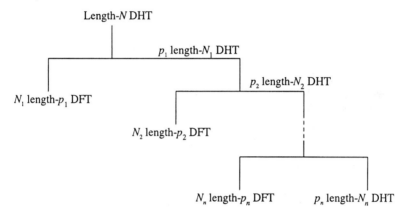

Figure 4.5   DHT decomposition steps for composite sequence length.

Similar to the DFT case, the non-radix fast algorithms for the DHT require irregular data index mapping (or data swapping) which makes in-place computation difficult. The presented PFA can solve this problem by properly generating indices of input data at each stage. Let us consider an example, as shown in Figure 4.6 for $N = 24$, without loss of generality. The input sequence is divided into 8 groups and each group contains 3 elements whose original input index is represented by $d_{n,l}$. We first compute 8 length 3 real-valued DFTs whose input indices are $d_{n,l}$ for $0 \leq n < 8$ and $0 \leq l < 3$. Then 3 length 8 DHTs whose inputs are the outputs of the DFTs are computed. Observing Figure 4.6, we can conclude that the input indices can be obtained by

$$d_{n,l} = \begin{cases} d_{n-1,p-1} + 1, & n = 1, \ldots, q-1, \quad l = 0 \\ d_{n-1,l-1} + 1, & n = 1, \ldots, q-1, \quad l = 1, \ldots, p-1, \end{cases} \tag{4.65}$$

where $d_{0,l}, l = 0, .., p-1$, can be pre-calculated. In Figure 4.6, a broken arrow indicates an increment-by-one operation required in (4.65). Therefore, modulo operation used in (4.61) is replaced by an increment operation. Once $d_{n,l}$ are obtained, the length 3 real-valued DFT can be computed. Thus, both data swapping and modulo operations are avoided. In-place computation can also be implemented by writing the outputs of the length 3 DFT into the data locations that are no longer used by the input data. The right side of Figure 4.6 shows that the outputs of each length 3 real-valued DFT can be placed in the data stores that are used for its input data. One important property achieved by this arrangement is that the input data for the next stage DHT computation are in order so that the same data indexing scheme can be used for further decomposition when N has more than one factor of odd integers. This property ensures that the presented algorithm can achieve in-place computation. In summary, the following steps can compute the length N DHT.

- compute input data indices according to (4.65) to regroup sequence $x(n)$ into $x'(n, l)$;

- compute length p DFT according to (4.62) for $n = 0, \ldots, q - 1$.

- compute length q DHT according to (4.57) and (4.59).

- obtain the final outputs by combining the outputs of length q DHT according to (4.59).

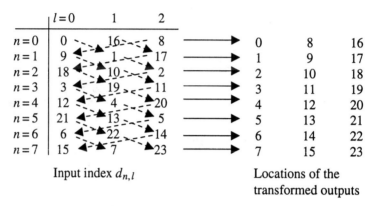

Input index $d_{n,l}$ Locations of the transformed outputs

Figure 4.6   Input index calculation and in-place computation.

By using the property $X(k) = X^*(N - k)$ $(k = 0, 1, \ldots, N/2)$ of the DFT for a real input sequence, (4.59) requires $p - 1$ length q DHTs. According to (4.57)–(4.59), the number of real multiplications required by the algorithm is

$$M_H(N) = pM_H(q) + qM_F(p) \tag{4.66}$$

and the required number of real additions is

$$A_H(N) = pA_H(q) + qA_F(p) + (p - 1)q, \tag{4.67}$$

where the subscripts H and F indicate the number of operations needed by the DHT and DFT computation, respectively. We now consider an optimization method to reduce the number of multiplications needed by (4.66). When $p = 3$, for example, the length 3 DFT of real data requires one real multiplication, 4 real additions and 1 shift operation. However, we note that $v(n, 0) = 0$ and

$$v(n, 1) = -v(n, 2) = \frac{\sqrt{3}}{2}[x(n + q) - x(n + 2q)]. \tag{4.68}$$

The multiplication required by (4.68) can be saved if the second term of (4.59) becomes a scaled length q DHT by a factor of $\sqrt{3}/2$. Such arrangements are also possible for other p values. For example, the imaginary parts of a length 5 DFT of real data can be expressed by

$$
\begin{aligned}
v(n, 1) &= -v(n, 4) \\
&= \sin(\frac{2\pi}{5})\left\{[x(n + q) - x(n + 4q)] + 2\cos(\frac{2\pi}{5})[x(n + 2q) - x(n + 3q)]\right\} \tag{4.69}
\end{aligned}
$$

and

$$v(n, 2) = -v(n, 3)$$
$$= \sin(\frac{2\pi}{5}) \left\{ 2\cos(\frac{2\pi}{5})[x(n+q) - x(n+4q)] - [x(n+2q) - x(n+3q)] \right\}. \quad (4.70)$$

When the scaling factors outside the braces in (4.69) and (4.70) are not counted, the length 5 DFT of real data uses 3 multiplications and 13 additions. For a length 9 DFT of real data, the output associated with a scaling factor is

$$v(n, 3) = -v(n, 6)$$
$$= \sin(\frac{4\pi}{9}) \left\{ \begin{array}{c} [x(n+q) - x(n+8q)] - [x(n+2q) - x(n+7q)] \\ +[x(n+4q) - x(n+5q)] \end{array} \right\}. \quad (4.71)$$

Therefore, the length 9 DFT of real data used by our algorithm requires 9 multiplications and 32 additions. In general, the reduction of multiplications obtained from rearranging the scaling factors is Kq, where K is the number of scaling factors associated with the imaginary parts of the length p DFT and q is the number of length p real-valued DFTs needed by (4.59). All these scaling factors are absorbed by a scaled length q DHT. For $N = 2^m$, the scaled DHT can be decomposed into a scaled length $N/2$ DHT and two standard length $N/4$ DHTs. To minimize the computational complexity for a scaled DHT, the scaling factor for the odd indexed transform outputs can be combined with the twiddle factors or, equivalently, the DHT computation uses scaled transform coefficients. The number of additions needed by a scaled length N DHT is the same as that needed by a standard length N DHT, but the required number of multiplications becomes

$$M_H^s(N) = M_H^s(\frac{N}{2}) + 2M_H(\frac{N}{4}) + N - 2, \quad N > 16, \quad (4.72)$$

where the superscript s indicates M_H^s is for a scaled DHT. The number of extra multiplications for a scaled DHT is small. For example, Table 4.3 shows that for $N = 1024$, only 34 extra multiplications are needed compared to a standard DHT. Based on (4.66) and (4.67) and the above discussion, Table 4.4 shows the number of additions and multiplications needed by a length $p \cdot q$ DHT for $p = 3, 5$ and 9.

One basic requirement of the PFA is that p and q are relatively prime. Various values of p and q can be selected to achieve the required sequence length. To minimize computational complexity, we choose p to be an odd integer that can be a prime number or a product of relatively prime numbers. In general a small value of p is preferred for an easy optimization of the length p DFT computation. For better computational efficiency, we choose q to be 2^m because DHTs of such sequence lengths need a smaller number of operations and a regular computational structure. Table 4.3 shows the computational complexity required by Sorensen's split-radix 2/4 and Duhamel's DHT/FFT hybrid algorithms. Based upon the split-radix 2/4 algorithm, the number of multiplications required by a scaled DHT is calculated, as shown in Table 4.3. By using Table 4.4 and the number of operations listed in Table 4.3, the number of multiplications and additions needed by the presented algorithm for $p = 3, 5$ and 9 can be calculated, as shown in Table 4.5. When p is a product of odd numbers, the same decomposition approach can also be used. For example, we choose $p = 3$ and $q = 5 \cdot 2^m$ for $N = 15 \cdot 2^m$. Figures 4.7 (a) and 4.7 (b) show a comparison of the required numbers of multiplications and additions per transform output for various sequence lengths. Figure 4.7 (c) shows that the algorithm requires substantially less computational complexity than Sorensen's split-radix 2/4 algorithm and Duhamel's algorithm.

Table 4.3   Computational complexity for length 2^r DHT.

N	Algorithm [2] $M(N)$	$A(N)$	Algorithm [12, 32] $M(N)$	$A(N)$	Scaled by $\sqrt{3}/2$ $M(N)$	$A(N)$
1	0	0	0	0	1	0
2	0	2	0	2	2	2
4	0	8	0	8	4	8
8	2	22	2	22	8	22
16	10	62	10	62	16	62
32	40	168	34	166	50	168
64	118	418	98	422	132	418
128	320	1008	258	1030	338	1008
256	806	2354	642	2438	828	2354
512	1952	5392	1538	5638	1978	5392
1024	4582	12146	3586	12806	4612	12146
2048	10528	27024	8194	28628	10562	27024

Table 4.4   Computational complexity for $p = 3$, 5 and 9.

	$M_H(N)$	$A_H(N)$
$p = 3$	$2M_H(q) + M_H^s(q)$	$3A_H(q) + 2N$
5	$3M_H(q) + 2M_H^s(q) + 3q$	$5A_H(q) + 12q + N$
9	$8M_H(q) + M_H^s(q) + N$	$9A_H(q) + 31q + N$

It is known that for $N = 2^m$, the number of multiplications needed by DHTs with $N = 1, 2$ and 4 is a zero constant. Our algorithm also reserves this desirable property for $p > 1$. As shown in Figure 4.7 (a), the number of multiplications per transform point for $N/p = 1$, 2 and 4, where $p = 3$, 5, 9 and 15, is a non-zero constant. Compared to the number of multiplications needed by the algorithms reported in [32] and [12], Figure 4.7 (a) shows that when N is large (e.g., > 100), our algorithm achieves a reduction of the multiplicative complexity. The described algorithm needs an additive complexity that is between those used by the algorithms reported in [12] and [32], as shown in Figure 4.7 (b). Finally, Figure 4.7 (c) shows that the presented algorithm for $p = 5$ uses about the same number of arithmetic operations (additions plus multiplications) as that needed by the algorithms reported in [12, 32]. For other p values, however, the presented algorithm is more computationally efficient when N is large.

For a comprehensive comparison with other algorithms, Figure 4.8 shows the number of arithmetic operations required by Sorensen's algorithm [32], Pei's mixed radix algorithm [28], Meher's algorithm [25], Lun's algorithm [21], the prime-factor real-valued DFT algorithm [15] and the presented algorithm for $p = 3$. It is clearly shown that the presented algorithm needs a smaller number of operations compared to all the algorithms for most cases. Although it is possible that other algorithms ([25] and [28], for example) require a smaller number of operations for some sequence lengths, they generally need many more operations for other sequence lengths compared to the presented algorithm. Figure 4.8 also shows that the algorithms in the non-radix type category generally need more operations than the split-radix 2/4 algorithm.

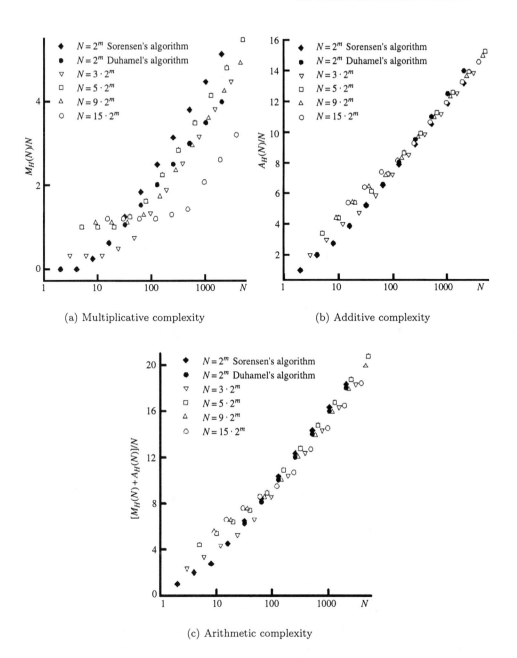

(a) Multiplicative complexity   (b) Additive complexity

(c) Arithmetic complexity

Figure 4.7   Comparison of computational complexity.

The avoidance of the data swapping operation makes an important improvement on computational structures so that in-place computation can be achieved. Therefore, the computational structure of the described algorithm is much simpler than those of all the above mentioned non-radix algorithms for composite sequence lengths. One further advantage is that the described algorithm allows using existing DHT subroutines which were designed for the sequence length being a power of two. One possible drawback of the algorithm is

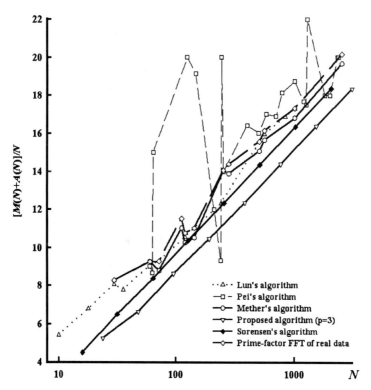

Figure 4.8   Comparison of arithmetic complexity required by non-radix type algorithms.

Table 4.5   The required number of operations for $p = 3, 5, 9$ and 15.

N/p	$p = 3$		$p = 5$		$p = 9$		$p = 15$	
	$M(N)$	$A(N)$	$M(N)$	$A(N)$	$M(N)$	$A(N)$	$M(N)$	$A(N)$
1	1	6	5	17	10	40	18	81
2	2	18	10	44	20	98	36	192
4	4	48	20	108	40	232	72	444
8	12	114	46	246	96	518	156	978
16	36	282	110	582	240	1198	348	2226
32	130	696	314	1384	658	2792	974	5112
64	368	1638	810	3178	1652	6322	2472	11454
128	978	3792	2020	7216	4050	14192	6114	25488
256	2440	8598	4842	16122	9580	31426	14592	56046
512	5882	19248	11348	35664	22202	69008	34110	122352
1024	13776	42582	26042	78138	50484	150274	78216	265134

using complex arithmetic for length p DFT computations. However, this requirement will be trivial when p is small since the length p DFT can be optimized and coded as subroutines.

4.5.2   Algorithm for type-II DHT

The transform outputs $X(k)$ of the type-II DHT, defined in (4.2), can be grouped into

$$X(kp) = \sum_{n=0}^{q-1} u(n) \mathrm{cas} \frac{\pi(2n+1)k}{q}, \qquad 0 \le k < q, \tag{4.73}$$

which is a length q type-II DHT. The input sequence $u(n)$ in (4.73) is given by

$$u(n,0) = \sum_{i=0}^{p-1} x(n+iq). \tag{4.74}$$

Similarly,

$$
\begin{aligned}
X(kp \pm mq) &= \sum_{n=0}^{N-1} x(n) \mathrm{cas} \frac{\pi(2n+1)(kp \pm mq)}{q} \\
&= \sum_{n=0}^{N-1} x(n) \left[\cos \frac{\pi(2n+1)m}{p} \mathrm{cas} \frac{\pi(2n+1)k}{q} \right. \\
&\quad \left. \pm \sin \frac{\pi(2n+1)m}{p} \mathrm{cas} \frac{-\pi(2n+1)k}{q} \right].
\end{aligned}
\tag{4.75}
$$

We regroup $X(kp + mq)$ and $X(kp - mq)$ into new sequences

$$
\begin{aligned}
F(k,m) &= \frac{X(kp+mq) + X(kp-mq)}{2} \\
&= \sum_{n=0}^{N-1} x(n) \cos \frac{\pi(2n+1)m}{p} \mathrm{cas} \frac{\pi(2n+1)k}{q} \\
&= \sum_{n=0}^{q-1} u(n,m) \mathrm{cas} \frac{\pi(2n+1)k}{q},
\end{aligned}
\tag{4.76}
$$

$$
\begin{aligned}
G(k,m) &= \frac{X(kp+mq) - X(kp-mq)}{2} \\
&= \sum_{n=0}^{N-1} x(n) \sin \frac{\pi(2n+1)m}{p} \mathrm{cas} \frac{-\pi(2n+1)k}{q} \\
&= \sum_{n=0}^{q-1} v(n,m) \mathrm{cas} \frac{-\pi(2n+1)k}{q},
\end{aligned}
\tag{4.77}
$$

where for both (4.76) and (4.77), $1 \le m \le (p-1)/2$ and $0 \le k \le q-1$. If (4.77) is computed by $-G(q-k,m)$ ($k = 0, 1, \ldots, q-1$), both (4.76) and (4.77) are length q type-II DHTs. It is noted that $G(q,m) = -G(0,m)$. The sequence $u(n,m)$ in (4.76) is given by

$$u(n,m) = \sum_{i=0}^{p-1} x(n+iq) \cos \frac{\pi(2n+2iq+1)m}{p} \tag{4.78}$$

and the sequence $v(n,m)$ in (4.77) is

$$
\begin{aligned}
v(n,m) &= \sum_{i=0}^{p-1} x(n+iq) \sin \frac{\pi(2n+2iq+1)m}{p} \\
&= (-1)^n \sum_{i=0}^{p-1} x(n+iq)(-1)^{iq} \cos \frac{\pi[2(n+iq)+1](p-2m)}{p}.
\end{aligned}
\tag{4.79}
$$

The final transform outputs are given by (4.73) and

$$X(kp + mq) = F(k, m) + G(k, m), \quad X(kp - mq) = F(k, m) - G(k, m)$$
$$0 \leq k \leq q - 1, \qquad\qquad\qquad 1 \leq m \leq (p-1)/2. \tag{4.80}$$

When $kp + mq > N$ or $kp - mq < 0$,

$$X(N + k) = -X(k), \quad X(-k) = -X(N - k). \tag{4.81}$$

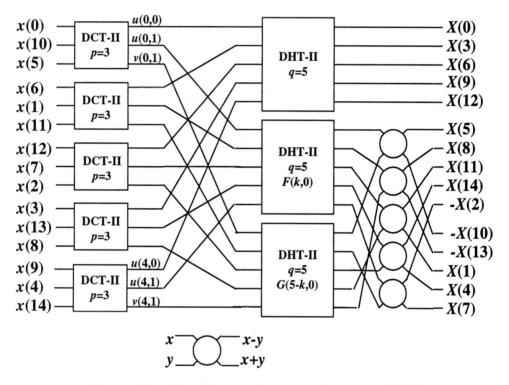

Figure 4.9   Signal flow graph of a length 15 type-II DHT.

Appendix F shows that the combination of (4.74), (4.78) and (4.79) can be converted into a length p type-II DCT. Therefore, the length N type-II DHT is decomposed into p length q type-II DHTs, given by (4.73), (4.76) and (4.77), and q length p type-II DCTs. Figure 4.9 shows the signal flow graph for $N = 15$. One desirable property of the described decomposition is that there is no twiddle factor in the entire computation. However, reordering the original input sequence is necessary, which will be discussed shortly. The required computational complexity is

$$A_H^{II}(N) = pA_H^{II}(q) + qA_C^{II}(p) + (p - 1)q \tag{4.82}$$
$$M_H^{II}(N) = pM_H^{II}(q) + qM_C^{II}(p), \tag{4.83}$$

where the superscripts indicate the type of the transform, and the subscripts H and C indicate that the associated functions are for DHT and DCT computations, respectively.

4.5.3 Algorithm for type-III DHT

The type-III DHT of $X(k)$, defined in (4.3), can be decomposed into

$$
x(n) = \sum_{k=0}^{q-1} X(pk) \operatorname{cas} \frac{\pi(2n+1)k}{q}
$$

$$
+ \sum_{m=1}^{(p-1)/2} \left[\sum_{k=0}^{q-1} Y(k,m) \operatorname{cas} \frac{\pi(2n+1)(pk+qm)}{N} \right.
$$

$$
\left. + \sum_{k=0}^{q-1} Z(k,m) \operatorname{cas} \frac{\pi(2n+1)(pk-qm)}{N} \right]
$$

$$
= a(n) + \sum_{m=1}^{(p-1)/2} \left[b(n,m) \cos \frac{\pi(2n+1)m}{p} + c(n,m) \sin \frac{\pi(2n+1)m}{p} \right], \quad (4.84)
$$

where

$$
a(n) = \sum_{k=0}^{q-1} X(pk) \operatorname{cas} \frac{\pi(2n+1)k}{q} \tag{4.85}
$$

$$
b(n,m) = \sum_{k=0}^{q-1} [Y(k,m) + Z(k,m)] \operatorname{cas} \frac{\pi(2n+1)k}{q} \tag{4.86}
$$

$$
c(n,m) = \sum_{k=0}^{q-1} [Y(k,m) - Z(k,m)] \operatorname{cas} \frac{-\pi(2n+1)k}{q}. \tag{4.87}
$$

Equations (4.85) – (4.87) are length q type-III DHTs. In (4.86) and (4.87), $Y(k,m)$ and $Z(k,m)$ are the reordered input sequences according to

$$
Y(k,m) = \begin{cases} X(kp+mq), & kp+mq < N \\ -X(kp+mq-N), & kp+mq > N \end{cases} \tag{4.88}
$$

and

$$
Z(k,m) = \begin{cases} X(kp-mq), & kp > mq \\ -X(N+kp-mq), & kp < mq \end{cases}. \tag{4.89}
$$

It can be proven that for $i = 0, 1, \ldots, p-1$, $a(n+iq) = a(n)$, $b(n+iq,m) = b(n,m)$ and $c(n+iq,m) = c(n,m)$; therefore, (4.84) becomes

$$
x(n+iq) = a(n) + \sum_{m=1}^{(p-1)/2} \left[b(n,m) \cos \frac{\pi(2n+2iq+1)m}{p} \right.
$$

$$
\left. + c(n,m) \sin \frac{\pi(2n+2iq+1)m}{p} \right] \tag{4.90}
$$

for $0 \le n \le q-1$ and $0 \le i \le p-1$. Equation (4.90) is a length p type-III DCT, as shown in Appendix G. Therefore, the length N type-III DHT is decomposed into p length q type-III DHTs and q length p type-III DCTs. The signal flow diagram of the decomposition process for a length 15 type-III DHT can be obtained by converting the inputs (outputs) of

Figure 4.9 into the outputs (inputs). As in the decomposition for the type-II DHT, there is no twiddle factor in the entire computation and reordering the original input sequence is necessary. It can be easily seen that the required computational complexity is

$$A_H^{III}(N) = pA_H^{III}(q) + qA_C^{III}(p) + (p-1)q \tag{4.91}$$
$$M_H^{III}(N) = pM_H^{III}(q) + qM_C^{III}(p). \tag{4.92}$$

4.5.4 Index mapping for type-II and -III DHTs

The PFAs for a type-II (or type-III) DHT require the original input (or output) sequence to be reordered according to

$$j = (n + iq) \mod p, \qquad 0 \le j, i \le p-1, \tag{4.93}$$

where $i = 0, 1, \ldots, p-1$ and $n + iq$ is the original input index. The mapping process is to select p data for each n according to the above equation. Similar to the approach used in the PFA reported in Chapter 3, the required data can be read directly from the original data array by using pre-calculated data indices.

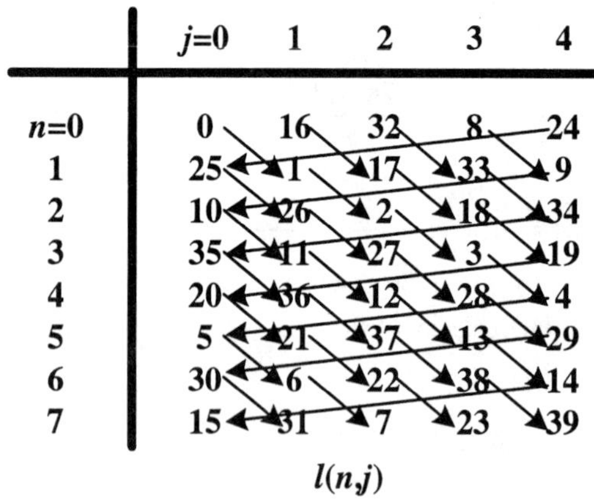

(a)

j=0	1	2	3	4
0	16	32	8	24
1	17	33	9	25
2	18	34	10	26
3	19	35	11	27
4	20	36	12	28
5	21	37	13	29
6	22	38	14	30
7	23	39	15	31

(b)

Figure 4.10 Arrangements for (a) input sequence reordering and (b) in-place computation ($p = 5$ and $q = 8$).

The input data associated with original indices expressed by $n + iq$ $(i = 0, 1, \ldots, p-1)$ for each n are grouped to be the input data of a length p type-II DCT computation. Figure 4.10 (a) shows an example for $p = 5$ and $q = 8$ in which for each $n, j = (n + iq) \mod p$ is arranged according to the increasing order. If we define the new input order $l(n, j)$, Figure 4.10 (a) shows that for $n > 0$,

$$l(n,j) = \begin{cases} l(n-1, j-1) + 1, & j > 0 \\ l(n-1, p-1) + 1, & j = 0 \end{cases}, \tag{4.94}$$

where for $n = 0, l(0, j) = (iq) \mod p$ $(i = 0, 1, \ldots, p-1)$, which can be pre-calculated and arranged into an increasing order. Figure 4.10 (a) shows that $l(n, j)$, pointed to by the arrow, can be obtained by increasing by one operation on the number from which the arrow leaves.

In-place computation can be achieved. Once the p input data for each n are used for the length p type-II DCT computation, the memory locations they occupied can be used for the outputs of the DCT computation. Note that a desirable property achieved from such an arrangement is that the outputs of the q length p type-II DCTs can be arranged into a natural order, as shown in Figure 4.10 (b). Therefore, the same procedure of data indexing can be applied for the next stage computation. In general, such a method for reordering of the input sequence can also be used for the PFAs of other transforms. For a type-III

Table 4.6   Computational complexity for $N = 2^m, 3 \cdot 2^m, 5 \cdot 2^m$ and $15 \cdot 2^m$.

Type-III DHT					
N	$M(N)$	$A(N)$	N	$M(N)$	$A(N)$
$N = 2^m$			$N = 3 \cdot 2^m$		
8	6	26	12	10	42
16	18	70	24	26	126
32	46	186	48	70	306
64	114	454	96	170	750
128	270	1082	192	406	1746
256	626	2502	384	938	4014
512	1422	5690	768	2134	9042
$N = 5 \cdot 2^m$			$N = 15 \cdot 2^m$		
20	30	98	30	49	192
40	70	266	60	98	414
80	170	622	120	226	1038
160	390	1474	240	542	2346
320	890	3358	480	1234	5382
640	1990	7586	960	2798	11994
1280	4410	16862	1920	6226	26598

DHT, trivial data reordering of the initial input sequence according to (4.88) and (4.89) is necessary when $pk + qm > N$ and $pk - qm < 0$. The mapping between the output indices $n + iq$ in (4.90) and the outputs of the type-III DCT should be carefully made for each n. This can be understood by considering the fact that the reordering process of the input sequence for a type-II DHT is equivalent to the reordering process of the output sequence for a type-III DHT. Therefore, the mapping process can be carried out in a similar way to that used in the reordering process for a type-II DHT.

4.5.5   Computational complexity for type-II and -III DHTs

The PFA is particularly useful when N contains a few small factors which are mutually prime. For large N, it is better to combine with other fast algorithms to minimize the overall computational complexity. For example, we can assume $N = p \cdot q$, where p is an odd integer and $q = 2^m$. The length N DHT can be decomposed into p length 2^m DHTs and 2^m length p DCTs by the prime factor decomposition. Then the length 2^m DHT can be decomposed by the radix-2 or split-radix algorithms. Of course, the order of using the two types of algorithms can be reversed to decompose the entire computation into 2^m length p

DHTs. In general, the former approach is preferred to minimize the overall computational complexity because the radix-2 or the split-radix algorithm has a smaller growth rate in terms of the number of operations per transform point, which can be demonstrated in Appendix E, Chapter 6. The computational complexity presented in this subsection is achieved from the former approach.

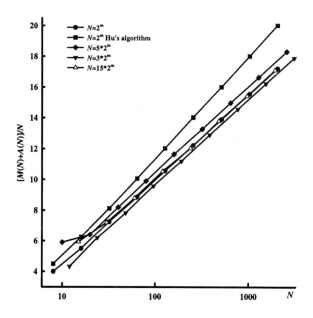

Figure 4.11   Comparison of computational complexity for type-III DHT.

Based on the computational complexity given in the previous subsections, Table 4.6 shows the numbers of additions and multiplications for the type-III DHT for $N = q \cdot 2^m$, where $q = 1$, 3, 5 and 15. The computational complexity for a type-II DHT is the same as that for a type-III DHT. Therefore any comment on the algorithms of the type-III DHT is also valid for the algorithm of the type-II DHT. Figures 4.11 compares the computational complexity needed by the presented algorithms and that needed by the algorithm in [38] which appears to use the minimum number of operations among the fast algorithms reported in the literature. Comparison with other PFAs for type-II and -III DHTs is not possible because such algorithms are hardly available in the published literature. For a type-III DHT, Figure 4.11 shows that the presented algorithms require a minimum computational complexity when $q = 3$ compared to that for other q values. The same computational complexity is needed for $q = 1$ and 15. However, the most computational complexity is needed for $q = 5$. The computational complexity required by the algorithms for all cases is substantially smaller than that achieved by the algorithm in [38] for $q = 1$ (see Table 4.7) When $N = 256$, for example, our algorithms need 70% additions and 93% multiplications for a type-III DHT compared with those needed by the algorithm in [38]. The savings are mainly achieved by eliminating twiddle factors in the prime factor decomposition and the efficient decomposition for the even sequence lengths.

Although the PFA is the most computationally efficient, the requirement that the sequence length must contain mutually prime factors is restrictive for some applications. The radix-q algorithm to be considered in the next section attempts to remove this condition.

Table 4.7 Computational complexity needed by the algorithm in [38].

$N =$	Type-III DHT	
	$M(N)$	$A(N)$
4	2	10
8	8	28
16	24	76
32	64	196
64	160	484
128	384	1156
256	896	2692

4.6 RADIX-q ALGORITHMS

Let us consider another decomposition approach that allows us to divide the entire computational task based on the odd factors. We generally assume that the sequence length N contains odd factors which do not have to be mutually prime.

4.6.1 Algorithm for type-II DHT

The type-II DHT defined in (4.2) can be expressed by

$$
\begin{aligned}
X(k) \;=\; & \sum_{n=0}^{N/q-1} x\!\left(qn + \frac{q-1}{2}\right)\mathrm{cas}\,\frac{\pi(2n+1)k}{N/q} \\
& + \sum_{m=0}^{(q-3)/2}\sum_{n=0}^{N/q-1} x(qn+m)\,\mathrm{cas}\,\frac{\pi[q(2n+1)-(q-1-2m)]k}{N} \\
& + \sum_{m=0}^{(q-3)/2}\sum_{n=0}^{N/q-1} x(qn+q-1-m)\,\mathrm{cas}\,\frac{\pi[q(2n+1)+(q-1-2m)]k}{N},
\end{aligned}
\tag{4.95}
$$

where q is an odd factor. By using the property $\mathrm{cas}(\alpha+\beta)=\mathrm{cas}(\alpha)\cos(\beta)+\mathrm{cas}(-\alpha)\sin(\beta)$, (4.95) can be expressed by

$$
\begin{aligned}
X(k) \;=\; & A(k) + \sum_{m=0}^{(q-3)/2} B(k,m)\cos\frac{\pi(q-1-2m)k}{N} \\
& + \sum_{m=0}^{(q-3)/2} C(k,m)\sin\frac{\pi(q-1-2m)k}{N},
\end{aligned}
\tag{4.96}
$$

where $0 \le k < N$. For $0 \le k \le N/q - 1$ and $0 \le m \le (q-3)/2$,

$$A(k) = \sum_{n=0}^{N/q-1} x(qn + \frac{q-1}{2}) \text{cas} \frac{\pi(2n+1)k}{N/q} \tag{4.97}$$

$$B(k,m) = \sum_{n=0}^{N/q-1} [x(qn+m) + x(qn+q-1-m)] \text{cas} \frac{\pi(2n+1)k}{N/q} \tag{4.98}$$

$$C(k,m) = \sum_{n=0}^{N/q-1} [x(qn+q-1-m) - x(qn+m)] \text{cas} \frac{-\pi(2n+1)k}{N/q}, \tag{4.99}$$

Therefore, $A(k), B(k,m)$ and $C(k,m)$ are the length N/q type-II DHTs. It can be easily proven that $A(k), B(k,m)$ and $C(k,m)$ have the following properties:

$$A(k + \frac{iN}{q}) = (-1)^i A(k) \tag{4.100}$$

$$B(k + \frac{iN}{q}, m) = (-1)^i B(k,m) \tag{4.101}$$

$$C(k + \frac{iN}{q}, m) = (-1)^i C(k,m) \tag{4.102}$$

for $0 \le k \le N/q - 1$ and $0 \le i \le q - 1$. By using the above properties, (4.96) can be written as

$$X(k + iN/q) = (-1)^i A(k) + \sum_{m=0}^{(q-3)/2} \left[y(k,m) \cos \frac{i\pi(2m+1)}{q} \right.$$
$$\left. + z(k,m) \sin \frac{i\pi(2m+1)}{q} \right], \tag{4.103}$$

where $0 \le k \le N/q - 1$, $0 \le i \le q - 1$, and

$$y(k,m) = B(k,m) \cos \frac{\pi(q-1-2m)k}{N} + C(k,m) \sin \frac{\pi(q-1-2m)k}{N} \tag{4.104}$$

$$z(y,m) = B(k,m) \sin \frac{\pi(q-1-2m)k}{N} - C(k,m) \cos \frac{\pi(q-1-2m)k}{N}. \tag{4.105}$$

The computation in (4.103) can be transformed into a length q type-III DCT, as shown in Appendix H. The entire length N DHT is therefore decomposed into N/q length q type-III DCTs and q length N/q type-II DHTs. If the subsequence length N/q in (4.97) – (4.99) also contains odd factors, the same decomposition approach can be recursively continued until the subsequence length becomes q or has no odd factor. The decomposition costs for the radix-q algorithm are:

- $(q-1)N/q$ additions for $B(k,m)$ and $C(k,m)$ in (4.98) and (4.99)

- $2(q-1)(N/q-1)$ multiplications and $(q-1)(N/q-1)$ additions for (4.104) and (4.105).

Therefore, the numbers of additions and multiplications required by the algorithm are

$$A_H^{II}(N) = qA_H^{II}(\frac{N}{q}) + (\frac{N}{q})A_C^{III}(q) + (q-1)(\frac{2N}{q} - 1) \tag{4.106}$$

$$M_H^{II}(N) = qM_H^{II}(\frac{N}{q}) + (\frac{N}{q})M_C^{III}(q) + 2(q-1)(\frac{N}{q} - 1), \tag{4.107}$$

where the superscripts and subscripts indicate the types of the DCT or DHT computation involved, respectively.

4.6.2 Algorithm for type-III DHT

The type-III DHT, defined in (4.3), can be decomposed into

$$x(nq + \frac{q-1}{2}) = \sum_{k=0}^{N-1} X(k)\text{cas}\frac{\pi(2n+1)k}{N/q} = \sum_{k=0}^{N/q-1} u(k)\text{cas}\frac{\pi(2n+1)k}{N/q}, \qquad (4.108)$$

where

$$u(k) = \sum_{i=0}^{q-1}(-1)^i X(k + \frac{iN}{q}), \quad 0 \le k \le \frac{N}{q} - 1. \qquad (4.109)$$

Similarly, we have

$$x(nq + m) = \sum_{k=0}^{N-1} X(k)\text{cas}\frac{\pi[(2nq+q)-(q-2m-1)]k}{N} \qquad (4.110)$$

$$x(nq + q - m - 1) = \sum_{k=0}^{N-1} X(k)\text{cas}\frac{\pi[(2nq+q)+(q-2m-1)]k}{N}, \qquad (4.111)$$

where for both (4.110) and (4.111), $0 \le n \le N/q - 1$. We further group $x(nq + m)$ and $x(nq + q - m - 1)$ into

$$F(n, m) = \frac{x(nq+q-m-1) + x(nq+m)}{2} \qquad (4.112)$$

$$= \sum_{k=0}^{N-1} X(k)\cos\frac{\pi(q-2m-1)k}{N}\text{cas}\frac{\pi(2n+1)k}{p}$$

$$= \sum_{k=0}^{p-1} v(k,m)\text{cas}\frac{\pi(2n+1)k}{p}, \qquad (4.113)$$

where

$$v(k) = \sum_{i=0}^{q-1}(-1)^i X(k + \frac{iN}{q})\cos\frac{\pi(q-2m-1)(k+iN/q)}{N} \qquad (4.114)$$

and

$$G(n, m) = \frac{x(nq+q-m-1) - x(nq+m)}{2}$$

$$= \sum_{k=0}^{N-1} X(k)\sin\frac{\pi(q-2m-1)k}{N}\text{cas}\frac{-\pi(2n+1)k}{p}$$

$$= \sum_{k=0}^{p-1} w(k,m)\text{cas}\frac{-\pi(2n+1)k}{p}, \qquad (4.115)$$

where

$$w(k) = \sum_{i=0}^{q-1} (-1)^i X(k + \frac{iN}{q}) \sin \frac{\pi(q - 2m - 1)(k + iN/q)}{N}. \tag{4.116}$$

The final transform outputs are achieved by (4.108) and

$$\begin{aligned}
x(nq + q - m - 1) &= F(n, m) + G(n, m) \\
x(nq + m) &= F(n, m) - G(n, m).
\end{aligned} \tag{4.117}$$

For (4.113), (4.115) and (4.117), $n = 0, 1, \ldots, p - 1$ and $m = 0, 1, \ldots, (q - 3)/2$. Therefore, the length N type-III DHT is decomposed into q length N/q type-III DHTs as shown in (4.108), (4.113) and (4.115). Furthermore (4.114) and (4.116) can be expressed by

$$v(k, m) = \cos \frac{\pi(q - 2m - 1)k}{N} A(k, m) + \sin \frac{\pi(q - 2m - 1)k}{N} B(k, m) \tag{4.118}$$

$$w(k, m) = \sin \frac{\pi(q - 2m - 1)k}{N} A(k, m) - \cos \frac{\pi(q - 2m - 1)k}{N} B(k, m), \tag{4.119}$$

where for $m = 0, 1, \ldots, (q - 3)/2$,

$$A(k, m) = \sum_{i-0}^{q-1} X(k + \frac{iN}{q}) \cos \frac{i\pi(2m + 1)}{q} \tag{4.120}$$

$$B(k, m) = \sum_{i-0}^{q-1} X(k + \frac{iN}{q}) \sin \frac{i\pi(2m + 1)}{q}. \tag{4.121}$$

Appendix I shows that the combination of (4.109), (4.120) and (4.121) can be transformed into a length q type-II DCT. Therefore, the entire type-III DHT computation is decomposed into q length N/q type-III DHTs and N/q length q type-II DCTs. Figure 4.12 shows the signal flow graph for a length 15 type-III DHT computation. The decomposition costs are

- $2(q - 1)(N/q - 1)$ multiplications and $(q - 1)(N/q - 1)$ additions for (4.119)

- $(q - 1)N/q$ additions in (4.117).

Therefore, the number of additions required by the radix-q algorithm for a type-III DHT is

$$A_H^{III}(N) = q A_H^{III}(\frac{N}{q}) + \frac{N}{q} A_C^{II}(q) + (q - 1)(\frac{2N}{q} - 1) \tag{4.122}$$

and the number of multiplications is

$$M_H^{III}(N) = q M_H^{III}(\frac{N}{q}) + \frac{N}{q} M_C^{II}(q) + 2(q - 1)(\frac{N}{q} - 1). \tag{4.123}$$

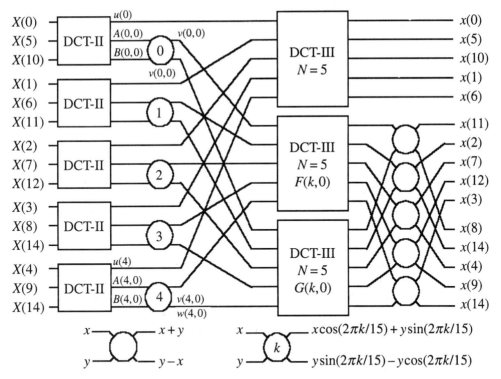

Figure 4.12   Signal flow graph of a length 15 type-III DHT.

4.6.3   Algorithm for type-IV DHT

The type-IV DHT of sequence $x(n)$ ($n = 0, 1, \ldots, N-1$) is defined by (4.4). It can be rewritten as

$$
\begin{aligned}
X(k) = {} & \sum_{n=0}^{N/q-1} x(qn + \frac{q-1}{2})\mathrm{cas}\frac{\pi(2n+1)(2k+1)}{2N/q} \\
& + \sum_{m=0}^{(q-3)/2} \left[\sum_{n=0}^{N/q-1} x(qn+m)\mathrm{cas}\frac{\pi[q(2n+1)-(q-1-2m)](2k+1)}{2N} \right. \\
& \left. + \sum_{n=0}^{N/q-1} x(qn+q-m-1)\mathrm{cas}\frac{\pi[q(2n+1)+(q-1-2m)](2k+1)}{2N} \right] . \quad (4.124)
\end{aligned}
$$

the second and third terms in (4.124) into

$$
X(k) = A(k) + \sum_{m=0}^{(q-3)/2} \left[B(k,m) \cos\frac{\pi\alpha(2k+1)}{2N} + C(k,m) \sin\frac{\pi\alpha(2k+1)}{2N} \right], \quad (4.125)
$$

where $\alpha = q - 2m - 1$ and

$$
\begin{aligned}
A(k) &= (-1)^i A(k + \frac{iN}{q}) \\
&= \sum_{n=0}^{N/q-1} x(qn + \frac{q-1}{2}) \text{cas} \frac{\pi(2n+1)(2k+1)}{2N/q}
\end{aligned}
\tag{4.126}
$$

$$
\begin{aligned}
B(k,m) &= (-1)^i B(k + \frac{iN}{q}) \\
&= \sum_{n=0}^{N/q-1} [x(qn + m) + x(qn + q - m - 1)] \text{cas} \frac{\pi(2n+1)(2k+1)}{2N/q}
\end{aligned}
\tag{4.127}
$$

$$
\begin{aligned}
C(k,m) &= (-1)^i C(k + \frac{iN}{q}) \\
&= \sum_{n=0}^{N/q-1} [x(qn + m) - x(qn + q - m - 1)] \text{cas} \frac{-\pi(2n+1)(2k+1)}{2N/q}
\end{aligned}
\tag{4.128}
$$

$$
0 \le m \le \frac{q-3}{2}, \quad 0 \le k \le p-1, \quad 0 \le i \le q-1,
$$

which are length N/q type-IV DHTs. Furthermore, (4.125) can be calculated by

$$
\begin{aligned}
X(k + \frac{iN}{q}) &= (-1)^i A(k) + \sum_{m=0}^{(q-3)/2} \Big[B(k,m) \cos \frac{\pi\alpha(2k + 2iN/q + 1)}{2N} \\
&\quad + C(k,m) \sin \frac{\pi\alpha(2k + 2iN/q + 1)}{2N} \Big]
\end{aligned}
\tag{4.129}
$$

$$
0 \le k \le N/q - 1, \quad 0 \le i \le q - 1,
$$

which can be rewritten as

$$
\begin{aligned}
X(k + \frac{iN}{q}) &= (-1)^i A(k) + \sum_{m=0}^{(q-3)/2} \Big[D(k,m) \cos \frac{\pi(2m+1)i}{q} \\
&\quad + E(k,m) \sin \frac{\pi(2m+1)i}{q} \Big],
\end{aligned}
\tag{4.130}
$$

where

$$
D(k,m) = B(k,m) \cos \frac{\pi\alpha(2k+1)}{2N} + C(k,m) \sin \frac{\pi\alpha(2k+1)}{2N}
\tag{4.131}
$$

$$
E(k,m) = B(k,m) \sin \frac{\pi\alpha(2k+1)}{2N} - C(k,m) \cos \frac{\pi\alpha(2k+1)}{2N}.
\tag{4.132}
$$

Appendix H can be used to show that (4.130) can be transformed into a length q type-III DCT. Therefore the length N type-IV DHT is decomposed into q length N/q type-IV DHTs and N/q length q type-III DCTs. Figure 4.13 shows the signal flow graph for a length 15 type-IV DHT. In general, the decomposition costs are

- $2(q-1)N/q$ multiplications and $(q-1)N/q$ additions for (4.131) and (4.132).

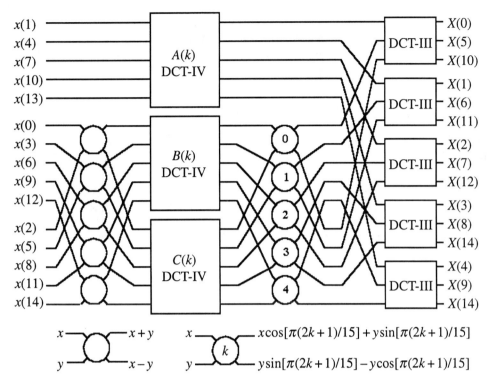

$$x \text{---} \overline{} \text{---} x+y \qquad x \text{---}$$

Figure 4.13   Signal flow graph of a length 15 type-IV DHT.

- $(q-1)N/q$ additions for (4.127) and (4.128).

The required total numbers of arithmetic operations are

$$M_H^{IV}(N) = qM_H^{IV}\left(\frac{N}{q}\right) + \frac{N}{q}M_C^{III}(q) + \frac{2(q-1)N}{q} \qquad (4.133)$$

$$A_H^{IV}(N) = qA_H^{IV}\left(\frac{N}{q}\right) + \frac{N}{q}A_C^{III}(q) + \frac{2(q-1)N}{q}. \qquad (4.134)$$

4.6.4   Computational complexity

Computation for $N = 3^m$

For $N = 3^m$, the type-II DHT requires a length 3 type-III DCT which uses 1 multiplication and 4 additions. However, the multiplication used in the DCT computation can be absorbed by the multiplications needed for twiddle factors in (4.104). Therefore, the number of multiplications needed by the radix-q algorithm for a length 3^m type-II DHT is

$$M_H^{II}(N) = 3M_H^{II}\left(\frac{N}{3}\right) + \frac{4N}{3} - 3. \qquad (4.135)$$

Similarly, for the length 3^m type-III DHT, the multiplication needed by the length 3 type-II DCT can also be combined with the twiddle factors in (4.119) for $k > 0$. Thus, the number of multiplications for the type-III DHT is

$$M_H^{III}(N) = 3M_H^{III}\left(\frac{N}{3}\right) + \frac{4N}{3} - 3. \qquad (4.136)$$

A similar arrangement can also be made to minimize the number of multiplications for a length 3^m type-IV DHT. For example, the multiplication needed by $\sin[\pi(2m+1)i/q]$ in (4.130) can be absorbed by the twiddle factor in (4.132). The number of multiplications needed by the radix-q algorithm is

$$M_H^{IV}(N) = 3M_H^{IV}(\frac{N}{3}) + \frac{4N}{3}. \tag{4.137}$$

The number of additions needed by the type-II, -III and -IV DHTs can be calculated by (4.106), (4.122) and (4.134) for $q = 3$.

Computation for $N = 5^m$

For $N = 5^m$, the radix-q algorithms require length 5 type-III and type-II DCTs which generally need 5 multiplications and 13 additions. A similar arrangement as shown for $N = 3^m$ can be made to save two multiplications. Therefore, the number of multiplications for the type-II DHT is

$$M_H^{II}(N) = 5M_H^{II}(\frac{N}{5}) + \frac{11N}{5} - 6 \tag{4.138}$$

for the type-III DHT is

$$M_H^{III}(N) = 5M_H^{III}(\frac{N}{5}) + \frac{11N}{5} - 6 \tag{4.139}$$

and for the type-IV DHT is

$$M_H^{IV}(N) = 5M_H^{IV}(\frac{N}{5}) + \frac{11N}{5}. \tag{4.140}$$

Again, the number of additions needed by the type-II, -III and -IV DHTs can be calculated by (4.106), (4.122) and (4.134) for $q = 5$.

Table 4.8   Computational complexity for $N = 2^m$.

| | Presented | | | | | | Algorithm in [17] | | | | | |
| | Type-III | | | Type-IV | | | Type-III | | | Type-IV | | |
N	M	A	Total	M	A	Total	M	A	Total	M	A	Total
4	2	6	8	4	12	16	2	10	12	6	10	16
8	6	26	32	12	28	40	8	28	36	16	32	48
16	18	70	88	28	84	112	24	76	100	40	88	128
32	46	186	232	68	204	272	64	196	260	96	224	320
64	114	454	568	156	500	656	160	484	644	224	544	768
128	270	1082	1352	356	1164	1520	384	1156	1540	512	1280	1792
256	626	2502	3128	736	2676	3412	896	2692	3588	1152	2944	4096

Comparison

Our main objective is to support DHT computation of various sequence lengths without increasing the computational complexity in comparison with other reported fast algorithms. In general, comparisons on the required computational complexity should be made on the basis of the same sequence lengths. However, this is not possible because other fast algorithms supporting composite sequence lengths are not widely reported in the literature. Instead, we first compare the computational complexity required by both the presented and other algorithms for the sequence length $N = 2^m$. By using the computational complexity for $N = 2^m$ as a reference, we make further comparisons on the required computational complexity for various sequence lengths.

Table 4.8 lists the number of arithmetic operations needed by the presented algorithms and that reported in [17] for type-III and type-IV DHTs. When $N = 256$, for example, the presented algorithm requires 626 multiplications and 2502 additions for type-III DHT and 736 multiplications and 2676 additions for type-IV DHT compared to the algorithm in [17] requiring 896 multiplications and 2692 additions for type-III DHT and 1152 multiplications and 2944 additions for type-IV DHT. For $N = 3^m$ and 5^m, however, the radix-*q* algorithms generally require more operations than that needed for $N = 2^m$, as shown in Tables 4.9 and 4.10.

Table 4.9   Computational complexity for type-III DHT.

$N = 3^m$				$N = 5^m$			
N	M	A	Total	N	M	A	Total
9	12	34	46	5	5	13	18
27	69	172	241	25	74	166	240
81	312	730	1042	125	639	1351	1990
243	1257	2836	4093	625	4564	9376	13940
729	4740	10450	15190	3125	29689	60001	89690

Table 4.10   Computational complexity for type-IV DHT.

$N = 3^m$				$N = 5^m$			
N	M	A	Total	N	M	A	Total
9	15	34	45	5	5	17	22
27	81	207	288	25	80	195	275
81	351	837	1188	125	675	1525	2200
243	1377	3159	4536	625	4750	10375	15125
729	5103	11421	1624	3125	30625	65625	96250

By jointly using the radix-*q* and the radix-2 algorithms, DHTs of arbitrarily composite sequence lengths can be efficiently computed. Table 4.11 lists the numbers of multiplications and additions needed for $N = 3 \cdot 2^m$ and $5 \cdot 2^m$. Figures 4.14 (a) and (b) compare the computational complexity required by the presented algorithms for type-III and -IV DHTs. It shows that for both type-III and -IV DHTs, more operations are needed for $N = 3^m$ and 5^m compared to that needed for $N = 2^m$. For the presented algorithm, the computational complexity for $N = 3 \cdot 2^m$ and $5 \cdot 2^m$ requires more operations than that needed for $N = 2^m$. Figures 4.14 (a) and 4.14 (b) compare the computational complexity needed by the presented and the other existing algorithms ([17], for example) for various sequence lengths.

The radix-*q* algorithms generally have a more complex computational structure than the radix-2 algorithm. In particular, the radix-*q* algorithms differ from the prime factor algo-

Table 4.11   Computational complexity for type-III and -IV DHTs.

	$N = 3 \cdot 2^m$					$N = 5 \cdot 2^m$			
	Type-III		Type-IV			Type-III		Type-IV	
$N =$	M	A	M	A	M	A	M	A	M
6	2	20	8	26	5	5	17	5	17
12	10	58	16	64	10	10	44	20	54
24	26	146	44	164	20	30	118	40	128
48	70	358	100	388	40	70	286	100	316
96	170	842	236	908	80	170	682	220	732
192	406	1942	532	2068	160	390	1574	500	1684
384	938	4394	1196	4652	320	890	3578	1100	3788
768	2134	9814	2644	10328	640	1990	8006	2420	8436
1536	4776	21674	5804	22700	1280	4410	17722	5260	18572
3072	10580	47446	12624	49492	2560	9670	38854	11380	40564

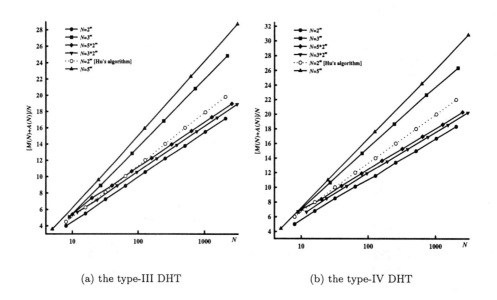

(a) the type-III DHT   (b) the type-IV DHT

Figure 4.14   Comparison of computational complexity.

rithm in that they allow the decomposition of sequence lengths containing any combination of odd factors. The prime factor algorithm can only be used when the factors are mutually prime. In-place computation can also be achieved because the memory space for input data can be reused by the partially processed data, as shown in Figures 4.12 and 4.13. Because of the close relation between the generalized DFT of real data and the DHT [3, 38], the algorithm can also be used to reduce the computational complexity for the generalized DFT computation of composite sequence lengths.

4.7   FAST ALGORITHMS USING TYPE-I DHT

This section presents another type of fast algorithm for type-II, -III and -IV DHTs based on the relationships among the different types of DHTs. The type-II and -III DHTs are decomposed into two length $N/2$ type-I DHTs and the type-IV DHT is converted into two length $N/2$ type-II or type -III DHTs. The algorithms achieve a simple computational structure and support a wide range of sequence lengths. The sequence length is assumed again to be $N = q \cdot 2^m$ throughout this subsection, where q is an odd positive integer.

4.7.1   Algorithm for type-II DHT

Before we decompose the type-II DHT, let us rearrange the input sequence $x(n)$ by

$$y(n) = \begin{cases} x(n), & 0 \leq n < N \\ x(n - N), & n > N - 1 \\ x(N + n), & n < 0 \end{cases} .$$
(4.141)

Based on (4.141), the type-II DHT of sequence $x(n)$ defined in (4.2) can be computed by

$$\begin{aligned} X(k) &= \sum_{n=0}^{N/2-1} y(2n + \frac{q-1}{2})\mathrm{cas}\frac{\pi(4n+q)k}{N} + \sum_{n=0}^{N/2-1} y(2n - \frac{q+1}{2})\mathrm{cas}\frac{\pi(4n-q)k}{N} \\ &= \cos\frac{\pi qk}{N}A(k) + \sin\frac{\pi qk}{N}B(k) \end{aligned}$$
(4.142)

$$X(k + \frac{N}{2}) = (-1)^{(q+1)/2}\left[\sin\frac{\pi pk}{N}A(k) - \cos\frac{\pi pk}{N}B(k)\right],$$
(4.143)

where for (4.142) and (4.143), $0 \leq k < N/2$, and

$$A(k) = A(k + \frac{N}{2}) = \sum_{n=0}^{N/2-1}\left[y(2n + \frac{q-1}{2}) + y(2n - \frac{q+1}{2})\right]\mathrm{cas}\frac{2\pi nk}{N/2}$$
(4.144)

$$B(\frac{N}{2} - k) = B(N - k) = \sum_{n=0}^{N/2-1}\left[y(2n + \frac{q-1}{2}) - y(2n - \frac{q+1}{2})\right]\mathrm{cas}\frac{2\pi nk}{N/2},$$
(4.145)

where $B(N/2) = B(N) = B(0)$. Both (4.144) and (4.145) are the length $N/2$ type-I DHTs that can be computed by the fast algorithms presented in Sections 4.2 and 4.3. When $N = 2q$, the required number of multiplications is

$$M_{II}(N) = 2M_I(\frac{N}{2}).$$
(4.146)

For $N > 2q$, the number of additions needed by such a decomposition is

$$A_{II}(N) = 2A_I(\frac{N}{2}) + 2N - 2q$$
(4.147)

and the number of multiplications is

$$M_{II}(N) = 2M_I(\frac{N}{2}) + 2N - 6q.$$
(4.148)

It can be seen that this algorithm is derived by using the decimation-in-time decomposition technique.

4.7.2   Algorithm for type-III DHT

We use the decimation-in-frequency decomposition technique for the type-III DHT. For example, the transform outputs defined in (4.3) can be expressed by

$$x[(2n + \frac{q-1}{2}) \bmod N] = \sum_{k=0}^{N-1} X(k) \cos \frac{\pi(4n+q)k}{N} \tag{4.149}$$

$$x[(N + 2n - \frac{q+1}{2}) \bmod N] = \sum_{k=0}^{N-1} X(k) \cos \frac{\pi(4n-q)k}{N}, \tag{4.150}$$

where $0 \leq n < N/2$. We combine (4.149) and (4.150) to form new sequences

$$
\begin{aligned}
f(n) &= \frac{x\{[n + (q-1)/2] \bmod N\} + x\{[N + n - (q+1)/2] \bmod N\}}{2} \\
&= \sum_{k=0}^{N/2-1} \left[X(k) \cos \frac{\pi qk}{N} + (-1)^{(q+1)/2} X(k + \frac{N}{2}) \sin \frac{\pi qk}{N} \right] \mathrm{cas} \frac{2\pi nk}{N/2} \tag{4.151}
\end{aligned}
$$

$$
\begin{aligned}
g(\frac{N}{2} - n) &= \frac{x\{[n + (q-1)/2] \bmod N\} - x\{[N + n - (q+1)/2] \bmod N\}}{2} \\
&= \sum_{k=0}^{N/2-1} \left[X(k) \sin \frac{\pi qk}{N} - (-1)^{(q+1)/2} X(k + \frac{N}{2}) \cos \frac{\pi qk}{N} \right] \mathrm{cas} \frac{2\pi nk}{N/2}, \tag{4.152}
\end{aligned}
$$

where $0 \leq n < N/2$. Both (4.151) and (4.152) are length $N/2$ type-I DHTs. The final outputs of the type-III DHT are obtained by

$$x \left[(2n + \frac{q-1}{2}) \bmod N \right] = f(n) + g(n) \tag{4.153}$$

$$x \left[(N + 2n - \frac{q+1}{2}) \bmod N \right] = f(n) - g(n), \tag{4.154}$$

where $0 \leq n < N/2$. In fact, the algorithm for the type-III DHT is based upon the decimation-in-frequency decomposition, and the algorithm for the type-II DHT is based on the decimation-in-time decomposition. It can be verified that the fast algorithms for both the type-II and -III DHTs require the same number of arithmetic operations. Therefore, (4.146) – (4.148) can be used to calculate the number of arithmetic operations for the type-III DHT.

4.7.3   Algorithm for type-IV DHT

Based on the decimation-in-frequency decomposition, the type-IV DHT of sequence $x(n)$ defined in (4.4) can be grouped into

$$
\begin{aligned}
F(k) &= \frac{X[2k + (q-1)/2] + X[N + 2k - (q+1)/2]}{2} \\
&= \sum_{n=0}^{N/2-1} \left[x(n) \cos \frac{\pi q(2n+1)}{2N} + (-1)^{(q+1)/2} x(n + \frac{N}{2}) \sin \frac{\pi q(2n+1)}{2N} \right] \\
&\quad \cdot \mathrm{cas} \frac{\pi(2n+1)k}{N/2} \tag{4.155}
\end{aligned}
$$

$$G(k) = \frac{X[2k + (q-1)/2] - X[2k - (q+1)/2]}{2}$$

$$= \sum_{n=0}^{N/2-1} \left[x(n) \sin \frac{\pi q(2n+1)}{2N} - (-1)^{(q+1)/2} x(n + \frac{N}{2}) \cos \frac{\pi q(2n+1)}{2N} \right]$$

$$\cdot \operatorname{cas} \frac{-\pi(2n+1)k}{N/2}. \tag{4.156}$$

The final outputs of the DHT are obtained from

$$X\left(2k + \frac{q-1}{2}\right) = F(k) + G(k), \quad X\left(2k - \frac{q+1}{2}\right) = F(k) - G(k), \tag{4.157}$$

where $0 \le k < N/2$. When $2k + (q-1)/2 \le N$ and $2k - (q+1)/2 < 0$, the transform outputs are achieved by

$$X(t) = \begin{cases} -X(t-N), & t \ge N \\ -X(t+N), & t < 0 \end{cases}. \tag{4.158}$$

Therefore, the type-IV DHT is decomposed into two length $N/2$ type-II DHTs. An equivalent decimation-in-time algorithm can be used to partition the type-IV DHT into two length $N/2$ type-III DHTs.

The required number of additions for such a decomposition is

$$A_{IV}(N) = 2A_{II}(\frac{N}{2}) + 2N, \quad N \ge q \tag{4.159}$$

and the number of multiplications is

$$M_{IV}(N) = 2M_{II}(\frac{N}{2}) + 2N, \quad N > q. \tag{4.160}$$

When $N = 2q$, $2q$ multiplications can be saved from trivial twiddle factors.

4.8   CHAPTER SUMMARY

This chapter deals with various fast algorithms for the four types of 1D DHTs. The generalized split-radix algorithms are simple in concept and can be easily implemented. With the odd factor q used in the sequence length, the split-radix algorithms can reduce the number of arithmetic operations needed by the twiddle factors. Our PFAs have made important improvements to simplify the procedure of the index mapping and make in-place computation possible. The radix-q algorithms can decompose the DHT computation, which is very useful when the factors of the sequence length are not mutually prime. It has been demonstrated that for an arbitrary composite length, a combination of the split-radix algorithm and the PFA (or the radix-q algorithm) is necessary. It is also shown that the computational complexity and the implementation complexity can often be reduced. Most of the concepts or decomposition approaches can be extended for the MD DHTs, which is the main subject of the next chapter.

Appendix A. Computation of odd indexed outputs

For $N = 16$, the odd indexed transform outputs are computed as follows:

Procedure DHT16(var xx:data) (* for odd outputs of length 16 DHT*)

```
Var a, b, c, d : real; i :  integer; v, t:  data;
begin
for i:=0 to 7 do v[i] := x[i]-x[i+8];        (* 8 additions *)
t[0] := v[0]-v[4];    t[1] := 1.414*v[2];    (* 1 addition 1 mul *)
t[2] := 1.414*v[6];   t[3] := t[0]+t[2];     (* 1 addition 1 mul *)
t[4] := t[0]-t[2];    t[0] := v[0]+v[4];     (* 2 additions *)
t[1] := t[0]+t[1];    t[2] := t[0]-t[2];     (* 2 additions *)
t[10]:= v[1]+v[3];    t[11]:= v[5]-v[7];     (* 2 additions *)
t[12]:= v[1]-v[3];    t[13]:= v[5]+v[7];     (* 2 additions *)
t[5] :=0.765t[11];    t[7] := 1.306(t[10]+t[11]); (* 1 addition 2 muls*)
t[6] := 1.8478t[10];t[5] := t[7]-t[5];       (* 1 addition 1 mul *)
t[6] := t[6]-t[7];    t[9] := 1.306(t[12]+t[13]); (* 2 additions 1 mul*)
t[7] := 0.7654t[13];t[8] := 1.8478t[12];     (* 2 muls *)
t[7] := t[9]-t[7];    t[8] := t[9]-t[8];     (* 2 additions *)
x[1] := t[1]+t[5];    x[9] := t[1]-t[5];     (* 2 additions *)
x[5] := t[2]+t[6];    x[13]:= t[2]-t[6];     (* 2 additions *)
x[3] := t[3]+t[7];    x[11]:= t[3]-t[7];     (* 2 additions *)
x[7] := t[4]+t[8];    x[15]:= t[4]-t[8];     (* 2 additions *)
end;
```

where $1.8478 = [\mathrm{cas}(3\pi/8) + \mathrm{cas}(-\pi/8)], 0.7654 = [\mathrm{cas}(3\pi/8) - \mathrm{cas}(-\pi/8)]$ and $1.3065 = \mathrm{cas}(3\pi/8)$. In total, 8 multiplications and 32 additions are used. For a complete length 16 DHT, $62 (= 8 + 22 + 32)$ additions and $10 (2 + 8)$ multiplications are necessary.

Appendix B. Subroutines for split-radix DHT algorithm

This appendix presents a number of subroutines which are needed for the computation of length $3 \cdot 2^m$ (not length 2^m) DHTs. It is assumed that both the input and output sequences are presented in a natural order. Rotation operations are realized by using 4 multiplications and 2 additions. The algorithm is implemented by a recursive subroutine, i.e., procedure DHT ( ), with terminating subroutines DHT12( ) and DHT24( ). To test them, a main program is required to generate the input sequence and to call DHT( ). The comments in the procedures show the count of operations required by the length 12 and -24 DHTs.

```
type  data=array[0..2048] of real; (* Data array to be proceeded *)
temp=array[0..12] of real;
(*Data stores for length 12 and 24 DHT routines *)
(**********DATA REORDERING********************)
Procedure swap(var x:data; S:integer; E:integer);
(* This procedure works only for N=3*2**m*)
var N, N1, N2, i,j, k : integer; t : real;
begin N:=E-S;        (*Sequence length to be processed*)
if N>6 then
begin N1:=N div 2;    N2:=N1 div 2;
for i:=S+N2 to S+N1-1 do
begin k:= N2+i;   t:=x[i];   x[i]:=x[k];   x[k]:=t;   end;
swap(x, S, S+N1);   (*for the first half sequence*)
swap(x, S+N1, E);   (*for the second half sequence*)
end  else if N=6 then
```

```
begin i:=S+1;   k:=S+2;   t:=x[i];   x[i]:=x[k];   x[k]:=t;
j:=S+3;   k:=S+4;   t:=x[j];   x[j]:=x[k];   x[k]:=t;
t:=x[i];   x[i]:=x[k];          x[k]:=t;
end; end;
```

(********** length 12 FHT **************)

```
Procedure DHT12(Var x :  data; F : integer);(* 52 adds and 4 muls *)
Var i, j, k :  integer;   y, t :  temp;
begin for i:=0 to 5 do    (* butterfly computation *)
begin k:=i+F;   j:=k+6;  y[i]:=x[k]+x[j];
y[i+6]:=x[k]-x[j];end; (* 12 adds *)
for i:=0 to 2 do
begin j:=i+3; t[i]:=y[i]+y[j]; t[j]:=y[i]-y[j]; end;
(* 6 adds *)
t[6]:=0.366025404*(t[4]+t[5]);(* 1 add, 1 mul *)
t[7]:=0.366025404*(t[2]-t[1]);(* 1 add, 1 mul *)
x[F]:=t[0]+t[1]+t[2]; x[F+6]:=t[3]+t[5]-t[4]; (* 4 additions *)
x[F+2]:=t[3]+t[4]+t[6]; x[F+8]:=t[0]-t[1]+t[7]; (* 4 additions *)
x[F+4]:=t[0]-t[2]-t[7]; x[F+10]:=t[3]-t[5]-t[6]; (* 4 additions *)
t[0]:=y[6]+y[9];     t[3]:=y[6]-y[9]; (* 2 additions *)
t[1]:=y[7]+y[8];       t[4]:=y[7]-y[8]; (* 2 additions *)
t[2]:=y[10]+y[11]; t[5]:=y[10]-y[11]; (* 2 additions *)
t[6]:=0.366025404*(t[1]+t[5]); (* 1 add, 1 mul*)
t[7]:=0.366025404*(t[2]-t[4]); (* 1 add, 1 mul*)
x[F+1]:=t[0]+t[1]+t[6]; x[F+7]:=t[3]+t[7]-t[4]; (* 4 additions *)
x[F+3]:=t[3]+t[2]+t[4]; x[F+9]:=t[0]-t[1]+t[5]; (* 4 additions *)
x[F+5]:=t[0]-t[6]-t[5]; x[F+11]:=t[3]-t[7]-t[2];(* 4 additions *)
end;
```

(********** length 24 FHT ***************)

```
Procedure DHT24(Var x :  data; F : integer);(*120 adds and 12 mul*)
Var  i, j, k, n :  integer; y, t :  temp;
begin for  i:=0 to 11 do
begin j:=F+i;   k:=j+12;   t[0]:=x[j];
x[j]:=t[0]+x[k];     x[k]:=t[0]-x[k];             (* 24 additions *)
end; DHT12(x, F); (*call length 12 DHT, 52 adds and 4 mults *)
i:=F+12;   j:=i+1;   k:=i+5;   t[0]:=x[j]+x[k];
t[1]:=x[j]-x[k];             (* 2 additions *)
j:=i+7;     k:=i+11;   t[2]:=x[j]-x[k];
t[3]:=x[j]+x[k];             (* 2 additions *)
j:=i+3;     k:=i+9;
t[4]:=0.707106781*(t[2]+2.0*x[j]);   (* 1 add, 1 mult, 1 shift *)
t[5]:=0.707106781*(t[1]-2.0*x[k]);   (* 1 add, 1 mult, 1 shift *)
t[6]:=1.224744871*t[3];   t[7]:=1.224744871*t[0];    (* 2 muls *)
y[0]:=t[4]+t[7];          y[6] := -y[0];            (* 1 addition *)
y[1]:=1.414213562*(t[1]+x[k]);   y[7] := -y[1];     (* 1 add, 1 mult *)
y[2]:=t[7]-t[4];          y[8] := -y[2];            (* 1 addition *)
y[3]:=t[5]+t[6];          y[9] := -y[3];            (* 1 addition *)
y[4]:=1.414213562*(x[j]-t[2]);   y[10]:=-y[4];      (* 1 add, 1 mult*)
y[5]:=t[6]-t[5];          y[11]:= -y[5];            (* 1 addition *)
j:=i+6;     t[0]:=x[i]+x[j];     t[3]:=x[i]-x[j];   (* 2 additions *)
j:=i+2;     k:=i+4;             t[1] := x[j]+x[k];
t[4]:=x[j]-x[k];                               (* 2 additions *)
j:=i+8;   k:=i+10;   t[2]:= x[j]+x[k];   t[5]:=x[j]-x[k];(* 2 adds *)
t[6]:=0.366025404*(t[1]+t[5]);                 (* 1 add, 1 mult *)
t[7]:=0.366025404*(t[4]-t[2]);                 (* 1 add, 1 mult *)
y[12]:=t[0]+t[1]+t[6];     y[18] := y[12];     (* 2 additions *)
y[13]:=t[2]+t[3]+t[4];     y[19] := y[13];     (* 2 additions *)
y[14]:=t[0]-t[6]-t[5];     y[20] := y[14];     (* 2 additions *)
y[15]:=t[3]-t[4]-t[7];     y[21] := y[15];     (* 2 additions *)
y[16]:=t[0]-t[1]+t[5];     y[22] := y[16];     (* 2 additions *)
```

```
y[17]:=t[3]-t[2]+t[7];    y[23] := y[17];              (* 2 additions *)
for i:=0 to 11 do
begin j := 11-i; k:=F+j; n := k+j;
x[n]:=x[k];          x[n+1]:=y[j]+y[j+12];             (* 12 additions *)
end;   end;


(*********RECURSIVE DHT*********************)
Procedure FHT (Var x:data; F:integer; L:integer);
(* x the input data, F the first data of x,  L the last data of x *)
Var N, N1, N2, N3, i, m, m1, m2, m3 :  integer;  a, a1, a2: real;
begin N:=L-F+1;    a:=6.283185307/N;   (*N the size of data sequence *)
if N > 24 then
begin N1:=N div 2;  N2:=N1 div 2;   N3:=N2 div 2;
for   i:=0 to N1-1 do                   (* butterfly computation *)
begin m1:=i+F;       m:=m1+N1;      a2:=x[m1];
x[m1]:=a2+x[m];    x[m]:=a2-x[m];
end; m2:=F+N1;                          (* rotation operations *)
for i:=1 to N3-1 do
begin a1:=a*i;      m1:=i+m2;       m:=m1+N2;
a2:=x[m1]*cos(a1)-x[m]*sin(a1);
x[m]:=x[m1]*sin(a1)+x[m]*cos(a1);
x[m1]:=a2;          a1:=a*(i+N3);
m1:=m1+N3;          m:=m + N3;
a2:=x[m1]*cos(a1)-x[m]*sin(a1);
x[m]:=x[m1]*sin(a1)+x[m]*cos(a1);  x[m1]:=a2;
end; m1:=N3+m2;     m:=m1 +N2;
a2:=(x[m1]-x[m])*0.707106781;       x[m]:=(x[m1]+x[m])*0.707106781;
x[m1]:=a2;          m3:=F+N1;
m:=m3-1;            FHT(x, F, m);       (*Length (L-F+1)/2 DHT*)
m:=m+N2;            FHT(x, m3,m);       (*Length (L-F+1)/4 DHT*)
m:=m+1;             FHT(x, m, L);       (*Length (L-F+1)/4 DHT*)
m3:=F+N;
for i:=1 to (N2 div 2)-1 do
begin m1:=m+i;       a2:=x[m1];       m2:= m3-i;
x[m1]:=x[m2]; x[m2] := a2;
end; m1:=F+N1;     m2:=m1+N2; a2:=x[m1]; (* butterfly computation *)
x[m1]:=a2+x[m2];    a:=a2-x[m2];
m1:=m1+1;           m2:=m2-1;
for i:=m1 to m2 do
begin a2:=x[i];     m:=i+N2;
x[i]:=a2+x[m];    x[m-1]:=a2-x[m];
end; x[L]:=a;
swap(x,F+N1,m3);    swap(x,F,m3);       (* reordering data*)
end else if N=24 then   DHT24(x, F)     (* terminating the calls*)
else if N=12 then DHT12(x,F);           (* terminating the calls*)
end;
```

Appendix C. Computation of $C(n,k)$ and $S(n,k)$

For convenience, we assume $u(m) = x(m) + x(9-m)(m = 1,2,3,4)$ and define

$$C(k) = \sum_{m=0}^{8} x(m) \cos \frac{2\pi mk}{9}, \quad k = 1,2,4. \tag{C.1}$$

It can be calculated as follows. In total, 3 multiplications and 14 additions are needed.

$$\alpha = 2\pi/9; \qquad t1=x(0)-0.6u(3);$$
$$t2=u(4)-u(1); \qquad t4=u(2)-u(4);$$
$$t2=t2-t3; \qquad t3=t3-t4;$$
$$X(1)=t1+t2; \qquad X(2)=t1+t3;$$
$$X(4)=t1-t2-t3;$$

Similarly, if $v(m) = x(m) - x(9 - m)(m = 1, 2, 3, 4)$, then the sequence

$$S(k) = \sum_{m=0}^{8} x(m) \sin \frac{2\pi mk}{9}, \quad k = 1, 2, 4 \qquad (C.2)$$

can be computed by the following routine. In total, 4 multiplications and 13 additions are needed.

$$\alpha = 2\pi/9; \qquad t2=v(3)\sin 3\alpha;$$
$$t2=[v(2)+v(4)]\sin 2\alpha; \qquad t3=[v(1)-u(4)]\sin\alpha;$$
$$t4=[v(1)+v(2)]\sin 4\alpha; \qquad t2=t2+t3;$$
$$t3=t3+t4; \qquad X(1)=t1+t2;$$
$$X(2)=t3-t1; \qquad X(1)=t1+t2;$$
$$X(2)=t1+t3; \qquad X(4)=t1-t2+t3;$$

Appendix D. length 9 DHT computation

Procedure DHT9(var x : data1); (* 40 additions, 10 muls and 2 shifts *)
var t1, t2, t3, t4 : real; u, y : data1;

```
u[0] := x[0];              u[1] := x[1]+x[8];         u[2] := x[2]+x[7];
u[3] := x[3]+x[6];         u[4] := x[4]+x[5];         t3 := u[1]-u[4];
t4 := u[2]-u[4];           t1:= 0.76604 t3;           t2 := t1+0.17364 t4;
t1 := 0.17364 t3;          t3 := 0.93969 t1-t4;       t1 := -t2-t3;
t4 := x[0] + u[3];         y[3] := u[1]+u[2]+u[4];     x[0] := t4+y[3];
y[3] := t4-0.5 y[3];       t4 := u[0]-0.5 u[3];        y[1] := t2+t4;
y[2] := t3+t4;             y[4] := t1+t4;             u[1]:=x[1]-x[8];
u[2] := x[2]-x[7];         u[3] := x[3]-x[6];         u[4] := x[4]-x[5];
t3:= u[1]+u[2];            t4 := u[2]+u[4];           t1 := 0.64278 t3;
t2 := t1+0.34202 t4;       t1 := 0.98481 t3;          t3 := 0.64278 t1-t4;
t1 := t3-t2;               t4 := 0.86602 u[3];        x[3]:=0.86602 (u[1]-u[2]+u[4]);
x[1]:=t2+t4;               x[2] := t3-t4;             x[4] := t1+t4;
x[8] := y[1] - x[1];       x[1] := x[1] + y[1];       x[7] := y[2] - x[2];
x[2]:=x[2]+y[2];           x[6] := y[3] - x[3];       x[3] := x[3] + y[3];
x[5] := y[4] - x[4];       x[4] := x[4] + y[4];
end;
```

Appendix E. length 15 DHT computation

The length 15 DHT can be calculated by the following equations.

$$X(3k) = \sum_{n=0}^{4} [x(n) + x(n+5) + x(n+10)] \mathrm{cas} \frac{2\pi nk}{5} \tag{E.1}$$

$$
\begin{aligned}
F(k) &= \frac{X(3k+5) + X(3k-5)}{2} \\
&= \sum_{n=0}^{4} \left[x(n) \cos \frac{2\pi n}{3} + x(n+5) \cos \frac{2\pi(n+5)}{3} \right. \\
&\quad \left. + x(n+10) \cos \frac{2\pi(n+10)}{3} \right] \mathrm{cas} \frac{2\pi nk}{5}
\end{aligned}
\tag{E.2}
$$

$$
\begin{aligned}
G(k) &= \frac{X(3k+5) - X(3k-5)}{2} \\
&= \sum_{n=0}^{4} \left[x(n) \sin \frac{2\pi n}{3} + x(n+5) \sin \frac{2\pi(n+5)}{3} \right. \\
&\quad \left. + x(n+10) \sin \frac{2\pi(n+10)}{3} \right] \mathrm{cas} \frac{-2\pi nk}{5}.
\end{aligned}
\tag{E.3}
$$

Equations (E.1) – (E.3), $k = 0$ to 4, are length 5 DHTs. The transformed outputs are obtained by

$$x[(3k+5) \bmod 15] = F(k) + G(k), \quad x[(3k+10) \bmod 15] = F(k) - G(k). \tag{E.4}$$

In (E.2), the twiddle factors are either –0.5 or 1.0. Thus, no multiplications are needed. Furthermore, (E.3) is a scaled length 5 DHT since the twiddle factors are either 0.0 or 0.866. The length 5 DHT uses 17 additions and 5 multiplications. However, the optimized scaled length 5 DHT needs 3 extra multiplications. Therefore, the length 15 DHT requires 81 additions and 18 multiplications. To compute a scaled length 15 DHT, (E.1) and (E.2) require 6 extra multiplications. However, (E.3) needs no extra multiplication since the scaling factor for the length 15 DHT can be combined with that for the scaled length 5 DHT.

Appendix F. Proof

If p and q are relatively prime, (4.74), (4.78) and (4.79) can be transformed into a length p type-II DCT. According to the Chinese remainder theorem, we can always find a set of integers l so that for any n

$$l = (n + iq) \bmod p, \quad 0 \le l, i \le p - 1. \tag{F.1}$$

Therefore, we have

$$u(n, m) = \sum_{l=0}^{p-1} x(l) \cos \frac{\pi(2l+1)m}{p}, \quad 0 \le m \le \frac{p-1}{2} \tag{F.2}$$

$$v(n, m) = \sum_{l=0}^{p-1} x(l) \sin \frac{\pi(2l+1)m}{p}, \quad 1 \le m \le \frac{p-1}{2}. \tag{F.3}$$

It can be shown that (F.2) corresponds to the computation associated with even rows and (F.3) corresponds to the computation associated with odd rows of the kernel matrix of the type-II DCT. The effect of variable n in (F.1) is to change the column indices of the DCT kernel matrix or equivalently to change the original input order.

Appendix G. Proof

Similar to the proof in Appendix F, according to the Chinese remainder theorem, the second term in (4.90) can be converted into

$$\sum_{m=1}^{(p-1)/2} b(l, m) \cos \frac{\pi(2l+1)2m}{2p}. \tag{G.1}$$

where l is defined in (F.1). It can be seen that (G.1) corresponds to the even columns of the type-III DCT kernel matrix. Similarly, we have

$$\sum_{m=1}^{(p-1)/2} c(l, m) \sin \frac{\pi(2l+1)2m}{2p} = \sum_{m=1}^{(p-1)/2} c(l, m)(-1)^l \cos \frac{\pi(2l+1)(p-2m)}{2p}, \tag{G.2}$$

which corresponds to the odd columns of the type-III DCT kernel matrix. The multiplication factor by $(-1)^l$ in (G.2) can be eliminated by changing the order of the matrix columns.

Appendix H. Derivation of type-III DCT

The length q type-III DCT is defined by

$$X(t) = \sum_{m=0}^{q-1} x(m) \cos \frac{\pi(2t+1)m}{2q}, \quad 0 \le t \le q-1 \tag{H.1}$$

which can be rewritten into

$$X(t + \frac{q-1}{2}) = \sum_{m=0}^{q-1} x(m) \cos \frac{\pi(2t+q)m}{2q}$$

$$= x(0) + (-1)^t \sum_{m=1}^{(q-1)/2} (-1)^m x(2m) \cos \frac{\pi(q-2m)t}{q}$$

$$+ \sum_{m=1}^{(q-1)/2} (-1)^m x(2m+1) \sin \frac{\pi(2m+1)t}{q}, \tag{H.2}$$

where $0 \leq t \leq (q-1)/2$. If we let $m = (q - 2n - 1)/2$ $(m = 1, \ldots, (q-1)/2)$ and $(n = 0, 1, \ldots, (q-3)/2)$, the second terms of (H.2) can be changed so that it becomes

$$x\left[\frac{q-1}{2} + t\right] = x(0) + (-1)^t \sum_{n=0}^{(q-3)/2} (-1)^{(q-2n-1)/2} x(q - 2n - 1) \cos \frac{\pi(2n+1)t}{q}$$

$$+ \sum_{m=0}^{(q-3)/2} (-1)^m x(2m+1) \sin \frac{\pi(2m+1)t}{q}. \tag{H.3}$$

Similarly, we have

$$x\left[\frac{q-1}{2} - t\right] = \sum_{m=0}^{q-1} x(m) \cos \frac{\pi(q - 2t)m}{2q}$$

$$= x(0) + (-1)^t \sum_{m=1}^{(q-1)/2} (-1)^m x(2m) \cos \frac{\pi(q - 2m)t}{q}$$

$$+ \sum_{m=1}^{(q-1)/2} (-1)^m x(2m+1) \sin \frac{\pi(2m+1)t}{q}. \tag{H.4}$$

In (H.3) and (H.4), the second term corresponds to the terms related to $x(2m)$ and the last term is associated with $x(2m+1)$. The format of both (H.3) and (H.4) is the same as that in (4.103) or (4.125) if the difference in sign (i.e., $(-1)^m$ and $(-1)^t$) is ignored. These sign differences can be easily dealt with at the input and output of the type-III DCT. Therefore, the computation of (4.103) and (4.125) can be performed by the type-III DCT.

Example: When $q = 5$, (4.103) becomes

$$X(k + \frac{iN}{5}) = (-1)^i A(k) + z(k,0) \sin \frac{i\pi}{5} + y(k,1) \cos \frac{3i\pi}{5} \tag{H.5}$$

$$+ z(k,1) \sin \frac{3i\pi}{5} + y(k,0) \cos \frac{i\pi}{5}$$

$$0 \leq k \leq \frac{N}{5} - 1, \quad 0 \leq i \leq 4, \tag{H.6}$$

which can be computed by

$$\begin{vmatrix} X(k + 2N/5) \\ X(k + 4N/5) \\ X(k) \\ -X(k + N/5) \\ -X(k + 3N/5) \end{vmatrix} = T_{DCT} \begin{vmatrix} A(k) \\ z(k,0) \\ -y(k,1) \\ -z(k,1) \\ y(k,0) \end{vmatrix}, \tag{H.7}$$

where $0 \leq k \leq N/5 - 1$ and T_{DCT} is the type-III kernel matrix.

Appendix I. Derivation of type-II DCT

The length q type-II DCT of sequence $x(i)$ is defined by

$$X(m) = \sum_{i=0}^{q-1} x(i)\cos\frac{\pi(2i+1)m}{2q}, \quad 0 \le m \le q-1. \tag{I.1}$$

If we let $a = q - 2m - 1$, (I.1) can be rewritten as

$$X(q - 2m - 1)$$

$$= \sum_{i=1}^{(q-1)/2} x(\frac{q-1}{2} - i) \cos\frac{\pi(q-2i)a}{2q} + \sum_{i=0}^{(q-1)/2} x(i+\frac{q-1}{2}) \cos\frac{\pi(q+2i)a}{2q}$$

$$= (-1)^{m+(q-1)/2} \left[\sum_{i=1}^{(q-1)/2} x(\frac{q-1}{2} - i) \cos\frac{\pi i a}{q} + \sum_{i=0}^{(q-1)/2} x(i+\frac{q-1}{2}) \cos\frac{\pi i a}{q} \right]$$

$$= (-1)^{m+(q-1)/2} x(\frac{q-1}{2})$$

$$+ (-1)^{m+(q-1)/2} \sum_{i=1}^{(q-1)/2} (-1)^i \left[x(\frac{q-1}{2} - i) + x(\frac{q-1}{2} + i) \right] \cos\frac{\pi i(2m+1)}{q}$$

$$0 \le m \le \frac{q-1}{2}, \tag{I.2}$$

which has the same format as that of (4.120) for each particular k if the differences in sign are ignored. Similarly,

$$X(2m + 1) = \sum_{i=1}^{(q-1)/2} x(\frac{q-1}{2} - i) \cos\frac{\pi(q-2i)(2m+1)}{2q}$$

$$+ \sum_{i=0}^{(q-1)/2} x(i+\frac{q-1}{2}) \cos\frac{\pi(q+2i)(2m+1)}{2q}$$

$$= (-1)^m \sum_{i=1}^{(q-1)/2} \left[x(\frac{q-1}{2} - i) - x(i+\frac{q-1}{2}) \right] \sin\frac{\pi i(2m+1)}{q} \tag{I.3}$$

$$0 \le m \le \frac{q-3}{2},$$

which is equivalent to (4.121) if the differences in sign are ignored. In fact, (4.120) and (4.121) are related to the computation corresponding to the even and odd rows of the type-II DCT kernel matrix, respectively. If the multiplication by (-1) for some terms in (I.2) and (I.3) are properly considered, (4.109), (4.120) and (4.121) form a length q type-II DCT.

REFERENCES

1. N. Anupindi, S. B. Narayanan and K. M. M. Prabhu, New radix-3 algorithm, *Electron. Lett,* vol. 26, no. 18, 1537–1539, 1990.

2. G. Bi, New split-radix algorithm for the discrete Hartley transform, *IEEE Trans. Signal Process.,* vol. 45, no. 2, 297–302, 1997.

3. G. Bi and Y. Chen, Fast generalised DFT and DHT algorithms, *Signal Processing,* vol. 65, 383–390, 1998.

4. G. Bi, On computation of the discrete W transform, *IEEE Trans. Signal Process.,* vol. 47, no. 5, 1450–1453, 1999.

5. G. Bi and J. Liu, Fast odd-factor algorithm for discrete W transform, *Electron. Lett.,* vol. 34, no. 5, 431–433, 1998.

6. G. Bi and C. Lu, Prime factor algorithms for generalised discrete Hartley transform, *Electron. Lett.,* vol. 35, no. 20, 1708–1710, 1999.

7. G. Bi, Y. Chen and Y. Zeng, Fast algorithms for generalised discrete Hartley transform of composite sequence lengths, *IEEE Trans. Signal Process.,* vol. 47, no. 9, 893–901, 2000.

8. G. Bi, Fast algorithms for generalised discrete Hartley transform, *Circuits, Systems, Computers,* vol. 10, no. 1, 77–83, 2000.

9. G. E. J. Bold, A comparison of the time involved in computing fast Hartley and fast Fourier transforms, *Proceedings of IEEE,* vol. 73, no. 12, 1863–1864, 1985.

10. S. C. Chan and K. L. Ho, Polynomial transform fast Hartley transform, *IEEE Int. Sym. Circuits and System,* 642–645, 1991.

11. Y. H. Chan and W. C. Siu, New fast discrete Hartley transform algorithm, *Electron. Lett.,* vol. 27, no. 4, 347–349, 1991.

12. P. Duhamel and M. Vetterli, Improved Fourier and Hartley transform algorithm: application to cyclic convolution of real data, *IEEE Trans. Acoust., Speech, Signal Process.,* vol. ASSP-35, no. 6, 818–824, 1987.

13. D. M. W. Evans, An improved digit-reversal permutation algorithm for the fast Fourier and Hartley transforms, *IEEE Trans. Acoust., Speech, Signal Process.,* 1120–1125, 1987.

14. I. J. Good, The relationship between two fast Fourier transforms, *IEEE Trans. Computers,* vol. C-20, 310–317, 1971.

15. M. T. Heideman, C. S. Burrus and H. W Johnson, Prime-factor FFT algorithm for real-valued series, *ICASSP,* 28A.7.1–28A.7.4, 1984.

16. H. S. Hou, The fast Hartley transform algorithm, *IEEE Trans. Computers,* vol. C-36, no. 2, 147–156, 1987.

17. N. C. Hu, H. I. Chang and O. K. Ersoy, Generalized discrete Hartley transforms, *IEEE Trans. Signal Process.,* vol. 40, no. 12, 2931–2940, 1992.

18. E. A. Jonckheere and C. W. Ma, Split-radix fast Hartley transform in one and two dimensions, *IEEE Trans. Signal Process.*, vol. 39, no. 2, 499–503, 1991.

19. J. C. Liu and T. P. Lin, Short-time Hartley transform, *IEE Proceedings-F*, vol. 140, no. 3, 171–174, 1993.

20. D. Liu, D. Wang and Z. Wang, Interpolation using Hartley transform, *Electron. Lett.*, vol. 28, no. 2, 209–210, 1992.

21. D. P. K. Lun and W. C. Siu, Fast radix-3/9 discrete Hartley transform, *IEEE Trans. Signal Process.*, vol. 41, no. 7, 2494–2499, 1993.

22. D. P. K. Lun and Wan-Chi Siu, On prime factor mapping for the discrete Hartley transform, *IEEE Trans. Signal Process.*, vol. 40, no. 6, 1399–1411, 1992.

23. H. S. Malvar, Fast computation of the discrete cosine transform and the discrete Hartley transform, *IEEE Trans. Acoust., Speech, Signal Process.*, vol. ASSP-35, no. 10, 1484–1485, 1987.

24. H. S. Malvar, *Signal Processing with Lapped Transforms*, Artech House, London, 1992.

25. P. K. Meher, J. K. Satapathy and G. Panda, New high-speed prime-factor algorithm for discrete Hartley transform, *IEE Proceedings-F*, vol. 140, no. 1, 63–70, 1993.

26. R. P. Millane, Analytic properties of the Hartley transform and their implications, *Proceedings of IEEE*, vol. 82, no. 3, 413–428, 1994.

27. H. J. Nussbaumer, *Fast Fourier Transform and Convolution Algorithms*, Springer-Verlag, New York, 1981.

28. Soo-Chang Pei and Sy-Been Jaw, In-place in-order mixed radix fast Hartley transforms, *Signal Processing*, vol. 48, 123–134, 1996.

29. Soo-Chang Pei and Ja-Ling Wu, Split-radix fast Hartley transform, *Electron. Lett.*, vol. 22, no. 1, 26–27, 1986.

30. M. Popovic and D. Sevic, A new look at the comparison of the fast Hartley and Fourier transforms, *IEEE Trans. Signal Process.*, vol. 42, no. 8, 2178 – 2182, 1994.

31. A. D. Poularikas, *The Transforms and Applications Handbook*, Chapter 4, IEEE Press, Wachington, D. C. 1996.

32. H. V. Sorensen, D. L. Jones, C. S. Burrus and M. T. Heideman, On computing the discrete Hartley transform, *IEEE Trans. Acoust., Speech, Signal Process.*, vol. ASSP-33, no. 4, 1231–1238, 1985.

33. P. R. Uniyal, Transforming real-valued sequences: fast Fourier versus fast Hartley transform algorithms, *IEEE Trans. Signal Process.*, vol. 42, no. 11, 3249–3254, 1994.

34. K. Vassiliadis, P. Agelidis and G Sergiadis, Single-channel demodulator and Hartley transform in MRI, *IEEE Trans. Medical Imaging*, vol. 10, no. 4, 638–641, 1991.

35. M. Vetterli and P. Duhamel, Split-radix algorithms for length pm DFT's, *IEEE Trans. Acoust., Speech, Signal Processing*, vol. 37, no. 1, 57–64, 1989.

36. Z. Wang, Harmonic analysis with a real frequency function. I. Aperiodic case, *Appl. Math. Comput.*, vol. 9, 53–73; II. Periodic and bounded cases, vol. 9, 153–163; III. Data sequence, vol. 9, 245–255, 1981.

37. Z. Wang, Comments on generalised discrete Hartley transform, *IEEE Trans. Signal Process.*, vol. 43, no. 7, 1711-1712, 1995.

38. Z. Wang, A prime factor fast W transform algorithm, *IEEE Trans. Signal Process.*, vol. 40, no. 9, 2361–2368, 1992.

39. Z. Wang, The generalised discrete W transform and its application to interpolation, *26th Asilomar Conference,* vol. 1, 241–245, 1992.

40. J. Xi and J. F. Chicharo, Shift properties of discrete W transforms and real time discrete W analyzers, *IEEE Trans. Circuits Syst. II,* vol. 44, no. 1, 41–45, 1997.

41. N. J. Zhao, In-place radix-3 fast Hartley transform algorithm, *Electron. Lett.,* vol. 28, no. 3, 319–321, 1992.

5

Fast Algorithms for MD Discrete Hartley Transform

This chapter presents a number of fast algorithms for the computation of multidimensional (MD) discrete Hartley transform (DHT).

- Section 5.2 extends the concept of the one dimensional (1D) split-radix algorithm for the two-dimensional (2D) type-I DHT. Both decimation-in-time and decimation-in-frequency decomposition algorithms are presented.

- Section 5.3 deals with fast algorithms for 2D type-II, -III and -IV DHTs. It is shown that these algorithms can be derived by decomposing the entire computation into even-even, even-odd, odd-even and odd-odd computational blocks

- Section 5.4 uses the 2D type-I DHT to derive fast algorithms for 2D type-II, -III and -IV DHTs. Extending the decomposition concepts for DHTs of higher dimensions is also illustrated.

  The derivation of the above algorithms is based on the symmetry and periodicity properties of the transform kernel. In the next few sections, we apply the polynomial transform (PT) to the computation of MD DHTs to reduce the number of dimensions of the computation task.

- Section 5.5 presents the PT-based algorithm for MD type-I DHTs whose dimensional sizes are a power of two.

- Section 5.6 presents the PT-based algorithms for MD type-II DHTs whose dimensional sizes are powers of two. Using similar techniques, the fast algorithm for MD type-III DHTs is also described.

- Section 5.7 presents the PT-based algorithm for MD type-I DHTs whose dimensional sizes are powers of a prime number.

- Section 5.8 presents the PT-based algorithm for MD type-II DHTs whose dimensional sizes are powers of a prime number.

Discussions on the computational complexity for each of these algorithms are also presented.

5.1   INTRODUCTION

Image transforms play an important role in digital image processing as a theoretical and implementation tool for numerous applications, notably in digital image filtering, restoration, encoding and analysis. In particular, the MD discrete Fourier transform (DFT) has been a highly acceptable tool for frequency analysis ([1, 12, 9, 21, 25], for example). To deal with real sequences, however, the DFT is not the best choice since complex operations are required. Alternatively, the generalized DHT (or the discrete W transform) is often employed for real input arrays [8, 16]. Dedicated processors with a large memory space may be needed for such applications because the computational complexity becomes prohibitive for processing tasks of more than one dimension.

A general theoretical framework [7] was reported to convert the MD DHT successively into the 1D DFTs. The computational complexity is equivalent to that required by the row-column method for computation of the MD transforms. Fast algorithms for the type-II and -III 2D DHTs were reported to reduce the computational complexity [17, 31]. They were based on decomposing the MD DHTs into computational tasks of lower dimensions. It was also shown that by using the PT, the rD type-II and -III DHTs can be decomposed into an $(r-1)$D PT and 1D DHTs. Substantial reduction of the computational complexity in terms of the number of additions and multiplications was achieved compared to that needed in [7]. However, such an algorithm needs a process to reorder the input sequence for the conversion process and to reorder the coefficients of the $(r-1)$D PT to obtain the final transform outputs. The Diophantine index equations between the input index and the output index are solved to convert the general 2D DHT into N 1D length N type-II DHTs [31].

In the literature, many fast algorithms have been reported for various transforms such as the DFT, discrete cosine transform (DCT) and DHT. In particular, the radix-type fast algorithms [3, 16] for these transforms have been popular because they require a balanced computational complexity and a regular computational structure. Desirable features such as simple data indexing by using bit-reversal techniques and in-place computation are critical for minimizing the overall implementation and computation costs. Because the computational complexity for the MD DHT is prohibitive, developing radix-type algorithms is extremely important. It is known that the DHT has a close relationship to the DFT of real data [13, 26]. In general, the fast DHT algorithms require the same multiplicative complexity as and a similar additive complexity to that needed by their fast Fourier transform (FFT) counterpart. For example, an indirect computation by using the fast DHT and real FFT algorithms achieves such a computational requirement [13].

For clarity, let us formally define the four types of 2D DHTs of a real-valued array $x(n_1, n_2)$ $(0 \leq n_1 < N_1, 0 \leq n_2 < N_2)$. DHTs of higher dimensions can be similarly defined as shown below.

Type-I DHT

$$X(k_1, k_2) = \sum_{n_1=0}^{N_1-1} \sum_{n_2=0}^{N_2-1} x(n_1, n_2) \text{cas} \left(\frac{2\pi n_1 k_1}{N_1} + \frac{2\pi n_2 k_2}{N_2} \right) \qquad (5.1)$$

Type-II DHT

$$X(k_1, k_2) = \sum_{n_1=0}^{N_1-1} \sum_{n_2=0}^{N_2-1} x(n_1, n_2) \text{cas} \left(\frac{\pi(2n_1+1)k_1}{N_1} + \frac{\pi(2n_2+1)k_2}{N_2} \right) \qquad (5.2)$$

Type-III DHT

$$X(k_1, k_2) = \sum_{n_1=0}^{N_1-1} \sum_{n_2=0}^{N_2-1} x(n_1, n_2) \mathrm{cas} \left(\frac{\pi n_1 (2k_1 + 1)}{N_1} + \frac{\pi n_2 (2k_2 + 1)}{N_2} \right) \qquad (5.3)$$

Type-IV DHT

$$X(k_1, k_2) = \sum_{n_1=0}^{N_1-1} \sum_{n_2=0}^{N_2-1} x(n_1, n_2) \mathrm{cas} \left(\frac{\pi (2n_1 + 1)(2k_1 + 1)}{2N_1} + \frac{\pi (2n_2 + 1)(2k_2 + 1)}{2N_2} \right), \qquad (5.4)$$

where $0 \le k_1 \le N_1 - 1$, $0 \le k_2 \le N_2 - 1$ and cas( )= cos( ) + sin( ). The different types of DHTs can be easily identified by the use of parameters in the cas( ) functions. It is noted that the dimension sizes N_1 and N_2 are not necessarily the same.

5.2   SPLIT-RADIX ALGORITHMS FOR 2D TYPE-I DHT

5.2.1   Decimation-in-frequency algorithm

Even-even decomposition
For the 2D type-I DHT, we assume that the dimensional size $N_1 = N_2 = N$, and $N = q \cdot 2^m$, where q is an odd integer and $m > 0$. By setting different q values, various array sizes can be accommodated by the algorithm. By using the symmetric property of the cas( ) function, the computation task can be divided into several smaller subtasks. The even-even indexed transform outputs are obtained by

$$X(2k_1, 2k_2) = \sum_{n_1=0}^{N/2-1} \sum_{n_2=0}^{N/2-1} y_0(n_1, n_2) \mathrm{cas} \frac{2\pi(n_1 k_1 + n_2 k_2)}{N/2}, \qquad (5.5)$$

where $0 \le k_1, k_2 < N/2$, and

$$y_0(n_1, n_2) = \left[x(n_1, n_2) + x(n_1, n_2 + \frac{N}{2}) \right] + \left[x(n_1 + \frac{N}{2}, n_2) + x(n_1 + \frac{N}{2}, n_2 + \frac{N}{2}) \right]. (5.6)$$

Equation (5.5) is an $N/2 \times N/2$ type-I DHT whose input sequence is defined by (5.6).

If $N = 2q$, the even-odd, odd-even and odd-odd indexed transform outputs can be computed by

$$\begin{aligned} X(2k_1 + r_1 q, 2k_2 + r_2 q) &= \sum_{n_1=0}^{2q-1} \sum_{n_2=0}^{2q-1} x(n_1, n_2)(-1)^{(n_1 r_1 + n_2 r_2)} \mathrm{cas} \frac{2\pi(n_1 k_1 + n_2 k_2)}{q} \\ &= \sum_{n_1=0}^{q-1} \sum_{n_2=0}^{q-1} y_{r_1, r_2}(n_1, n_2)(-1)^{(n_1 r_1 + n_2 r_2)} \\ &\quad \cdot \mathrm{cas} \frac{2\pi(n_1 k_1 + n_2 k_2)}{q}, \end{aligned} \qquad (5.7)$$

where $0 \le k_1, k_2 < q$, and

$$\begin{aligned} y_{r_1, r_2}(n_1, n_2) &= x(n_1, n_2) + (-1)^{r_1 q} x(n_1 + q, n_2) + (-1)^{r_2 q} x(n_1, n_2 + q) \\ &\quad + (-1)^{r_1 q + r_2 q} x(n_1 + q, n_2 + q). \end{aligned} \qquad (5.8)$$

In (5.7) and (5.8), an even (or odd) value of r_1 or r_2 indicates an even (or odd) index of the transform outputs. By setting $r_1 = 0$ and $r_2 = 1$, for example, (5.7) computes $X(2k_1, 2k_2+q)$ which gives the even-odd indexed transform outputs. One desirable feature of (5.7) is that multiplications for twiddle factors are not needed because all the twiddle factors are trivial. When $N > 2q$, we have

$$
\begin{aligned}
X(4k_1 \pm r_1 q, 4k_2 \pm r_2 q) &= \sum_{n_1=0}^{N-1} \sum_{n_2=0}^{N-1} x(n_1, n_2) \text{cas} \left(\frac{2\pi(n_1 k_1 + n_2 k_2)}{N/4} \pm \frac{2\pi(n_1 r_1 + n_2 r_2)}{N/q} \right) \\
&= \sum_{n_1=0}^{N-1} \sum_{n_2=0}^{N-1} x(n_1, n_2) \left(\cos \frac{2\pi(n_1 r_1 + n_2 r_2)}{N/q} \text{cas} \frac{2\pi(n_1 k_1 + n_2 k_2)}{N/4} \right. \\
&\quad \left. \pm \sin \frac{2\pi(n_1 r_1 + n_2 r_2)}{N/q} \text{cas} \frac{-2\pi(n_1 k_1 + n_2 k_2)}{N/4} \right),
\end{aligned}
\tag{5.9}
$$

where $0 \le k_1, k_2 < N/4$. When $N = 4q$ in (5.9), the twiddle factors are again trivial so that a number of multiplications and additions can be saved. For simplicity we define $a = n_1 + N/4, b = n_2 + N/4$ and $\beta = 8\pi(n_1 k_1 + n_2 k_2)/N$. The computation can be further decomposed into twelve $N/4 \times N/4$ DHTs.

Even-odd decomposition

By setting $r_2 = 1$ in (5.9), we form an auxiliary sequence defined by

$$
\begin{aligned}
F_{r_1,1}(k_1, k_2) &= \frac{X(4k_1 + r_1 q, 4k_2 + q) + X(4k_1 + r_1 q, \; 4k_2 - q)}{2} \\
&= \sum_{n_1=0}^{N-1} \sum_{n_2=0}^{N-1} x(n_1, n_2) \cos \frac{2\pi(r_1 n_1 + r_2)}{N/q} \text{cas} \frac{2\pi(n_1 k_1 + n_2 k_2)}{N/4} \\
&= \sum_{n_1=0}^{N/4-1} \sum_{n_2=0}^{N/4-1} \left\{ \left[y_1(n_1, n_2) + (-1)^{\frac{r_1}{2}} y_1(a, n_2) \right] \cos \frac{2\pi(r_1 n_1 + n_2)}{N/q} \right. \\
&\quad \left. + (-1)^{\frac{q+1}{2}} \left[y_2(n_1, b) + (-1)^{\frac{r_1}{2}} y_2(a, b) \right] \sin \frac{2\pi(r_1 n_1 + n_2)}{N/q} \right\} \text{cas}(\beta)
\end{aligned}
\tag{5.10}
$$

and similarly,

$$
\begin{aligned}
G_{r_1,1}(k_1, k_2) &= \frac{X(4k_1 + r_1 q, 4k_2 + q) - X(4k_1 + r_1 q, \; 4k_2 - q)}{2} \\
&= \sum_{n_1=0}^{N/4-1} \sum_{n_2=0}^{N/4-1} \left\{ \left[y_1(n_1, n_2) + (-1)^{\frac{r_1}{2}} y_1(a, n_2) \right] \sin \frac{2\pi(r_1 n_1 + n_2)}{N/q} \right. \\
&\quad \left. - (-1)^{\frac{q+1}{2}} \left[y_2(n_1, b) + (-1)^{\frac{r_1}{2}} y_2(a, b) \right] \cos \frac{2\pi(r_1 n_1 + n_2)}{N/q} \right\} \text{cas}(-\beta),
\end{aligned}
\tag{5.11}
$$

where for (5.10) and (5.11), $0 \le k_1, k_2 < N/4$, $r_1 = 0$ or 2 and

$$
\begin{aligned}
y_1(n_1, n_2) &= \left[x(n_1, n_2) - x(n_1, n_2 + \frac{N}{2}) \right] + \left[x(n_1 + \frac{N}{2}, n_2) \right. \\
&\quad \left. - x(n_1 + \frac{N}{2}, n_2 + \frac{N}{2}) \right].
\end{aligned}
\tag{5.12}
$$

Therefore, the even-odd decomposition derives four $N/4 \times N/4$ type-I DHTs whose inputs are rotated by twiddle factors.

Odd-even decomposition

By setting $r_1 = 1$ in (5.9), we have

$$
\begin{aligned}
F_{1,r_2}(k_1, k_2) &= \frac{X(4k_1 + q, 4k_2 + r_2q) + X(4k_1 - q, \, 4k_2 + r_2q)}{2} \\
&= \sum_{n_1=0}^{N-1} \sum_{n_2=0}^{N-1} x(n_1, n_2) \cos \frac{2\pi(n_1 + r_2n_2)}{N/q} \, \text{cas} \, \frac{2\pi(n_1 k_1 + n_2 k_2)}{N/4} \\
&= \sum_{n_1=0}^{N/4-1} \sum_{n_2=0}^{N/4-1} \left\{ [y_2(n_1, n_2) + (-1)^{\frac{r_2}{2}} y_2(n_1, b)] \cos \frac{2\pi(n_1 + r_2n_2)}{N/q} \right. \\
&\quad \left. + (-1)^{\frac{q+1}{2}} [y_2(a, n_2) + (-1)^{\frac{r_2}{2}} y_2(a, b)] \sin \frac{2\pi(n_1 + r_2n_2)}{N/q} \right\} \text{cas}\,(\beta) \quad (5.13)
\end{aligned}
$$

and

$$
\begin{aligned}
G_{1,r_2}(k_1, k_2) &= \frac{X(4k_1 + q, 4k_2 + r_2q) - X(4k_1 - q, \, 4k_2 + r_2q)}{2} \\
&= \sum_{n_1=0}^{N/4-1} \sum_{n_2=0}^{N/4-1} \left\{ [y_2(n_1, n_2) + (-1)^{\frac{r_2}{2}} y_2(n_1, b)] \sin \frac{2\pi(n_1 + r_2n_2)}{N/q} \right. \\
&\quad \left. - (-1)^{\frac{q+1}{2}} [y_2(a, n_2) + (-1)^{\frac{r_2}{2}} y_2(a, b)] \cos \frac{2\pi(n_1 + r_2n_2)}{N/q} \right\} \text{cas}\,(-\beta), (5.14)
\end{aligned}
$$

where for (5.13) and (5.14), $0 \leq k_1, k_2 < N/4$, $r_2 = 0, 2$ and

$$
\begin{aligned}
y_2(n_1, n_2) &= \left[x(n_1, n_2) + x(n_1, n_2 + \frac{N}{2}) \right] - \left[x(n_1 + \frac{N}{2}, n_2) \right. \\
&\quad \left. + x(n_1 + \frac{N}{2}, n_2 + \frac{N}{2}) \right].
\end{aligned} \quad (5.15)
$$

Odd-odd decomposition

Similarly, the odd-odd decomposition can be achieved by setting $r_2 = 1$ in (5.9) to achieve

$$
\begin{aligned}
F_{r_1,1}(k_1, k_2) &= \frac{X(4k_1 + r_1q, 4k_2 + q) + X(4k_1 - r_1q, \, 4k_2 - q)}{2} \\
&= \sum_{n_1=0}^{N-1} \sum_{n_2=0}^{N-1} x(n_1, n_2) \cos \frac{2\pi(r_1n_1 + n_2)}{N/q} \, \text{cas} \, \frac{2\pi(n_1 k_1 + n_2 k_2)}{N/4} \\
&= \sum_{n_1=0}^{N/4-1} \sum_{n_2=0}^{N/4-1} \left\{ [y_3(n_1, n_2) - r_1 y_3(a, b)] \cos \frac{2\pi(r_1n_1 + n_2)}{N/q} \right. \\
&\quad \left. + (-1)^{\frac{q+1}{2}} [y_3(n_1, b) + r_1 y_3(a, n_2)] \sin \frac{2\pi(r_1n_1 + n_2)}{N/q} \right\} \text{cas}(\beta) \quad (5.16)
\end{aligned}
$$

and

$$G_{r_1,1}(k_1, k_2) = \frac{X(4k_1 + r_1q, 4k_2 + q) - X(4k_1 - r_1q, \ 4k_2 - q)}{2}$$

$$= \sum_{n_1=0}^{N/4-1} \sum_{n_2=0}^{N/4-1} \left\{ [y_3(n_1, n_2) - r_1 y_3(a, b)] \sin \frac{2\pi(r_1 n_1 + n_2)}{N/q} \right.$$

$$\left. - (-1)^{\frac{q+1}{2}} [y_3(n_1, b) + r_1 y_3(a, n_2)] \cos \frac{2\pi(r_1 n_1 + n_2)}{N/q} \right\} \text{cas}(-\beta), \quad (5.17)$$

where for (5.16) and (5.17), $0 \leq k_1, k_2 < N/4$, $r_1 = 1, -1$ and

$$y_3(n_1, n_2) = \left[x(n_1, n_2) - x\left(n_1, n_2 + \frac{N}{2}\right) \right] - \left[x\left(n_1 + \frac{N}{2}, n_2\right) \right.$$

$$\left. - x\left(n_1 + \frac{N}{2}, n_2 + \frac{N}{2}\right) \right]. \quad (5.18)$$

The entire $N \times N$ DHT is decomposed into one $N/2 \times N/2$ type-I DHT defined in (5.5) and twelve $N/4 \times N/4$ type-I DHTs given in (5.10), (5.11), (5.13), (5.14), (5.16) and (5.17). The final even-odd, odd-even and odd-odd indexed transform outputs are obtained by

$$X(4k_1 + r_1q, 4k_2 + r_2q) = F_{r_1,r_2}(k_1, k_2) \pm G_{r_1,r_2}(k_1, k_2), \quad (5.19)$$

where $0 \leq k_1, k_2 < N/4$ and r_1 and r_2 are the values used in the decomposition process. When (5.19) is used, the following properties should be applied:

$$X(k_1, \ N - k_2) = X(k_1, \ -k_2)$$
$$X(N - k_1, \ k_2) = X(-k_1, \ k_2)$$
$$X(N - k_1, \ N - k_2) = X(-k_1, \ -k_2). \quad (5.20)$$

Figure 5.1 shows the computation of the even-odd, odd-even and odd-odd indexed transform outputs for $q = 1$. This algorithm is derived by directly using the properties of the trigonometric identities rather than using matrix manipulation or some intermediate variables in the decomposition process as in [20, 29]. One major difference from the decomposition in [29] is that the property of (5.20) is used for the negative indices, which allows the parameter q to be associated with the twiddle factors. Such an arrangement provides substantial savings on computational complexity and a wider range of choices on transform sizes.

5.2.2   Computational complexity

Now let us consider the computational complexity needed by the algorithm in terms of the numbers of multiplications and additions. We assume that a rotation operation is implemented by 3 multiplications and 3 additions. From (5.5) and (5.7), the required number of multiplications for $N = 2q$ is

$$M(2q) = 4M(q) \quad (5.21)$$

and the number of additions is

$$A(2q) = 4A(q) + 8q^2. \quad (5.22)$$

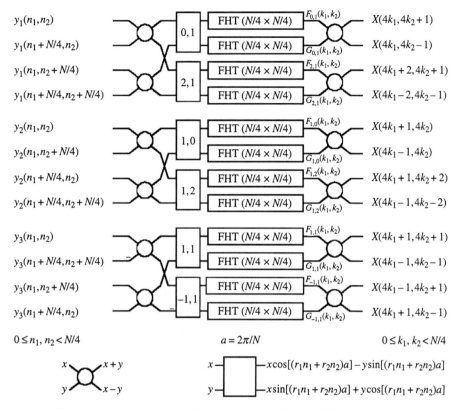

Figure 5.1 Computation of even-odd, odd-even and odd-odd indexed outputs for $q = 1$.

When $N = 4q$, the twiddle factors in (5.10), (5.11), (5.13), (5.14), (5.16) and (5.17) are either zero or one which needs no multiplication. Therefore, the number of multiplications needed by the algorithm is

$$M(4q) = 4M(2q) + 12M(q) \tag{5.23}$$

and the number of additions is

$$A(4q) = 4A(2q) + 12A(q) + 56q^2. \tag{5.24}$$

Equations (5.21) and (5.23) show that when $N = 2q$ and $4q$, no multiplications are needed by the twiddle factors, which is similar to Wu's algorithms in [29] which do not need any multiplications for 2×2 and 4×4 DHTs. According to (5.10) – (5.20), the decomposition costs for $N > 4$ are:

- $24(N/4)^2 - M_t$ multiplications and $24(N/4)^2 - A_t$ additions for twiddle factors in (5.10), (5.11), (5.13), (5.14), (5.16) and (5.17), where $M_t = 9qN$ and $A_t = 3qN$ are the number of multiplications and additions saved from trivial twiddle factors.

- $2N^2$ additions for (5.6), (5.12), (5.15) and (5.18).

- $12(N/4)^2$ additions for (5.19).

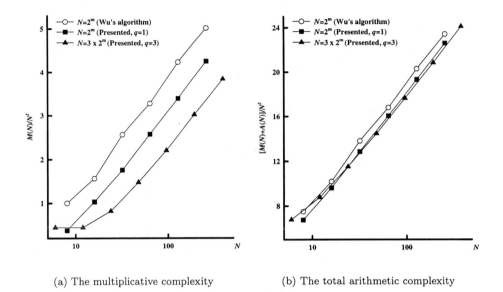

(a) The multiplicative complexity   (b) The total arithmetic complexity

Figure 5.2   Computational complexity for $q = 1$ and 3.

Therefore, the total number of multiplications required by the above algorithm is

$$M(N) = 4M(\frac{N}{2}) + 12M(\frac{N}{4}) + \frac{3N^2}{2} - 9qN, \quad N > 4 \tag{5.25}$$

and the total number of additions is

$$A(N) = 4A(\frac{N}{2}) + 12A(\frac{N}{4}) + \frac{17N^2}{4} - 3qN \quad N > 4. \tag{5.26}$$

Table 5.1   Comparison of computational complexity for $q = 1$ and 3.

Wu's algorithm [29]			Presented algorithm					
$q = 1$			$q = 1$			$q = 3$		
N	M	A	N	M	A	N	M	A
8	64	416	8	24	408	6	16	228
16	400	2208	16	264	2216	12	64	1200
32	2624	11512	32	1800	11368	24	472	6168
64	13424	55344	64	10536	55176	48	3400	29928
128	69216	262984	128	55560	260840	96	20296	142248
256	328080	1205376	256	277992	1200712	192	111208	656328

The initial values, i.e., the computational complexity for a $q \times q$ DHT, are $M(1 \times 1) = 0$, $A(1 \times 1) = 0$ for $q = 1$ and $M(3 \times 3) = 4$ and $A(3 \times 3) = 39$ for $q = 3$. Table 5.1 shows the computational complexity required by Wu's algorithm [29] and the presented one for $q = 1$

and 3. Figure 5.2 (a) shows that the presented algorithm significantly reduces the number of multiplications for both $q = 1$ and 3, and Figure 5.2 (b) compares the total number of arithmetic operations (additions plus multiplications) required by these fast algorithms. Figure 5.2 (b) also shows that the algorithm achieves a better computational efficiency for $q = 3$ than that for $q = 1$. Compared to the hybrid FDHT/RFFT algorithm [13] and real FFT algorithm [23], the presented one uses exactly the same number of multiplications and a few more additions. Because our algorithm computes the DHT from lower order DHTs, the computational structure is simpler than that of the hybrid FDHT/RFFT algorithm.

5.2.3   Decimation-in-time algorithm

For simplicity, let us define $a = 4\pi(n_1 k_1 + n_2 k_2)/N$, $\rho = 2\pi q/N$ and $\beta = 8\pi(n_1 k_1 + n_2 k_2)/N$. With the decimation-in-time decomposition, the type-I 2D DHT in (5.1) can be given as

$$
\begin{aligned}
X(k_1, k_2) = & \sum_{n_1=0}^{N_1/2-1} \sum_{n_2=0}^{N_2/2-1} x(2n_1, 2n_2)\mathrm{cas}\,(\alpha) \\
& + \sum_{n_1=0}^{N_1/4-1} \sum_{n_2=0}^{N_2/4-1} \left\{ \sum_{l=0}^{1} x[4n_1 + (-1)^l q, 4n_2]\mathrm{cas}[\beta + (-1)^l \rho k_1] \right. \\
& + \sum_{l=0}^{1} x[4n_1, 4n_2 + (-1)^l q]\mathrm{cas}[\beta + (-1)^l \rho k_2] \\
& + \sum_{l=0}^{1} x[4n_1 + (-1)^l q, 4n_2 + (-1)^l q]\mathrm{cas}[\beta + (-1)^l \rho(k_1 + k_2)] \\
& + \sum_{l=0}^{1} x[4n_1 + (-1)^l q, 4n_2 - (-1)^l q]\mathrm{cas}[\beta + (-1)^l \rho(k_1 - k_2)] \\
& + \sum_{l=0}^{1} x[4n_1 + (-1)^l q, 4n_2 + (-1)^l 2q]\mathrm{cas}[\beta + (-1)^l \rho(k_1 + 2k_2)] \\
& \left. + \sum_{l=0}^{1} x[4n_1 + (-1)^l 2q, 4n_2 + (-1)^l q]\mathrm{cas}[\beta + (-1)^l \rho(2k_1 + k_2)] \right\}, \quad (5.27)
\end{aligned}
$$

where $N_1 = N_2 = N = q \cdot 2^m$. Each term of the inner sum in (5.27) can be further decomposed into two terms. By rearranging these terms, (5.27) can be expressed into

$$
\begin{aligned}
X(K_1, K_2) = & A(k_1, k_2) + (-1)^i[\cos(\rho k_1)B_0(k_1, k_2) + \sin(\rho k_1)B_1(k_1, k_2)] \\
& + (-1)^j \cos(\rho k_2)\, C_0(k_1, k_2) + \sin(\rho k_2)C_1(k_1, k_2) \\
& + (-1)^{i+j} \{\cos[\rho(k_1 + k_2)]D_0(k_1, k_2) + \sin[\rho(k_1 + k_2)]D_1(k_1, k_2)\} \\
& + (-1)^{i-j} \{\cos[\rho(k_1 - k_2)]E_0(k_1, k_2) + \sin[\rho(k_1 - k_2)]E_1(k_1, k_2)\} \\
& + (-1)^{i} \{\cos[\rho(k_1 + 2k_2)]F_0(k_1, k_2) + \sin[\rho(k_1 + 2k_2)]F_1(k_1, k_2)\} \\
& + (-1)^{j} \{\cos[\rho(2k_1 + k_2)]G_0(k_1, k_2) + \sin[\rho(2k_1 + k_2)]G_1(k_1, k_2)\}, (5.28)
\end{aligned}
$$

where $K_1 = k_1 + \frac{iN}{2}$, $K_2 = k_2 + \frac{iN}{2}$, and $(i, j) = (0,0), (1,0), (0,1)$ and $(1,1)$.

We further define $y(4n_1+l_1q, 4n_2+l_2q, l) = x(4n_1+l_1q, 4n_2+l_2q)+(-1)^l x(4n_1-l_1q, 4n_2-l_2q)$, where l_1 or $l_2 = 1, 0, -1$, and $l = 0, 1$. For any positive integers i and j, we have

$$A(k_1, k_2) = A(k_1 + \frac{iN}{2}, k_2 + \frac{jN}{2}) = \sum_{n_1=0}^{N_1/2-1} \sum_{n_2=0}^{N_2/2-1} x(2n_1, 2n_2)\mathrm{cas}\,(\alpha) \qquad (5.29)$$

$$B_l(k_1, k_2) = B_l(k_1 + \frac{iN}{4}, k_2 + \frac{jN}{4}) = \sum_{n_1=0}^{N_1/4-1} \sum_{n_2=0}^{N_2/4-1} y(4n_1 + q, 4n_2, l)\mathrm{cas}[(-1)^l \beta] \quad (5.30)$$

$$C_l(k_1, k_2) = C_l(k_1 + \frac{iN}{4}, k_2 + \frac{jN}{4})$$
$$= \sum_{n_1=0}^{N_1/4-1} \sum_{n_2=0}^{N_2/4-1} y(4n_1, 4n_2 + q, l)\mathrm{cas}[(-1)^l \beta] \qquad (5.31)$$

$$D_l(k_1, k_2) = D_l(k_1 + \frac{iN}{4}, k_2 + \frac{jN}{4})$$
$$= \sum_{n_1=0}^{N_1/4-1} \sum_{n_2=0}^{N_2/4-1} y(4n_1 + q, 4n_2 + q, l)\mathrm{cas}[(-1)^l \beta] \qquad (5.32)$$

$$E_l(k_1, k_2) = E_l(k_1 + \frac{iN}{4}, k_2 + \frac{jN}{4})$$
$$= \sum_{n_1=0}^{N_1/4-1} \sum_{n_2=0}^{N_2/4-1} y(4n_1 + q, 4n_2 - q, l)\mathrm{cas}[(-1)^l \beta] \qquad (5.33)$$

$$F_l(k_1, k_2) = F_l(k_1 + \frac{iN}{4}, k_2 + \frac{jN}{4})$$
$$= \sum_{n_1=0}^{N_1/4-1} \sum_{n_2=0}^{N_2/4-1} y(4n_1 + q, 4n_2 + 2q, l)\mathrm{cas}[(-1)^l \beta] \qquad (5.34)$$

$$G_l(k_1, k_2) = G_l(k_1 + \frac{iN}{4}, k_2 + \frac{jN}{4})$$
$$= \sum_{n_1=0}^{N_1/4-1} \sum_{n_2=0}^{N_2/4-1} y(4n_1 + 2q, 4n_2 + q, l)\mathrm{cas}[(-1)^l \beta], \qquad (5.35)$$

where for (5.30) – (5.35), $l = 0, 1$, and $0 \le k_1, k_2 < N/4$. When the input index of $x(\ )$ in (5.27) is smaller than 0 or larger than $N - 1$, we have

$$x(t_1, t_2) = \begin{cases} x[(t_1) \bmod N, (t_2) \bmod N], & t_1 \ge N \text{ and/or } t_2 \ge N \\ x[(N + t_1) \bmod N, (n + t_2) \bmod N], & t_1 \le 0 \text{ and/or } t_2 \le 0 \end{cases} \qquad (5.36)$$

Based on (5.28), we have the transform outputs $X(k_1 + N/4, k_2)$

$$X(k_1 + \frac{N}{4}, k_2) = A(k_1 + \frac{N}{4}, k_2)$$
$$+ (-1)^{(q+1)/2}[\sin(\rho k_1)B_0(k_1, k_2) - \cos(\rho k_1)B_1(k_1, k_2)]$$
$$+ [\cos(\rho k_2)\, C_0(k_1, k_2) + \sin(\rho k_2)C_1(k_1, k_2)]$$

$$+ (-1)^{(q+1)/2}\{\sin[\rho(k_1 + k_2)]D_0(k_1, k_2) - \cos[\rho(k_1 + k_2)]D_1(k_1, k_2)\}$$
$$+ (-1)^{(q+1)/2}\{\sin[\rho(k_1 - k_2)]E_0(k_1, k_2) - \cos[\rho(k_1 - k_2)]E_1(k_1, k_2)\}$$
$$+ (-1)^{(q+1)/2}\{\sin[\rho(k_1 + 2k_2)]F_0(k_1, k_2) - \cos[\rho(k_1 + 2k_2)]F_1(k_1, k_2)\}$$
$$- \{\cos[\rho(2k_1 + k_2)]G_0(k_1, k_2) + \sin[\rho(2k_1 + k_2)]G_1(k_1, k_2)\}. \tag{5.37}$$

The other transform outputs $X(k_1, k_2 + N/4)$ and $X(k_1 + N/4, k_2 + N/4)$ can be similarly derived; the results are not presented here for simplicity. Therefore, the entire type-I DHT is decomposed into one $N/2 \times N/2$ type-I DHT in (5.29) and twelve $N/4 \times N/4$ type-I DHTs in (5.30) – (5.35), which are combined, as shown in (5.28) and (5.37), to produce the final transformed outputs.

Table 5.2   Computational complexity required by the presented algorithm.

N	$A(N)$
$2q$	$A(2q) = 4A(q) + 8q^2$
$4q$	$A(4q) = A(2q) + 12A(q) + 56q^2$
$N > 4q$	$A(N) = A(\frac{N}{2}) + 12A(\frac{N}{4}) + \frac{17N^2}{4} - 3qN$
N	$M(N)$
$2q$	$M(2q) = 4M(q)$
$4q$	$M(4q) = M(2q) + 12M(q)$
$N > 4q$	$M(N) = M(\frac{N}{2}) + 12M(\frac{N}{4}) + \frac{3N^2}{2} - 9qN$

5.2.4   Computational complexity

Now we consider the computational complexity of the fast algorithms. It is assumed that one rotation operation is realized by using four multiplications and two additions. When $N = 2q$, the $N \times N$ type-I DHT can be computed by

$$X(k_1, k_2) = \sum_{n_1=0}^{N/2-1} \sum_{n_2=0}^{N/2-1} x(2n_1, 2n_2)\text{cas}(\alpha)$$
$$+ (-1)^{k_1} \sum_{n_1=0}^{N/2-1} \sum_{n_2=0}^{N/2-1} [x(2n_1 + q, 2n_2) + x(2n_1 - q, 2n_2)]\text{cas}(\alpha)$$
$$+ (-1)^{k_2} \sum_{n_1=0}^{N/2-1} \sum_{n_2=0}^{N/2-1} [x(2n_1, 2n_2 + q) + x(2n_1, 2n_2 - q)]\text{cas}(\alpha)$$
$$+ (-1)^{k_1+k_2} \sum_{n_1=0}^{N/2-1} \sum_{n_2=0}^{N/2-1} [x(2n_1 + q, 2n_2 + q) + x(2n_1 - q, 2n_2 - q)]\text{cas}(\alpha), \tag{5.38}$$

which contains four $q \times q$ type-I DHTs. It is also noted that for $N = 4q$, all twiddle factors such as those in (5.28) and (5.37) become trivial and do not need multiplications. Table 5.2 lists the equations calculating the number of multiplications and additions required by the algorithm. The computational complexity for a $q \times q$ DHT are $A(1) = M(1) = 0$ for $q = 1$, and $A(3) = 39$ and $M(3) = 4$ for $q = 3$.

In general, extra arithmetic operations are needed for dimension sizes other than a power of two. In contrast, the presented algorithms achieve savings on the number of operations for $q > 1$ because the number of trivial twiddle factors is proportional to q. Based on Table 5.2, we can calculate the number of additions and multiplications needed by the presented algorithm.

5.3   FAST ALGORITHMS FOR 2D TYPE-II, -III AND -IV DHTs

5.3.1   Algorithm for type-II DHT

If the sizes N_1 and N_2 of both dimensions are even, (5.2) can be decomposed into

$$X(2k_1, 2k_2) = \sum_{n_1=0}^{N_1/2-1} \sum_{n_2=0}^{N_2/2-1} y_0(n_1, n_2) \mathrm{cas}\left(\frac{\pi(2n_1+1)k_1}{N_1/2} + \frac{\pi(2n_2+1)k_2}{N_2/2}\right) \quad (5.39)$$

$$X(2k_1, 2k_2+1) = \sum_{n_1=0}^{N_1/2-1} \sum_{n_2=0}^{N_2/2-1} y_1(n_1, n_2)$$
$$\cdot \mathrm{cas}\left(\frac{\pi(2n_1+1)k_1}{N_1/2} + \frac{\pi(2n_2+1)(2k_2+1)}{2(N_2/2)}\right) \quad (5.40)$$

$$X(2k_1+1, 2k_2) = \sum_{n_1=0}^{N_1/2-1} \sum_{n_2=0}^{N_2/2-1} y_2(n_1, n_2)$$
$$\cdot \mathrm{cas}\left(\frac{\pi(2n_1+1)(2k_1+1)}{2(N_1/2)} + \frac{\pi(2n_2+1)k_2}{N_2/2}\right) \quad (5.41)$$

$$X(2k_1+1, 2k_2+1) = \sum_{n_1=0}^{N_1/2-1} \sum_{n_2=0}^{N_2/2-1} y_3(n_1, n_2)$$
$$\cdot \mathrm{cas}\left(\frac{\pi(2n_1+1)(2k_1+1)}{2(N_1/2)} + \frac{\pi(2n_2+1)(2k_2+1)}{2(N_2/2)}\right), \quad (5.42)$$

where the input sequences for (5.39) – (5.42) are

$$\begin{bmatrix} y_0(n_1,n_2) \\ y_1(n_1,n_2) \\ y_2(n_1,n_2) \\ y_3(n_1,n_2) \end{bmatrix} = \begin{bmatrix} 1 & 1 & 1 & 1 \\ 1 & 1 & -1 & -1 \\ 1 & -1 & 1 & -1 \\ 1 & -1 & -1 & 1 \end{bmatrix} \begin{bmatrix} x(n_1,n_2) \\ x(n_1+\frac{N_1}{2},n_2) \\ x(n_1,n_2+\frac{N_2}{2}) \\ x(n_1+\frac{N_1}{2},n_2+\frac{N_2}{2}) \end{bmatrix}. \quad (5.43)$$

According to their indices, (5.39) – (5.42) are known as the type-II even-even, even-odd, odd-even and odd-odd computation arrays, respectively. The even-even array is the $N_1/2 \times N_2/2$ type-II DHT, and the odd-odd array is the $N_1/2 \times N_2/2$ type-IV DHT. The even-even array can be further decomposed in the same way as shown above. In the rest of this subsection, decomposition of the even-odd and odd-even arrays is considered. Computation of the odd-odd array is considered in the next subsection.

The type-II even-odd decomposition

For simplicity, let us formally define the $N_1 \times N_2$ type-II even-odd array to be

$$X_{EO}(k_1, k_2) = \sum_{n_1=0}^{N_1-1} \sum_{n_2=0}^{N_2-1} x(n_1, n_2) \text{cas}\left(\frac{\pi(2n_1+1)k_1}{N_1} + \frac{\pi(2n_2+1)(2k_2+1)}{2N_2}\right), \quad (5.44)$$

which can be grouped into

$$\begin{aligned}
F_0(k_1, k_2) &= \frac{X_{EO}(2k_1, 2k_2) + X_{EO}(2k_1, 2k_2-1)}{2} \\
&= \sum_{n_1=0}^{N_1-1} \sum_{n_2=0}^{N_2-1} 0.5x(n_1, n_2) \left[\text{cas}\left(\frac{\pi(2n_1+1)k_1}{N_1/2} + \frac{\pi(2n_2+1)(4k_2+1)}{2N_2}\right) \right. \\
&\quad \left. + \text{cas}\left(\frac{\pi(2n_1+1)k_1}{N_1/2} + \frac{\pi(2n_2+1)(4k_2-1)}{2N_2}\right)\right] \\
&= \sum_{n_1=0}^{N_1/2-1} \sum_{n_2=0}^{N_2/2-1} f_0(n_1, n_2) \text{cas}\left(\frac{\pi(2n_1+1)k_1}{N_1/2} + \frac{\pi(2n_2+1)k_2}{N_2/2}\right), \quad (5.45)
\end{aligned}$$

where

$$f_0(n_1, n_2) = u_0(n_1, n_2) \cos\frac{\pi(2n_2+1)}{2N_2} - v_0(n_1, n_2) \sin\frac{\pi(2n_2+1)}{2N_2} \quad (5.46)$$

$$u_0(n_1, n_2) = x(n_1, n_2) + x(n_1 + \frac{N_1}{2}, n_2) \quad (5.47)$$

$$v_0(n_1, n_2) = x(n_1, n_2 + \frac{N_2}{2}) + x(n_1 + \frac{N_1}{2}, n_2 + \frac{N_2}{2}). \quad (5.48)$$

Similarly, we have

$$\begin{aligned}
G_0(k_1, k_2) &= \frac{X_{EO}(2k_1, 2k_2) - X_{EO}(2k_1, 2k_2-1)}{2} \\
&= \sum_{n_1=0}^{N_1/2-1} \sum_{n_2=0}^{N_2/2-1} g_0(n_1, n_2) \text{cas}\left(\frac{-\pi(2n_1+1)k_1}{N_1/2} - \frac{\pi(2n_2+1)k_2}{N_2/2}\right), \quad (5.49)
\end{aligned}$$

where

$$g_0(n_1, n_2) = u_0(n_1, n_2) \sin\frac{\pi(2n_2+1)}{2N_2} + v_0(n_1, n_2) \cos\frac{\pi(2n_2+1)}{2N_2}. \quad (5.50)$$

Both (5.45) and (5.49) are the $N_1/2 \times N_2/2$ type-II DHT. Similarly, we have

$$\begin{aligned}
F_1(k_1, k_2) &= \frac{X_{EO}(2k_1+1, 2k_2) + X_{EO}(2k_1+1, 2k_2-1)}{2} \\
&= \sum_{n_1=0}^{N_1/2-1} \sum_{n_2=0}^{N_2/2-1} f_1(n_1, n_2) \text{cas}\left(\frac{\pi(2n_1+1)(2k_1+1)}{2(N_1/2)} + \frac{\pi(2n_2+1)k_2}{N_2/2}\right) \quad (5.51)
\end{aligned}$$

$$\begin{aligned}
G_1(k_1, k_2) &= \frac{X_{EO}(2k_1+1, 2k_2) - X_{EO}(2k_1+1, 2k_2-1)}{2} \\
&= \sum_{n_1=0}^{N_1/2-1} \sum_{n_2=0}^{N_2/2-1} g_1(n_1, n_2) \text{cas}\left(\frac{-\pi(2n_1+1)(2k_1+1)}{2(N_1/2)} - \frac{\pi(2n_2+1)k_2}{N_2/2}\right),
\end{aligned}$$

$$(5.52)$$

where

$$f_1(n_1, n_2) = u_1(n_1, n_2) \cos \frac{\pi(2n_2 + 1)}{2N_2} - v_1(n_1, n_2) \sin \frac{\pi(2n_2 + 1)}{2N_2} \qquad (5.53)$$

$$g_1(n_1, n_2) = u_1(n_1, n_2) \sin \frac{\pi(2n_2 + 1)}{2N_2} + v_1(n_1, n_2) \cos \frac{\pi(2n_2 + 1)}{2N_2} \qquad (5.54)$$

$$u_1(n_1, n_2) = x(n_1, n_2) - x(n_1 + \frac{N_1}{2}, n_2) \qquad (5.55)$$

$$v_1(n_1, n_2) = x(n_1, n_2 + \frac{N_2}{2}) - x(n_1 + \frac{N_1}{2}, n_2 + \frac{N_2}{2}). \qquad (5.56)$$

Both (5.51) and (5.52) are the $N_1/2 \times N_2/2$ type-II odd-even sequences which are to be considered in the next subsection. The final outputs of (5.44) can be obtained by

$$X_{EO}(2k_1, 2k_2) = F_0(k_1, k_2) + G_0(k_1, k_2) \qquad (5.57)$$

$$X_{EO}(2k_1, 2k_2 - 1) = F_0(k_1, k_2) - G_0(k_1, k_2) \qquad (5.58)$$

$$X_{EO}(2k_1 + 1, 2k_2) = F_1(k_1, k_2) + G_1(k_1, k_2) \qquad (5.59)$$

$$X_{EO}(2k_1 + 1, 2k_2 - 1) = F_1(k_1, k_2) - G_1(k_1, k_2), \qquad (5.60)$$

where for (5.57) – (5.60), $0 \le k_1 < N_1/2, 1 \le k_2 < N_2/2$ and $X_{EO}(k_1, -1) = -X_{EO}(k_1, N_2 - 1)$.

The type-II odd-even decomposition
The $N_1 \times N_2$ odd-even sequence is defined by

$$X_{OE}(k_1, k_2) = \sum_{n_1=0}^{N_1-1} \sum_{n_2=0}^{N_2-1} x(n_1, n_2) \text{cas} \left(\frac{\pi(2n_1 + 1)(2k_1 + 1)}{2N_1} + \frac{\pi(2n_2 + 1)k_2}{N_2} \right), \qquad (5.61)$$

where $0 \le k_1 < N_1/2, 1 \le k_2 < N_2/2$. Equation (5.61) can be decomposed into

$$\begin{aligned} F_2(k_1, k_2) &= \frac{X_{OE}(2k_1, 2k_2) + X_{OE}(2k_1 - 1, 2k_2)}{2} \\ &= \sum_{n_1=0}^{N_1/2-1} \sum_{n_2=0}^{N_2/2-1} f_2(n_1, n_2) \text{cas} \left(\frac{\pi(2n_1 + 1)k_1}{N_1/2} + \frac{\pi(2n_2 + 1)k_2}{N_2/2} \right) \end{aligned} \qquad (5.62)$$

$$\begin{aligned} G_2(k_1, k_2) &= \frac{X_{OE}(2k_1, 2k_2) - X_{OE}(2k_1 - 1, 2k_2)}{2} \\ &= \sum_{n_1=0}^{N_1/2-1} \sum_{n_2=0}^{N_2/2-1} g_2(n_1, n_2) \text{cas} \left(\frac{-\pi(2n_1 + 1)k_1}{N_1/2} - \frac{\pi(2n_2 + 1)k_2}{N_2/2} \right) \end{aligned} \qquad (5.63)$$

$$\begin{aligned} F_3(k_1, k_2) &= \frac{X_{OE}(2k_1, 2k_2 + 1) + X_{OE}(2k_1 - 1, 2k_2 + 1)}{2} \\ &= \sum_{n_1=0}^{N_1/2-1} \sum_{n_2=0}^{N_2/2-1} f_3(n_1, n_2) \text{cas} \left(\frac{\pi(2n_1 + 1)k_1}{N_1/2} + \frac{\pi(2n_2 + 1)(2k_2 + 1)}{2(N_2/2)} \right) \end{aligned} \qquad (5.64)$$

$$G_3(k_1, k_2) = \frac{X_{OE}(2k_1, 2k_2 + 1) - X_{OE}(2k_1 - 1, 2k_2 + 1)}{2}$$

$$= \sum_{n_1=0}^{N_1/2-1} \sum_{n_2=0}^{N_2/2-1} g_3(n_1, n_2) \text{cas} \left(\frac{-\pi(2n_1 + 1)k_1}{N_1/2} - \frac{\pi(2n_2 + 1)(2k_2 + 1)}{2(N_2/2)} \right). \quad (5.65)$$

For (5.62) – (5.65), the input sequences are

$$f_2(n_1, n_2) = u_2(n_1, n_2) \cos \frac{\pi(2n_1 + 1)}{2N_1} - v_2(n_1, n_2) \sin \frac{\pi(2n_1 + 1)}{2N_1} \quad (5.66)$$

$$g_2(n_1, n_2) = u_2(n_1, n_2) \sin \frac{\pi(2n_1 + 1)}{2N_1} + v_2(n_1, n_2) \cos \frac{\pi(2n_1 + 1)}{2N_1} \quad (5.67)$$

$$f_3(n_1, n_2) = u_3(n_1, n_2) \cos \frac{\pi(2n_1 + 1)}{2N_1} - v_3(n_1, n_2) \sin \frac{\pi(2n_1 + 1)}{2N_1} \quad (5.68)$$

$$g_3(n_1, n_2) = u_3(n_1, n_2) \sin \frac{\pi(2n_1 + 1)}{2N_1} + v_3(n_1, n_2) \cos \frac{\pi(2n_1 + 1)}{2N_1} \quad (5.69)$$

$$u_2(n_1, n_2) = x(n_1, n_2) + x(n_1, n_2 + \frac{N_2}{2}) \quad (5.70)$$

$$v_2(n_1, n_2) = x(n_1 + \frac{N_1}{2}, n_2) + x(n_1 + \frac{N_1}{2}, n_2 + \frac{N_2}{2}) \quad (5.71)$$

$$u_3(n_1, n_2) = x(n_1, n_2) - x(n_1, n_2 + \frac{N_2}{2}) \quad (5.72)$$

$$v_3(n_1, n_2) = x(n_1 + \frac{N_1}{2}, n_2) - x(n_1 + \frac{N_1}{2}, n_2 + \frac{N_2}{2}). \quad (5.73)$$

Both (5.62) and (5.63) are the $N_1/2 \times N_2/2$ type-II DHT, and (5.64) and (5.65) are the $N_1/2 \times N_2/2$ type-II even-odd sequence which can be computed as shown in the last subsection. The final outputs of (5.61) can be obtained by

$$X_{OE}(2k_1, 2k_2) = F_2(k_1, k_2) + G_2(k_1, k_2) \quad (5.74)$$
$$X_{OE}(2k_1 - 1, 2k_2) = F_2(k_1, k_2) - G_2(k_1, k_2) \quad (5.75)$$
$$X_{OE}(2k_1, 2k_2 + 1) = F_3(k_1, k_2) + G_3(k_1, k_2) \quad (5.76)$$
$$X_{OE}(2k_1 - 1, 2k_2 + 1) = F_3(k_1, k_2) - G_3(k_1, k_2), \quad (5.77)$$

where for (5.74) – (5.77), $0 \leq k_1 \leq N_1/2 - 1, 1 \leq k_2 \leq N_2/2 - 1$ and $X_{OE}(-1, k_2) = -X_{OE}(N_1 - 1, k_2)$ for any positive integer k_2.

5.3.2 Algorithm for type-III DHT

Decomposition techniques similar to those used in the last subsections can be applied to decompose the type-III DHT which is defined in (5.3). Instead of giving the details of the derivation, the necessary equations are listed to describe the decomposition process. The type-III DHT can be computed by

$$X_{III}(k_1, k_2) = A(k_1, k_2) + B(k_1, k_2) + C(k_1, k_2) + D(k_1, k_2), \quad (5.78)$$

where

$$A(k_1, k_2) = A(k_1, k_2 + \frac{N_2}{2}) = A(k_1 + \frac{N_1}{2}, k_2) = A(k_1 + \frac{N_1}{2}, k_2 + \frac{N_2}{2})$$

$$= \sum_{n_1=0}^{N_1/2-1} \sum_{n_2=0}^{N_2/2-1} x(2n_1, 2n_2) \mathrm{cas} \left(\frac{\pi n_1(2k_1+1)}{N_1/2} + \frac{\pi n_2(2k_2+1)}{N_2/2} \right) \quad (5.79)$$

$$B(k_1, k_2) = B(k_1, k_2 + \frac{N_2}{2}) = -B(k_1 + \frac{N_1}{2}, k_2) = -B(k_1 + \frac{N_1}{2}, k_2 + \frac{N_2}{2})$$

$$= \sum_{n_1=0}^{N_1/2-1} \sum_{n_2=0}^{N_2/2-1} x(2n_1 + 1, 2n_2)$$

$$\cdot \mathrm{cas} \left(\frac{\pi(2n_1+1)(2k_1+1)}{2(N_1/2)} + \frac{\pi n_2(2k_2+1)}{N_2/2} \right) \quad (5.80)$$

$$C(k_1, k_2) = -C(k_1, k_2 + \frac{N_2}{2}) = C(k_1 + \frac{N_1}{2}, k_2) = -C(k_1 + \frac{N_1}{2}, k_2 + \frac{N_2}{2})$$

$$= \sum_{n_1=0}^{N_1/2-1} \sum_{n_2=0}^{N_2/2-1} x(2n_1, 2n_2 + 1)$$

$$\cdot \mathrm{cas} \left(\frac{\pi n_1(2k_1+1)}{N_1/2} + \frac{\pi(2n_2+1)(2k_2+1)}{2(N_2/2)} \right) \quad (5.81)$$

$$D(k_1, k_2) = -D(k_1, k_2 + \frac{N_2}{2}) = -D(k_1 + \frac{N_1}{2}, k_2) = C(k_1 + \frac{N_1}{2}, k_2 + \frac{N_2}{2})$$

$$= \sum_{n_1=0}^{N_1/2-1} \sum_{n_2=0}^{N_2/2-1} x(2n_1 + 1, 2n_2 + 1)$$

$$\cdot \mathrm{cas} \left(\frac{\pi(2n_1+1)(2k_1+1)}{2(N_1/2)} + \frac{\pi(2n_2+1)(2k_2+1)}{2(N_2/2)} \right), \quad (5.82)$$

where for (5.79) – (5.82), $0 \le k_1 < N_1/2$ and $1 \le k_2 < N_2/2$. It is noted that (5.79) is a type-III DHT, and (5.82) is a type-IV DHT which can be further decomposed as shown in the next subsection. Furthermore, (5.80) and (5.81) are the type-III odd-even and even-odd arrays. Because they have a symmetric relation, we show only the process of decomposing (5.80) in the rest of this subsection.

Let us define the $N_1 \times N_2$ type-III even-odd array to be

$$X_{EO}(k_1, k_2) = \sum_{n_1=0}^{N_1-1} \sum_{n_2=0}^{N_2-1} x(n_1, n_2) \mathrm{cas} \left(\frac{\pi n_1(2k_1+1)}{N_1} + \frac{\pi(2n_2+1)(2k_2+1)}{2N_2} \right) \quad (5.83)$$

which can be expressed by

$$X_{EO}(k_1, k_2) = \cos \frac{\pi(2k_2+1)}{2N_2} E_0(k_1, k_2) + \sin \frac{\pi(2k_2+1)}{2N_2} E_1(k_1, k_2), \quad (5.84)$$

where

$$
\begin{aligned}
E_0(k_1, k_2) =\ & \sum_{n_1=0}^{N_1/2-1} \sum_{n_2=0}^{N_2/2-1} u_0(n_1, n_2)\mathrm{cas}\left(\frac{\pi n_1(2k_1+1)}{N_1/2} + \frac{\pi n_2(2k_2+1)}{N_2/2}\right) \\
& + \sum_{n_1=0}^{N_1/2-1} \sum_{n_2=0}^{N_2/2-1} u_1(n_1, n_2) \\
& \cdot \mathrm{cas}\left(\frac{\pi(2n_1+1)(2k_1+1)}{2(N_1/2)} + \frac{\pi n_2(2k_2+1)}{N_2/2}\right)
\end{aligned} \tag{5.85}
$$

in which the first term is an $N_1/2 \times N_2/2$ type-III DHT and the second term is the $N_1/2 \times N_2/2$ type-III odd-even sequence. The input sequences in (5.85) are

$$
\begin{aligned}
u_0(n_1, n_2) &= x(2n_1, 2n_2) + x(2n_1, 2n_2 - 1) \\
u_1(n_1, n_2) &= x(2n_1+1, 2n_2) + x(2n_1+1, 2n_2 - 1).
\end{aligned} \tag{5.86}
$$

Similarly we have

$$
\begin{aligned}
E_1(k_1, k_2) =\ & \sum_{n_1=0}^{N_1/2-1} \sum_{n_2=0}^{N_2/2-1} v_0(n_1, n_2)\mathrm{cas}\left(\frac{-\pi n_1(2k_1+1)}{N_1/2} - \frac{\pi n_2(2k_2+1)}{N_2/2}\right) \\
& + \sum_{n_1=0}^{N_1/2-1} \sum_{n_2=0}^{N_2/2-1} v_1(n_1, n_2) \\
& \cdot \mathrm{cas}\left(\frac{-\pi(2n_1+1)(2k_1+1)}{2(N_1/2)} - \frac{\pi n_2(2k_2+1)}{N_2/2}\right),
\end{aligned} \tag{5.87}
$$

where the input sequences are

$$
\begin{aligned}
v_0(n_1, n_2) &= x(2n_1, 2n_2) - x(2n_1, 2n_2 - 1) \\
v_1(n_1, n_2) &= x(2n_1+1, 2n_2) - x(2n_1+1, 2n_2 - 1),
\end{aligned} \tag{5.88}
$$

where in both (5.86) and (5.88), $x(n_1, -1) = -x(n_1, N_2 - 1)$ and n_1 is a positive integer. The $N_1/2 \times N_2/2$ type-III odd-even arrays in (5.85) and (5.87) can be further decomposed into two $N_1/4 \times N_2/4$ type-III DHTs and two $N_1/4 \times N_2/4$ type-III odd-even arrays.

5.3.3   Algorithm for type-IV DHT

The type-IV DHT defined in (5.4) can be decomposed into four $N_1/2 \times N_2/2$ type-II DHTs, as described below.

$$
\begin{aligned}
F_4(k_1, k_2) &= \frac{X_{IV}(2k_1, 2k_2) + X_{IV}(2k_1 - 1, 2k_2 - 1)}{2} \\
&= \sum_{n_1=0}^{N_1/2-1} \sum_{n_2=0}^{N_2/2-1} f_4(n_1, n_2)\mathrm{cas}\left(\frac{\pi(2n_1+1)k_1}{N_1/2} + \frac{\pi(2n_2+1)k_2}{N_2/2}\right)
\end{aligned} \tag{5.89}
$$

$$
\begin{aligned}
G_4(k_1, k_2) &= \frac{X_{IV}(2k_1, 2k_2) - X_{IV}(2k_1 - 1, 2k_2 - 1)}{2} \\
&= \sum_{n_1=0}^{N_1/2-1} \sum_{n_2=0}^{N_2/2-1} g_4(n_1, n_2)\mathrm{cas}\left(\frac{-\pi(2n_1+1)k_1}{N_1/2} - \frac{\pi(2n_2+1)k_2}{N_2/2}\right)
\end{aligned} \tag{5.90}
$$

$$F_5(k_1, k_2) = \frac{X_{IV}(2k_1, 2k_2 - 1) + X_{IV}(2k_1 - 1, 2k_2)}{2}$$

$$= \sum_{n_1=0}^{N_1/2-1} \sum_{n_2=0}^{N_2/2-1} f_5(n_1, n_2) \text{cas}\left(\frac{\pi(2n_1 + 1)k_1}{N_1/2} + \frac{\pi(2n_2 + 1)k_2}{N_2/2}\right) \quad (5.91)$$

$$G_5(k_1, k_2) = \frac{X_{IV}(2k_1, 2k_2 - 1) - X_{IV}(2k_1 - 1, 2k_2)}{2}$$

$$= \sum_{n_1=0}^{N_1/2-1} \sum_{n_2=0}^{N_2/2-1} g_5(n_1, n_2) \text{cas}\left(\frac{-\pi(2n_1 + 1)k_1}{N_1/2} - \frac{\pi(2n_2 + 1)k_2}{N_2/2}\right), \quad (5.92)$$

where for (5.89) – (5.92), $0 \le k_1 < N_1/2$, and $1 \le k_2 < N_2/2$. The input arrays for (5.89) – (5.92) are

$$f_4(n_1, n_2) = u_4(n_1, n_2) \cos\left(\frac{\pi(2n_1 + 1)}{2N_1} + \frac{\pi(2n_2 + 1)}{2N_2}\right)$$

$$- v_4(n_1, n_2) \sin\left(\frac{\pi(2n_1 + 1)}{2N_1} + \frac{\pi(2n_2 + 1)}{2N_2}\right) \quad (5.93)$$

$$g_4(n_1, n_2) = u_4(n_1, n_2) \sin\left(\frac{\pi(2n_1 + 1)}{2N_1} + \frac{\pi(2n_2 + 1)}{2N_2}\right)$$

$$+ v_4(n_1, n_2) \cos\left(\frac{\pi(2n_1 + 1)}{2N_1} + \frac{\pi(2n_2 + 1)}{2N_2}\right) \quad (5.94)$$

$$f_5(n_1, n_2) = u_5(n_1, n_2) \cos\left(\frac{\pi(2n_1 + 1)}{2N_1} - \frac{\pi(2n_2 + 1)}{2N_2}\right)$$

$$- v_5(n_1, n_2) \sin\left(\frac{\pi(2n_1 + 1)}{2N_1} - \frac{\pi(2n_2 + 1)}{2N_2}\right) \quad (5.95)$$

$$g_5(n_1, n_2) = u_5(n_1, n_2) \sin\left(\frac{\pi(2n_1 + 1)}{N_1} - \frac{\pi(2n_2 + 1)}{N_2}\right)$$

$$+ v_5(n_1, n_2) \cos\left(\frac{\pi(2n_1 + 1)}{N_1} - \frac{\pi(2n_2 + 1)}{N_2}\right), \quad (5.96)$$

where

$$u_4(n_1, n_2) = x(n_1, n_2) - x(n_1 + \frac{N_1}{2}, n_2 + \frac{N_2}{2}) \quad (5.97)$$

$$v_4(n_1, n_2) = x(n_1 + \frac{N_1}{2}, n_2) + x(n_1, n_2 + \frac{N_2}{2}) \quad (5.98)$$

$$u_5(n_1, n_2) = x(n_1, n_2) + x(\frac{N_1}{2}, n_2 + \frac{N_2}{2}) \quad (5.99)$$

$$v_5(n_1, n_2) = x(n_1 + \frac{N_1}{2}, n_2) - x(n_1, n_2 + \frac{N_2}{2}). \quad (5.100)$$

The final transform outputs can be obtained by

$$X_{IV}(2k_1, 2k_2) = F_4(k_1, k_2) + G_4(k_1, k_2) \quad (5.101)$$

$$X_{IV}(2k_1 - 1, 2k_2 - 1) = F_4(k_1, k_2) - G_4(k_1, k_2) \quad (5.102)$$

$$X_{IV}(2k_1, 2k_2 - 1) = F_5(k_1, k_2) + G_5(k_1, k_2) \quad (5.103)$$

$$X_{IV}(2k_1 - 1, 2k_2) = F_5(k_1, k_2) - G_5(k_1, k_2), \quad (5.104)$$

where for (5.101) – (5.104), $0 \le k_1 \le N_1/2 - 1$, $0 \le k_2 \le N_2/2 - 1$, and $X_{IV}(-1, -1) = X_{IV}(N_1 - 1, N_2 - 1)$, $X_{IV}(k_1, -1) = -X_{IV}(k_1, N_2 - 1)$, $X_{IV}(-1, k_2) = -X_{IV}(N_1 - 1, k_2)$, where k_1 and k_2 are any positive integers.

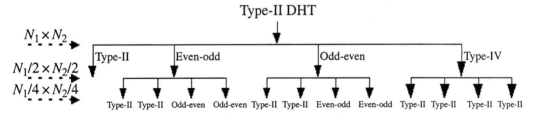

Figure 5.3   Decomposition process for the type-II DHT.

5.3.4   Computational complexity

In summary, the entire $N_1 \times N_2$ type-II DHT can be decomposed into a type-II DHT, an even-odd array, an odd-even array and a type-IV DHT whose transform size is $N_1/2 \times N_2/2$. Figure 5.3 shows the first two steps of the decomposition process for the type-II DHT. Because the type-II has a symmetric relation to the type-III DHT, we conclude that the decomposition of both types of DHT requires the same computational complexity. Therefore, the comments made for the type-II DHT are also valid for the type-III DHT.

The number of additions needed by the presented algorithm for the type-II DHT is

$$A_{II}(N_1 \times N_2) = A_{II}(\frac{N_1}{2} \times \frac{N_2}{2}) + A_{IV}(\frac{N_1}{2} \times \frac{N_2}{2}) + A_{EO}(\frac{N_1}{2} \times \frac{N_2}{2})$$
$$+ A_{OE}(\frac{N_1}{2} \times \frac{N_2}{2}) + 2N_1N_2, \tag{5.105}$$

where the last term is the number of additions needed by (5.43), and the number of multiplications is

$$M_{II}(N_1 \times N_2) = M_{II}(\frac{N_1}{2} \times \frac{N_2}{2}) + M_{IV}(\frac{N_1}{2} \times \frac{N_2}{2}) + M_{EO}(\frac{N_1}{2} \times \frac{N_2}{2})$$
$$+ M_{OE}(\frac{N_1}{2} \times \frac{N_2}{2}), \tag{5.106}$$

The $N_1 \times N_2$ type-IV DHT can be decomposed into four $N_1/2 \times N_2/2$ type-II DHTs. The required number of additions is

$$A_{IV}(N_1 \times N_2) = 4A_{II}(\frac{N_1}{2} \times \frac{N_2}{2}) + 3N_1N_2, \tag{5.107}$$

where the last term is the number of additions needed by (5.93) – (5.104), and the number of multiplications is

$$M_{IV}(N_1 \times N_2) = 4M_{II}(\frac{N_1}{2} \times \frac{N_2}{2}) + 2N_1N_2, \tag{5.108}$$

where the last term is the number of multiplications needed by the twiddle factors in (5.93) – (5.96). In general, it is possible to have trivial twiddle factors so that a number of operations can be saved. When $N = N_1 = N_2$, for example, $6N$ multiplications and $2N$ additions can be saved.

It can be easily verified from the decomposition process that the type-II (or type-III) even-odd and type-II (or type-III) odd-even arrays require the same number of additions and multiplications. Therefore, only the number of operations required by the even-odd type-II array is given below:

$$A_{EO}(N_1 \times N_2) = 2A_{OE}(\frac{N_1}{2} \times \frac{N_2}{2}) + 2A_{II}(\frac{N_1}{2} \times \frac{N_2}{2}) + 3N_1N_2 \qquad (5.109)$$

and the number of multiplications is

$$M_{EO}(N_1 \times N_2) = 2M_{OE}(\frac{N_1}{2} \times \frac{N_2}{2}) + 2M_{II}(\frac{N_1}{2} \times \frac{N_2}{2}) + \frac{3N_1N_2}{2}. \qquad (5.110)$$

Table 5.3 The required computational complexity.

N	Type-II DHT		Type-IV DHT		EO or OE	
	$M(N \times N)$	$A(N \times N)$	$M(N \times N)$	$A(N \times N)$	$M(N \times N)$	$A(N \times N)$
2	0	8	0	8	4	4
4	8	40	8	56	32	64
8	64	352	112	336	160	400
16	496	2016	672	2144	832	2272
32	2832	10752	3840	11072	4704	11648
64	16080	53312	19136	55168	21216	57088
128	77648	255424	96320	262144	99168	269952
256	372304	1188544	440128	1217792	451936	1247360

The last term of (5.110) is the number of multiplications needed by the twiddle factors as shown in (5.46), (5.50), (5.53) and (5.54). When (5.46) and (5.50) are substituted into (5.43), $0.5N_1N_2$ multiplications are needed. Based on (5.105) – (5.110), Table 5.3 lists the number of arithmetic operations needed for the computation of type-II, -IV and the even-odd type-II array whose transform size is $N = N_1 = N_2$. The number of arithmetic operations for $N = 2$ and 4 are achieved by an optimized computation.

The main improvement made by the algorithms is the regularity of the computational structure. It has all the desirable features, including in-place computation and regular data reordering, performed by the bit-reversal method, which have been achieved by the radix-type algorithms for other transform computations. The simplicity of the algorithms enables an easy implementation. Further reduction of the total computation time and implementation cost can be achieved. The techniques used in deriving the fast algorithms can easily be applied to derive fast algorithms for higher dimensional DHTs. For processing of high dimensional data, these issues will become the main challenges to obtain high processing throughput.

5.4   FAST ALGORITHMS BASED ON TYPE-I DHT

Fast algorithms for the type-II, -III and -IV DHTs can also be derived from the use of the type-I DHT. These algorithms are similar to the radix-2 algorithm in the sense that the entire computation of $N_1 \times N_2$, where $N_1 = q_1 \cdot 2^{m_1}$ and $N_2 = q_2 \cdot 2^{m_2}$, type-II and -III DHTs is decomposed into four $N_1/2 \times N_2/2$ type-I DHTs. We also make use of factors q_1 or q_2 in the transform sizes to reduce the number of multiplications and additions from the trivial twiddle factors. In a similar way, the type-IV 2D DHT can be divided into four type-II DHTs whose size is $N_1/2 \times N_2/2$. For simplicity of presentation, the fast algorithms are illustrated by 2D type-II, -III and -IV DHTs. The decomposition approach can be easily extended for the computation of higher dimensions.

5.4.1 Algorithm for 2D type-II DHT

The $N_1 \times N_2$ type-II DHT of input array $x(n_1, n_2)$ defined in (5.2) can be decomposed into

$$
\begin{aligned}
X(k_1, k_2) &= \sum_{n_1=0}^{N_1/2-1} \sum_{n_2=0}^{N_2/2-1} x\left(2n_1 + \frac{q_1 - 1}{2}, 2n_2 + \frac{q_2 - 1}{2}\right) \\
&\quad \cdot \mathrm{cas}\left(\frac{\pi(4n_1 + q_1)k_1}{N_1} + \frac{\pi(4n_2 + q_2)k_2}{N_2}\right) \\
&+ \sum_{n_1=0}^{N_1/2-1} \sum_{n_2=0}^{N_2/2-1} x\left(2n_1 - \frac{q_1 + 1}{2}, 2n_2 + \frac{q_2 - 1}{2}\right) \\
&\quad \cdot \mathrm{cas}\left(\frac{\pi(4n_1 - q_1)k_1}{N_1} + \frac{\pi(4n_2 + q_2)k_2}{N_2}\right) \\
&+ \sum_{n_1=0}^{N_1/2-1} \sum_{n_2=0}^{N_2/2-1} x\left(2n_1 + \frac{q_1 - 1}{2}, 2n_2 - \frac{q_2 + 1}{2}\right) \\
&\quad \cdot \mathrm{cas}\left(\frac{\pi(4n_1 + q_1)k_1}{N_1} + \frac{\pi(4n_2 - q_2)k_2}{N_2}\right) \\
&+ \sum_{n_1=0}^{N_1/2-1} \sum_{n_2=0}^{N_2/2-1} x\left(2n_1 - \frac{q_1 + 1}{2}, 2n_2 - \frac{q_2 + 1}{2}\right) \\
&\quad \cdot \mathrm{cas}\left(\frac{\pi(4n_1 - q_1)k_1}{N_1} + \frac{\pi(4n_2 - q_2)k_2}{N_2}.\right)
\end{aligned} \tag{5.111}
$$

By using the property $\mathrm{cas}(\alpha+\beta) = \cos(\alpha)\mathrm{cas}(\beta)+\sin(\alpha)\mathrm{cas}(-\beta)$, (5.111) can be decomposed into eight terms which can be grouped into

$$
\begin{aligned}
X(k_1, k_2) &= [\cos(\alpha_1 k_1 + \alpha_2 k_2)A(k_1, k_2) + \sin(\alpha_1 k_1 + \alpha_2 k_2)B(k_1, k_2)] \\
&\quad + [\cos(\alpha_1 k_1 - \alpha_2 k_2)C(k_1, k_2) + \sin(\alpha_1 k_1 - \alpha_2 k_2)D(k_1, k_2)]
\end{aligned} \tag{5.112}
$$

$$
\begin{aligned}
X(k_1 &+ \frac{N_1}{2}, k_2 + \frac{N_2}{2}) \\
&= (-1)^{\frac{q_1+q_2}{2}} [\sin(\alpha_1 k_1 + \alpha_2 k_2)A(k_1, k_2) - \cos(\alpha_1 k_1 + \alpha_2 k_2)B(k_1, k_2)] \\
&\quad + (-1)^{\frac{q_1-q_2}{2}} [\sin(\alpha_1 k_1 - \alpha_2 k_2)C(k_1, k_2) - \cos(\alpha_1 k_1 - \alpha_2 k_2)D(k_1, k_2)]
\end{aligned} \tag{5.113}
$$

$$X(k_1 + \frac{N_1}{2}, k_2)$$
$$= (-1)^{\frac{q_1+1}{2}}[\sin(\alpha_1 k_1 + \alpha_2 k_2)A(k_1, k_2) - \cos(\alpha_1 k_1 + \alpha_2 k_2)B(k_1, k_2)]$$
$$+ (-1)^{\frac{q_1+1}{2}}[\sin(\alpha_1 k_1 - \alpha_2 k_2)C(k_1, k_2) - \cos(\alpha_1 k_1 - \alpha_2 k_2)D(k_1, k_2)] \quad (5.114)$$

$$X(k_1, k_2 + \frac{N_2}{2})$$
$$= (-1)^{\frac{q_2+1}{2}}[\sin(\alpha_1 k_1 + \alpha_2 k_2)A(k_1, k_2) - \cos(\alpha_1 k_1 + \alpha_2 k_2)B(k_1, k_2)]$$
$$- (-1)^{\frac{q_2+1}{2}}[\sin(\alpha_1 k_1 - \alpha_2 k_2)C(k_1, k_2) - \cos(\alpha_1 k_1 - \alpha_2 k_2)D(k_1, k_2)], \quad (5.115)$$

where for $(5.112) - (5.115)$, $a_1 = \pi q_1/N_1$, $a_2 = \pi q_2/N_2$, $0 \le k_1 < N_1/2$, $0 \le k_2 < N_2/2$ and

$$A(k_1, k_2) = A(k_1 + \frac{i_1 N_1}{2}, k_2 + \frac{i_2 N_2}{2})$$
$$= \sum_{n_1=0}^{N_1/2-1} \sum_{n_2=0}^{N_2/2-1} y_0(n_1, n_2)\mathrm{cas}\left(\frac{2\pi n_1 k_1}{N_1/2} + \frac{2\pi n_2 k_2}{N_2/2}\right) \quad (5.116)$$

$$B(k_1, k_2) = B(k_1 + \frac{i_1 N_1}{2}, k_2 + \frac{i_2 N_2}{2})$$
$$= \sum_{n_1=0}^{N_1/2-1} \sum_{n_2=0}^{N_2/2-1} y_1(n_1, n_2)\mathrm{cas}\left(\frac{-2\pi n_1 k_1}{N_1/2} - \frac{2\pi n_2 k_2}{N_2/2}\right) \quad (5.117)$$

$$C(k_1, k_2) = C(k_1 + \frac{i_1 N_1}{2}, k_2 + \frac{i_2 N_2}{2})$$
$$= \sum_{n_1=0}^{N_1/2-1} \sum_{n_2=0}^{N_2/2-1} y_2(n_1, n_2)\mathrm{cas}\left(\frac{2\pi n_1 k_1}{N_1/2} + \frac{2\pi n_2 k_2}{N_2/2}\right) \quad (5.118)$$

$$D(k_1, k_2) = D(k_1 + \frac{i_1 N_1}{2}, k_2 + \frac{i_2 N_2}{2})$$
$$= \sum_{n_1=0}^{N_1/2-1} \sum_{n_2=0}^{N_2/2-1} y_3(n_1, n_2)\mathrm{cas}\left(\frac{-2\pi n_1 k_1}{N_1/2} - \frac{2\pi n_2 k_2}{N_2/2}\right), \quad (5.119)$$

where $i_1, i_2 = (0,1), (1,0), (1,1)$. The input sequences in $(5.116) - (5.119)$ are respectively

$$y_0(n_1, n_2) = x(2n_1 + \frac{q_1-1}{2}, 2n_2 + \frac{q_2-1}{2}) + x(2n_1 - \frac{q_1+1}{2}, 2n_2 - \frac{q_2+1}{2}) \quad (5.120)$$

$$y_1(n_1, n_2) = x(2n_1 + \frac{q_1-1}{2}, 2n_2 + \frac{q_2-1}{2}) - x(2n_1 - \frac{q_1+1}{2}, 2n_2 - \frac{q_2+1}{2}) \quad (5.121)$$

$$y_2(n_1, n_2) = x(2n_1 + \frac{q_1-1}{2}, 2n_2 - \frac{q_2+1}{2}) + x(2n_1 - \frac{q_1+1}{2}, 2n_2 + \frac{q_2-1}{2}) \quad (5.122)$$

$$y_3(n_1, n_2) = x(2n_1 + \frac{q_1-1}{2}, 2n_2 - \frac{q_2+1}{2}) - x(2n_1 - \frac{q_1+1}{2}, 2n_2 + \frac{q_2-1}{2}). \quad (5.123)$$

When $n_1 = 0$ and/or $n_2 = 0$, the index of the input in (5.123) becomes negative. The data with negative index can be substituted by

$$x(t_1, t_2) = \begin{cases} x(N_1 + t_1, t_2), & t_1 < 0, t_2 > 0 \\ x(t_1, N_2 + t_2), & t_1 > 0, t_2 < 0 \\ x(N_1 + t_1, N_2 + t_2), & t_1 < 0, t_2 < 0 \end{cases} \quad . \quad (5.124)$$

The negative sign of $B(k_1, k_2)$ and $D(k_1, k_2)$ in (5.116) – (5.119) can be eliminated by substituting k_1 and k_2 with $N_1/2 - k_1$ and $N_2/2 - k_2$, respectively. Therefore, the 2D type-II DHT is decomposed into four $N_1/2 \times N_2/2$ type-I DHTs which can be computed by the fast algorithm presented previously. According to (5.112) – (5.115), the required number of multiplications is

$$M_{II}(N_1 \times N_2) = 4M_I(\frac{N_1}{2} \times \frac{N_2}{2}) + 2N_1N_2 - T - T_m \tag{5.125}$$

and the number of additions is

$$A_{II}(N_1 \times N_2) = 4A_I(\frac{N_1}{2} \times \frac{N_2}{2}) + 3N_1N_2 - T_a. \tag{5.126}$$

The last terms T_m in (5.125) and T_a in (5.126) are the number of multiplications and additions, respectively, saved from trivial twiddle factors in (5.112) – (5.115). When $N = N_1 = N_2 > 2$ and $q = q_1 = q_2$, for example, T_a and T_m are $2qN$ and $14q(N - 2q)$, respectively. It is shown that the split-radix algorithms in [3] can be decomposed into one $N_1/2 \times N_2/2$ and twelve $N_1/4 \times N_2/4$ DHTs with some twiddle factors. Because both $A(k_1, k_2)$ and $C(k_1, k_2)$ in (5.112) – (5.115) are the same type of DHT, they require the same twiddle factors. Therefore, half of the multiplications used by the twiddle factors for the DHT computation can be saved if appreciate arrangements are made to utilize this property. Similar savings can be achieved from $B(k_1, k_2)$ and $D(k_1, k_2)$ in (5.112) – (5.115). In (5.125) the third term T is the number of multiplications saved from these twiddle factors. For $N = N_1 = N_2 > 8$, $T = 3N^2/4 - 9qN$.

5.4.2   Algorithm for 2D type-III DHT

In the last subsection, the decomposition is based on regrouping the input data, which corresponds to the decimation-in-time decomposition. The decimation-in-frequency approach can be used to regroup the transform outputs. For example, (5.3) can be decomposed into

$$x(2n_1 + \frac{q_1 - 1}{2}, 2n_2 + \frac{q_2 - 1}{2})$$
$$= \sum_{k_1=0}^{N_1-1} \sum_{k_2=0}^{N_2-1} X(k_1, k_2) \text{cas}\left(\frac{\pi(4n_1 + q_1)k_1}{N_1} + \frac{\pi(4n_2 + q_2)k_2}{N_2}\right) \tag{5.127}$$

$$x(2n_1 - \frac{q_1 - 1}{2}, 2n_2 - \frac{q_2 - 1}{2})$$
$$= \sum_{k_1=0}^{N_1-1} \sum_{k_2=0}^{N_2-1} X(k_1, k_2) \text{cas}\left(\frac{\pi(4n_1 - q_1)k_1}{N_1} + \frac{\pi(4n_2 - q_2)k_2}{N_2}\right) \tag{5.128}$$

$$x(2n_1 + \frac{q_1 - 1}{2}, 2n_2 - \frac{q_2 - 1}{2})$$
$$= \sum_{k_1=0}^{N_1-1} \sum_{k_2=0}^{N_2-1} X(k_1, k_2) \text{cas}\left(\frac{\pi(4n_1 + q_1)k_1}{N_1} + \frac{\pi(4n_2 - q_2)k_2}{N_2}\right) \tag{5.129}$$

$$x(2n_1 - \frac{q_1 - 1}{2}, 2n_2 + \frac{q_2 - 1}{2})$$
$$= \sum_{k_1=0}^{N_1-1} \sum_{k_2=0}^{N_2-1} X(k_1, k_2) \text{cas}\left(\frac{\pi(4n_1 - q_1)k_1}{N_1} + \frac{\pi(4n_2 + q_2)k_2}{N_2}\right). \tag{5.130}$$

If (5.127) and (5.130) are combined by addition and subtraction, we have

$$
\begin{aligned}
a(n_1, n_2) &= \frac{x(2n_1 + \frac{q_1-1}{2}, 2n_2 + \frac{q_2-1}{2}) + x(2n_1 - \frac{q_1-1}{2}, 2n_2 - \frac{q_2-1}{2})}{2} \\
&= \sum_{k_1=0}^{N_1/2-1} \sum_{k_2=0}^{N_2/2-1} \left\{ \left[X(k_1, k_2) + (-1)^{\frac{q_1+q_2}{2}} X(k_1 + \frac{N_1}{2}, k_2 + \frac{N_2}{2}) \right] \right. \\
&\quad \cdot \cos(\alpha_1 k_1 + \alpha_2 k_2) + \left[(-1)^{(q_1+1)/2} X(k_1 + \frac{N_1}{2}, k_2) \right. \\
&\quad \left. + (-1)^{(q_2+1)/2} X(k_1, k_2 + \frac{N_2}{2}) \right] \sin(\alpha_1 k_1 + \alpha_2 k_2) \Bigg\} \\
&\quad \cdot \cos\left(\frac{2\pi n_1 k_1}{N_1/2} + \frac{2\pi n_2 k_2}{N_2/2} \right)
\end{aligned}
\tag{5.131}
$$

$$
\begin{aligned}
b(n_1, n_2) &= \frac{x(2n_1 + \frac{q_1-1}{2}, 2n_2 + \frac{q_2-1}{2}) - x(2n_1 - \frac{q_1-1}{2}, 2n_2 - \frac{q_2-1}{2})}{2} \\
&= \sum_{k_1=0}^{N_1/2-1} \sum_{k_2=0}^{N_2/2-1} \left\{ \left[X(k_1, k_2) + (-1)^{\frac{q_1+q_2}{2}} X(k_1 + \frac{N_1}{2}, k_2 + \frac{N_2}{2}) \right] \right. \\
&\quad \cdot \sin(\alpha_1 k_1 + \alpha_2 k_2) + \left[(-1)^{\frac{q_1-1}{2}} X(k_1 + \frac{N_1}{2}, k_2) \right. \\
&\quad \left. + (-1)^{\frac{q_2-1}{2}} X(k_1, k_2 + \frac{N_2}{2}) \right] \cos(\alpha_1 k_1 + \alpha_2 k_2) \Bigg\} \\
&\quad \cdot \cos\left(\frac{-2\pi n_1 k_1}{N_1/2} - \frac{2\pi n_2 k_2}{N_2/2} \right)
\end{aligned}
\tag{5.132}
$$

$$
\begin{aligned}
c(n_1, n_2) &= \frac{x(2n_1 + \frac{q_1-1}{2}, 2n_2 - \frac{q_2+1}{2}) + x(2n_1 - \frac{q_1+1}{2}, 2n_2 + \frac{q_2-1}{2})}{2} \\
&= \sum_{k_1=0}^{N_1/2-1} \sum_{k_2=0}^{N_2/2-1} \left\{ \left[X(k_1, k_2) + (-1)^{\frac{q_1-q_2}{2}} X(k_1 + \frac{N_1}{2}, k_2 + \frac{N_2}{2}) \right] \right. \\
&\quad \cdot \cos(\alpha_1 k_1 - \alpha_2 k_2) - \left[(-1)^{\frac{q_1-1}{2}} X(k_1 + \frac{N_1}{2}, k_2) \right. \\
&\quad \left. - (-1)^{\frac{q_2-1}{2}} X(k_1, k_2 + \frac{N_2}{2}) \right] \sin(\alpha_1 k_1 - \alpha_2 k_2) \Bigg\} \\
&\quad \cdot \cos\left(\frac{2\pi n_1 k_1}{N_1/2} + \frac{2\pi n_2 k_2}{N_2/2} \right)
\end{aligned}
\tag{5.133}
$$

$$
\begin{aligned}
d(n_1, n_2) &= \frac{x(2n_1 + \frac{q_1-1}{2}, 2n_2 - \frac{q_2+1}{2}) - x(2n_1 - \frac{q_1+1}{2}, 2n_2 + \frac{q_2-1}{2})}{2} \\
&= \sum_{k_1=0}^{N_1/2-1} \sum_{k_2=0}^{N_2/2-1} \left\{ \left[X(k_1, k_2) + (-1)^{\frac{q_1-q_2}{2}} X(k_1 + \frac{N_1}{2}, k_2 + \frac{N_2}{2}) \right] \right.
\end{aligned}
$$

$$\cdot \sin{(\alpha_1 k_1 - \alpha_2 k_2)} + \left[(-1)^{\frac{q_1-1}{2}} X(k_1 + \frac{N_1}{2}, k_2) \right.$$

$$\left. - (-1)^{\frac{q_2-1}{2}} X(k_1, k_2 + \frac{N_2}{2}) \right] \cos{(\alpha_1 k_1 - \alpha_2 k_2)} \Big\}$$

$$\cdot \mathrm{cas}\left(-\frac{2\pi n_1 k_1}{N_1/2} - \frac{2\pi n_2 k_2}{N_2/2} \right). \tag{5.134}$$

Therefore, the type-III DHT is decomposed into four $N_1/2 \times N_2/2$ type-I DHTs. The final transform outputs are given by

$$x(2n_1 + \frac{q_1-1}{2}, 2n_2 + \frac{q_2-1}{2}) = a(n_1, n_2) + b(n_1, n_2) \tag{5.135}$$

$$x(2n_1 - \frac{q_1+1}{2}, 2n_2 - \frac{q_2+1}{2}) = a(n_1, n_2) - b(n_1, n_2) \tag{5.136}$$

$$x(2n_1 + \frac{q_1-1}{2}, 2n_2 - \frac{q_2+1}{2}) = c(n_1, n_2) + d(n_1, n_2) \tag{5.137}$$

$$x(2n_1 - \frac{q_1+1}{2}, 2n_2 + \frac{q_2-1}{2}) = c(n_1, n_2) - d(n_1, n_2) \tag{5.138}$$

for $0 \le n_1 < N_1/2$, and $0 \le n_2 < N_2/2$. When the array indices in (5.135) – (5.138) are negative or larger than N_1 or N_2, we have

$$x(t_1, t_2) = \begin{cases} x[(t_1) \bmod N_1, (t_2) \bmod N_2], & t_1 \le N_1 \text{ and/or } t_2 \le N_2 \\ x[(N_1 + t_1) \bmod N_1, (N_1 + t_2) \bmod N_2], & t_1 < 0 \text{ and/or } t_2 < 0 \end{cases} \tag{5.139}$$

It can be easily seen that the algorithm for the type-III DHT is based on the decimation-in-frequency decomposition and that for the type-II DHT is based on the decimation-in-time decomposition. However, the techniques used in the decomposition process are the same. Therefore, the numbers of additions and multiplications required by the presented algorithm for the type-III DHT are the same as those used by the algorithm for the type-II DHT. By changing the subscript II into III, (5.125) and (5.126) can be used to calculate the numbers of multiplications and additions for the type-III DHT.

5.4.3   Algorithm for 2D type-IV DHT

The 2D type-IV DHT of array $X(k1, k2)$ defined in (5.4) can be converted into four type-II DHTs, as shown below:

$$A(k_1, k_2) = \frac{X(2k_1 + \frac{q_1-1}{2}, 2k_2 + \frac{q_2-1}{2}) + X(2k_1 - \frac{q_1+1}{2}, 2k_2 - \frac{q_2+1}{2})}{2}$$

$$= \sum_{n_1=0}^{N_1/2-1} \sum_{n_2=0}^{N_2/2-1} \left\{ \left[x(n_1, n_2) + (-1)^{\frac{q_1+q_2}{2}} x(n_1 + \frac{N_1}{2}, n_2 + \frac{N_2}{2}) \right] \right.$$

$$\cdot \cos[\alpha_1(n_1 + 0.5) + \alpha_2(n_2 + 0.5)] - \left[(-1)^{\frac{q_1-1}{2}} x(n_1 + \frac{N_1}{2}, n_2) \right.$$

$$\left. + (-1)^{\frac{q_2-1}{2}} x(n_1, n_2 + \frac{N_2}{2}) \right] \sin[\alpha_1(n_1 + 0.5) + \alpha_2(n_2 + 0.5)] \Big\}$$

$$\cdot \mathrm{cas}\left(\frac{\pi(2n_1 + 1)k_1}{N_1/2} + \frac{\pi(2n_2 + 1)k_2}{N_2/2} \right) \tag{5.140}$$

$$B(k_1, k_2) = \frac{X(2k_1 + \frac{q_1-1}{2}, 2k_2 + \frac{q_2-1}{2}) - X(2k_1 - \frac{q_1+1}{2}, 2k_2 - \frac{q_2+1}{2})}{2}$$

$$= \sum_{n_1=0}^{N_1/2-1} \sum_{n_2=0}^{N_2/2-1} \left\{ \left[x(n_1, n_2) + (-1)^{\frac{q_1+q_2}{2}} x(n_1 + \frac{N_1}{2}, n_2 + \frac{N_2}{2}) \right] \right.$$

$$\cdot \sin\left[\alpha_1(n_1 + 0.5) + \alpha_2(n_2 + 0.5) \right] + \left[(-1)^{\frac{q_1-1}{2}} x(n_1 + \frac{N_1}{2}, n_2) \right.$$

$$\left. + (-1)^{\frac{q_2-1}{2}} x(n_1, n_2 + \frac{N_2}{2}) \right] \cos\left[\alpha_1(n_1 + 0.5) + \alpha_2(n_2 + 0.5) \right] \right\}$$

$$\cdot \text{cas}\left(-\frac{\pi(2n_1+1)k_1}{N_1/2} - \frac{\pi(2n_2+1)k_2}{N_2/2} \right) \tag{5.141}$$

$$C(k_1, k_2) = \frac{X(2k_1 + \frac{q_1-1}{2}, 2k_2 - \frac{q_2+1}{2}) + X(2k_1 - \frac{q_1+1}{2}, 2k_2 + \frac{q_2-1}{2})}{2}$$

$$= \sum_{n_1=0}^{N_1/2-1} \sum_{n_2=0}^{N_2/2-1} \left\{ \left[x(n_1, n_2) + (-1)^{\frac{q_1-q_2}{2}} x(n_1 + \frac{N_1}{2}, n_2 + \frac{N_2}{2}) \right] \right.$$

$$\cdot \cos\left[\alpha_1(n_1 + 0.5) - \alpha_2(n_2 + 0.5) \right] - \left[(-1)^{\frac{q_1-1}{2}} x(n_1 + \frac{N_1}{2}, n_2) \right.$$

$$\left. - (-1)^{\frac{q_2-1}{2}} x(n_1, n_2 + \frac{N_2}{2}) \right] \sin\left[\alpha_1(n_1 + 0.5) - \alpha_2(n_2 + 0.5) \right] \right\}$$

$$\cdot \text{cas}\left(\frac{\pi(2n_1+1)k_1}{N_1/2} + \frac{\pi(2n_2+1)k_2}{N_2/2} \right) \tag{5.142}$$

$$D(k_1, k_2) = \frac{X(2k_1 + \frac{q_1-1}{2}, 2k_2 - \frac{q_2+1}{2}) - X(2k_1 - \frac{q_1+1}{2}, 2k_2 + \frac{q_2-1}{2})}{2}$$

$$= \sum_{n_1=0}^{N_1/2-1} \sum_{n_2=0}^{N_2/2-1} \left\{ \left[x(n_1, n_2) + (-1)^{\frac{q_1-q_2}{2}} x(n_1 + \frac{N_1}{2}, n_2 + \frac{N_2}{2}) \right] \right.$$

$$\cdot \sin\left[\alpha_1(n_1 + 0.5) - \alpha_2(n_2 + 0.5) \right] + \left[(-1)^{\frac{q_1-1}{2}} x(n_1 + \frac{N_1}{2}, n_2) \right.$$

$$\left. - (-1)^{\frac{q_2-1}{2}} x(n_1, n_2 + \frac{N_2}{2}) \right] \cos\left[\alpha_1(n_1 + 0.5) - \alpha_2(n_2 + 0.5) \right] \right\}$$

$$\cdot \text{cas}\left(-\frac{\pi(2n_1+1)k_1}{N_1/2} - \frac{\pi(2n_2+1)k_2}{N_2/2}. \right) \tag{5.143}$$

The final transform outputs are given by

$$X(2k_1 + \frac{q_1-1}{2}, 2k_2 + \frac{q_2-1}{2}) = A(k_1, k_2) + B(k_1, k_2) \tag{5.144}$$

$$X(2k_1 - \frac{q_1+1}{2}, 2k_2 - \frac{q_2+1}{2}) = A(k_1, k_2) - B(k_1, k_2) \tag{5.145}$$

$$X(2k_1 + \frac{q_1-1}{2}, 2k_2 - \frac{q_2+1}{2}) = C(k_1, k_2) + D(k_1, k_2) \tag{5.146}$$

$$X(2k_1 - \frac{q_1+1}{2}, 2k_2 + \frac{q_2-1}{2}) = C(k_1, k_2) - D(k_1, k_2), \tag{5.147}$$

where $0 \leq k_1 < N_1/2$, and $0 \leq k_2 < N_2/2$. Similar to (5.139), the following properties have to be applied when the indices of the transform outputs in (5.144) – (5.147) become negative or larger than N_1 and/or N_2:

$$X(t_1, t_2) = \begin{cases} -X[(t_1) \bmod N_1, (t_2) \bmod N_2], & t_1 \leq N_1 \quad \text{or} \quad t_2 \leq N_2 \\ X[(t_1) \bmod N_1, (t_2) \bmod N_2], & t_1 \leq N_1 \quad \text{and} \quad t_2 \leq N_2 \\ -X[(N_1 + t_1) \bmod N_1, (N_1 + t_2) \bmod N_2], & t_1 < 0 \quad \text{or} \quad t_2 < 0 \\ X[(N_1 + t_1) \bmod N_1, (N_1 + t_2) \bmod N_2], & t_1 < 0 \quad \text{and} \quad t_2 < 0. \end{cases} \quad (5.148)$$

Therefore, the type-IV DHT is converted into four $N_1/2 \times N_2/2$ type-II DHTs. The required number of multiplications is

$$M_{IV}(N_1 \times N_2) = 4M_{II}(\frac{N_1}{2} \times \frac{N_2}{2}) + 2N_1 N_2 - T_m \quad (5.149)$$

and the number of additions is

$$A_{IV}(N_1 \times N_2) = 4A_{II}(\frac{N_1}{2} \times \frac{N_2}{2}) + 3N_1 N_2 - T_a, \quad (5.150)$$

where $T_m = 6qN$ and $T_a = 2qN$ for $N = N_1 = N_2$.

5.4.4 Algorithm for 3D type-II DHT

In this subsection, we use the 3D type-II DHT as an example to show the decomposition approach to be applied to a higher dimensional DHT. The 3D type-II DHT of array $x(n_1, n_2, n_3)$ is defined by

$$X(k_1, k_2, k_3) = \sum_{n_1=0}^{N_1-1} \sum_{n_2=0}^{N_2-1} \sum_{n_3=0}^{N_3-1} x(n_1, n_2, n_3)$$
$$\cdot \text{cas}\left(\frac{\pi(2n_1 + 1)k_1}{N_1} + \frac{\pi(2n_2 + 1)k_2}{N_2} + \frac{\pi(2n_3 + 1)k_3}{N_3}\right), \quad (5.151)$$

where $N_1 = q_1 \cdot 2^{m_1}, N_2 = q_2 \cdot 2^{m_2}$ and $N_3 = q_3 \cdot 2^{m_3}$. For simplicity, we assume the odd factors $q = q_1 = q_2 = q_3$ without loss of generality. The 3D type-II DHT can be rewritten into

$$X(k_1, k_2, k_3) = \sum_{l_1=0}^{1} \sum_{l_2=0}^{1} \sum_{l_3=0}^{1} w(k_1, k_2, k_3, l_1, l_2, l_3), \quad (5.152)$$

where

$$w(k_1, k_2, k_3, l_1, l_2, l_3) = \sum_{n_1=0}^{N_1/2-1} \sum_{n_2=0}^{N_2/2-1} \sum_{n_3=0}^{N_3/2-1} y(l_1, l_2, l_3)$$
$$\cdot \text{cas}\left[4\pi\left(\frac{n_1 k_1}{N_1} + \frac{n_2 k_2}{N_2} + \frac{n_3 k_3}{N_3}\right) + \alpha(l_1, l_2, l_3)\right] \quad (5.153)$$

$$y(l_1, l_2, l_3) = x\left[2n_1 + (-1)^{l_1}(\frac{q-1}{2} + l_1), 2n_2 + (-1)^{l_2}(\frac{q-1}{2} + l_2),\right.$$
$$\left. 2n_3 + (-1)^{l_3}(\frac{q-1}{2} + l_3)\right] \quad (5.154)$$

and

$$\alpha(l_1, l_2, l_3) = q\pi \left[(-1)^{l_1}\frac{k_1}{N_1} + (-1)^{l_2}\frac{k_2}{N_2} + (-1)^{l_3}\frac{k_3}{N_3}\right]. \tag{5.155}$$

With the techniques similar to those used in the previous subsections, (5.153) can be decomposed into 16 terms which are then combined to rewrite (5.152) into

$$\begin{aligned}
X(k_1, k_2, k_3) = \ &\cos\left[\alpha(0,0,0)\right] \sum_{n_1=0}^{N_1/2-1} \sum_{n_2=0}^{N_2/2-1} \sum_{n_3=0}^{N_3/2-1} [y(0,0,0) + y(1,1,1)]\mathrm{cas}(\beta) \\
&+ \sin\left[\alpha(0,0,0)\right] \sum_{n_1=0}^{N_1/2-1} \sum_{n_2=0}^{N_2/2-1} \sum_{n_3=0}^{N_3/2-1} [y(0,0,0) - y(1,1,1)]\,\mathrm{cas}(-\beta) \\
&+ \cos\left[\alpha(0,0,1)\right] \sum_{n_1=0}^{N_1/2-1} \sum_{n_2=0}^{N_2/2-1} \sum_{n_3=0}^{N_3/2-1} [y(0,0,1) + y(1,1,0)]\,\mathrm{cas}(\beta) \\
&+ \sin\left[\alpha(0,0,1)\right] \sum_{n_1=0}^{N_1/2-1} \sum_{n_2=0}^{N_2/2-1} \sum_{n_3=0}^{N_3/2-1} [y(0,0,1) - y(1,1,0)]\,\mathrm{cas}(-\beta) \\
&+ \cos\left[\alpha(0,1,0)\right] \sum_{n_1=0}^{N_1/2-1} \sum_{n_2=0}^{N_2/2-1} \sum_{n_3=0}^{N_3/2-1} [y(0,1,0) + y(1,0,1)]\,\mathrm{cas}(\beta) \\
&+ \sin\left[\alpha(0,1,0)\right] \sum_{n_1=0}^{N_1/2-1} \sum_{n_2=0}^{N_2/2-1} \sum_{n_3=0}^{N_3/2-1} [y(0,1,0) - y(1,0,1)]\,\mathrm{cas}(-\beta) \\
&+ \cos\left[\alpha(0,1,1)\right] \sum_{n_1=0}^{N_1/2-1} \sum_{n_2=0}^{N_2/2-1} \sum_{n_3=0}^{N_3/2-1} [y(0,1,1) + y(1,0,0)]\,\mathrm{cas}(\beta) \\
&+ \sin\left[\alpha(0,1,1)\right] \sum_{n_1=0}^{N_1/2-1} \sum_{n_2=0}^{N_2/2-1} \sum_{n_3=0}^{N_3/2-1} [y(0,1,1) - y(1,0,0)]\,\mathrm{cas}(-\beta),
\end{aligned} \tag{5.156}$$

where

$$\beta = 4\pi \left[n_1\frac{k_1}{N_1} + n_2\frac{k_2}{N_2} + n_3\frac{k_3}{N_3}\right].$$

Therefore, the 3D type-II DHT is decomposed into eight $N_1/2 \times N_2/2 \times N_3/2$ type-I DHTs which are rotated by twiddle factors. Similar to the 2D type-II DHT, the transform outputs $X(k_1+i_1N_1/2, k_2+i_2N_2/2, k_3+i_3N_3/2)$, where $0 \le k_1 < N_1/2, 0 \le k_2 < N_2/2, 0 \le k_3 < N_3/2$, $(i_1, i_2, i_3) = (0,0,0)$ to $(1,1,1)$, can be obtained by combining the eight rotated $N_1/2 \times N_2/2 \times N_3/2$ type-II DHTs, which is not detailed here for simplicity of presentation. In general, for each (i_1, i_2, i_3), the twiddle factors cos( ) and sin( ) in (5.156) may change sign, or cos( ) and sin( ) are swapped with possible sign changes. Table 5.4 lists the twiddle factors in rows 0 and 5 and the associated signs for each (i_1, i_2, i_3). The twiddle factors in column j, $(j = 1, 2, \ldots, 8)$ are corresponding to the $(j-1)$th term in (5.156). Therefore, for each (i_1, i_2, i_3), the twiddle factor associated with the jth term is the twiddle factor in the

Table 5.4   Signs and twiddle factors for the 3D type-II DHT.

i_1, i_2, i_3	$\cos[\alpha(0,0,0)]$	$\sin[\alpha(0,0,0)]$	$\cos[\alpha(0,0,1)]$	$\sin[\alpha(0,0,1)]$
0,0,0	1	1	1	1
0,1,1	-1	-1	1	1
1,0,1	-1	-1	1	1
1,1,0	-1	-1	-1	-1
i_1, i_2, i_3	$\cos[\alpha(0,1,0)]$	$\sin[\alpha(0,1,0)]$	$\cos[\alpha(0,1,1)]$	$\sin[\alpha(0,1,1)]$
0,0,0	1	1	1	1
0,1,1	1	1	-1	-1
1,0,1	-1	-1	1	1
1,1,0	1	1	1	1
i_1, i_2, i_3	$\sin[\alpha(0,0,0)]$	$\cos[\alpha(0,0,0)]$	$\sin[\alpha(0,0,1)]$	$\cos[\alpha(0,0,1)]$
0,0,1	$(-1)^{\frac{(q+1)}{2}}$	$(-1)^{\frac{(q-1)}{2}}$	$(-1)^{\frac{(q-1)}{2}}$	$(-1)^{\frac{(q+1)}{2}}$
0,1,0	$(-1)^{\frac{(q+1)}{2}}$	$(-1)^{\frac{(q-1)}{2}}$	$(-1)^{\frac{(q+1)}{2}}$	$(-1)^{\frac{(q-1)}{2}}$
1,0,0	$(-1)^{\frac{(q+1)}{2}}$	$(-1)^{\frac{(q-1)}{2}}$	$(-1)^{\frac{(q+1)}{2}}$	$(-1)^{\frac{(q-1)}{2}}$
1,1,1	$(-1)^{\frac{(q-1)}{2}}$	$(-1)^{\frac{(q+1)}{2}}$	$(-1)^{\frac{(q+1)}{2}}$	$(-1)^{\frac{(q-1)}{2}}$
i_1, i_2, i_3	$\sin[\alpha(0,1,0)]$	$\cos[\alpha(0,1,0)]$	$\sin[\alpha(0,1,1)]$	$\cos[\alpha(0,1,1)]$
0,0,1	$(-1)^{\frac{(q+1)}{2}}$	$(-1)^{\frac{(q-1)}{2}}$	$(-1)^{\frac{(q-1)}{2}}$	$(-1)^{\frac{(q+1)}{2}}$
0,1,0	$(-1)^{\frac{(q-1)}{2}}$	$(-1)^{\frac{(q+1)}{2}}$	$(-1)^{\frac{(q-1)}{2}}$	$(-1)^{\frac{(q+1)}{2}}$
1,0,0	$(-1)^{\frac{(q+1)}{2}}$	$(-1)^{\frac{(q-1)}{2}}$	$(-1)^{\frac{(q+1)}{2}}$	$(-1)^{\frac{(q-1)}{2}}$
1,1,1	$(-1)^{\frac{(q+1)}{2}}$	$(-1)^{\frac{(q-1)}{2}}$	$(-1)^{\frac{(q-1)}{2}}$	$(-1)^{\frac{(q+1)}{2}}$

jth column with the sign located at the same column. Based on Table 5.4 and (5.156), the number of multiplications needed by the algorithm is

$$M_{II}(N_1 \times N_2 \times N_3) = 8M_I(\frac{N_1}{2} \times \frac{N_2}{2} \times \frac{N_3}{2}) + 2N_1N_2N_3 - T_m \qquad (5.157)$$

and the number of additions is

$$A_{II}(N_1 \times N_2 \times N_3) = 8A_I(\frac{N_1}{2} \times \frac{N_2}{2} \times \frac{N_3}{2}) + 3N_1N_2N_3 - T_a, \qquad (5.158)$$

where the last terms in (5.157) and (5.158) are the number of operations saved from the twiddle factors. In general, there are three categories of trivial twiddle factors according to their values and the ways in which arithmetic operations can be saved. The values of the twiddle factors in the first category are either 0, which needs no addition or multiplication, or ± 1 which needs no multiplication. Those in the second category have the same value. If they are used in the same summation such as (5.156), for example, a number of multiplications can be saved by performing summation before multiplication. The third category includes twiddle factors whose values are a power of two so that the associated multiplication can be replaced by a shift operation in the fixed point number system. For high dimensional discrete transforms, there are many trivial twiddle factors that belong to the second category. For large transform sizes, however, they are not regularly related to the data index. If savings on operations for twiddle factors are to be made, the computational structure of the algorithm becomes irregular and requires a substantial overhead of computation time because too many special arrangements in implementation are made. However, the trivial twiddle factors in the first category can be used because special arrangements in programming can easily be made. When the twiddle factors in the first category are used to minimize the number of operations, for example, $T_a = 2qN$ and $T_m = 4qN$ for $N = N_1 = N_2 = N_3$.

Table 5.5   The required computational complexity for the type-II DHT.

Algorithm in [31]			Presented algorithm				
$q = 1$			$q = 1$			$q = 3$	
N	$M(N)$	$A(N)$	$M(N)$	$A(N)$	N	$M(N)$	$A(N)$
8	64	384	44	432	6	16	228
16	384	2176	364	2368	12	100	1272
32	2048	11264	2204	11872	24	652	6348
64	10240	55296	12028	57632	48	4300	31296
128	49152	262144	62012	269600	96	23932	146787
256	229376	1212416	302908	1239456	192	124636	671772
512	1048576	5505024	1437116	5588256	384	623644	3051804

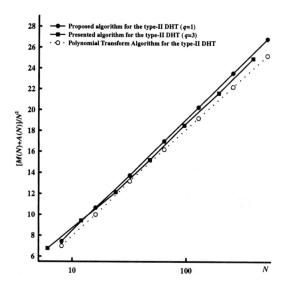

Figure 5.4   Comparison of number of arithmetic operations needed for type-II DHT.

5.4.5   Discussion

The fast algorithms presented in this section have a highly regular computational structure for low implementation complexity and can be easily extended to algorithms for higher dimensional processing. They possess all the desirable properties which have been available from the radix-type algorithms reported for various transforms ([3, 23], for example). An additional feature of these algorithms is that a parameter q is used in the computation. One advantage is that a wide range of transform sizes can be flexibly selected by setting the desirable value of q. In general, computation of a transform whose array size is other than a power of two by the radix-type fast algorithms requires extra arithmetic operations. In contrast, the derived algorithms achieve savings in the number of operations because the number of arithmetic operations saved from trivial twiddle factors proportionally increases with q. Table 5.5 lists the number of arithmetic operations needed by the 2D type-II DHT algorithms for $q = 1$ and 3.

Fast algorithms based on the PT for the type-II and -III DHTs were reported in [31]. Reduction of the number of operations was achieved by converting the MD type-II and -III DHTs into lower dimensional computations. The number of operations needed by the PT algorithm is given in Table 5.5. Figure 5.4 compares the computational complexity needed by the presented and the PT algorithms in terms of the number of arithmetic operations per transform output. It shows that the presented algorithm for both $q = 1$ and 3 uses more operations than that needed by the PT. Figure 5.4 also shows that the presented algorithm for $q = 3$ requires fewer operations than that needed for $q = 1$.

5.5 PT-BASED RADIX-2 ALGORITHM FOR MD TYPE-I DHT

In this section, we will discuss the PT-based algorithms for the MD type-I DHT [14, 32, 33]. We assume that the number of dimensions is r, where $r > 1$, and the dimension size N_i $(i = 1, 2, \ldots, r)$ is a power of two. Based on the index mapping and MD PT, the rD DHT is converted into a series of 1D type-II DHTs and PTs. The main advantages of the PT-based algorithms are structural regularity and reduction of the number of operations compared to other fast algorithms. In general, it is difficult to provide an algorithm that can be used for any number of dimensions. The only available algorithm in the literature is the row-column algorithm whose computational complexity is generally acceptable when both the number of dimensions and the dimensional sizes are small. In comparison, the PT-based algorithm is much more computationally efficient than the row-column algorithm.

5.5.1 Algorithm

The rD DHT is defined by

$$X(k_1, k_2, \ldots, k_r) = \sum_{n_1=0}^{N_1-1} \cdots \sum_{n_r=0}^{N_r-1} x(n_1, n_2, \ldots, n_r) \, \text{cas} \left(\frac{2\pi n_1 k_1}{N_1} + \cdots + \frac{2\pi n_r k_r}{N_r} \right), \quad (5.159)$$

where $k_i = 0, 1, \ldots, N_i - 1$, $i = 1, 2, \ldots, r$ and the dimensional sizes N_i are powers of two. Without loss of generality, we assume that $N_r \geq N_i$ $(i = 1, 2, \ldots, r - 1)$ which leads to $N_r = 2^t$ and $N_r/N_i = 2^{l_i}$, where $l_i \geq 0$. By dividing the inner sum into two parts, we have

$$X(k_1, k_2, \ldots, k_r) = \sum_{n_1=0}^{N_1-1} \cdots \sum_{n_{r-1}=0}^{N_{r-1}-1} \sum_{n_r=0}^{N_r/2-1} x(n_1, \ldots, n_{r-1}, 2n_r + 1)$$

$$\cdot \text{cas} \left(\frac{2\pi n_1 k_1}{N_1} + \cdots + \frac{2\pi n_{r-1} k_{r-1}}{N_{r-1}} + \frac{2\pi (2n_r + 1) k_r}{N_r} \right)$$

$$+ \sum_{n_1=0}^{N_1-1} \cdots \sum_{n_{r-1}=0}^{N_{r-1}-1} \sum_{n_r=0}^{N_r/2-1} x(n_1, \ldots, n_{r-1}, 2n_r)$$

$$\cdot \text{cas} \left(\frac{2\pi n_1 k_1}{N_1} + \cdots + \frac{2\pi n_{r-1} k_{r-1}}{N_{r-1}} + \frac{2\pi n_r k_r}{N_r/2} \right)$$

$$= Y(k_1, k_2, \ldots, k_r) + W(k_1, k_2, \ldots, k_r), \tag{5.160}$$

where

$$Y(k_1, k_2, \ldots, k_r + \frac{N_r}{2}) = -Y(k_1, k_2, \ldots, k_r) \tag{5.161}$$

$$W(k_1, k_2, \ldots, k_r + \frac{N_r}{2}) = W(k_1, k_2, \ldots, k_r). \tag{5.162}$$

In both (5.161) and (5.162), the index k_r is from 0 to $N_r/2 - 1$. Since $W(k_1, k_2, \ldots, k_r)$ is an rD type-I DHT with size $N_1 \times \cdots \times N_{r-1} \times (N_r/2)$, it can be further decomposed in the same way as shown above.

Let us consider the computation of $Y(k_1, k_2, \ldots, k_r)$ with the following lemma to convert $Y(k_1, k_2, \ldots, k_r)$ into an $(r-1)$D polynomial transform.

Lemma 4 *Let $p_i(n_r)$ be the least non-negative remainder of $(2n_r + 1)p_i$ modulo N_i, and $A = \{(n_1, \ldots, n_{r-1}, 2n_r + 1) \mid 0 \leq n_i \leq N_i - 1, 1 \leq i \leq r - 1, 0 \leq n_r \leq N_r/2 - 1\}$, $B = \{(p_1(n_r), p_2(n_r), \ldots, p_{r-1}(n_r), 2n_r + 1) \mid 0 \leq p_i \leq N_i - 1, 1 \leq i \leq r - 1, 0 \leq n_r \leq N_r/2 - 1\}$. Then $A = B$.*

Proof. The same lemma is used in Chapter 3 for developing the FPT algorithm for the MD DFT.

The first term in (5.160) is a multiple sum in which the indices belong to A. Since $A = B$, the indices also belong to B. Therefore we can convert it into

$$Y(k_1, k_2, \ldots, k_r) = \sum_{p_1=0}^{N_1-1} \cdots \sum_{p_{r-1}=0}^{N_{r-1}-1} \sum_{n_r=0}^{N_r/2-1} x(p_1(n_r), \ldots, p_{r-1}(n_r), 2n_r + 1)$$

$$\cdot \text{cas} \left(\frac{2\pi p_1(n_r)k_1}{N_1} + \cdots + \frac{2\pi p_{r-1}(n_r)k_{r-1}}{N_{r-1}} + \frac{2\pi(2n_r + 1)k_r}{N_r} \right). \tag{5.163}$$

Because $p_i(n_r) \equiv p_i(2n_r + 1) \mod N_i$, and the function cas( ) is periodic with period 2π, (5.163) becomes

$$Y(k_1, k_2, \ldots, k_r) = \sum_{p_1=0}^{N_1-1} \cdots \sum_{p_{r-1}=0}^{N_{r-1}-1} \sum_{n_r=0}^{N_r/2-1} x(p_1(n_r), \ldots, p_{r-1}(n_r), 2n_r + 1)$$

$$\cdot \text{cas} \frac{\pi(2n_r + 1)(p_1 k_1 2^{l_1} + \cdots + p_{r-1}k_{r-1}2^{l_{r-1}} + k_r)}{N_r/2}. \tag{5.164}$$

Let

$$V_{p_1, \ldots, p_{r-1}}(m) = \sum_{n_r=0}^{N_r/2-1} x(p_1(n_r), \ldots, p_{r-1}(n_r), 2n_r + 1) \text{cas} \frac{\pi(2n_r + 1)m}{N_r/2}, \tag{5.165}$$

where $p_i = 0, 1, \ldots, N_i - 1$, $i = 1, 2, \ldots, r - 1$ and $m = 0, 1, \ldots, N_r/2 - 1$. Equation (5.165) defines $N_1 N_2 \cdots N_{r-1}$ 1D type-II DHTs whose dimensional size is $N_r/2$. Therefore, (5.164) becomes

$$Y(k_1, k_2, \ldots, k_r) = \sum_{p_1=0}^{N_1-1} \sum_{p_2=0}^{N_2-1} \cdots \sum_{p_{r-1}=0}^{N_{r-1}-1}$$
$$V_{p_1,\ldots,p_{r-1}}(p_1 k_1 2^{l_1} + \cdots + p_{r-1} k_{r-1} 2^{l_{r-1}} + k_r). \tag{5.166}$$

Since $V_{p_1,\ldots,p_{r-1}}(m)$ has the property of

$$V_{p_1,\ldots,p_{r-1}}(m + uN_r/2) = (-1)^u V_{p_1,\ldots,p_{r-1}}(m), \tag{5.167}$$

(5.166) can be computed if $V_{p_1,\ldots,p_{r-1}}(m)$ is known for $m = 0, 1, \ldots, N_r/2 - 1$. Direct computation of (5.166) requires many additions. To reduce the required number of additions, let us compute (5.166) by using the PT. We define the generating polynomial of $Y(k_1, \ldots, k_r)$ by

$$Y_{k_1,\ldots,k_{r-1}}(z) = \sum_{k_r=0}^{N_r/2-1} Y(k_1, \ldots, k_r) z^{k_r}. \tag{5.168}$$

From (5.166) we have

$$Y_{k_1,\ldots,k_{r-1}}(z) \equiv \sum_{p_1=0}^{N_1-1} \cdots \sum_{p_{r-1}=0}^{N_{r-1}-1} \sum_{k_r=0}^{N_r/2-1} V_{p_1,\ldots,p_{r-1}}(p_1 k_1 2^{l_1} + \cdots + p_{r-1} k_{r-1} 2^{l_{r-1}} + k_r)$$
$$\cdot z^{k_r} \bmod (z^{N_r/2} + 1). \tag{5.169}$$

Because $V_{p_1,\ldots,p_{r-1}}(\lambda + k_r) z^{k_r} \bmod (z^{N_r/2} + 1)$ is periodic with period $N_r/2$ for variable k_r, (5.169) becomes

$$Y_{k_1,\ldots,k_{r-1}}(z) \equiv \sum_{p_1=0}^{N_1-1} \cdots \sum_{p_{r-1}=0}^{N_{r-1}-1} \sum_{k_r=0}^{N_r/2-1} V_{p_1,\ldots,p_{r-1}}(k_r)$$
$$\cdot z^{k_r - p_1 k_1 2^{l_1} - \cdots - p_{r-1} k_{r-1} 2^{l_{r-1}}} \bmod (z^{N_r/2} + 1). \tag{5.170}$$

If we define

$$U_{p_1,\ldots,p_{r-1}}(z) = \sum_{m=0}^{N_r/2-1} V_{p_1,\ldots,p_{r-1}}(m) z^m, \tag{5.171}$$

then (5.170) becomes

$$Y_{k_1,\ldots,k_{r-1}}(z) \equiv \sum_{p_1=0}^{N_1-1} \sum_{p_2=0}^{N_2-1} \cdots \sum_{p_{r-1}=0}^{N_{r-1}-1} U_{p_1,\ldots,p_{r-1}}(z)$$
$$\cdot z_1^{p_1 k_1} \cdots z_{r-1}^{p_{r-1} k_{r-1}} \bmod (z^{N_r/2} + 1), \tag{5.172}$$

where $z_i \equiv z^{-2^{l_i}} \bmod (z^{N_r/2} + 1)$ $(i = 1, 2, \ldots, r - 1)$. Equation (5.172) is an $(r - 1)$D PT. Once $Y_{k_1,\ldots,k_{r-1}}(z)$ is computed according to (5.172), $Y(k_1, \ldots, k_r)$ is obtained from the coefficients of $Y_{k_1,\ldots,k_{r-1}}(z)$.

The main steps of the PT-based algorithms for the type-I DHT whose dimensional sizes are powers of two are summarized below and illustrated in Figure 5.5.

Algorithm 5 *PT-based radix-2 algorithm for MD type-I DHT*

Step 1.   *Compute $N_1 N_2 \cdots N_{r-1}$ 1D type-II DHTs of length $N_r/2$ according to (5.165) and generate $U_{p_1,\dots,p_{r-1}}(z)$ according to (5.171).*

Step 2.   *Compute the $(r-1)$D polynomial transform of the polynomial sequence $U_{p_1,\dots,p_{r-1}}(z)$ according to (5.172) to get $Y_{k_1,\dots,k_{r-1}}(z)$ whose coefficients are $Y(k_1,\dots,k_r)$.*

Step 3.   *Compute $W(k_1,\dots,k_r)$, that is, the second term in (5.160), which is an MD type-I DHT with $k_r = 0,\dots,N_r/2 - 1$. The same method used in Steps 1 and 2 can be used for further decomposition.*

Step 4.   *Compute $X(k_1,\dots,k_r)$ from $Y(k_1,\dots,k_r)$ and $W(k_1,\dots,k_r)$ as described in (5.160).*

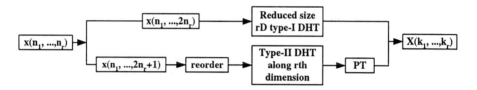

Figure 5.5   Flow chart of the PT-based radix-2 algorithm for MD type-I DHT.

5.5.2   Computational complexity

Now let us consider the number of arithmetic operations needed by the derived algorithm. We compute $N_1 N_2 \cdots N_{r-1}$ 1D type-II DHTs with length $N_r/2$ in Step 1 and an $(r-1)$D PT with size $N_1 \times \cdots \times N_{r-1}$ in Step 2. The known algorithms in [27] or Chapter 4 for computing a 1D type-II DHT of length $N_r/2$ require $\frac{1}{4}N_r \log_2 N_r - \frac{1}{2}N_r$ multiplications and $\frac{3}{4}N_r \log_2 N_r - \frac{3}{2}N_r$ additions, respectively. The $(r-1)$D PT in (5.172) requires $\frac{N}{2} \log_2(N_1 \cdots N_{r-1})$ additions (see Chapter 2 or [6, 18, 19, 24]), where $N = N_1 N_2 \cdots N_r$. Therefore, the total number of multiplications and additions can be expressed, respectively, by

$$M_u(N_1,\dots,N_{r-1},N_r) = M_u(N_1,\dots,N_{r-1},\frac{N_r}{2}) + \frac{N}{4} \log_2 N_r - \frac{N}{2} \qquad (5.173)$$

and

$$A_d(N_1,\dots,N_{r-1},N_r) = A_d(N_1,\dots,N_{r-1},\frac{N_r}{2}) + \frac{N}{2} \log_2 N + \frac{N}{4} \log_2 N_r - \frac{N}{2}. \qquad (5.174)$$

Based on (5.173) and (5.174), the computational complexity can be given by a very complex expression. For simplicity, let us consider the transform whose dimensional sizes are identical, that is, $N_1 = N_2 = \cdots = N_r = 2^t$. By recursively using (5.173) and (5.174), we obtain

$$M_u(2^t) = \frac{1}{2}(1 - \frac{1}{2^r})2^{rt}t - (1 - \frac{1}{2^r})2^{rt} + M_u(2^{t-1}) \qquad (5.175)$$

$$A_d(2^t) = (1 - \frac{1}{2^r})r2^{rt}t - (2 - \frac{r+2}{2^r})2^{rt} + \frac{1}{2}(1 - \frac{1}{2^r})2^{rt}t + A_d(2^{t-1}), \qquad (5.176)$$

where in (5.175), $M_u(2^t)$ represents the number of multiplications needed by the transform whose dimensional sizes are 2^t. Similarly, $A_d(2^t)$ in (5.176) represents the required number of additions. We can repeatedly use (5.175) and (5.176) until the dimensional size becomes 2 or the MD type-I DHT whose size is $2 \times \cdots \times 2$. It can easily be verified that $M_u(2) = 0$ and $A_d(2) = r2^r$. Therefore, (5.175) and (5.176) become

$$
\begin{aligned}
M_u(2^t) &= \frac{1}{2}(1 - \frac{1}{2^r}) \sum_{i=2}^{t} 2^{ri} i - (1 - \frac{1}{2^r}) \sum_{i=2}^{t} 2^{ri} + M_u(2) \\
&= \frac{1}{2}Nt - \frac{2^{r+1} - 1}{2^{r+1} - 2}N + \frac{2^{r+1} - 1}{2^{r+1} - 2} + \frac{1}{2}(2^r - 1)
\end{aligned}
\tag{5.177}
$$

$$
\begin{aligned}
A_d(2^t) &= r(1 - \frac{1}{2^r}) \sum_{i=2}^{t} 2^{ri} i - (2 - \frac{r+2}{2^r}) \sum_{i=2}^{t} 2^{ri} + \frac{1}{2}(1 - \frac{1}{2^r}) \sum_{i=2}^{t} 2^{ri} i + A_d(2) \\
&= (r + 1/2)Nt - \frac{2^{r+2} - 3}{2^{r+1} - 2}N + \frac{2^{r+2} - 3}{2^{r+1} - 2} + \frac{3}{2}(2^r - 1).
\end{aligned}
\tag{5.178}
$$

In general, the commonly used fast algorithms for the rD type-I DHT convert (5.159) into a transform whose kernel is separable before the row-column method is applied [7, 22]. The conversion process requires rN additions. Therefore, such an algorithm needs rN/N_1 1D type-I DHTs of length N_1 plus rN additions. If each 1D DHT with length N_1 needs $\frac{1}{2}N_1 \log_2 N_1 - \frac{3}{2}N_1 + 2$ multiplications and $\frac{3}{2}N_1 \log_2 N_1 - \frac{5}{2}N_1 + 6$ additions, which is the case for some best known algorithms [11, 19, 27], the number of multiplications and additions of the row-column method can be expressed, respectively, by

$$
\overline{M_u} = \frac{N}{2} \log_2 N - \frac{3rN}{2} + \frac{2rN}{N_1}
\tag{5.179}
$$

$$
\overline{A_d} = \frac{N}{2} \log_2 N - \frac{3rN}{2} + \frac{6rN}{N_1}.
\tag{5.180}
$$

In comparison, the PT-based algorithm substantially reduces the number of arithmetic operations compared to the row-column method. Approximately, the relationship between the computational complexities needed by the PT-based and the row-column algorithms can be expressed by

$$
M_u = \frac{1}{r}\overline{M_u}, \; A_d = \frac{2r+1}{3r}\overline{A_d}
$$

The reduction of arithmetic operations increases with the number of dimensions r.

Compared with some recently reported algorithms which are different from the row-column method [3, 17], this algorithm also has advantages. The algorithm in [17] has a complex structure and can only deal with a 2D type-I DHT with size $2^l \times 2^l$. Compared with the algorithm introduced in [3], the PT-based algorithm considerably reduces the computational complexity. Table 5.6 shows a detailed comparison of the number of operations for an $N_1 \times N_1$ 2D type-I DHT among the PT-based algorithm and the ones in [3] and [17]. This algorithm can be used to compute DHTs of different dimensions whose sizes are not necessarily the same.

The PT-based algorithm has a relatively simple structure. The main part of the algorithm is in Step 1, in which a number of 1D type-II DHTs are computed, and in Step 2, which

Table 5.6   Comparison of computational complexity needed by three 2D type-I DHT algorithms.

	PT-based		Algorithm in [3]		Algorithm in [17]	
N_1	M_u	A_d	M_u	A_d	M_u	A_d
8	26	354	24	408	24	408
16	218	2018	264	2216	216	2168
32	1370	10594	1800	11368	1368	10936
64	7514	52578	10536	55176	7512	53304
128	38234	251234	55560	260840	38232	252728
256	185690	1168738	277992	1200712	185688	1171768

computes an $(r-1)$D PT. In Step 1, an index mapping or sequence reordering is needed when these 1D DHTs are computed. The index mapping can be pre-calculated and stored in an array of $N/2$ integers. Extra computation is needed for on-line computation of data indices for sequence reordering. For example, if an array $x_1(n_1, \ldots, n_r)$ of size $N/2$ is used to store the new data indices, the index mapping is realized according to

$$x_1(p_1, \ldots, p_{r-1}, n_r) = x(p_1(n_r), \ldots, p_{r-1}(n_r), 2n_r + 1) \qquad (5.181)$$

whose indices $p_i(n_r)$ $(i = 1, 2, \ldots, r-1)$ are computed according to $(2n_r+1)p_i \bmod N_i$, which requires $\frac{N}{2} - \frac{N}{N_r}$ additions, $\frac{N}{2} - \frac{N}{N_r}$ multiplications and $\frac{N}{2} - \frac{N}{N_r}$ modulo operations (shift operations are not included). This computational overhead, which is only a small fraction of the total computation time, is the price for reduction of the multiplicative complexity by r times compared to that needed by the row-column algorithm.

5.6   PT-BASED RADIX-2 ALGORITHM FOR MD TYPE-II DHT

This section derives a PT-based fast algorithm for the MD type-II DHT [31].

5.6.1   Algorithm

We assume that $N_r = 2^t$, $N_r/N_i = 2^{l_i}$ $(i = 1, 2, \ldots, r-1)$, where $l_i \geq 0$ is an integer. An rD type-II DHT with size $N_1 \times N_2 \times \cdots \times N_r$ is defined by

$$X(k_1, k_2, \ldots, k_r) = \sum_{n_1=0}^{N_1-1} \sum_{n_2=0}^{N_2-1} \cdots \sum_{n_r=0}^{N_r-1} x(n_1, n_2, \ldots, n_r)$$
$$\cdot \operatorname{cas}\left[\frac{\pi(2n_1+1)k_1}{N_1} + \cdots + \frac{\pi(2n_r+1)k_r}{N_r}\right], \qquad (5.182)$$

where $k_i = 0, 1, \ldots, N_i - 1$ and $i = 1, 2, \ldots, r$. We shall change the order of the multiple sums to eliminate the redundant operations in the computation.

Lemma 5 *Let $p_i(n_r)$ be the least non-negative remainder of $(2p_i+1)n_r + p_i$ module N_i, and $A = \{(n_1, n_2, \ldots, n_r) \mid 0 \leq n_i \leq N_i - 1, 1 \leq i \leq r\}$, $B = \{(p_1(n_r), p_2(n_r), \ldots, p_{r-1}(n_r), n_r) \mid 0 \leq p_i \leq N_i - 1, 1 \leq i \leq r-1, 0 \leq n_r \leq N_r - 1\}$. Then $A = B$.*

Proof. It is sufficient to prove that the elements in B are different from each other. Let $(p_1(n_r), p_2(n_r), \ldots, p_{r-1}(n_r), n_r)$ and $(p'_1(n'_r), \ldots, p'_{r-1}(n'_r), n'_r)$ be two elements in B. If they are the same, then

$$p_i(n_r) = p'_i(n'_r), \quad i = 1, 2, \ldots, r-1, \quad n_r = n'_r.$$

From the definition of $p_i(n_r)$ we see that

$$(2p_i + 1)n_r + p_i \equiv (2p'_i + 1)n_r + p'_i \bmod N_i.$$

Hence

$$(2n_r + 1)(p_i - p'_i) \equiv 0 \bmod N_i, \quad i = 1, 2, \ldots, r-1.$$

Since $2n_r + 1$ is relatively prime with N_i, we have $p_i \equiv p'_i \bmod N_i$, or $p_i = p'_i$.

Equation (5.182) is a sum with its indices belonging to A. Since $A = B$, (5.182) is also a sum with its indices belonging to B. Therefore,

$$
\begin{aligned}
X(k_1, k_2, \ldots, k_r) = & \sum_{p_1=0}^{N_1-1} \cdots \sum_{p_{r-1}=0}^{N_{r-1}-1} \sum_{n_r=0}^{N_r-1} x(p_1(n_r), \ldots, p_{r-1}(n_r), n_r) \\
& \cdot \operatorname{cas}\left[\frac{\pi(2p_1(n_r) + 1)k_1}{N_1} + \cdots + \frac{\pi(2p_{r-1}(n_r) + 1)k_{r-1}}{N_{r-1}}\right. \\
& \left. + \frac{\pi(2n_r + 1)k_r}{N_r}\right].
\end{aligned}
$$

Because

$$2p_i(n_r) + 1 \equiv 2(2p_i + 1)n_r + 2p_i + 1 \equiv (2p_i + 1)(2n_r + 1) \bmod 2N_i$$

and the function cas( ) is periodic with period 2π, we have

$$
\begin{aligned}
X(k_1, k_2, \ldots, k_r) = & \sum_{p_1=0}^{N_1-1} \cdots \sum_{p_{r-1}=0}^{N_{r-1}-1} \sum_{n_r=0}^{N_r-1} x(p_1(n_r), \ldots, p_{r-1}(n_r), n_r) \\
& \cdot \operatorname{cas}\frac{\pi(2n_r + 1)[(2p_1 + 1)k_1 2^{l_1} + \cdots + (2p_{r-1} + 1)k_{r-1} 2^{l_{r-1}} + k_r]}{N_r}.
\end{aligned}
$$

Let

$$V_{p_1, \ldots, p_{r-1}}(m) = \sum_{n_r=0}^{N_r-1} x(p_1(n_r), \ldots, p_{r-1}(n_r), n_r)\operatorname{cas}\frac{\pi(2n_r + 1)m}{N_r}, \qquad (5.183)$$

where $p_i = 0, 1, \ldots, N_i - 1$, $i = 1, 2, \ldots, r-1$, and $m = 0, 1, \ldots, N_r - 1$. These are $N_1 N_2 \cdots N_{r-1}$ type-II DHT of length N_r. Then

$$
\begin{aligned}
X(k_1, k_2, \ldots, k_r) = & \sum_{p_1=0}^{N_1-1} \sum_{p_2=0}^{N_2-1} \cdots \sum_{p_{r-1}=0}^{N_{r-1}-1} \\
& V_{p_1, \ldots, p_{r-1}}[(2p_1 + 1)k_1 2^{l_1} + \cdots + (2p_{r-1} + 1)k_{r-1} 2^{l_{r-1}} + k_r]. \quad (5.184)
\end{aligned}
$$

Because

$$V_{p_1, \ldots, p_{r-1}}(m + uN_r) = (-1)^u V_{p_1, \ldots, p_{r-1}}(m),$$

(5.184) can be computed if $V_{p_1,\ldots,p_{r-1}}(m)$ is known. A direct computation of (5.184) requires too many additions. It is possible to use the PT to substantially reduce the number of additions that are needed to compute (5.184). Let

$$X_{k_1,\ldots,k_{r-1}}(z) = \sum_{k_r=0}^{N_r-1} X(k_1,\ldots,k_r)z^{k_r}.$$

From (5.184) we see that

$$X_{k_1,\ldots,k_{r-1}}(z) \equiv \sum_{p_1=0}^{N_1-1} \cdots \sum_{p_{r-1}=0}^{N_{r-1}-1} \sum_{k_r=0}^{N_r-1} V_{p_1,\ldots,p_{r-1}}[(2p_1+1)k_1 2^{l_1}$$
$$+ \cdots + (2p_{r-1}+1)k_{r-1} 2^{l_{r-1}} + k_r]z^{k_r} \bmod (z^{N_r}+1).$$

Since $V_{p_1,\ldots,p_{r-1}}(\lambda+k_r)z^{k_r} \bmod (z^{N_r}+1)$ is periodic with period N_r for variable k_r, we have

$$X_{k_1,\ldots,k_{r-1}}(z) \equiv \sum_{p_1=0}^{N_1-1} \cdots \sum_{p_{r-1}=0}^{N_{r-1}-1} \sum_{k_r=0}^{N_r-1} V_{p_1,\ldots,p_{r-1}}(k_r)$$
$$\cdot z^{k_r-(2p_1+1)k_1 2^{l_1} - \cdots - (2p_{r-1}+1)k_{r-1} 2^{l_{r-1}}} \bmod (z^{N_r}+1).$$

Let

$$U_{p_1,\ldots,p_{r-1}}(z) = \sum_{k_r=0}^{N_r-1} V_{p_1,\ldots,p_{r-1}}(k_r)z^{k_r}$$

and

$$Y_{k_1,\ldots,k_{r-1}}(z) \equiv \sum_{p_1=0}^{N_1-1} \sum_{p_2=0}^{N_2-1} \cdots \sum_{p_{r-1}=0}^{N_{r-1}-1} U_{p_1,\ldots,p_{r-1}}(z)$$
$$\cdot z_1^{p_1 k_1} \cdots z_{r-1}^{p_{r-1} k_{r-1}} \bmod (z^{N_r}+1), \tag{5.185}$$

where $z_i \equiv z^{-2^{l_i+1}} \bmod (z^{N_r}+1)$ $(i=1,2,\ldots,r-1)$. It can be seen that (5.185) is an $(r-1)$D PT. After getting $Y_{k_1,\ldots,k_{r-1}}(z)$ from (5.185), we have

$$X_{k_1,\ldots,k_{r-1}}(z) \equiv Y_{k_1,\ldots,k_{r-1}}(z)z^{-k_1 2^{l_1} - \cdots - k_{r-1} 2^{l_{r-1}}} \bmod (z^{N_r}+1),$$

that is, we can get $X_{k_1,\ldots,k_{r-1}}(z)$ (therefore, $X(k_1,\ldots,k_r)$) by reordering the coefficients of $Y_{k_1,\ldots,k_{r-1}}(z)$. The PT-based algorithm for the rD type-II DHT can be summarized into the following steps. Figure 5.6 shows the flow chart of the algorithm.

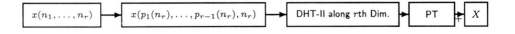

Figure 5.6   Steps of the PT-based radix-2 algorithm for MD type-II DHT.

Algorithm 6 *PT-based radix-2 algorithm for MD type-II DHT*

Step 1.   *Compute $N_1 \cdots N_{r-1}$ 1D type-II DHT of length N_r.*

$$V_{p_1,\ldots,p_{r-1}}(m) = \sum_{n_r=0}^{N_r-1} x(p_1(n_r),\ldots,p_{r-1}(n_r),n_r)\mathrm{cas}\frac{\pi(2n_r+1)m}{N_r},$$

$$p_i = 0,1,\ldots,N_i-1, \ i=1,2,\ldots,r-1, \ m=0,1,\ldots,N_r-1.$$

Generate $U_{p_1,\ldots,p_{r-1}}(z)$ as

$$U_{p_1,\ldots,p_{r-1}}(z) = \sum_{m=0}^{N_r-1} V_{p_1,\ldots,p_{r-1}}(m)z^m.$$

Step 2.   *Compute the $(r-1)D$ PT of the polynomial sequence $U_{p_1,\ldots,p_{r-1}}(z)$.*

$$Y_{k_1,\ldots,k_{r-1}}(z) \equiv \sum_{p_1=0}^{N_1-1}\sum_{p_2=0}^{N_2-1}\cdots\sum_{p_{r-1}=0}^{N_{r-1}-1} U_{p_1,\ldots,p_{r-1}}(z)$$
$$\cdot z_1^{p_1 k_1}\cdots z_{r-1}^{p_{r-1}k_{r-1}} \bmod (z^{N_r}+1),$$

where $k_i = 0,1,\ldots,N_i-1$, $i=1,2,\ldots,r-1$, and z_i is defined as before.

Step 3.   *Reorder the coefficients of the polynomials $Y_{k_1,\ldots,k_{r-1}}(z)$ to obtain $X_{k_1,\ldots,k_{r-1}}(z)$.*

$$X_{k_1,\ldots,k_{r-1}}(z) \equiv Y_{k_1,\ldots,k_{r-1}}(z)z^{-k_1 2^{l_1}-\cdots-k_{r-1}2^{l_{r-1}}} \bmod (z^{N_r}+1).$$

The coefficients of $X_{k_1,\ldots,k_{r-1}}(z)$ are $X(k_1,\ldots,k_r)$.

5.6.2 Computational complexity

The presented algorithm computes $N_1 \cdots N_{r-1}$ 1D type-II DHTs of length N_r and an $(r-1)D$ PT with module $z^{N_r}+1$. Assuming that one of some best known algorithms in [11, 19, 27] or Chapter 4 is used, the 1D type-II DHT of length N_r needs $\frac{1}{2}N_r\log_2 N_r - \frac{1}{2}N_r$ multiplications and $\frac{3}{2}N_r\log_2 N_r - \frac{3}{2}N_r$ additions. The $(r-1)D$ PT needs $N\log_2(N_1\cdots N_{r-1})$ additions (see Chapter 2 or [6, 18, 19, 24]). Therefore, the total computational complexity is

$$M_u = \frac{N}{2}\log_2 N_r - \frac{N}{2} \tag{5.186}$$

$$A_d = \frac{3N}{2}\log_2 N_r - \frac{3N}{2} + N\log_2(N_1\cdots N_{r-1}), \tag{5.187}$$

where $N = N_1\cdots N_r$.

The rD type-II DHT can be converted into a transform with a separable kernel, which is then computed by the row-column method. It can be easily found that the numbers of multiplications and additions needed by this approach are

$$\overline{M}_u = \frac{N}{2}\log_2 N - \frac{rN}{2}$$

and

$$\overline{A}_d = \frac{3N}{2} \log_2 N - \frac{rN}{2}.$$

In comparison, the PT-based algorithm is much more computationally efficient than the row-column-based algorithm. Especially when $N_1 = N_2 = \cdots = N_r$, we have

$$M_u = \frac{1}{r}\overline{M}_u, \quad A_d \approx \frac{2r+1}{3r}\overline{A}_d$$

which shows that the reduction is proportional to r, the number of dimensions. Tables 5.7 and 5.8 compare the computational complexity needed by the two algorithms for 2D and 3D type-II DHTs, respectively.

Table 5.7   Comparison of number of operations for 2D type-II DHT.

N_1	PT-based algorithm			Row-column-based algorithm		
	M_u	A_d	$M_u + A_d$	$\overline{M}_u$	$\overline{A}_d$	$\overline{M}_u + \overline{A}_d$
4	8	56	64	16	80	96
8	64	384	448	128	512	640
16	384	2176	2560	768	2816	3584
32	2048	11264	13312	4096	14336	18432
64	10240	55296	65536	20480	69632	90112
128	49152	262144	311296	98304	327680	425984
256	229376	1212416	1441792	458752	1507328	1966080
512	1048576	5505024	6553600	2097152	6815744	8912896
1024	4718592	24641536	29360128	9437184	30408704	39845888

Table 5.8   Comparison of number of operations for 3D type-II DHT.

N_1	PT-based algorithm			Row-column-based algorithm		
	M_u	A_d	$M_u + A_d$	$\overline{M}_u$	$\overline{A}_d$	$\overline{M}_u + \overline{A}_d$
4	32	352	384	96	480	576
8	512	4608	5120	1536	6144	7680
16	6144	51200	57344	18432	67584	86016
32	65536	524288	589824	196608	688128	884736
64	655360	5111808	5767168	1966080	6684672	8650752
128	6291456	48234496	54525952	18874368	62914560	81788928

The PT-based algorithm has a simple computational structure. Step 1 and Step 2 are the main parts of the algorithm. Step 1 computes a number of 1D type-II DHTs and Step 2 is an $(r-1)$D PT which can be efficiently implemented. In Step 1, the input sequence $x(n_1, \ldots, n_r)$ should be reordered.

5.6.3 PT-based radix-2 algorithm for MD type-III DHT

An rD type-III DHT with size $N_1 \times N_2 \times \cdots \times N_r$ is

$$X(k_1, k_2, \ldots, k_r) = \sum_{n_1=0}^{N_1-1} \sum_{n_2=0}^{N_2-1} \cdots \sum_{n_r=0}^{N_r-1} x(n_1, n_2, \ldots, n_r)$$
$$\cdot \operatorname{cas}\left(\frac{\pi(2k_1+1)n_1}{N_1} + \cdots + \frac{\pi(2k_r+1)n_r}{N_r}\right),$$

where $k_i = 0, 1, \ldots, N_i - 1$ and $i = 1, 2, \ldots, r$. We again assume that the size of each dimension is a power of two, that is, $N_r = 2^t$, $N_r/N_i = 2^{l_i}$ for a non-negative integer l_i $(i = 1, 2, \ldots, r-1)$.

Since the rD type-III DHT is the inversion of the rD type-II DHT, its PT algorithm can be easily derived from the PT algorithm of the rD type-II DHT. Here we just present the main steps of the algorithm without providing the details.

Algorithm 7 *PT-based radix-2 algorithm for rD type-III DHT*

Step 1.   *Generate $x_{n_1,\ldots,n_{r-1}}(z)$ as*

$$x_{n_1,\ldots,n_{r-1}}(z) = \sum_{n_r=0}^{N_r-1} x(n_1, \ldots, n_r) z^{n_r}$$

and reorder the coefficients of the polynomials $x_{n_1,\ldots,n_{r-1}}(z)$ to get $\overline{x}_{n_1,\ldots,n_{r-1}}(z)$ as

$$\overline{x}_{n_1,\ldots,n_{r-1}}(z) \equiv x_{n_1,\ldots,n_{r-1}}(z) z^{n_1 2^{l_1} + \cdots + n_{r-1} 2^{l_{r-1}}} \bmod (z^{N_r} + 1)$$

for $n_i = 0, 1, \ldots, N_i - 1$ and $i = 1, 2, \ldots, r-1$.

Step 2.   *Use the FPT algorithm to compute the $(r-1)$D PT of the polynomial sequence $\overline{x}_{n_1,\ldots,n_{r-1}}(z)$*

$$Y_{k_1,\ldots,k_{r-1}}(z) \equiv \sum_{n_1=0}^{N_1-1} \sum_{n_2=0}^{N_2-1} \cdots \sum_{n_{r-1}=0}^{N_{r-1}-1} \overline{x}_{n_1,\ldots,n_{r-1}}(z)$$
$$\cdot z_1^{n_1 k_1} \cdots z_{r-1}^{n_{r-1} k_{r-1}} \bmod (z^{N_r} + 1),$$

where $k_i = 0, 1, \ldots, N_i - 1$, and $z_i \equiv z^{2^{l_i+1}} \bmod (z^{N_r} + 1)$. Let

$$Y_{k_1,\ldots,k_{r-1}}(z) = \sum_{k_r=0}^{N_r-1} y(k_1, \ldots, k_r) z^{k_r}.$$

Step 3.   *Compute $N_1 \cdots N_{r-1}$ 1D type-III DHTs of length N_r*

$$X(p_1(l), \ldots, p_{r-1}(l), l) = \sum_{m=0}^{N_r-1} y(p_1, \ldots, p_{r-1}, m) \operatorname{cas} \frac{\pi(2l+1)m}{N_r},$$

where $p_i = 0, 1, \ldots, N_i - 1$, $l = 0, 1, \ldots, N_r - 1$, and $p_i(l)$ is the least non-negative remainder of $(2p_i + 1)l + p_i$ modulo N_i.

The number of operations needed by the algorithm is the same as that needed by the algorithm for the rD type-II DHT.

5.7 PT-BASED RADIX-q ALGORITHM FOR MD TYPE-I DHT

In this section, a PT-based algorithm is presented for the rD type-I DHT whose dimensional size is $q^{l_1} \times q^{l_2} \times \cdots \times q^{l_r}$, where q is an odd prime number and $r > 1$ is the number of dimensions. It shows that the rD type-I DHT is decomposed into a number of reduced 1D type-I DHTs and an $(r-1)$D PT. Computation of the reduced 1D DHT is also discussed. Analysis and comparisons of the required computational complexity are also presented.

5.7.1 Algorithm

Let us assume that the dimension sizes are $N_r = q^t$, $N_r/N_i = q^{l_i}$ $(i = 1, 2, \ldots, r-1)$, where $l_i \geq 0$ is an integer and q is an odd prime number. An rD type-I DHT with size $N_1 \times N_2 \times \cdots \times N_r$ is

$$
X(k_1, k_2, \ldots, k_r) = \sum_{n_1=0}^{N_1-1} \sum_{n_2=0}^{N_2-1} \cdots \sum_{n_r=0}^{N_r-1} x(n_1, n_2, \ldots, n_r)
$$
$$
\cdot \mathrm{cas}\left(\frac{2\pi n_1 k_1}{N_1} + \cdots + \frac{2\pi n_r k_r}{N_r} \right),
$$

where $k_i = 0, 1, \ldots, N_i - 1$ and $i = 1, 2, \ldots, r$. It can be easily shown that the rD type-I DHT can be decomposed into two parts, that is,

$$
X(k_1, k_2, \ldots, k_r) = A(k_1, k_2, \ldots, k_r) + B(k_1, k_2, \ldots, k_r), \tag{5.188}
$$

where

$$
A(k_1, k_2, \ldots, k_r) = \sum_{n_1=0}^{N_1-1} \cdots \sum_{n_{r-1}=0}^{N_{r-1}-1} \sum_{q|n_r}^{N_r-1} x(n_1, \ldots, n_{r-1}, n_r)
$$
$$
\cdot \mathrm{cas}\left(\frac{2\pi n_1 k_1}{N_1} + \cdots + \frac{2\pi n_r k_r}{N_r} \right) \tag{5.189}
$$

$$
B(k_1, k_2, \ldots, k_r) = \sum_{n_1=0}^{N_1-1} \cdots \sum_{n_{r-1}=0}^{N_{r-1}-1} \sum_{q \nmid n_r}^{N_r-1} x(n_1, \ldots, n_{r-1}, n_r)
$$
$$
\cdot \mathrm{cas}\left(\frac{2\pi n_1 k_1}{N_1} + \cdots + \frac{2\pi n_r k_r}{N_r} \right), \tag{5.190}
$$

where in (5.189), the symbol $q \mid n_r$ associated with the inner sum represents the n_r $(0 \leq n_r < N_r)$ values that are divisible by q, and in (5.190), the symbol $q \nmid n_r$ associated with the inner sum represents the n_r $(0 \leq n_r < N_r)$ values that are not divisible by q. It is seen that (5.189) becomes an rD type-I DHT with size $N_1 \times \cdots \times N_{r-1} \times N_r/q$

$$
A(k_1, k_2, \ldots, k_r) = \sum_{n_1=0}^{N_1-1} \cdots \sum_{n_{r-1}=0}^{N_{r-1}-1} \sum_{n_r=0}^{N_r/q-1} x(n_1, \ldots, n_{r-1}, qn_r)
$$
$$
\cdot \mathrm{cas}\left(\frac{2\pi n_1 k_1}{N_1} + \cdots + \frac{2\pi n_{r-1} k_{r-1}}{N_{r-1}} + \frac{2\pi n_r k_r}{N_r/q} \right), \tag{5.191}
$$

where $k_i = 0, 1, \ldots, N_i - 1$, $i = 1, 2, \ldots, r - 1$ and $k_r = 0, 1, \ldots, N_r/q - 1$. It can be easily proven that $A(k_1, k_2, \ldots, k_r + u\frac{N_r}{q}) = A(k_1, k_2, \ldots, k_r)$ for an arbitrary integer u.

Now let us convert (5.190) into a series of reduced 1D type-I DHTs of length N_r plus an $(r-1)$D PT. Let $p_i(n_r)$ be the least non-negative remainder of $p_i n_r$ modulo N_i ($i = 1, 2, \ldots, r - 1$). Then

$$
\begin{aligned}
B(k_1, k_2, \ldots, k_r) &= \sum_{p_1=0}^{N_1-1} \cdots \sum_{p_{r-1}=0}^{N_{r-1}-1} \sum_{q \nmid n_r}^{N_r-1} x(p_1(n_r), \ldots, p_{r-1}(n_r), n_r) \\
&\quad \cdot \text{cas}\left(\frac{2\pi p_1(n_r) k_1}{N_1} + \cdots + \frac{2\pi p_{r-1}(n_r) k_{r-1}}{N_{r-1}} + \frac{2\pi n_r k_r}{N_r}\right) \\
&= \sum_{p_1=0}^{N_1-1} \cdots \sum_{p_{r-1}=0}^{N_{r-1}-1} \sum_{q \nmid n_r}^{N_r-1} x(p_1(n_r), \ldots, p_{r-1}(n_r), n_r) \\
&\quad \cdot \text{cas}\frac{2\pi n_r (p_1 k_1 q^{l_1} + \cdots + p_{r-1} k_{r-1} q^{l_{r-1}} + k_r)}{N_r}.
\end{aligned}
$$

Let

$$
V_{p_1, \ldots, p_{r-1}}(l) = \sum_{q \nmid n_r}^{N_r-1} x(p_1(n_r), \ldots, p_{r-1}(n_r), n_r) \text{cas}\frac{2\pi n_r l}{N_r}, \tag{5.192}
$$

where $p_i = 0, 1, \ldots, N_i - 1$, $i = 1, 2, \ldots, r - 1$ and $l = 0, 1, \ldots, N_r - 1$. We define the reduced 1D type-I DHT of length M as

$$
W(l) = \sum_{q \nmid m}^{M-1} F(m) \text{cas}\frac{2\pi ml}{M}, \quad l = 0, 1, \ldots, M - 1, \ M = q^t, \tag{5.193}
$$

which is a summation derived from a 1D type-I DHT without the terms satisfying $m \equiv 0 \bmod q$. Therefore, (5.192) defines $N_1 N_2 \cdots N_{r-1}$ reduced 1D type-I DHTs of length N_r. Then

$$
\begin{aligned}
B(k_1, k_2, \ldots, k_r) &= \sum_{p_1=0}^{N_1-1} \sum_{p_2=0}^{N_2-1} \cdots \sum_{p_{r-1}=0}^{N_{r-1}-1} \\
&\quad \cdot V_{p_1, \ldots, p_{r-1}}(p_1 k_1 q^{l_1} + \cdots + p_{r-1} k_{r-1} q^{l_{r-1}} + k_r). \tag{5.194}
\end{aligned}
$$

Since $V_{p_1, \ldots, p_{r-1}}(l + uN_r) = V_{p_1, \ldots, p_{r-1}}(l)$ for an arbitrary integer u, we see that (5.194) can be computed if the reduced 1D type-I DHTs $V_{p_1, \ldots, p_{r-1}}(l)$ defined in (5.192) are available. We use the PT to compute (5.194) to minimize the number of additions for combining the outputs of the reduced 1D type-I DHTs.

Let $B_{k_1,\ldots,k_{r-1}}(z) = \sum_{k_r=0}^{N_r-1} B(k_1,\ldots,k_r) z^{k_r}$, then

$$
\begin{aligned}
B_{k_1,\ldots,k_{r-1}}(z) &\equiv \sum_{p_1=0}^{N_1-1} \cdots \sum_{p_{r-1}=0}^{N_{r-1}-1} \sum_{k_r=0}^{N_r-1} V_{p_1,\ldots,p_{r-1}}(p_1 k_1 q^{l_1} + \cdots + p_{r-1}k_{r-1}q^{l_{r-1}} + k_r) \\
&\quad \cdot z^{k_r} \bmod (z^{N_r} - 1) \\
&\equiv \sum_{p_1=0}^{N_1-1} \cdots \sum_{p_{r-1}=0}^{N_{r-1}-1} \sum_{k_r=0}^{N_r-1} V_{p_1,\ldots,p_{r-1}}(k_r) \\
&\quad \cdot z^{k_r - p_1 k_1 q^{l_1} - \cdots - p_{r-1}k_{r-1}q^{l_{r-1}}} \bmod (z^{N_r} - 1) \\
&\equiv \sum_{p_1=0}^{N_1-1} \cdots \sum_{p_{r-1}=0}^{N_{r-1}-1} U_{p_1,\ldots,p_{r-1}}(z) \\
&\quad \cdot z_1^{p_1 k_1} \cdots z_{r-1}^{p_{r-1}k_{r-1}} \bmod (z^{N_r} - 1), \quad\quad (5.195)
\end{aligned}
$$

where for $i = 1, 2, \ldots, r-1$,

$$
z_i \equiv z^{-q^{l_i}} \bmod (z^{N_r} - 1)
$$

and

$$
U_{p_1,\ldots,p_{r-1}}(z) = \sum_{l=0}^{N_r-1} V_{p_1,\ldots,p_{r-1}}(l) z^l.
$$

It is seen that (5.195) is like an $(r-1)$D PT. These decomposition steps are summarized below and illustrated in Figure 5.7.

Algorithm 8 *PT-based radix-q algorithm for rD type-I DHT*

Step 1.   *Compute $N_1 N_2 \cdots N_{r-1}$ 1D type-I reduced DHTs according to (5.192).*

Step 2.   *Compute the $(r-1)$D pseudo PT according to (5.195) to get $B(k_1, \ldots, k_r)$, which are the coefficients of $B_{k_1,\ldots,k_{r-1}}(z)$.*

Step 3.   *Compute the rD type-I DHT with size $N_1 \times \cdots \times N_{r-1} \times N_r/q$ in (5.191) by recursively using Step 1 and Step 2.*

Step 4.   *Compute $X(k_1, k_2, \ldots, k_r)$ from (5.188).*

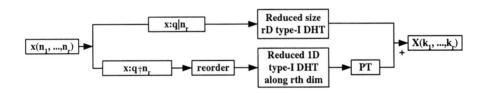

Figure 5.7   Flow chart of the PT-based radix-q algorithm for rD type-I DHT.

5.7.2   Computation of reduced type-I DHT

In general, many existing algorithms for length q^l 1D type-I DHTs, such as those in [15, 32, 35], can be amended for the reduced DHT defined in (5.193). Based on the algorithm in

[32, 35], it can be shown that the reduced 1D type-I DHT of length M can be recursively decomposed into q reduced DHTs of length M/q.

Algorithm 9 *Algorithm for the reduced 1D type-I DHT*

Step 1.   *Compute $\overline{C}_j(m)$ and $\overline{S}_j(m)$ $(j = 0, 1, \ldots, (q-1)/2)$ according to*

$$\overline{C}_j(m) = \sum_{i=0}^{q-1} F(iM/q + m) \cos \frac{2\pi ij}{q}$$

$$\overline{S}_j(m) = \sum_{i=1}^{q-1} F(iM/q + m) \sin \frac{2\pi ij}{q}$$

$$0 \leq m \leq M/q - 1, \quad m \not\equiv 0 \bmod q.$$

Step 2.   *Compute $d_0(m)$, $d_j(m)$ and $d'_j(m)$ $(j = 1, 2, \ldots, (q-1)/2)$ according to*

$$d_0(m) = \overline{C}_0(m)$$

$$d_j(m) = \overline{C}_j(m) 2 \cos \frac{2\pi mj}{M} - \overline{S}_j(m) 2 \sin \frac{2\pi mj}{M}$$

$$d'_j(m) = \overline{C}_j(m) 2 \sin \frac{2\pi mj}{M} + \overline{S}_j(m) 2 \cos \frac{2\pi mj}{M}$$

$$0 \leq m \leq M/q - 1, \quad m \not\equiv 0 \bmod q.$$

Step 3.   *Compute $D_j(l)$ and $D'_j(l)$ which are the length M/q reduced 1D type-I DHTs of sequences $d_j(m)$ and $d'_j(m)$.*

Step 4.   *Obtain $W(ql + j) = C_j(l)$ $(j = 1, 2, \ldots, (q-1)/2)$ by*

$$C_j(l) = \frac{D_j(l) + D'_j(M/q - l)}{2}, \quad l = 0, 1, \ldots, \frac{M}{q} - 1$$

$$C_{q-j}(l-1) = \frac{D_j(l) - D'_j(M/q - l)}{2}, \quad l = 1, 2, \ldots, \frac{M}{q},$$

where $D_j(M/q) = D_j(0)$, $D'_j(M/q) = D'_j(0)$ and $W(ql) = D_0(l)$.

We assume that the computation in Step 1 for a fixed m needs $\alpha(q)$ multiplications and $\beta(q)$ additions. The total computational complexity required by Step 1, Step 2 and Step 4 is $(\alpha(q) + 2q - 2)(\frac{M}{q} - \frac{M}{q^2})$ multiplications and $(\beta(q) + q - 1)(\frac{M}{q} - \frac{M}{q^2}) + (q-1)\frac{M}{q}$ additions. Let $\varphi(M)$ and $\psi(M)$ represent the number of multiplications and additions, respectively, for the computation of the length M transform, then

$$\varphi(M) = (\alpha(q) + 2q - 2)(1 - \frac{1}{q})\frac{M}{q} + q\varphi(\frac{M}{q})$$

$$\psi(M) = (\beta(q) + q - 1)(1 - \frac{1}{q})\frac{M}{q} + (q-1)\frac{M}{q} + q\psi(\frac{M}{q}). \tag{5.196}$$

Some values of $\alpha(q)$ and $\beta(q)$ for small q are given in [32, 35]. For example,

$$\alpha(3) = 0, \quad \beta(3) = 4, \quad \alpha(5) = 4, \quad \beta(5) = 14.$$

The computational process for Step 1 when $q = 3$ is

$$\overline{C}_0(m) = F(m) + \left[F(\frac{M}{3} + m) + F(\frac{2M}{3} + m)\right]$$

$$\overline{C}_1(m) = F(m) - \frac{1}{2}\left[F(\frac{M}{3} + m) + F(\frac{2M}{3} + m)\right]$$

$$\overline{C}_2(m) = \frac{\sqrt{3}}{2}\left[F(\frac{M}{3} + m) - F(\frac{2M}{3} + m)\right].$$

Since the factor $\frac{\sqrt{3}}{2}$ can be incorporated into Step 2, the multiplication involving this factor is not included in $\alpha(3)$.

For the computation in Step 3, we use the same technique to decompose the reduced 1D type-I DHTs of length M/q into smaller ones. However, when the length is small enough, such as $M = q$ or $M = q^2$, we usually use the cyclic convolution algorithm [6, 19]. Therefore, we have

$$\varphi(3^t) = (8t - 5)3^{t-2}, \qquad \psi(3^t) = (2t - \frac{2}{3})3^t$$

$$\varphi(5^t) = (48t - 23)5^{t-2}, \qquad \psi(5^t) = (92t - 17)5^{t-2}.$$

5.7.3   Computational complexity

The presented algorithm computes $N_1 N_2 \cdots N_{r-1}$ reduced 1D type-I DHTs of length N_r and an $(r - 1)$D pseudo PT with size $N_1 \times \cdots \times N_{r-1}$. The $(r - 1)$D pseudo PT requires $(q - 1)N \log_q(N_1 N_2 \cdots N_{r-1})$ additions (see Chapter 2), where $N = N_1 N_2 \cdots N_r$. The number of multiplications is

$$M_u(N_1, \ldots, N_{r-1}, N_r) = \frac{N}{N_r}\varphi(N_r) + M_u(N_1, \ldots, N_{r-1}, \frac{N_r}{q}) \qquad (5.197)$$

and the number of additions is

$$A_d(N_1, \ldots, N_{r-1}, N_r) = \frac{N}{N_r}\psi(N_r) + (q - 1)N \log_q \frac{N}{N_r} + N$$

$$+ A_d(N_1, \ldots, N_{r-1}, \frac{N_r}{q}), \qquad (5.198)$$

where $\varphi(N_r)$ and $\psi(N_r)$ are the numbers of multiplications and additions, respectively, for a reduced 1D type-I DHT of length N_r. It is difficult to get a closed expression for arbitrary N_i. However, we can obtain a closed form expression when $N_1 = N_2 = \cdots = N_r = q^t$. For simplicity, we define $M_u(q^t) = M_u(q^t, \ldots, q^t, q^t)$ and $A_d(q^t) = A_d(q^t, \ldots, q^t, q^t)$ and assume that $\varphi(q^t) = a_1 t q^t + a_2 q^t + a_3$ and $\psi(q^t) = b_1 t q^t + b_2 q^t + b_3$. Based on (5.197) and (5.198) we can deduce that

$$M_u(q^t) = a_1 \frac{q}{q - 1} t q^{rt} + (a_2 - \frac{a_1}{q^r - 1})\frac{q}{q - 1}(q^{rt} - 1)$$

$$+ a_3 \frac{q^r - 1}{(q - 1)(q^{r-1} - 1)}(q^{(r-1)t} - 1) \qquad (5.199)$$

$$A_d(q^t) = (b_1 \frac{q}{q - 1} + (r - 1)q)t q^{rt} + (b_2 + 1 - \frac{b_1}{q^r - 1} - \frac{q^r - q}{q^r - 1})\frac{q}{q - 1}(q^{rt} - 1)$$

$$+ b_3 \frac{q^r - 1}{(q - 1)(q^{r-1} - 1)}(q^{(r-1)t} - 1). \qquad (5.200)$$

The row-column method has to convert the entire computation task into a transform whose kernel is separable. The conversion process generally needs rN additions [7, 10, 32]. If we assume that the number of multiplications and additions for a 1D type-I DHT of length q^t is $\tau(q^t) = c_1 t q^t + c_2 q^t + c_3$ and $\rho(q^t) = d_1 t q^t + d_2 q^t + d_3$, respectively, the computational complexity required by the row-column method is

$$\overline{M}_u(q^t) = (c_1 t + c_2) r q^{rt} + c_3 r q^{(r-1)t}$$
$$\overline{A}_d(q^t) = (d_1 t + d_2 + 1) r q^{rt} + d_3 r q^{(r-1)t}.$$

If we assume that $a_1 = \frac{q-1}{q} c_1$ and $a_2 \approx \frac{q-1}{q} c_2$ (which is generally true), then $M_u \approx \frac{1}{r} \overline{M}_u$, which indicates that the number of multiplications needed by the algorithm is only $1/r$ times that needed by the row-column method.

It is more difficult to compare the additive complexity. Let us consider two cases when $q = 3$ and $q = 5$. Using the method described above, we have

$$M_u(3^t) = 4t3^{rt-1} - \frac{5 \cdot 3^r + 3}{6(3^r - 1)}(3^{rt} - 1)$$

$$A_d(3^t) = rt3^{rt+1} - 3^{rt} + 1 \tag{5.201}$$

$$M_u(5^t) = 12t5^{rt-1} - \left(\frac{23}{20} + \frac{12}{5(5^r - 1)}\right)(5^{rt} - 1)$$

$$A_d(5^t) = \left(5r - \frac{2}{5}\right)t5^{rt} - \frac{17 \cdot 5^{r-1} - 5}{4(5^r - 1)}(5^{rt} - 1). \tag{5.202}$$

The numbers of operations for the DHT (see Chapter 4 or [32, 35]) are

$$\tau(3^t) = \left(\frac{4t}{3} - 1\right)3^t, \quad \rho(3^t) = (8t - 1)3^{t-1}$$

$$\tau(5^t) = 12t5^{t-1}, \quad \rho(5^t) = 22t5^{t-1}.$$

Then the total number of multiplications and additions needed by the row-column method is

$$\overline{M}_u(3^t) = \left(\frac{4t}{3} - 1\right)r3^{rt}, \quad \overline{A}_d(3^t) = 2(4t + 1)r3^{rt-1} \tag{5.203}$$

$$\overline{M}_u(5^t) = 12rt5^{rt-1}, \quad \overline{A}_d(5^t) = \left(\frac{22t}{5} + 1\right)r5^{rt}. \tag{5.204}$$

It can be seen from these two cases that the number of multiplications is approximately $\frac{1}{r}$ times that used by the row-column method, while the number of additions is slightly increased. In particular, the total number of arithmetic operations (additions plus multiplications) is reduced considerably, as seen from (5.201)–(5.204). Table 5.9 lists the number of operations for $q = 3$ and $r = 3$.

5.8   PT-BASED RADIX-q ALGORITHM FOR MD TYPE-II DHT

In this section, we consider a PT-based algorithm for the rD type-II DHT whose dimensional size is $q^{l_1} \times q^{l_2} \times \cdots \times q^{l_r}$ [34], where q is an odd prime number. The rD type-II DHT can be converted into a series of reduced 1D type-II DHTs by using the MD PT. Our

Table 5.9   Comparison of the number of operations needed by two algorithms.

t	PT-based algorithm			Row-column based algorithm		
	M_u	A_d	$M_u + A_d$	$\overline{M}_u$	$\overline{A}_d$	$\overline{M}_u + \overline{A}_d$
2	1300	12394	13694	3645	13122	16767
3	61321	511759	573080	177147	511758	688905
4	2364232	18600436	20964668	6908733	18068994	24977767
5	82966117	631341909	7143180260	243931419	602654094	846585513

discussion includes the details of algorithm derivation, computation steps and the required computational complexity. For easy understanding, we first consider the algorithm for the 2D type-II DHT to illustrate some basic concepts. The computation methods for the reduced 1D type-II DHT are also discussed, which are useful in the computation of the rD type-II DHT. By extending these concepts, the fast algorithm for the rD, where $r \geq 3$, type-II DHT is obtained. The detailed analysis of the computational complexity required by the algorithm is finally discussed.

5.8.1   PT-based radix-q algorithm for 2D type-II DHT

We assume that the dimensional sizes are M and N which are powers of an odd prime number q. If $M \geq N$, we can write $M = q^t$ and $M/N = q^L$, where $t > 0$ and $L \geq 0$, respectively. When $M < N$, we can simply swap M and N to satisfy the above assumption. The 2D type-II DHT is defined as

$$X(k,l) = \sum_{n=0}^{N-1} \sum_{m=0}^{M-1} x(n,m) \mathrm{cas}\left(\frac{\pi(2n+1)k}{N} + \frac{\pi(2m+1)l}{M}\right),$$

where $k = 0, 1, \ldots, N-1$ and $l = 0, 1, \ldots, M-1$. The 2D type-II DHT can be decomposed into

$$X(k,l) = A(k,l) + B(k,l), \tag{5.205}$$

where

$$A(k,l) = \sum_{n=0}^{N-1} \sum_{m \equiv (q-1)/2}^{M-1} x(n,m) \mathrm{cas}\left(\frac{\pi(2n+1)k}{N} + \frac{\pi(2m+1)l}{M}\right) \tag{5.206}$$

$$B(k,l) = \sum_{n=0}^{N-1} \sum_{m \not\equiv (q-1)/2}^{M-1} x(n,m) \mathrm{cas}\left(\frac{\pi(2n+1)k}{N} + \frac{\pi(2m+1)l}{M}\right). \tag{5.207}$$

In (5.206), $\displaystyle\sum_{m \equiv (q-1)/2}^{M-1}$ means to accumulate the terms associated with index m such that $(m \bmod q) = (q-1)/2$ $(m = 0, 1, \ldots, M-1)$. In (5.207), $\displaystyle\sum_{m \not\equiv (q-1)/2}^{M-1}$ means to accumulate the terms associated with index m such that $(m \bmod q) \neq (q-1)/2$ $(m = 0, 1, \ldots, M-1)$. These two notations will be used throughout this section. In fact, (5.206) can be expressed as a 2D type-II DHT with size $N \times M/q$:

$$A(k,l) = (-1)^u A(k, l + \frac{uM}{q})$$

$$= \sum_{n=0}^{N-1} \sum_{m=0}^{M/q-1} x(n, mq + \frac{q-1}{2}) \text{cas} \left(\frac{\pi(2n+1)k}{N} + \frac{\pi(2m+1)l}{M/q} \right), \quad (5.208)$$

where $k = 0, 1, \ldots, N-1$, $l = 0, 1, \ldots, M/q - 1$ and u is an arbitrary integer. The decomposition process can be continued if necessary.

Now let us study the computation of (5.207). If we assume that $p(m)$ is the least non-negative residue of $(2p+1)m + p$ modulo N, the following lemma can be easily proved.

Lemma 6 *Let* $S_1 = \{(n,m) \mid 0 \leq n \leq N-1, 0 \leq m \leq M-1 \text{ and } m \not\equiv \frac{q-1}{2} \bmod q\}$, $S_2 = \{(p(m), m) \mid 0 \leq p \leq N-1, 0 \leq m \leq M-1 \text{ and } m \not\equiv \frac{q-1}{2} \bmod q\}$. *Then* $S_1 = S_2$.

Based on the lemma, (5.207) becomes

$$B(k,l) = \sum_{p=0}^{N-1} \sum_{m \not\equiv (q-1)/2}^{M-1} x(p(m), m) \text{cas} \left(\frac{\pi(2p(m)+1)k}{N} + \frac{\pi(2m+1)l}{M} \right). \quad (5.209)$$

From the definition of $p(m)$, (5.207) can be expressed by

$$B(k,l) = \sum_{p=0}^{N-1} \sum_{m \not\equiv (q-1)/2}^{M-1} x(p(m), m) \text{cas} \frac{\pi(2m+1)((2p+1)kq^L + l)}{M}$$

$$= \sum_{p=0}^{N-1} V_p((2p+1)kq^L + l), \quad (5.210)$$

where $k = 0, 1, \ldots, N-1$, $l = 0, 1, \ldots, M-1$, and

$$V_p(i) = \sum_{m \not\equiv (q-1)/2}^{M-1} x(p(m), m) \text{cas} \frac{\pi(2m+1)i}{M} \quad (5.211)$$

for $i = 0, 1, \ldots, M-1$ and $p = 0, 1, \ldots, N-1$. Let us define a reduced type-II DHT of length M as

$$W(l) = \sum_{m \not\equiv (q-1)/2}^{M-1} F(m) \text{cas} \frac{\pi(2m+1)l}{M}, \quad l = 0, 1, \ldots, M-1, \quad (5.212)$$

which is a length M type-II DHT without the terms satisfying $(m \bmod q) = (q-1)/2$. Therefore, (5.211) specifies N reduced type-II DHTs of length M. The computation of (5.212) will be discussed shortly. Since $V_p(i + uM) = (-1)^u V_p(i)$, (5.210) can be computed if the reduced type-II DHTs defined in (5.211) are available.

In general, a direct computation of (5.210) requires $NM(N-1)$ additions. Based on the following lemma, however, the polynomial transform can be used to achieve considerable savings on the number of additions.

Lemma 7 *Let* $B_k(z)$ *be the generating polynomial of* $B(k,l)$, *that is,*

$$B_k(z) = \sum_{l=0}^{M-1} B(k,l) z^l$$

and

$$U_p(z) \equiv \sum_{l=0}^{M-1} V_p(l)z^l \bmod (z^M + 1).$$

(5.213)

Assume that the pseudo PT of $U_p(z)$ is $Y_k(z)$, that is,

$$Y_k(z) \equiv \sum_{p=0}^{N-1} U_p(z)\hat{z}^{pk} \bmod (z^M + 1),$$

(5.214)

where

$$\hat{z} \equiv z^{-2q^L} \bmod (z^M + 1).$$

Then we have

$$B_k(z) \equiv Y_k(z)z^{-kq^L} \bmod (z^M + 1).$$

(5.215)

Proof. Based on the property $V_p(i + uM) = (-1)^u V_p(i)$ and (5.210), we have

$$
\begin{aligned}
B_k(z) &\equiv \sum_{p=0}^{N-1}\sum_{l=0}^{M-1} V_p((2p+1)kq^L + l)z^l \bmod (z^M + 1) \\
&\equiv \sum_{p=0}^{N-1}\sum_{l=0}^{M-1} V_p(l)z^{l-(2p+1)kq^L} \bmod (z^M + 1) \\
&\equiv z^{-kq^L}\sum_{p=0}^{N-1} U_p(z)\hat{z}^{pk} \bmod (z^M + 1) \\
&\equiv Y_k(z)z^{-kq^L} \bmod (z^M + 1).
\end{aligned}
$$

Since $\hat{z}^N \equiv 1 \bmod (z^M + 1)$, the simple FFT-like algorithm (known as the fast polynomial transform or FPT) described in Chapter 2 (also see [6, 18, 19, 24]) can be used to compute (5.214), which generally needs $(q-1)NM\log_q N$ additions. The fast algorithm for the 2D type-II DHT can be summarized by the following steps.

Algorithm 10 *PT-based radix-q algorithm for 2D type-II DHT*

Step 1. *Compute N reduced 1D type-II DHTs according to (5.211).*

Step 2. *Compute a pseudo PT defined in (5.214) and reorder the coefficients of the polynomials according to (5.215) to get $B_k(z)$ whose coefficients are $B(k,l)$.*

Step 3. *Compute the 2D type-II DHT with size $N \times M/q$ defined in (5.208) by recursively using Steps 1 and 2.*

Step 4. *Compute $X(k,l)$ according to (5.205).*

5.8.2   Radix-q algorithm for reduced type-II DHT

In general, many fast algorithms [4, 5, 11, 19, 28, 32, 35] for the type-II DHT can be modified for the reduced type-II DHT. We shall modify the algorithm in [32, 35] for the reduced type-II DHT. The outputs $W(l)$ in (5.212) can be decomposed according to $C_j(l) = W(ql + j)$

($j = 0, 1, \ldots, q - 1$, $l = 0, 1, \ldots, M/q - 1$). Then $C_j(l)$ can be converted into the reduced length M/q type-II DHT. We have

$$
\begin{aligned}
C_0(l) &= W(ql) = \sum_{\substack{m \not\equiv (q-1)/2}}^{M-1} F(m) \mathrm{cas} \frac{\pi(2m+1)l}{M/q} \\
&= \sum_{\substack{m \not\equiv (q-1)/2}}^{M/q-1} \sum_{i=0}^{q-1} F(iM/q + m) \mathrm{cas} \frac{\pi(2m+1)l}{M/q} \\
&= \sum_{\substack{m \not\equiv (q-1)/2}}^{M/q-1} d_0(m) \mathrm{cas} \frac{\pi(2m+1)l}{M/q},
\end{aligned}
\tag{5.216}
$$

which becomes the reduced length M/q type-II DHT of the input sequence

$$
d_0(m) = \sum_{i=0}^{q-1} F(\frac{iM}{q} + m)
\tag{5.217}
$$

for $m = 0, 1, \ldots, M/q - 1$ and $m \not\equiv (q-1)/2 \bmod q$. Let $D_j(l) = C_j(l) + C_{q-j}(l-1) = W(ql + j) + W(ql - j)$ ($l = 0, 1, \ldots, M/q - 1$), then

$$
\begin{aligned}
D_j(l) &= \sum_{\substack{m \not\equiv (q-1)/2}}^{M-1} 2F(m) \cos \frac{\pi(2m+1)j}{M} \mathrm{cas} \frac{\pi(2m+1)l}{M/q} \\
&= \sum_{\substack{m \not\equiv (q-1)/2}}^{M/q-1} d_j(m) \mathrm{cas} \frac{\pi(2m+1)l}{M/q},
\end{aligned}
\tag{5.218}
$$

which is the reduced length M/q type-II DHT of input sequence

$$
d_j(m) = \sum_{i=0}^{q-1} 2F(\frac{iM}{q} + m) \cos \frac{\pi(2iM/q + 2m + 1)j}{M}.
\tag{5.219}
$$

Furthermore, we have

$$
\begin{aligned}
C_j(l) - C_{q-j}(l-1) &= W(ql + j) - W(ql - j) \\
&= \sum_{\substack{m \not\equiv (q-1)/2}}^{M-1} 2F(m) \sin \frac{\pi(2m+1)j}{M} \mathrm{cas} \frac{-\pi(2m+1)l}{M/q}.
\end{aligned}
$$

If we define

$$
d_j'(m) = \sum_{i=0}^{q-1} 2F(\frac{iM}{q} + m) \sin \frac{\pi(2iM/q + 2m + 1)j}{M}
\tag{5.220}
$$

and let the reduced type-II DHT of $d_j'(m)$ be $D_j'(l)$ ($l = 0, 1, \ldots, M/q - 1$),

$$
D_j'(l) = \sum_{\substack{m \not\equiv (q-1)/2}}^{M/q-1} d_j'(m) \mathrm{cas} \frac{\pi(2m+1)l}{M/q},
\tag{5.221}
$$

then

$$C_{q-j}(l-1) - C_j(l) = D'_j(M/q - l), \tag{5.222}$$

where $D'_j(M/q) = -D'_j(0)$. Once $D_j(l)$ and $D'_j(l)$ are available, $C_j(l)$, where $j = 1, 2, \ldots, q-1$ and $l = 0, 1, \ldots, M/q - 1$, can be calculated according to

$$
\begin{aligned}
W(ql + j) &= C_j(l) = \frac{D_j(l) - D'_j(M/q - l)}{2} \\
W(ql - j) &= C_{q-j}(l - 1) = \frac{D_j(l) + D'_j(M/q - l)}{2},
\end{aligned}
\tag{5.223}
$$

where $l = 1, 2, \ldots, M/q$, $j = 1, 2, \ldots, (q-1)/2$ and $D_j(M/q) = -D_j(0)$.

Based on the trigonometric properties, $d_j(m)$ and $d'_j(m)$ can be simplified. For example, we can express (5.217), (5.219) and (5.220) as

$$
\begin{aligned}
d_0(m) &= \overline{C}_0(m) \\
d_j(m) &= \overline{C}_j(m)2\cos\frac{\pi(2m+1)j}{M} - \overline{S}_j(m)2\sin\frac{\pi(2m+1)j}{M} \\
d'_j(m) &= \overline{C}_j(m)2\sin\frac{\pi(2m+1)j}{M} + \overline{S}_j(m)2\cos\frac{\pi(2m+1)j}{M},
\end{aligned}
\tag{5.224}
$$

where

$$
\begin{aligned}
\overline{C}_0(m) &= \sum_{i=0}^{q-1} F(\frac{iM}{q} + m) \\
\overline{C}_j(m) &= \sum_{i=0}^{q-1} F(\frac{iM}{q} + m)\cos\frac{2\pi ij}{q} \\
\overline{S}_j(m) &= \sum_{i=1}^{q-1} F(\frac{iM}{q} + m)\sin\frac{2\pi ij}{q},
\end{aligned}
\tag{5.225}
$$

where for both (5.224) and (5.225), $j = 1, 2, \ldots, (q-1)/2$, $0 \le m \le M/q - 1$ and $m \not\equiv \frac{q-1}{2} \bmod q$. Therefore, a reduced type-II DHT of length M is decomposed into q reduced type-II DHTs of length M/q. The decomposition process can be summarized by the following steps.

Algorithm 11 *Radix-q algorithm for reduced type-II DHT*

Step 1.   *Compute $\overline{C}_j(m)$ and $\overline{S}_j(m)$ defined in (5.225).*

Step 2.   *Compute $d_j(m)$ and $d'_j(m)$ defined in (5.224).*

Step 3.   *Compute the reduced type-II DHTs of $d_j(m)$ and $d'_j(m)$ to get $C_0(l)$, $D_j(l)$ and $D'_j(l)$.*

Step 4.   *Compute $C_j(l)$ according to (5.223).*

We assume that the computation in Step 1 for a fixed m needs $\alpha(q)$ multiplications and $\beta(q)$ additions. For some small q, the values of $\alpha(q)$ and $\beta(q)$ are given in [32, 35]. For example,

$$\alpha(3) = 0, \ \beta(3) = 4, \ \alpha(5) = 4, \ \beta(5) = 14$$

The algorithm for Step 1 when $q = 3$ has also been given in Section 5.7. Then the total computational complexity needed by Steps 1, 2 and 4 is $(\alpha(q) + 2q - 2)(\frac{M}{q} - \frac{M}{q^2})$ multiplications and $(\beta(q) + q - 1)(\frac{M}{q} - \frac{M}{q^2}) + (q - 1)\frac{M}{q}$ additions. The number of multiplications, $\phi(M)$, and the number of additions, $\psi(M)$, needed by the computation of the length M reduced transform are then

$$\phi(M) = (\alpha(q) + 2q - 2)(1 - \frac{1}{q})\frac{M}{q} + q\phi(\frac{M}{q})$$

$$\psi(M) = (\beta(q) + q - 1)(1 - \frac{1}{q})\frac{M}{q} + (q - 1)\frac{M}{q} + q\psi(\frac{M}{q}). \tag{5.226}$$

It is known that some arithmetic operations can be saved due to the existence of trivial twiddle factors. The required computational complexity of our calculation includes these operations that can be saved if appropriate methods are taken.

The decomposition process can be continued until the subsequence lengths become q or q^2. When the length is small enough, we usually use the cyclic convolution algorithm to compute the reduced type-II DHT transform, which will be discussed in the next subsection.

5.8.3 Cyclic convolution algorithm for reduced type-II DHT

The reduced type-II DHT in (5.212) can also be computed by cyclic convolution. From the number theory (see the primitive theorem in Chapter 2 or [2]), it is known that there exists an odd primitive root g modulo q^t, that is, integer l ($1 \le l \le q^t - 1$, $l \not\equiv 0 \bmod q$) can be written as $(g^i \bmod q^t)$ ($i = 0, 1, \ldots, q^{t-1}(q-1) - 1$). Since g is odd, it is easy to show that g is also a primitive root modulo $2q^t$, that is, $(2m + 1)$ ($0 \le m \le q^t - 1$, $m \not\equiv (q-1)/2 \bmod q$) can also be written as $(g^i \bmod 2q^t)$ ($i = 0, 1, \ldots, q^{t-1}(q - 1) - 1$). For example, $g = 5$ is a primitive root for both $M = 3^t = 9$ and $2M = 18$, and $g = 3$ is a primitive root for both 7 and 14. In (5.212), $W(l)$ can be decomposed into $W(qk)$ and $W(l)$ ($l \not\equiv 0 \bmod q$). We have

$$W(qk) = \sum_{\substack{m \not\equiv (q-1)/2}}^{M/q-1} F^0(m) \mathrm{cas} \frac{\pi(2m+1)k}{M/q}, \quad k = 0, 1, \ldots, M/q - 1, \tag{5.227}$$

where for $0 \le m \le M/q - 1$ and $m \not\equiv (q - 1)/2 \bmod q$

$$F^0(m) = \sum_{i=0}^{q-1} F(\frac{iM}{q} + m). \tag{5.228}$$

By using the primitive root g discussed above for index permutation, we can convert $W(l)$ ($l \not\equiv 0 \bmod q$) into a cyclic convolution of length $q^{t-1}(q - 1)$.

Lemma 8 *Let $\langle n \rangle$ be the least non-negative residue of an integer n modulo M and $\langle g^j \rangle = g^j + \lambda(j)M$. Also, we define $\bar{W}(j) = (-1)^{\lambda(j)}W(\langle g^j \rangle)$, $\bar{F}(i) = F(\langle \frac{g^{-i}-1}{2} \rangle)$ and $H(i) = \mathrm{cas}(\frac{\pi g^i}{M})$. Then we have*

$$\bar{W}(j) = \sum_{i=0}^{q^{t-1}(q-1)-1} \bar{F}(i)H(j - i), \quad j = 0, 1, \ldots, q^{t-1}(q - 1) - 1, \tag{5.229}$$

which is a cyclic convolution of length $q^{t-1}(q - 1)$.

Proof. We have

$$
\begin{aligned}
W(\langle g^j \rangle) &= \sum_{m \not\equiv \frac{q-1}{2}} F(m) \mathrm{cas} \frac{\pi(2m+1)\langle g^j \rangle}{M} \\
&= (-1)^{\lambda(j)} \sum_{m \not\equiv \frac{q-1}{2}} F(m) \mathrm{cas} \frac{\pi(2m+1)g^j}{M} \\
&= (-1)^{\lambda(j)} \sum_{i=0}^{q^{t-1}(q-1)-1} F(\langle \frac{g^{-i}-1}{2} \rangle) \mathrm{cas} \frac{\pi g^{j-i}}{M}.
\end{aligned}
$$

Furthermore, since g is also a primitive root of $2M$ ($M = q^t$), we know that for arbitrary integer l,

$$
\begin{aligned}
H(j - i + lq^{t-1}(q-1)) &= \mathrm{cas} \frac{\pi g^{j-i+lq^{t-1}(q-1)}}{M} \\
&= \mathrm{cas} \frac{\pi(g^{j-i} + u2M)}{M} \\
&= H(j - i).
\end{aligned}
$$

Hence, (5.229) is a cyclic convolution that can be computed by the fast algorithms [6, 19] for cyclic convolution according to the values of q and t. For large M/q, (5.227) can be further decomposed in the same way. When the length becomes q, the reduced DHT-II can be computed by a cyclic convolution of length $q-1$ plus the computation of $W(0) = \sum_{m \neq (q-1)/2}^{q-1} F(m)$. The computation of $W(0)$ needs at most $q-2$ additions, and usually some or all additions can be incorporated into the computation of the cyclic convolution [6, 19]. The algorithm can be summarized as follows.

Algorithm 12 *Cyclic convolution algorithm for reduced type-II DHT*

Step 1.   *Compute a cyclic convolution according to (5.229).*

Step 2.   *Compute $F^0(m)$ according to (5.228).*

Step 3.   *Compute the reduced type-II DHT of length M/q according to (5.227). The same technique used in Steps 1 and 2 can be applied for further decomposition.*

Let us now consider a few examples.

- A length 3 reduced type-II DHT can be calculated by

$$
W(0) = F(0) + F(2) \tag{5.230}
$$

$$
W(1) = \frac{1}{2}[F(0) + F(2)] + \frac{\sqrt{3}}{2}[F(0) - F(2)] \tag{5.231}
$$

$$
W(2) = -\frac{1}{2}[F(0) + F(2)] + \frac{\sqrt{3}}{2}[F(0) - F(2)], \tag{5.232}
$$

which needs only 1 multiplication, one shift and 4 additions.

- A length 5 reduced type-II DHT can be calculated by a cyclic convolution of length 4, which needs 5 multiplications and 15 additions.

- A length 7 reduced type-II DHT can be calculated by a cyclic convolution of length 6, which needs 8 multiplications and 31 additions.

- A length 9 reduced type-II DHT can be decomposed into a cyclic convolution of length 6 and a length 3 reduced type-II DHT at the cost of 4 additions. Since the cyclic convolution of length 6 needs 8 multiplications and 31 additions [6, 19], the length 9 reduced DHT-II requires 9 multiplications and 39 additions. By using the decomposition described in Section 5.8.2, the length 9 reduced DHT-II needs 11 multiplications and 34 additions.

Based on (5.226), the computational complexity for the length 3^t and 5^t reduced type-II DHT is

$$\phi(3^t) = (8t - 5)3^{t-2}, \qquad \psi(3^t) = (2t - \frac{3}{2})3^t \tag{5.233}$$

$$\phi(3^t) = (8t - 7)3^{t-2}, \qquad \psi(3^t) = (2t + \frac{1}{3})3^t \tag{5.234}$$

$$\phi(5^t) = (48t - 23)5^{t-2}, \qquad \psi(5^t) = (92t - 17)5^t. \tag{5.235}$$

5.8.4   PT-based radix-q algorithm for MD type-II DHT

Let us generalize the decomposition concepts which were used for the 2D type-II DHT in the last subsection. For simplicity, we only give the major steps of computation without presenting details. We assume $N_r = q^t$, $N_r/N_i = q^{l_i}$ $(i = 1, 2, \ldots, r-1)$, where l_i and t are non-negative integers, and q is an odd prime number. An rD type-II DHT of size $N_1 \times N_2 \times \cdots \times N_r$ is

$$X(k_1, k_2, \ldots, k_r) = \sum_{n_1=0}^{N_1-1} \sum_{n_2=0}^{N_2-1} \cdots \sum_{n_r=0}^{N_r-1} x(n_1, n_2, \ldots, n_r)$$
$$\cdot \mathrm{cas}\left(\frac{\pi(2n_1+1)k_1}{N_1} + \cdots + \frac{\pi(2n_r+1)k_r}{N_r}\right),$$

where $k_i = 0, 1, \ldots, N_i - 1$ and $i = 1, 2, \ldots, r$. Let $p_i(n_r)$ be the least non-negative remainder of $(2p_i + 1)n_r + p_i$ modulo N_i $(i = 1, 2, \ldots, r-1)$. A combined PT and radix-q algorithm for the rD type-II DHT is given below.

Algorithm 13 *PT-based radix-q algorithm for rD type-II DHT*

Step 1.   *Compute $N_1 N_2 \cdots N_{r-1}$ reduced type-II DHTs with length N_r*

$$V_{p_1, \ldots, p_{r-1}}(m) = \sum_{\substack{n_r=0 \\ n_r \not\equiv (q-1)/2}}^{N_r-1} x(p_1(n_r), \ldots, p_{r-1}(n_r), n_r) \mathrm{cas} \frac{\pi(2n_r+1)m}{N_r},$$

where $p_i = 0, 1, \ldots, N_i - 1$, $i = 1, 2, \ldots, r-1$ and $m = 0, 1, \ldots, N_r - 1$.

Step 2.   *Compute the $(r-1)D$ PT*

$$Y_{k_1, \ldots, k_{r-1}}(z) \equiv \sum_{p_1=0}^{N_1-1} \sum_{p_2=0}^{N_2-1} \cdots \sum_{p_{r-1}=0}^{N_{r-1}-1} U_{p_1, \ldots, p_{r-1}}(z)$$
$$\cdot z_1^{p_1 k_1} \cdots z_{r-1}^{p_{r-1} k_{r-1}} \bmod (z^{N_r} + 1),$$

where $z_i \equiv z^{-2q^{l_i}} \bmod (z^{N_r} + 1)$ and

$$U_{p_1,\ldots,p_{r-1}}(z) = \sum_{m=0}^{N_r-1} V_{p_1,\ldots,p_{r-1}}(m) z^m.$$

Reorder the coefficients of the polynomials to get $B_{k_1,\ldots,k_{r-1}}(z)$ according to

$$B_{k_1,\ldots,k_{r-1}}(z) \equiv Y_{k_1,\ldots,k_{r-1}}(z) z^{-k_1 q^{l_1} - \cdots - k_{r-1} q^{l_{r-1}}} \bmod (z^{N_r} + 1).$$

Let $B(k_1,\ldots,k_r)$ be the coefficients of $B_{k_1,\ldots,k_{r-1}}(z)$.

Step 3.   *Compute an rD type-II DHT with size $N_1 \times \cdots \times N_{r-1} \times N_r/q$ defined by*

$$A(k_1, k_2, \ldots, k_r) = \sum_{n_1=0}^{N_1-1} \cdots \sum_{n_{r-1}=0}^{N_{r-1}-1} \sum_{n_r=0}^{N_r/q-1} x(n_1,\ldots,n_{r-1}, q n_r + \frac{q-1}{2})$$

$$\cdot \text{cas} \left(\frac{\pi(2n_1+1)k_1}{N_1} + \cdots + \frac{\pi(2n_{r-1}+1)k_{r-1}}{N_{r-1}} \right.$$

$$\left. + \frac{\pi(2n_r+1)k_r}{N_r/q} \right),$$

$$k_i = 0, 1, \ldots, N_i - 1, \quad i = 1, 2, \ldots, r-1, \quad k_r = 0, 1, \ldots, \frac{N_r}{q},$$

which can be computed by Step 1 and Step 2 recursively.

Step 4.   *Compute $X(k_1,\ldots,k_r) = A(k_1, k_2, \ldots, k_r) + B(k_1, k_2, \ldots, k_r)$.*

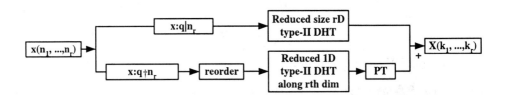

Figure 5.8   Flow chart of the PT-based radix-q algorithm for rD type-II DHT.

Figure 5.8 presents the flow chart of the algorithm. Since the rD type-III DHT is the inversion of the rD type-II DHT, its fast algorithm can be derived easily from the algorithm of the rD type-II DHT.

5.8.5   Computational complexity

Now we analyze the number of operations required by the fast algorithm for the rD type-II DHT. The algorithm computes $N_1 N_2 \cdots N_{r-1}$ reduced type-II DHTs of length N_r and an $(r-1)D$ PT of size $N_1 \times \cdots \times N_{r-1}$. The $(r-1)D$ PT needs $(q-1)N \log_q(N_1 N_2 \cdots N_{r-1})$ additions (see Chapter 2 or [6, 18, 19, 24]), where $N = N_1 N_2 \cdots N_r$. Therefore, the required numbers of multiplications and additions are

$$M_u(N_1, \ldots, N_{r-1}, N_r) = \frac{N}{N_r}\phi(N_r) + M_u(N_1, \ldots, N_{r-1}, \frac{N_r}{q})$$

$$A_d(N_1, \ldots, N_{r-1}, N_r) = \frac{N}{N_r}\psi(N_r) + (q-1)N\log_q \frac{N}{N_r} + N$$

$$+ A_d(N_1, \ldots, N_{r-1}, \frac{N_r}{q}), \tag{5.236}$$

where $\phi(N_r)$ and $\psi(N_r)$ are the number of multiplications and additions, respectively, for a reduced type-II DHT of length N_r. Although it is difficult to derive a closed expression of the number of operations for arbitrary N_i, a closed form expression can be obtained when $N_1 = N_2 = \cdots = N_r = q^t$. By defining $M_u(q^t)$ and $A_d(q^t)$ to represent $M_u(q^t, \ldots, q^t, q^t)$ and $A_d(q^t, \ldots, q^t, q^t)$, from (5.236) we have

$$M_u(q^t) = \frac{(q^r - 1)q^{(r-1)(t-1)}}{q - 1}\phi(q^t) + M_u(q^{t-1})$$

$$A_d(q^t) = \frac{(q^r - 1)q^{(r-1)(t-1)}}{q - 1}\psi(q^t) + (r-1)t(q^r - 1)q^{rt-r+1}$$

$$+ rq^{rt-r+1} + A_d(q^{t-1}). \tag{5.237}$$

Without loss of generality, we assume $\phi(q^t) = a_1 t q^t + a_2 q^t + a_3$ and $\psi(q^t) = b_1 t q^t + b_2 q^t + b_3$. From (5.237) we achieve

$$M_u(q^t) = \frac{a_1 q t q^{rt}}{q - 1} + (a_2 - \frac{a_1}{q^r - 1})\frac{q(q^{rt} - 1)}{q - 1} + a_3\frac{(q^r - 1)(q^{(r-1)t} - 1)}{(q - 1)(q^{r-1} - 1)}$$

$$A_d(q^t) = (\frac{b_1 q}{q - 1} + (r-1)q)t q^{rt} + (b_2 + 1 - \frac{b_1}{q^r - 1} - \frac{q^r - q}{q^r - 1})\frac{q(q^{rt} - 1)}{q - 1}$$

$$+ \frac{b_3(q^r - 1)(q^{(r-1)t} - 1)}{(q - 1)(q^{r-1} - 1)}. \tag{5.238}$$

On the other hand, the row-column approach needs rN additions to convert the rD type-II DHT into a form that has a separable kernel [7, 10, 32]. Let $\tau(q^t)$ and $\rho(q^t)$ represent the number of multiplications and additions, respectively, for computing a 1D type-II DHT of length q^t. We generally assume that $\tau(q^t) = c_1 t q^t + c_2 q^t + c_3$ and $\rho(q^t) = d_1 t q^t + d_2 q^t + d_3$. Therefore, the computational complexity of the row-column method for an rD type-II DHT with size $q^t \times q^t \times \cdots \times q^t$ is

$$\overline{M}_u(q^t) = (c_1 t + c_2)rq^{rt} + c_3 rq^{(r-1)t}$$

$$\overline{A}_d(q^t) = (d_1 t + d_2 + 1)rq^{rt} + d_3 rq^{(r-1)t}.$$

The number of multiplications needed by the PT-based algorithm is remarkably smaller than that needed by the row-column method. For most algorithms, $a_1 = \frac{q-1}{q}c_1$ and $a_2 \approx \frac{q-1}{q}c_2$, we have

$$M_u \approx \frac{1}{r}\overline{M}_u.$$

However, it is much more difficult to compare the number of additions. We consider two cases for $q = 3$ and $q = 5$. From (5.233), (5.235) and (5.238), we get

$$M_u(3^t) = \frac{4}{3}t3^{rt} - \frac{5 \cdot 3^r + 3}{6(3^r - 1)}(3^{rt} - 1)$$

$$A_d(3^t) = 3rt3^{rt} - 3^{rt} + 1$$

$$M_u(5^t) = \frac{12}{5}t5^{rt} - (\frac{23}{20} + \frac{12}{5(5^r - 1)})(5^{rt} - 1)$$

$$A_d(5^t) = (5r - \frac{2}{5})t5^{rt} - \frac{17 \cdot 5^{r-1} - 5}{4(5^r - 1)}(5^{rt} - 1). \tag{5.239}$$

The numbers of operations of the algorithms for the type-II DHT [32, 35] are

$$\tau(3^t) = (\frac{4t}{3} - 1)3^t, \quad \rho(3^t) = (8t - 1)3^{t-1}$$

$$\tau(5^t) = 12t5^{t-1}, \quad \rho(5^t) = 22t5^{t-1}.$$

For the row-column method, we have

$$\overline{M}_u(3^t) = (\frac{4t}{3} - 1)r3^{rt}$$

$$\overline{A}_d(3^t) = (8t + 2)r3^{rt-1}$$

$$\overline{M}_u(5^t) = 12rt5^{rt-1}$$

$$\overline{A}_d(5^t) = (\frac{22t}{5} + 1)r5^{rt}. \tag{5.240}$$

In these two cases, the number of multiplications required by the PT-based algorithm is approximately $\frac{1}{r}$ times that needed by the row-colum method. Although the number of additions is slightly increased, our algorithm considerably reduces the total number of operations (multiplications plus additions), which can be seen from (5.239) and (5.240). Table 5.10 lists the number of operations for $q = 3$ and $r = 3$.

Table 5.10   Comparison of the number of operations.

	PT-based algorithm			Row-column-based algorithm		
t	M_u	A_d	$M_u + A_d$	$\overline{M}_u$	$\overline{A}_d$	$\overline{M}_u + \overline{A}_d$
2	1300	12394	13694	3645	13122	16767
3	61321	511759	573080	177147	511758	688905
4	2364232	18600436	20964668	6908733	18068994	24977767
5	82966117	631341909	714318026	243931419	602654094	846585513

In general, the PT-based algorithms for an rD type-II DHT can handle transforms with size $q^{l_1} \times q^{l_2} \times \cdots \times q^{l_r}$, where q is an odd prime number. The dominant part using additions is for the computation of the $(r-1)$D PT, for which we use the simplest FPT algorithm. A considerable savings in the number of multiplications is achieved. The structure of the algorithm is also simple because it uses only the 1D DHT and the PT, which can be easily implemented. Although the number of additions increases slightly compared to the row-column method, the PT-based algorithms achieve substantial savings in terms of the total number of arithmetic operations. However, it is an open problem to find a suitable method to improve the FPT algorithm. The presented algorithm can also be easily extended to deal with dimension size $2^{k_1} q^{l_1} \times 2^{k_2} q^{l_2} \times \cdots \times 2^{k_r} q^{l_r}$. However, it seems impossible to extend the idea for an rD type-II DHT with size $q_1^{l_1} \times q_2^{l_2} \times \cdots \times q_r^{l_r}$, where q_i are different prime numbers.

5.9 CHAPTER SUMMARY

This chapter presents several fast algorithms for the computation of various types of MD DHTs. The split-radix algorithms, the recursive algorithm and the algorithms using the type-I DHT are developed by extending the decomposition concepts used for 1D DHTs to MD DHTs. These algorithms provide efficient computation and support various dimensional sizes. In particular, the PT-based fast algorithms reduce the number of operations considerably. The computational complexities of these fast algorithms are analyzed and compared with those of other reported ones.

REFERENCES

1. *Proceedings of IEEE (Special issue on Hartley transform),* vol. 82, no. 3, 1994.

2. T. M. Apostol, *Introduction to Analytic Number Theory,* Springer-Verlag, New York, p. 210, 1986.

3. G. Bi, Split-radix algorithm for 2D discrete Hartley transform, *Signal Processing,* vol. 63, no. 1, 45–53, 1997.

4. G. Bi and J. Liu, Fast odd-factor algorithm for discrete W transform, *Electron. Lett.,* vol. 34, no. 5, 431–433, 1998.

5. G. Bi and C. Lu, Prime-factor algorithms for generalized discrete Hartley transform, *Electron. Lett.,* vol. 35, no. 20, 1708–1710, 1999.

6. R. E. Blahut, *Fast Algorithms for Digital Signal Processing,* Addison-Wesley, Reading, MA, 1984.

7. T. Bortfeld and W. Dinter, Calculation of multidimensional Hartley transform using one dimensional Fourier transform, *IEEE Trans. Signal Process.,* vol. 42, no. 5, 1306–1310, 1995.

8. S. Boussakta and A. G. J. Holt, Fast multidimensional discrete Hartley transform using Fermat number transform, *IEE Proc. G,* vol. 135, 253–257, 1988.

9. R. N. Bracewell, *The Hartley Transform,* Oxford University Press, London, 1986.

10. R. N. Bracewell and O. Buneman, Fast two-dimensional Hartley transform, *Proc. IEEE,* vol. 74, no.5, 1282–1283, 1986.

11. V. Britanak and K. R. Rao, The fast generalized discrete Fourier transforms: a unified approach to the discrete sinusoidal transforms computation, *Signal Processing,* vol. 79, no. 2, 135–150, 1999.

12. O. Buneman, Multidimensional Hartley transform, *Proc. IEEE,* vol. 75, no. 2, 267, 1987.

13. S. C. Chan and K. L. Ho, Split vector-radix fast Fourier algorithm, *IEEE Trans. Signal Process.,* vol. 40, no. 8, 2029–2039, 1992.

14. S. C. Chan and K. L. Ho, Polynomial transform fast Hartley transform, *Electron. Lett.,* vol. 26, no. 21, 1914–1916, 1990.

15. C. Dubost, S. Desclos and A. Zerguerras, New radix-3 FHT algorithm, *Electron. Lett.,* vol. 26, no. 18, 1537–1538, 1990.

16. H. Hao and R. N. Bracewell, A three dimensional DFT algorithm using fast Hartley transform, *Proc. IEEE,* vol. 75, no. 2, 264–266, 1987.

17. N.-C. Hu, H.-I. Chang and F. F. Lu, Fast computation of the two dimensional generalized discrete Hartley transforms, *IEE Proc. Vis. Image Signal Process.,* vol. 142, no. 1, 35–39, 1995.

18. Z. R. Jiang and Y. H. Zeng, *Polynomial transform transform and its applications*, National University of Defense Technology Press, P. R. China, 1989.

19. Z. R. Jiang, Y. H. Zeng and P. N. Yu, *Fast Algorithms*, National University of Defense Technology Press, P. R. China, 1994.

20. E. A. Jonckheere and C. W. Ma, Split-radix fast Hartley transform in one and two dimensions, *IEEE Trans. Signal Process.*, vol. 39, no. 2, 499–503, 1991.

21. A. Kojima, N. Sakurai and J. Kishigami, Motion detection using 3D-FFT spectrum, *IEEE ICASSP*, V213–216, 1993

22. P. K. Meher, J. K. Satapathy and G. Panda, New high-speed prime-factor algorithm for discrete Hartley transform, *IEE Proceedings F*, vol. 140, no. 1, 63–70, 1993.

23. Z. J. Mou and P. Duhamel, In-place butterfly-style FFT of 2-D real sequences, *IEEE Trans. Acustics, Speech, Signal Process.*, ASSP-36, no. 10, 1642–1650, 1988.

24. H. J. Nussbaumer, *Fast Fourier Transform and Convolutional Algorithms*, Springer-Verlag, New York, 1981.

25. B. Porat and B. Friedlander, A frequency domain algorithm for multiframe detection and estimation of dim targets, *IEEE Trans. Pattern Analysis Machine Intelligence*, vol. 12, no. 4, 398–401, 1990.

26. H. V. Sorensen, D. L. Jones, C. S. Burrus and M. T. Heideman, On computing the discrete Hartley transform, *IEEE Trans. Acustics, Speech, Signal process.*, ASSP-33, no. 4, 1231–1238, 1985.

27. Z. Wang, The fast W transform-algorithms and programs, *Science in China (series A)*, vol. 32, 338–350, 1989.

28. Z. Wang and B. R. Hunt, The discrete W transform, *Appl. Math. Comput.*, vol. 16, 19–48, 1985.

29. J. L. Wu and S. C. Pei, The vector split-radix algorithm for 2D DHT, *IEEE Trans. Signal Process.*, vol. 41, no. 2, 960–965, 1993.

30. D. Yang, New fast algorithm to compute two-dimensional discrete Hartley transform, *Electron. Lett.*, vol. 25, no. 25, 1705–1706, 1989.

31. Y. H. Zeng and X. Li, Multidimensional polynomial transform algorithm for multidimensional discrete W transform, *IEEE Trans. Signal Process.*, vol. 47, vol. 7, 2050–2053, 1999.

32. Y. H. Zeng, L. Z. Cheng and M. Zhou, *Parallel Algorithms for Digital Signal Processing*, National University of Defense Technology Press, Changsha, P. R. China 1989. (in Chinese).

33. Y. H. Zeng, G. Bi and A. R. Leyman, Polynomial transform algorithms for multidimensional discrete Hartley transform, *Proc. IEEE International Symposium Circuits Systems*, Geneva, Switzerland, V517–520, 2000.

34. Y. H. Zeng, G. Bi and A. C. Kot, Combined polynomial transform and radix-q algorithm for MD discrete W transform, *IEEE Trans. Signal Process.*, vol. 49, no.3, 634–641, 2001.

35. Y. H. Zeng and Z. R. Jiang, A unified fast algorithm for the discrete W transform with arbitrary length, *Chinese J. Num. Math. & Appl.*, *Allerton Press*, New York, vol. 18, no. 4, 85–92, 1996.

6

Fast Algorithms for 1D Discrete Cosine Transform

This chapter presents fast algorithms for the four types of one-dimensional (1D) discrete cosine transforms (DCTs).

- Section 6.2 presents radix-2 algorithms for the type-II, -III and -IV DCTs.

- Section 6.3 derives prime factor algorithms for the type-II and -III DCTs.

- Section 6.4 describes radix-q algorithms for the type-II and -III DCTs.

- Section 6.5 introduces algorithms based on the relationship among the four types of DCTs.

Details of the mathematical derivations and comparisons on the computational complexity and structures are also provided.

6.1 INTRODUCTION

The DCT has been widely used in many applications because it is asymptotically equivalent to the Karhunen–Loève transform (KLT) of Markov-1 signals and a better approximation to the KLT than the discrete Fourier transform (DFT). A few books were published to provide excellent materials on the fast algorithms and applications of the DCT ([12] and [13], for example). This chapter presents several recently developed fast algorithms for various types of 1D DCTs. The fast algorithms for multidimensional (MD) DCTs will be dealt with in the next chapter. There exist four types of DCTs, as classified by Wang [18]. If the normalization factors are ignored, they are defined by

Type-I:

$$X(k) = \sum_{n=0}^{N} x(n) \cos \frac{\pi n k}{N}, \quad 0 \le k \le N \tag{6.1}$$

Type-II:

$$X(k) = \sum_{n=0}^{N-1} x(n) \cos \frac{\pi(2n+1)k}{2N}, \quad 0 \le k < N \tag{6.2}$$

Type-III:

$$X(k) = \sum_{n=0}^{N-1} x(n) \cos \frac{\pi n(2k+1)}{2N}, \quad 0 \le k < N \tag{6.3}$$

Type-IV:

$$X(k) = \sum_{n=0}^{N-1} x(n) \cos \frac{\pi(2n+1)(2k+1)}{4N}, \quad 0 \le k < N, \tag{6.4}$$

where the sequence length N can be an arbitrary composite number that is expressed by

$$N = 2^{r_0} q_1^{r_1}, \ldots, q_n^{r_n}, \tag{6.5}$$

where q_i $(i = 1, 2, \ldots, n)$ are odd integers. Our task is to find fast algorithms to efficiently compute DCTs of such sequence lengths.

In general the type-II and -III DCTs are considered to be a transform pair since the kernel matrix of the type-III is obviously the transpose of the kernel matrix of the type-II DCT. Furthermore, the type-IV DCT is the shifted version of the type-I DCT. Some useful properties of these DCTs can be found in [19, 21]. The type-II or -III DCTs are widely used in many applications while the type-I DCT is used for interpolation [17] and computations of other types of DCTs. The type-IV DCT is used in the lapped transform [11, 14] and is often required in the computation of the type-II and -III DCTs.

After many years of development, there exist numerous algorithms for all the types of DCTs. Based on a particular radix number q, the radix-type algorithms recursively divide the entire computation task into q length N/q DCTs. This type of fast algorithm particularly supports the sequence length $N = q^m$, where m is a positive integer. For example, the radix-2 algorithm [7] for $N = 2^m$ is the most popular and widely used. Other radix-type algorithms are the mixed radix algorithm for $N = 3^m 2^n$ [5] and the radix-q algorithms for $N = q^m$, where q is an odd integer [1, 2]. In particular, the radix-q algorithms can be combined with the radix-2 algorithm to support the composite sequence length without increasing the overall computational complexity compared to the radix-2 algorithm. The radix-type algorithms possess several desirable properties, including regular input/output indexing and in-place computation, that are very useful in reducing the overall implementation complexity. The fast algorithms for the DCTs whose sequence lengths are other than 2^m generally need a higher computational complexity and a more complex structure [5, 20]. The algorithms in the non-radix category, i.e., the prime factor algorithms and the polynomial-based fast algorithms, usually require a relatively small computational complexity. They generally need an index mapping process, and with these algorithms it is difficult to achieve in-place computation.

The main objective of this chapter is to present recently developed algorithms that support DCTs of composite sequence length, as defined in (6.5), without substantially increasing computational and structural complexity. One way to deal with such computations is to decompose the entire computation task according to the odd factors into a number of DCTs whose lengths do not contain odd factors. Then these DCTs can be computed by the radix-2 algorithm. It is also possible to decompose the entire computation task into a number of smaller DCTs whose lengths contain odd factors only. These small DCTs can be further decomposed by the fast algorithms that particularly deal with odd factors. It can be proven that the latter approach is generally preferred. Both approaches require fast algorithms that decompose the DCT computation according to odd factors in an efficient manner. Fast algorithms dealing with all the different odd factors have been found only recently [2, 3].

6.2 RADIX-2 ALGORITHMS

6.2.1 Algorithm for type-II DCT

The radix-2 algorithms are particularly suited to DCTs of even sequence length. By using the symmetric properties of the transform kernel function, the entire length N DCT is decomposed into two length $N/2$ DCTs. This process can be continued as far as the sequence length remains even. There exist a few versions of the radix-2 algorithm ([7, 10], for example). We now consider a radix-2 algorithm that has a recursive computational structure. Based on the decimation-in-frequency decomposition and the symmetric property of the transform kernel, the type-II DCT defined in (6.2) can be decomposed into

$$X(2k) = \sum_{n=0}^{N/2-1} [x(n) + x(N-n-1)] \cos \frac{\pi(2n+1)k}{2(N/2)}, \quad 0 \le k < \frac{N}{2}, \tag{6.6}$$

which is a length $N/2$ DCT of sequence $[x(n) + x(N-n-1)]$. We can combine the odd indexed DCT outputs to form a new sequence

$$
\begin{aligned}
F(k) &= \frac{X(2k+1) + X(2k-1)}{2} \\
&= \sum_{n=0}^{N/2-1} [x(n) - x(N-n-1)] \cos \frac{\pi(2n+1)}{2N} \cos \frac{\pi(2n+1)k}{2(N/2)},
\end{aligned} \tag{6.7}
$$

where $0 \le k < N/2$. $F(k)$ is again a length $N/2$ DCT of the sequence $[x(n) - x(N-n-1)] \cos \frac{\pi(2n+1)}{2N}$. By using the property $\cos(\alpha) = \cos(-\alpha)$, it can be easily proved that the odd indexed DCT $X(2k+1)$ can be recursively calculated by

$$X(2k+1) = 2F(k) - X(2k-1), \quad 1 \le k \le \frac{N}{2} - 1, \tag{6.8}$$

where the initial value $X(1) = F(0)$. This algorithm can be generally applied to sequence length $N = q \cdot 2^m$, where q is an odd integer. The decomposition process can be continued until a length q DCT is reached. Figure 6.1 shows the signal flow graph of this algorithm after the first decomposition step. This algorithm has a simple computational structure that leads to an easy implementation. An efficient hardware structure for sequential input data was reported to use this radix-2 algorithm [15]. It should be pointed out that the fast algorithm based on the decimation-in-time decomposition technique can also be derived.

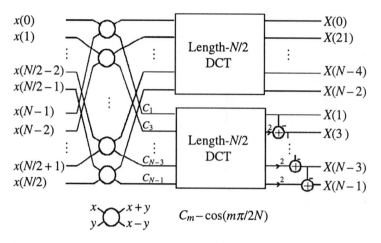

Figure 6.1   First decomposition step for length $q \cdot 2^m$ DCT.

6.2.2   Algorithm for type-III DCT

Based on the decimation-in-time decomposition, the type-III DCT, defined in (6.3), of a sequence $X(k)$ can be expressed by

$$x(n) = \sum_{k=0}^{N-1} X(k) \cos \frac{\pi(2n+1)k}{2N}, \qquad 0 \le n \le N-1, \tag{6.9}$$

which can be decomposed into

$$x(n) = A(n) + B(n) \tag{6.10}$$
$$x(N-n-1) = A(n) - B(n), \tag{6.11}$$

where for both (6.10) and (6.11), $0 \le n \le N/2 - 1$, and

$$A(n) = \sum_{k=0}^{N/2-1} X(2k) \cos \frac{\pi(2n+1)k}{2(N/2)} \tag{6.12}$$

$$B(n) = \sum_{k=0}^{N/2-1} X(2k+1) \cos \frac{\pi(2n+1)(2k+1)}{4(N/2)}, \tag{6.13}$$

It is noted that (6.12) is a length $N/2$ type-III DCT and (6.13) is a length $N/2$ type-IV DCT. The length $N/2$ type-III DCT can be similarly decomposed and the computation of the length $N/2$ type-IV DCT is given in the next subsection.

6.2.3   Algorithm for type-IV DCT

Now let us consider the computation of the type-IV DCT. Based on the definition of (6.4), we define a new sequence

$$F(k) = \frac{X(k) + X(k-1)}{2} = \cos \frac{\pi(2k+1)}{4N} \sum_{n=0}^{N-1} x(n) \cos \frac{\pi n(2k+1)}{2N},$$

which is a length N type-III DCT of input sequence $x(n)$, whose outputs are rotated by the twiddle factors $\cos \frac{\pi(2k+1)}{4N}$. By using the property $X(0) = X(-1)$, $X(k)$ can be obtained by

$$X(k) = 2F(k) - X(k-1), \quad 1 \le k \le N-1, \tag{6.14}$$

where the initial value $X(0) = F(0)$. The above algorithm is very simple if the fast algorithms for the type-III DCT are available. Therefore, the joint use of this decomposition and the one presented in the last subsection provides fast algorithms for both type-III and -IV DCTs.

The type-IV DCT can also be computed based on a decimation-in-time decomposition approach, as shown below.

$$
\begin{aligned}
X(k) &= \sum_{n=0}^{N/2-1} x(2n) \cos \frac{\pi[2(2n+1)-1](2k+1)}{4N} \\
&\quad + \sum_{n=0}^{N/2-1} x(2n+1) \cos \frac{\pi[2(2n+1)+1](2k+1)}{4N} \\
&= \cos \frac{\pi(2k+1)}{4N} \sum_{n=0}^{N/2-1} [x(2n) + x(2n+1)] \cos \frac{\pi(2n+1)(2k+1)}{2N} \\
&\quad + \sin \frac{\pi(2k+1)}{4N} \sum_{n=0}^{N/2-1} [x(2n) - x(2n+1)] \sin \frac{\pi(2n+1)(2k+1)}{2N} \\
&= \cos \frac{\pi(2k+1)}{4N} A(k) + \sin \frac{\pi(2k+1)}{4N} B(k), \tag{6.15}
\end{aligned}
$$

where $0 \le k \le N-1$, and

$$
A(k) = -A(N-k-1) = \sum_{n=0}^{N/2-1} [x(2n) + x(2n+1)] \cos \frac{\pi(2n+1)(2k+1)}{4(N/2)},
$$

which is a length $N/2$ type-IV DCT, and

$$
B(k) = B(N-k-1) = \sum_{n=0}^{N/2-1} [x(2n) - x(2n+1)] \sin \frac{\pi(2n+1)(2k+1)}{4(N/2)}.
$$

It is noted that $B(k)$ can also be computed by a length $N/2$ type-IV DCT because

$$
B(\frac{N}{2} - k - 1) = \sum_{n=0}^{N/2-1} [x(2n) - x(2n+1)](-1)^n \cos \frac{\pi(2n+1)(2k+1)}{4(N/2)}. \tag{6.16}
$$

Based on the properties of $A(k)$ and $B(k)$, we have

$$X(k) = \cos \frac{\pi(2k+1)}{4N} A(k) + \sin \frac{\pi(2k+1)}{4N} B(k) \tag{6.17}$$

$$X(N-k-1) = -\sin \frac{\pi(2k+1)}{4N} A(k) + \cos \frac{\pi(2k+1)}{4N} B(k), \tag{6.18}$$

where $0 \le k \le N/2 - 1$.

6.2.4   Computational complexity

Based on the decomposition procedures presented for the type-II, -III and -IV DCTs, we now consider the computational complexity needed by the radix-2 fast algorithms. The length N type-II DCT is decomposed into two length $N/2$ type-II DCTs, as seen in (6.6) and (6.7). In particular, the input sequence in (6.7) is rotated by a twiddle factor. The odd indexed DCT outputs are achieved by a recursive relation in (6.8). Therefore the computational complexity in terms of the numbers of additions and multiplications is

$$M_{II}(N) = 2M_{II}(\frac{N}{2}) + \frac{N}{2} \tag{6.19}$$

$$A_{II}(N) = 2A_{II}(\frac{N}{2}) + \frac{3N}{2} - 1, \tag{6.20}$$

where $M_{II}(N)$ and $A_{II}(N)$ are the numbers of multiplications and additions needed by the length N type-II DCT, respectively.

If the type-III DCT is used, as shown in (6.14), the computational complexity of the length N type-IV DCT becomes

$$M_{IV}(N) = M_{III}(N) + N \tag{6.21}$$
$$A_{IV}(N) = A_{III}(N) + N - 1. \tag{6.22}$$

If the computation described in (6.15) is used, the computational complexity becomes

$$M_{IV}(N) = 2M_{IV}(\frac{N}{2}) + 2N \tag{6.23}$$

$$A_{IV}(N) = 2A_{IV}(\frac{N}{2}) + 2N. \tag{6.24}$$

The computational complexity for the type-III DCT is the same as that for the type-II DCT, which can be easily proven.

6.3   PRIME FACTOR ALGORITHMS

Prime factor algorithms (PFAs) [4, 6, 7, 9, 16, 20] have been reported to be computationally efficient because twiddle factors are not needed between computational blocks. The basic principle of PFA is to convert a 1D DCT into a two-dimensional (2D) computation by using some properties of the transform kernel. Therefore, the 1D input sequence has to be converted into a 2D input array before the 2D computation. The outputs of the 2D computation have to be converted into a 1D output sequence according to some indexing rules. The conversion process (known as the index mapping process) for the input/output sequence is generally undesirable and sometimes has substantially undesirable effects on the computational efficiency of the fast algorithm. This happens because the mapping process often results in a complex computational structure and requires modulo and other arithmetic operations. Such requirements generally impose a substantial computational overhead which often offsets the savings achieved from removing the twiddle factors. Another drawback of most PFAs is that in-place computation is not possible, and thus much more memory space is required to implement the algorithm. By using table or matrix manipulation, several methods have been reported to simplify the mapping procedure [4, 8, 16]. These methods are not straightforward and are difficult to use particularly when the sequence length is large. This subsection describes fast algorithms based on the prime factor decomposition for the type-II and -III DCTs. It can be seen that these problems we have mentioned can be easily avoided.

6.3.1 Algorithm for type-II DCT

We assume that the sequence length $N = pq$, where p and q are mutually prime. The type-II DCT defined in (6.2) can be expressed as

$$
\begin{aligned}
X(|q\mu k + pm|) &= \sum_{n=0}^{N-1} x(n) \cos \frac{\pi(2n+1)(q\mu k + pm)}{2N} \\
&= A(k, m) - B(\mu k, m),
\end{aligned}
\tag{6.25}
$$

where

$$
A(k, m) = \sum_{n=0}^{N-1} x(n) \cos \frac{\pi(2n+1)m}{2q} \cos \frac{\pi(2n+1)k}{2p}
\tag{6.26}
$$

$$
B(\mu k, m) = \mu \sum_{n=0}^{N-1} x(n) \sin \frac{\pi(2n+1)m}{2q} \sin \frac{\pi(2n+1)k}{2p},
\tag{6.27}
$$

where for (6.25) – (6.27), $0 \le k < p, 0 \le m < q$, and μ is defined as

$$
\mu = \begin{cases} 1, & qk + mp < N \\ -1, & qk + mp > N \end{cases} .
\tag{6.28}
$$

Because $q\mu k + pm$ in (6.25) can be negative, the absolute value of $q\mu k + pm$ is used as the output indices. It is easy to prove that (6.26) and (6.27) are related by

$$
B(k, 0) = 0, \quad B(\mu k, m) = \mu A(p - k, q - m).
\tag{6.29}
$$

Therefore, (6.27) can be obtained from (6.26), or vice versa. Now, let us consider the computation of $A(k, m)$. Without presenting a detailed proof, $A(k, m)$ can be rewritten into a length p DCT:

$$
\begin{aligned}
A(k, m) = \sum_{n=0}^{p-1} \Bigg\{ &\sum_{i=0}^{(q-1)/2} x(2ip + n) \cos \frac{\pi(4ip + 2n + 1)m}{2q} \\
&+ \sum_{i=0}^{(q-1)/2} x(2ip - n - 1) \cos \frac{\pi(4ip - 2n - 1)m}{2q} \Bigg\} \cos \frac{\pi(2n+1)k}{2p}.
\end{aligned}
\tag{6.30}
$$

By using the property $\cos(a) = \cos(-a)$ in (6.30), the two terms in the braces can be rewritten into a closed form,

$$
\sum_{i=-(q-1)/2}^{(q-1)/2} y(i, n) \cos \frac{\pi(4ip + 2n + 1)m}{2q},
\tag{6.31}
$$

where

$$
y(i, n) = \begin{cases} x(2ip + n), & i \ge 0 \\ x(-2ip - n - 1), & i < 0 \end{cases} .
\tag{6.32}
$$

It is noted that in (6.31), $(4ip + 2n + 1)$ is odd and $2q$ is even, which are mutually prime. According to the Chinese remainder theorem, we can always find q different odd integers for each n so that

$$\cos \frac{\pi(2l + 1)m}{2q} \equiv \cos \frac{\pi(4ip + 2N + 1)m}{2q}. \tag{6.33}$$

It can be easily verified that this relation can be equivalently expressed by

$$2l + 1 = \begin{cases} ||4ip + 2n + 1| - 4q|, & |4ip + 2n + 1| > 2q \\ |4ip + 2n + 1|, & |4ip + 2n + 1| < 2q \end{cases}. \tag{6.34}$$

where $0 \leq n < p$, $-(q - 1)/2 \leq i \leq (q - 1)/2$ and $0 \leq l < q$. Therefore, (6.31) can be converted to

$$\sum_{l=0}^{q-1} y'(l, n) \cos \frac{\pi(2l + 1)m}{2q} \tag{6.35}$$

where l is derived from (6.34), and $y'(l, n)$ is a 2D array converted from the original input

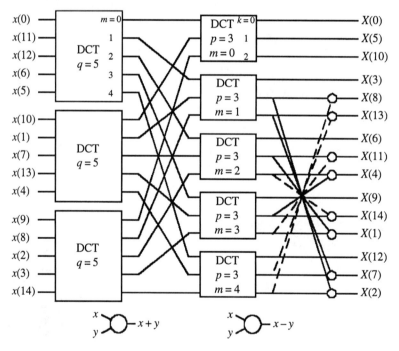

Figure 6.2   Signal flow graph of PFA ($qp = 15$).

sequence $x(n)$ according to the relation given in (6.33) or (6.34). Therefore, the length N DCT is decomposed into p length q and q length p DCTs, as indicated by (6.30). Figure 6.2 shows the signal flow graph of the PFA for a length 15 type-II DCT. This figure shows that three length 5 DCTs at the first stage and five length 3 DCTs at the second stage are needed. No twiddle factors are required between stages.

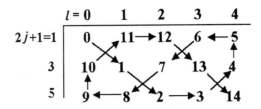

Figure 6.3   Input mapping graph for $q = 5$ and $p = 3$.

Table 6.1   Mapping of 1D input sequence into a 2D array.

Indices of	$j = 0$	1	2	2	1	0	0
$y'(j, l)$	$l = 0$	1	2	3	4	4	3
$y'(j, l) =$	$x(0)$	$x(1)$	$x(2)$	$x(3)$	$x(4)$	$x(5)$	$x(6)$
$j = 1$	2	2	1	0	0	1	2
$l = 2$	1	0	0	1	2	3	4
$x(7)$	$x(8)$	$x(9)$	$x(10)$	$x(11)$	$x(12)$	$x(13)$	$x(14)$

6.3.2   Index mapping

Index mapping is an important issue for the PFA. As mentioned earlier, the index mapping process often requires a considerable amount of computational overhead and complicates the computational structure to make in-place computation difficult. Based on our decomposition procedure, a very simple and straightforward method can be used for the index mapping of the PFA [3], as shown in Figure 6.3 for an example of a length 15 type-II DCT ($q = 5$ and $p = 3$). In the figure, $2j + 1$ and l are the indices used in (6.35), the numbers in the array are the original input indices and the arrows indicate that the next input index is achieved by the increment by one operation.

Let $l, 0 \leq l < q$, and $j, 0 \leq j < p$, be the 2D array indices and $n, 0 \leq n < N$, be the index of the original input sequence. In general, the mapping process can be performed by the following steps:

Step 1.   starting from 0, the index j of the 2D array is incremented by one until $q - 1$, then from $q - 1$, index j is decremented by one until 0. This process is continued for all input indices.

Step 2.   index l is generated in the same way as n except that the range is from 0 to $p - 1$ for increment and from $p - 1$ to 0 for decrement.

Step 3.   starting from 0 to $N - 1$, the index n of the original input sequence is increased by one for each increment/decrement operation described above. The index n of the original input sequence is mapped to indices (j, l) of the 2D array.

The mapping process is also shown in Table 6.1. Each column in the table specifies the relation between (j, l) and the original input index n. Such a mapping procedure is extremely simple, as shown in Appendix A. The subroutine of the mapping process requires only addition/subtraction and comparison.

Such a mapping procedure also makes in-place computation possible. After the computation at the first stage is completed, the inputs of each computational block (the length

5 DCT in Figure 6.2, for example) are no longer needed and the associated memory space can be reused for the outputs of the computational block. It is interesting to see that the transform outputs are also arranged into a natural order. The expression $|q\mu k + pm|$ is used to relate each computed output to the indices $k, 0 \leq k < p$, and $m, 0 \leq m < q$. This process can be easily implemented with operations of comparison, addition and subtraction.

6.3.3   Algorithm for type-III DCT

The type-III DCT, as defined in (6.3), of input sequence $X(k)$ can be expressed by

$$
\begin{aligned}
x(n) &= \sum_{k=0}^{q-1}\sum_{m=0}^{p-1} X(|\mu qk + pm|)\cos\frac{\pi(2n+1)(\mu qk + pm)}{2N} \\
&= \sum_{k=0}^{q-1}\sum_{m=0}^{p-1} X(|q\mu k + pm|)\left(\cos\frac{\pi(2n+1)k}{2q}\cos\frac{\pi(2n+1)m}{2p}\right. \\
&\quad \left. -\sin\frac{\pi(2n+1)k}{2q}\sin\frac{\pi(2n+1)m}{2p}\right) \\
&= \sum_{k=0}^{q-1}\sum_{m=0}^{p-1} X'(k,m)\cos\frac{\pi(2n+1)k}{2q}\cos\frac{\pi(2n+1)m}{2p},
\end{aligned}
\tag{6.36}
$$

which becomes a double summation whose input is

$$
X'(k,m) = \begin{cases} X(\mu qk + pm), & km = 0 \\ X(\mu qk + pm) + X[\mu p(q-k) + q(p-m)], & km > 0 \end{cases}.
\tag{6.37}
$$

In the second and third lines of (6.36), the term $\sin\frac{\pi(2n+1)k}{2q}\sin\frac{\pi(2n+1)m}{2p}$ is equivalent to $\cos\frac{\pi(2n+1)(q-k)}{2q}\cos\frac{\pi(2n+1)(p-m)}{2p}$. We can easily combine $\cos\frac{\pi(2n+1)k}{2q}\cos\frac{\pi(2n+1)m}{2p}$ and $\sin\frac{\pi(2n+1)k}{2q}\sin\frac{\pi(2n+1)m}{2p}$ to achieve the last line of (6.36). The combination details can be seen from the indices of the terms in (6.37).

The next step is to convert the 1D output sequence $x(n)$ into a 2D array $x'(n, l)$ so that (6.37) fully becomes a $q \times p$ 2D type-III DCT. The conversion process can be performed according to (6.34), Figure 6.3 or Table 6.1. We can rewrite (6.36) as

$$
x'(n, l) = \sum_{k=0}^{q-1}\sum_{m=0}^{p-1} X'(k,m)\cos\frac{\pi(2l+1)k}{2q}\cos\frac{\pi(2n+1)m}{2p},
\tag{6.38}
$$

where $0 \leq n < p$ and $0 \leq l < q$. It can be easily verified that the signal flow graph for the type-III DCT is the mirror image of Figure 6.2.

6.3.4   Computational complexity

One of the main advantages of the PFA is that no twiddle factor is needed between the computational stages, as seen in Figure 6.2. This simplifies the computational and structural complexity. It can be easily seen that the computational complexity needed by the PFA for both the type-II and -III DCTs is

$$
M(N) = pM(q) + qM(p)
\tag{6.39}
$$
$$
A(N) = pA(q) + qA(p) + (q-1)(p-1).
\tag{6.40}
$$

Compared to [8] and [9], this algorithm requires the same number of arithmetic operations. However, it requires $(q+1)(p-1)$ fewer arithmetic operations than the one in [16]. The main differences between this algorithm and others are the simplicity of the index mapping scheme and the availability of in-place computation. Similar to other reported PFAs, the derived one also deals with the sequence length containing mutually prime factors only. This limitation can be removed in the next section which considers fast algorithms that are able to deal with sequence lengths containing arbitrarily odd factors.

6.4 RADIX-q ALGORITHMS

In contrast to the PFAs, whose sequence lengths must contain mutually prime factors, the radix-q algorithm is used for a DCT whose sequence length N contains arbitrarily odd factors. By combining the radix-q algorithms with the radix-2 algorithms, we can compute DCTs of arbitrarily composite sequence length. In this section, the radix-q algorithm and its computational complexity for both the type-II and -III DCTs are described.

6.4.1 Radix-q algorithm for type-II DCT

Let us assume that N is divisible by an odd integer q. The type-II DCT defined in (6.2) can be partitioned into q length N/q DCTs, as described below.

$$
\begin{aligned}
X(k) = & \sum_{n=0}^{N/p-1} x(qn + \frac{q-1}{2}) \cos \frac{\pi(2n+1)k}{2N/q} \\
& + \sum_{m=0}^{\frac{q-3}{2}-1} \sum_{n=0}^{N/p-1} x(qn+m) \cos \frac{\pi[q(2n+1)-(q-1-2m)]k}{2N} \\
& + \sum_{m=0}^{\frac{q-3}{2}-1} \sum_{n=0}^{N/p-1} x(qn+q-m-1) \cos \frac{\pi[q(2n+1)+(q-1-2m)]k}{2N}.
\end{aligned}
\tag{6.41}
$$

By using the property $\cos(\alpha + \beta) = \cos\alpha\cos\beta - \sin\alpha\sin\beta$ and combining the second and third terms of (6.41), we can achieve

$$
\begin{aligned}
X(k) = & A(k) + \sum_{m=0}^{\frac{q-3}{2}-1} C_m(k) \cos \frac{\pi(q-1-2m)k}{2N} \\
& + \sum_{m=0}^{\frac{q-3}{2}-1} S_m(k) \sin \frac{\pi(q-1-2m)k}{2N},
\end{aligned}
\tag{6.42}
$$

where

$$
A(k) = \sum_{n=0}^{N/p-1} x(qn + \frac{q-1}{2}) \cos \frac{\pi(2n+1)k}{2N/q}
\tag{6.43}
$$

$$
C_m(k) = \sum_{n=0}^{N/p-1} [x(qn+m) + x(qn+q-m-1)] \cos \frac{\pi(2n+1)k}{2N/q}
\tag{6.44}
$$

$$S_m(k) = \sum_{n=0}^{N/p-1} [x(qn+m) - x(qn+q-m-1)] \sin \frac{\pi(2n+1)k}{2N/q} \qquad (6.45)$$

or

$$S_m(\frac{N}{q} - k) = \sum_{n=0}^{N/p-1} [x(qn+m) - x(qn+q-m-1)](-1)^n \cos \frac{\pi(2n+1)k}{2N/q}, \qquad (6.46)$$

where for (6.43) – (6.46), $0 \le m \le (q-3)/2$ and $0 \le k \le N/q-1$. Equation (6.42) contains q length N/q DCTs, $A(k)$, $C_m(k)$, and $S_m(k), 0 \le m \le (q-3)/2$, as defined in (6.43) – (6.46). For any integer j, $C_m(k)$ and $S_m(k)$ have the following properties:

$$C_m(\frac{jN}{q} \pm k) = \begin{cases} (-1)^{\frac{j}{2}} C_m(k), & (j \bmod 4) \text{ is even} \\ \pm(-1)^{\frac{j+1}{2}} C_m(\frac{N}{q} - k), & (j \bmod 4) \text{ is odd} \end{cases} \qquad (6.47)$$

$$S_m(\frac{jN}{q} \pm k) = \begin{cases} \pm(-1)^{\frac{j}{2}} S_m(k), & (j \bmod 4) \text{ is even} \\ (-1)^{\frac{j-1}{2}} S_m(\frac{N}{q} - k), & (j \bmod 4) \text{ is odd} \end{cases} . \qquad (6.48)$$

According to these properties, $X(k)$ can be further partitioned into $X(2jN/q + k)$ and $X(2jN/q - k)$, which can be regrouped to form new sequences:

$$F_{2j}(k) = \frac{X(2jN/q + k) + X(2jN/q - k)}{2}$$

$$= (-1)^j A(k) + (-1)^j \sum_{m=0}^{(q-3)/2} \left[C_m(k) + S_m(k) \tan \frac{\pi(q-1-2m)k}{2N} \right]$$

$$\cdot \cos \frac{\pi(q-1-2m)k}{2N} \cos \frac{2\pi(q-1-2m)j}{q} \qquad (6.49)$$

$$F_{2j+1}(k) = \frac{X[(2j+1)N/q + k] + X[(2j+1)N/q - k]}{2}$$

$$= (-1)^j \sum_{m=0}^{(q-3)/2} \left[C_m(\frac{N}{q} - k) \tan \frac{\pi(q-1-2m)k}{2N} + S_m(\frac{N}{q} - k) \right]$$

$$\cdot \cos \frac{\pi(q-1-2m)k}{2N} \cos \frac{\pi(q-1-2m)(2j+1)}{q}, \qquad (6.50)$$

where for (6.49) and (6.50), $0 \le j \le (q-3)/2$ and $0 \le k \le N/q - 1$. In (6.49) and (6.50), the twiddle factor $\cos[2\pi(q-1-2m)k/(2N)]$ is moved outside the square brackets. This allows each result obtained within the square brackets to be multiplied by the product of two trigonometric identities. Finally the transform outputs are

$$X(k) = F_0(k), \quad 0 \le k \le \frac{N}{q} - 1$$

$$X(\frac{jN}{q}) = F_j(0), \quad 1 \le j \le (q-1) \qquad (6.51)$$

$$X(\frac{jN}{q} + k) = 2F_j(k) - X(\frac{jN}{q} - k), \quad 1 \le j \le (q-1) \ 0 \le k \le \frac{N}{q} - 1.$$

Because $C_m(k)$ and $S_m(k)$ are the length N/q DCTs, they can be similarly decomposed if their sequence length N/q contains factors of q or any other odd factors. A radix-q

decomposition is performed if the same q value is used in the entire process. Similarly, a mixed radix decomposition is carried out if different q values are used. Thus, the limitation of the PFA is completely removed. By way of example, Figure 6.4 shows a signal flow graph of a length 27 DCT after the first step decomposition.

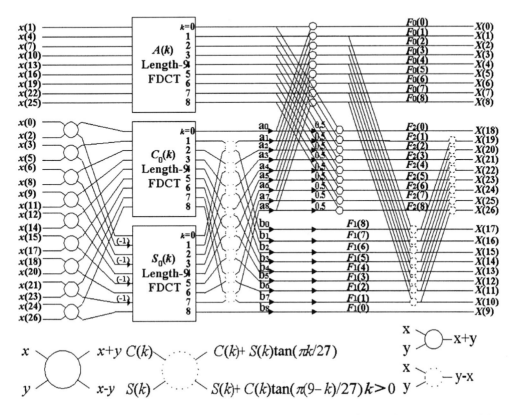

Figure 6.4   Signal flow graph of length 27 type-II DCT ($a = \pi/27$ and $b = 2\pi/3$).

6.4.2   Radix-q algorithm for type-III DCT

The decimation-in-time decomposition technique was used in the last subsection. For the type-III DCT, the decimation-in-frequency technique is to be used. With the same assumption on the sequence length N as that in the last subsection, the type-III DCT defined in (6.3) can be decomposed into

$$x\left(qn + \frac{q-1}{2}\right) = \sum_{k=0}^{N-1} X(k) \cos \frac{\pi(2n+1)k}{2(N/q)} \tag{6.52}$$

$$x(qn + m) = \sum_{k=0}^{N-1} X(k) \cos \frac{[\pi q(2n+1) - (q-1-2m)]k}{2(N/q)} \tag{6.53}$$

$$x(qn + q - m - 1) = \sum_{k=0}^{N-1} X(k) \cos \frac{[\pi q(2n+1) + (q-1-2m)]k}{2(N/q)}, \tag{6.54}$$

where for (6.52) – (6.54), $0 \leq n \leq N/q - 1$ and $0 \leq m \leq (q-3)/2$. Equation (6.52) can be rewritten as

$$
\begin{aligned}
x(qn + \frac{q-1}{2}) &= \sum_{k=1}^{N/q-1} \left\{ \sum_{i=1}^{(q-1)/2} X(\frac{2iN}{q} - k) \cos \frac{\pi(2n+1)(2iN/q+k)}{2(N/q)} \right. \\
&\quad \left. + \sum_{i=1}^{(q-1)/2} X(\frac{2iN}{q} - k) \cos \frac{\pi(2n+1)(2iN/q-k)}{2(N/q)} \right\} \\
&\quad + \sum_{k=1}^{N/q-1} X(k) \cos \frac{\pi(2n+1)k}{2(N/q)} + \sum_{i=0}^{(q-1)/2} X(\frac{2iN}{q}) \cos \frac{\pi(2n+1)(2iN/q)}{2(N/q)} \\
&= \sum_{k=1}^{N/q-1} \left\{ X(k) + \sum_{i=1}^{(q-1)/2} (-1)^i [X(\frac{2iN}{q} + k) + X(\frac{2iN}{q} - k)] \right\} \\
&\quad \cdot \cos \frac{\pi(2n+1)k}{2(N/q)} + \sum_{i=0}^{(q-1)/2} (-1)^k X(\frac{2iN}{q}) \\
&= \sum_{i=0}^{N/q-1} U(k) \cos \frac{\pi(2n+1)k}{2(N/q)}.
\end{aligned}
\tag{6.55}
$$

Because $\cos[\pi(2i+1)/2] = 0$, we note that input $X[(2i+1)N/q]$ is excluded from (6.55). By defining

$$
\begin{aligned}
S_i(k) &= X(\frac{2iN}{q} + k) + X(\frac{2iN}{q} - k) \\
T_i(k) &= X(\frac{2iN}{q} + k) - X(\frac{2iN}{q} - k),
\end{aligned}
$$

where $i = 1, \ldots, (q-1)/2$, we have

$$
U(k) = \begin{cases}
X(k) + \sum_{i=1}^{(q-1)/2} (-1)^i S_i(k), & k = 1, 2, \ldots, \frac{N}{q} - 1 \\
\sum_{i=0}^{(q-1)/2} (-1)^i X(\frac{2iN}{q}), & k = 0
\end{cases}
\tag{6.56}
$$

Therefore, (6.56) can be computed by a length N/q type-III DCT. New sequences are formed by combining (6.53) and (6.54),

$$
\begin{aligned}
F(n, m) &= \frac{x(qn+m) + x(qn+q-m-1)}{2} \\
&= \sum_{k=0}^{N-1} X(k) \cos \frac{\pi(q-1-2m)k}{2N} \cos \frac{\pi(2n+1)k}{2(N/q)}
\end{aligned}
\tag{6.57}
$$

$$
\begin{aligned}
G(n, m) &= \frac{x(qn+m) - x(qn+q-m-1)}{2} \\
&= \sum_{k=0}^{N-1} X(k) \sin \frac{\pi(q-1-2m)k}{2N} \sin \frac{\pi(2n+1)k}{2(N/q)}.
\end{aligned}
\tag{6.58}
$$

If we define $\alpha = \pi(q - 1 - 2m)$ for simplicity, (6.57) can be further decomposed into

$$
F(n,m) = \sum_{k=1}^{N/q-1} X(k) \cos \frac{\alpha k}{2N} \cos \frac{\pi(2n+1)k}{2(N/q)} + \sum_{k=1}^{N/q-1}
$$

$$
\left\{ \sum_{i=1}^{(q-1)/2} X(\frac{2iN}{q} + k) \cos \frac{\alpha(2iN/q - k)}{2N} \cos \frac{\pi(2n+1)(2iN/q + k)}{2(N/q)} \right.
$$

$$
\left. + \sum_{i=1}^{(q-1)/2} X(\frac{2iN}{q} - k) \cos \frac{\alpha(2iN/q - k)}{2N} \cos \frac{\pi(2n+1)(2iN/q - k)}{2(N/q)} \right\}
$$

$$
+ \sum_{i=0}^{(q-1)/2)} X(\frac{2iN}{q}) \cos \frac{i\alpha}{q} \cos \frac{\pi(2n+1)(2iN/q)}{2(N/q)} \tag{6.59}
$$

$$
= \sum_{k=1}^{N/q-1} \sum_{i=1}^{(q-1)/2} (-1)^i \left\{ S_i(k) \cos \frac{\alpha i}{q} \cos \frac{\alpha k}{2N} - T_i(k) \sin \frac{\alpha i}{q} \sin \frac{\alpha k}{2N} \right\}
$$

$$
\cdot \cos \frac{\pi(2n+1)k}{2(N/q)} + \sum_{i=0}^{(q-1)/2} (-1)^i X(\frac{2iN}{q}) \cos \frac{\alpha i}{q}
$$

$$
= \sum_{k=0}^{N/q-1} V(k,m) \cos \frac{\pi(2n+1)k}{2(N/q)},
$$

which is a length N/q type-III DCT whose input sequence is calculated by

$$
V(k,m) = \begin{cases} X(k) \cos \frac{\alpha k}{2N} + \sum_{i=1}^{(q-1)/2}(-1)^i \left\{ S_i(k) \cos \frac{\alpha i}{q} \cos \frac{\alpha k}{2N} \right. \\ \left. -T_i(k) \sin \frac{\alpha i}{q} \sin \frac{\alpha k}{2N} \right\}, & 1 \le k < \frac{N}{q}, \\ \sum_{i=0}^{(q-1)/2}(-1)^i X(\frac{2iN}{q}) \cos \frac{\alpha i}{q}, & k = 0 \end{cases} \tag{6.60}
$$

A similar approach can be used to rewrite (6.58). Without presenting the details of the derivation, we have

$$
G(m,n) = \sum_{k=1}^{N/q} W(k,m) \sin \frac{\pi(2n+1)k}{2(N/q)}, \tag{6.61}
$$

where

$$
W(k,m) = \begin{cases} \left\{ X(k) + \sum_{i=1}^{(q-1)/2}(-1)^i S_i(k) \cos \frac{\alpha i}{q} \right\} \sin \frac{\alpha k}{2N} \\ + \left\{ \sum_{i=1}^{(q-1)/2}(-1)^i T_i(k) \sin \frac{\alpha i}{q} \right\} \cos \frac{\alpha k}{2N}, & 1 \le k < \frac{N}{q}. \\ \sum_{i=0}^{(q-1)/2}(-1)^i X(\frac{2i-1}{q}) \sin \frac{\alpha(2i-1)}{2q}, & k = \frac{N}{q} \end{cases} \tag{6.62}
$$

We can compute (6.61) by

$$
G(n,m) = (-1)^n \sum_{k=0}^{N/q-1} W(\frac{N}{q} - k, m) \cos \frac{\pi(2n+1)k}{2(N/q)}, \tag{6.63}
$$

which is a length N/q type-III DCT. The final outputs can be obtained by (6.55) and

$$
\begin{aligned}
x(qn+m) &= F(n,m) + G(n,m) \\
x(qn+q-m-1) &= F(n,m) - G(n,m) \\
0 \le m \le \frac{q-3}{2}, \qquad 0 &\le n \le \frac{N}{q} - 1.
\end{aligned}
\tag{6.64}
$$

Therefore, the length N type-III DCT is decomposed into q length N/q type-III DCTs whose inputs are defined by (6.56), (6.60) and (6.62). If the subsequence length also contains factors of odd integers, further decomposition can be carried out in the same way. This algorithm is general and particularly suited to sequence lengths containing any possible combination of odd factors. Figure 6.5 shows the signal flow graph for $N = 27$.

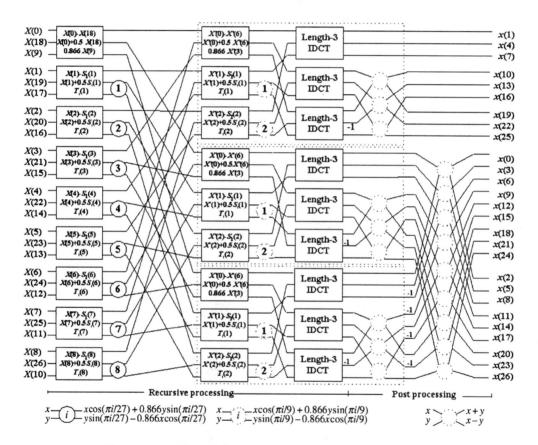

Figure 6.5   Signal flow graph of length 27 type-III DCT.

6.4.3   Computational complexity

This subsection considers the computational complexity required by the radix-q algorithm. Because the same computational complexity is needed for both the type-II and -III DCTs, our discussion is mainly based on the radix-q algorithm for the type-II DCT. Comparisons are made among a few algorithms in terms of the number of additions, multiplications and arithmetic operations (i.e., additions plus multiplications) for different q values.

For each $k = 1$ to $N/q - 1$, the decomposition costs of the radix-q algorithm are:

- $(q - 1)$ multiplications are used inside the brackets of (6.49) and (6.50).

- $q(q - 1)/2$ multiplications are used outside the brackets of (6.49) and (6.50).

- $(q^2 - 1)/2$ additions are used for (6.49) and (6.50).

- $(q - 1)$ additions are used to combine the inputs in (6.45) and (6.46).

Therefore, the number of multiplications required by the radix-q algorithm is

$$M(N) = qM(\frac{N}{q}) + \frac{(q^2 + q - 2)N}{2q} - R_m \qquad (6.65)$$

and the number of additions is

$$A(N) = qA(\frac{N}{q}) + \frac{(q + 5)(q - 1)N}{2q} - R_a, \qquad (6.66)$$

where $R_m = 3(q - 1)/2$ and $R_a = 2(q - 1)$ are the numbers of multiplications and additions saved from trivial twiddle factors for $k = 0$ in (6.49) and (6.50). When N is divisible by different q values, (6.65) and (6.66) can be recursively used to calculate the required computational complexity. When $N = q^m$, (6.65) and (6.66) can be more explicitly expressed by

$$M(N) = \frac{(q + 2)(q - 1)}{2q} N \log_q N + \frac{2M(q) - q^2 - q - 1}{2q} + 1.5 \qquad (6.67)$$

$$A(N) = \frac{(q + 5)(q - 1)}{2q} N \log_q N + \frac{2A(q) - q^2 - 4q + 1}{2q} + 2. \qquad (6.68)$$

In general, a lower computational efficiency than that given in (6.67) and (6.68) can be achieved by using the properties of twiddle factors. Furthermore, optimization of the small length q DCT can reduce the overall computational complexity. For example, additional properties of twiddle factors for $q = 3$, 5 and 9 can be utilized to achieve a lower computational complexity.

- **Computational complexity for $N = 3^m$**
For $q = 3$, there is a trivial factor $\cos(4\pi/3) = -0.5$ in $F_2(k)$, as seen in (6.49). According to (6.65) and (6.66), the number of nontrivial multiplications needed by the algorithm is

$$M(N) = 3M(\frac{N}{3}) + \frac{4N}{3} - 3, \quad N > 3 \qquad (6.69)$$

and the number of additions is

$$A(N) = 3A(\frac{N}{3}) + \frac{8N}{3} - 4, \quad N > 3, \qquad (6.70)$$

where the initial values are $M(9) = 8$ and $A(9) = 34$, which are the numbers of multiplications and additions needed to implement an optimized length 9 DCT, as shown in Appendix B. Compared with that in [5], the presented length 9 DCT computation saves two multiplications. Table 6.2 lists the number of operations needed by the algorithm for various values of q. A lower number of operations is achieved compared with that in [5].

Table 6.2   Computational complexity of the radix-q DCT algorithm.

$q = 3$			
	$M(N)$	$A(N)$	Total
$N = 9$	8	34	42
27	57	170	227
81	276	722	998
243	1149	2810	3959
729	4416	10370	14786
2187	16161	36938	53099
$q = 5$			
$N = 5$	4	13	17
25	69	162	231
125	614	1327	1941
625	4439	9252	13691
3125	29064	59377	88441
$q = 7$			
$N = 7$	18	24	42
49	306	408	714
343	3456	4608	8064
2401	33444	44592	78036
$q = 9$			
$N = 9$	8	34	42
81	264	740	1004
729	4224	10694	14918
6561	54768	132680	187448

• **Computational complexity for $N = 5^m$**

For $q = 5$, we can use the relationship $\cos(4\pi/5) + 0.5 = -\cos(2\pi/5) = -\cos(8\pi/5) = -0.309$ in (6.49) for $j = 1$ and 2. Therefore, we have

$$F_2(k) = -\{A(k) + 0.309[A_0(k) - A_1(k)] - 0.5A_1(k)\} \tag{6.71}$$

and

$$F_4(k) = A(k) - 0.309[A_0(k) - A_1(k)] + 0.5A_0(k), \tag{6.72}$$

where $A(k)$ is defined by (6.43), and

$$A_j(k) = C_j(k) \cos \frac{\pi(2-j)k}{N} + s_j(k) \sin \frac{\pi(2-j)k}{N}, \quad j = 1, 2. \tag{6.73}$$

Equations (6.71) and (6.72) require only five additions and one nontrivial multiplication instead of four additions and four multiplications otherwise. Hence, the computational complexity for $N = 5^m$ is

$$M(N) = 5M(\frac{N}{5}) + \frac{11N}{5} - 6, \quad N > 5 \tag{6.74}$$

Table 6.3   Computational complexity for $q = 2, 3, 5, 7$ and 9.

q	$M(N)$		$A(N)$	
2	$\frac{N}{2}\log_2 N$	$M(2) = 1$	$\frac{3N}{2}\log_2 N - N + 1$	$A(2) = 2$
3	$\frac{4N}{3}\log_3 N - \frac{35N}{18} + 1.5$	$M(9) = 8$	$\frac{8N}{3}\log_3 N - \frac{16N}{9} + 2$	$A(9) = 34$
5	$\frac{11N}{5}\log_5 N - \frac{17N}{10} + 1.5$	$M(5) = 4$	$\frac{21N}{5}\log_5 N - 2N + 2$	$A(5) = 13$
7	$\frac{27N}{7}\log_7 N - \frac{3N}{2} + 1.5$	$M(7) = 18$	$\frac{36N}{7}\log_7 N - 2N + 2$	$A(7) = 24$
9	$\frac{23N}{9}\log_9 N - \frac{15N}{8} + \frac{15}{8}$	$M(9) = 8$	$\frac{50N}{9}\log_9 N - 2N + 2$	$A(9) = 34$

and the number of additions is

$$A(N) = 5A(\frac{N}{5}) + \frac{21N}{5} - 8, \quad N > 5, \tag{6.75}$$

where $A(5) = 13$ and $M(5) = 4$, as shown in Appendix C. The number of operations required for $N = 5^m$ is also listed in Table 6.2.

• Computational complexity for $N = 9^m$

Optimization of the algorithm for $q = 9$ is also possible, as shown in Appendix D. Substantial savings in the numbers of both multiplications and additions are achieved by using a trivial property of the twiddle factors. The number of multiplications needed by the radix-q algorithm is

$$M(N) = 9M(\frac{N}{9}) + \frac{23N}{9} - 15, \quad N > 9 \tag{6.76}$$

and the number of additions is

$$A(N) = 9A(\frac{N}{9}) + \frac{50N}{9} - 16, \quad N > 9 \tag{6.77}$$

with initial values $A(9) = 34$ and $M(9) = 8$. The computational complexity of the radix-9 algorithm, as shown in Table 6.2, is a few operations more than that needed by the radix-3 algorithm. However, the property used in Appendix D is not available for general cases, and the same computational complexity cannot be achieved for other q values.

For various values of q, Table 6.3 provides a summary of the computational complexity in a closed form and their associated initial values. Appendix E presents the formulas for converting the recursive form into the closed form of the computational complexity. In particular, the algorithm for $q = 3$ uses $M(9)$ and $A(9)$ to be the initial values of (6.69) and (6.70) since the optimized length 9 DCT computation can effectively reduce the overall computational complexity. Figure 6.6 compares the computational efficiency for $N = q^m$, where $q = 3, 5, 7$ and 9, in terms of the total number of arithmetic operations per transform point, i.e., $[M(N) + A(N)]/N$. For $q = 5$ and 9, the radix-q algorithm is more computationally efficient than the radix-2 algorithm when N is small (i.e., less than 100) because of the optimized computation of small DCTs.

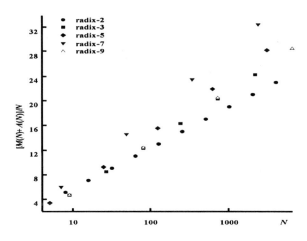

Figure 6.6   Computational complexity of radix-q DCT algorithm.

6.4.4   Algorithm for composite sequence lengths

Let the sequence length N be arbitrarily composite and expressed by (6.5). The length of the DCT can be decomposed in the following steps.

Step 1.   Decompose the length N DCT by a radix-2 algorithm until the subsequence length is not divisible by 2.

Step 2.   The length of the subsequence is now a product of odd integers. They can be further decomposed by the radix-q algorithm with a specific q value until the subsequence length is not divisible by the same q value.

Step 3.   The decomposition step is repeated for other q values.

In principle, the decomposition process can start with any radix number. A lower count of operations is obtained if the decomposition process starts with the smallest radix number, which is shown in Appendix E. This algorithm can be generally applied to sequence length $N = q \cdot 2^m$, where q is an odd integer. The decomposition process can be continued until a length q DCT is reached. For $q = 1$, Figure 6.1 becomes the traditional radix-2 algorithm and requires the same computational complexity as that reported in [7, 10, 13].

The decomposition approach for length $q \cdot 2^m$ has the same computational structure as that of the traditional radix-2 algorithm. The only additional requirement to implement the algorithm is to incorporate the length q DCTs into the subroutine of the radix-2 algorithm. The mixed radix algorithm in [5] is valid only for $N = 2^m 3^n$, which is a subset covered by our algorithm. Furthermore, the former generally needs a more complicated decomposition procedure since a decision has to be made on which of the two decomposition methods is to be used for a particular sequence length.

Table 6.4   Computational complexity.

N	$M(N)$	$A(N)$	Total	N	$M(N)$	$A(N)$	Total
	$N = 3 \cdot 2^m$				$N = 5 \cdot 2^m$		
12	14	49	63	10	13	40	53
24	40	133	173	20	36	109	145
48	104	337	441	40	92	277	369
96	256	817	1073	80	224	673	897
192	608	1921	2529	160	528	1585	2113
384	1408	4417	5825	320	1216	3649	4865
768	3200	9985	13185	640	2752	8257	11009
1536	7168	22273	29441	1280	6144	18433	24577
	$N = 7 \cdot 2^m$				$N = 9 \cdot 2^m$		
14	43	68	111	18	25	94	119
28	100	177	277	36	68	241	309
56	228	437	665	72	172	589	761
112	512	1041	1553	144	416	1393	1809
224	1136	2417	3553	288	976	3217	4193
448	2496	5505	8001	576	2240	7297	9537
896	5440	12353	17793	1152	5056	16321	21377
1792	11776	27393	39169	2304	11264	36097	47361
	$N = 15 \cdot 2^m$				$N = 27 \cdot 2^m$		
15	29	75	104	27	57	170	227
30	73	194	267	54	141	420	561
60	176	477	653	108	336	1001	1337
120	412	1133	1545	216	780	2325	3105
240	944	2625	3569	432	1776	5297	7073
480	2128	5969	8097	864	3984	11889	15873
960	4736	13377	18113	1728	8832	26369	35201

The computational complexity of the radix-q algorithm is determined by the decomposition cost (or the number of operations needed by twiddle factors) and the initial values of the recursive equations (or the computational complexity of small DCTs). Figure 6.6 shows that for small N, the radix-q algorithm is more efficient than the radix-2 algorithm for some values of q. On the other hand, the radix-2 algorithm has the smallest growth rate of computational complexity, as shown in the figure. A proof presented in Appendix E shows that the number of operations needed by twiddle factors has a major effect on the growth rate of computational complexity, whereas the number of operations needed by the length q DCT has a minor influence. For a better efficiency, it is necessary to combine the lower computational complexity of the small DCTs provided by the radix-q algorithm and the low growth rate of the radix-2 algorithm.

In order to achieve the lowest growth rate in computational complexity, the decomposition process should be started by the radix-2 method until the length of subsequences is not divisible by 2. Then, the subsequences are further decomposed by the radix-q algorithm.

The initial values of (6.19) and (6.20) are the numbers of multiplications and additions needed by the length q DCTs. Table 6.4 lists the numbers of additions and multiplications for $q = 3, 5, 7, 9, 15$ and 27. Figure 6.7 (a) shows that the additive complexity for length $q \cdot 2^m$ is slightly higher than that for length 2^m. Figure 6.7 (b) presents the multiplicative

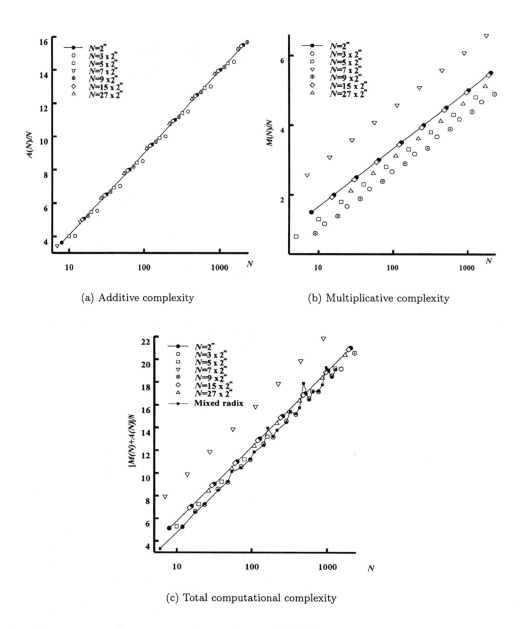

(a) Additive complexity

(b) Multiplicative complexity

(c) Total computational complexity

Figure 6.7   Computational complexity for DCTs of various sequence lengths.

complexity. Except for $m = 7$, significant savings in the number of multiplications are achieved compared with that for length 2^m. The multiplicative complexity for $q = 15$ is the same as that for length 2^m. The improvement is made by exploiting the additional properties of the twiddle factors and using optimized commutation of small DCTs. In contrast, the multiplicative complexity for $q = 7$ is much higher than the others because no additional property is used to reduce the number of operations needed by the twiddle factors.

It is important to compare the overall computational complexity when the time required for an addition is comparable with that for a multiplication. For example, it is a common feature that most high speed signal processing chips perform multiplication and addition at the same speed. Figure 6.7 (c) compares the overall computational complexity needed by the presented algorithm for various lengths of DCTs. It shows again that for all the q values considered (except for $q = 7$), the computation efficiency is better than or the same as that of length 2^m. Figure 6.7 (c) also compares the computational complexity required by the described algorithm and the mixed radix algorithm [5]. It shows that for $N = 3 \cdot 2^m$ and $9 \cdot 2^m$, the two algorithms require the same computational complexity. For other sequence lengths, however, higher computational complexity is needed by both algorithms.

6.5 FAST ALGORITHMS BASED ON TYPE-I DCT

It was suggested earlier that the decomposition for composite sequence length should be started by using the radix-2 algorithm and then the radix-q algorithm for a low growth rate of the total computational complexity. Following this guideline, this section presents another type of fast algorithm that allows such decompositions to be performed based on the odd factors of the composite sequence. These algorithms are very simple and flexible when used for sequence length $N = q \cdot 2^m$, where q is an odd factor. Further savings on arithmetic operations can be achieved from a large number of trivial twiddle factors. In general, these algorithms decompose the DCTs into type-I DCTs and discrete sine transforms (DSTs) which can be easily computed by their corresponding fast algorithms.

6.5.1 Algorithm for type-I DCT

We group the input sequence as follows. For $q = 1$ and 3, we define

$$y(n) = \begin{cases} x(4n+q), & 0 \le n < \frac{N}{4} \\ 0, & n = \frac{N}{4} \end{cases} \tag{6.78}$$

$$z(n) = \begin{cases} 0, & n = 0 \\ x(4n-q) & 1 \le n \le \frac{N}{4} \end{cases} \tag{6.79}$$

and for $q > 3$,

$$y(n) = \begin{cases} x(4n+q), & 0 \le n < \frac{N}{4} - t \\ x(2N - 4n - q), & \frac{N}{4} - t \le n < \frac{N}{4} \\ 0, & n = \frac{N}{4} \end{cases} \tag{6.80}$$

$$z(n) = \begin{cases} 0, & n = 0 \\ x(q - 4n) & 1 \le n \le t \\ x(4n - q) & t+1 \le n \le \frac{N}{4} \end{cases}, \tag{6.81}$$

where t is the integer part of $q/4$. Based on the decimation-in-time decomposition, the type-I DCT of input sequence $x(n)$, as defined in (6.1), can be expressed as

$$\begin{aligned} X_I(k) &= \sum_{n=0}^{N/2} x(2n) \cos \frac{\pi n k}{N/2} + \sum_{n=0}^{N/4} y(n) \cos \frac{\pi(4n+q)k}{N} + \sum_{n=0}^{N/4} z(n) \cos \frac{\pi(4n-q)k}{N} \\ &= A_c(k) + \cos \frac{\pi q k}{N} C_c(k) + \sin \frac{\pi q k}{N} S_c(k), \end{aligned} \tag{6.82}$$

where the subscript I indicates the type-I DCT, and

$$A_c(k) = A_c(N-k) = \sum_{n=0}^{N/2} x(2n) \cos \frac{\pi nk}{N/2}, \qquad 0 \le k \le \frac{N}{2} \tag{6.83}$$

$$C_c(k) = C_c(\frac{N}{2} - k) = C_c(\frac{N}{2} + k) = C_c(N - k)$$

$$= \sum_{n=0}^{N/4} [y(n) + z(n)] \cos \frac{\pi nk}{N/4}, \qquad 0 \le k \le \frac{N}{4} \tag{6.84}$$

$$S_c(k) = -S_c(\frac{N}{2} - k) = S_c(\frac{N}{2} + k) = -S_c(N - k)$$

$$= \sum_{n=1}^{N/4-1} [z(n) - y(n)] \sin \frac{\pi nk}{N/4}, \qquad 1 \le k < \frac{N}{4}. \tag{6.85}$$

Based on the above equations, we rewrite (6.82) into

$$X_I(k) = A_c(k) + \left[\cos \frac{\pi qk}{N} C_c(k) + \sin \frac{\pi qk}{N} S_c(k) \right], \qquad 0 \le k \le \frac{N}{4} \tag{6.86}$$

$$X_I(N-k) = A_c(k) - \left[\cos \frac{\pi qk}{N} C_c(k) + \sin \frac{\pi qk}{N} S_c(k) \right], \qquad 0 \le k \le \frac{N}{4} \tag{6.87}$$

$$X_I(\frac{N}{2} - k) = A_c(\frac{N}{2} - k)$$

$$+ (-1)^{\frac{q-1}{2}} \left[\sin \frac{\pi qk}{N} C_c(k) - \cos \frac{\pi qk}{N} S_c(k) \right], \qquad 0 \le k < \frac{N}{4} \tag{6.88}$$

$$X_I(\frac{N}{2} + k) = -A_c(\frac{N}{2} - k)$$

$$+ (-1)^{\frac{q-1}{2}} \left[\sin \frac{\pi qk}{N} C_c(k) - \cos \frac{\pi qk}{N} S_c(k) \right], \qquad 1 \le k < \frac{N}{4}. \tag{6.89}$$

Therefore, the length N type-I DCT is decomposed into a length $N/2$ type-I DCT, a length $N/4$ type-I DCT and a lengh-$N/4$ type-I DST, as defined in (6.83), (6.84) and (6.85), respectively. The decomposition is similar to the well-known radix-2/4 decomposition that is used in many fast algorithms for DFT and discrete Hartley transform (DHT) computations. Based on the decomposition approach, it can be found that

- $N/2 - (q+1)$ additions are needed by the summations inside the square brackets of (6.86) – (6.89)

- $N/2 + 2$ additions are needed by the operations outside the square brackets of (6.86) and (6.87), and $N/2 - 2$ additions are needed by the operations outside the square brackets of (6.88) and (6.89)

- $N/2 - 2$ additions are needed by (6.84) and (6.85) to combine $y(n)$ and $z(n)$.

The total number of additions needed by the algorithm is

$$A_I^c(N) = A_I^c(\frac{N}{2}) + A_I^c(\frac{N}{4}) + A_I^s(\frac{N}{4}) + 2N - q - 3. \tag{6.90}$$

Similarly, we have

- $N/2 - 1.5q + 0.5$ multiplications are needed by (6.86) or (6.87)

- $N/2 - 1.5q - 0.5$ multiplications are needed by (6.88) or (6.89).

The number of multiplications is

$$M_I^c(N) = M_I^c(\frac{N}{2}) + M_I^c(\frac{N}{4}) + M_I^s(\frac{N}{4}) + N - 3q. \tag{6.91}$$

In both (6.90) and (6.91), the subscript indicates the type of the DCT and the superscript c or s defines the cosine or sine transform.

6.5.2 Algorithm for type-I DST

Similarly, we define a length N type-I DST of input sequence $x(n)$ by

$$Y_I(k) = \sum_{n=1}^{N-1} x(n) \sin \frac{\pi n k}{N}, \quad 0 < k < N, \tag{6.92}$$

where $N = q \cdot 2^m, m > 1$ and q is an odd integer. For $q = 1$ and 3, we define

$$y(n) = \begin{cases} x(4n+q), & 0 \le n < \frac{N}{4} - t \\ 0, & n = \frac{N}{4} \end{cases} \tag{6.93}$$

$$z(n) = \begin{cases} 0, & n = 0 \\ x(4n-q), & t+1 \le n \le \frac{N}{4} \end{cases} \tag{6.94}$$

and for $q > 3$

$$y(n) = \begin{cases} x(4n+q), & 0 \le n < \frac{N}{4} - t \\ -x(2N - 4n - q), & \frac{N}{4} - t \le n < \frac{N}{4} \\ 0, & n = \frac{N}{4} \end{cases} \tag{6.95}$$

$$z(n) = \begin{cases} 0, & n = 0 \\ -x(q - 4n), & 1 \le n \le t \\ x(4n-q), & t+1 \le n \le \frac{N}{4} \end{cases}, \tag{6.96}$$

where t is the integer part of $q/4$. The type-I DST can be decomposed into

$$
\begin{aligned}
Y_I(k) &= \sum_{n=1}^{N/2-1} x(2n) \sin \frac{\pi n k}{N/2} + \sum_{n=1}^{N/4-1} x(4n+q) \sin \frac{\pi (4n+q)k}{N} \\
&\quad + \sum_{n=1}^{N/4-1} x(4n-q) \sin \frac{\pi (4n-q)k}{N} \\
&= A_s(k) + \cos \frac{\pi q k}{N} S_s(k) + \sin \frac{\pi q k}{N} C_s(k), \quad 0 < k < N,
\end{aligned} \tag{6.97}
$$

where

$$A_s(k) = -A_s(N-k) = \sum_{n=1}^{N/2-1} x(2n) \sin \frac{\pi nk}{N/2}, \qquad 0 < k < \frac{N}{2} \qquad (6.98)$$

$$S_s(k) = \sum_{n=1}^{N/4-1} [y(n) + z(n)] \sin \frac{\pi nk}{N/4}, \qquad 0 < k < \frac{N}{4} \qquad (6.99)$$

$$C_s(k) = \sum_{n=0}^{N/4} [y(n) - z(n)] \cos \frac{\pi nk}{N/4}, \qquad 0 \le k \le \frac{N}{4}. \qquad (6.100)$$

The DST defined in (6.92) can be expressed by

$$Y_I(k) = A_s(k) + \left[\cos \frac{\pi qk}{N} S_s(k) + \sin \frac{\pi qk}{N} C_s(k) \right], \qquad 1 \le k \le \frac{N}{4} \quad (6.101)$$

$$Y_I(N-k) = -A_s(k) + \left[\cos \frac{\pi qk}{N} S_s(k) + \sin \frac{\pi qk}{N} C_s(k) \right], \qquad 1 \le k \le \frac{N}{4} \quad (6.102)$$

$$Y_I(\frac{N}{2} - k) = A_s(\frac{N}{2} - k)$$
$$- (-1)^{\frac{q-1}{2}} \left[\sin \frac{\pi qk}{N} S_s(k) - \cos \frac{\pi qk}{N} C_s(k) \right], \quad 0 \le k < \frac{N}{4} \quad (6.103)$$

$$Y_I(\frac{N}{2} + k) = -A_s(\frac{N}{2} - k)$$
$$- (-1)^{\frac{q-1}{2}} \left[\sin \frac{\pi qk}{N} S_s(k) - \cos \frac{\pi qk}{N} C_s(k) \right], \quad 1 \le k < \frac{N}{4}. \quad (6.104)$$

Because the decomposition procedure is about the same as that used for the type-I DCT, the computational complexity can be similarly determined. However, two additions are saved when $k = 0$. The total number of additions needed by the fast algorithm is

$$A_I^s(N) = A_I^s(\frac{N}{2}) + A_I^c(\frac{N}{4}) + A_I^s(\frac{N}{4}) + 2N - q - 5. \qquad (6.105)$$

The total number of multiplications is

$$M_I^s(N) = M_I^s(\frac{N}{2}) + M_I^c(\frac{N}{4}) + M_I^s(\frac{N}{4}) + N - 3q. \qquad (6.106)$$

The type-I DCT and DST can be decomposed by other methods. When N is even, for example, the DCT can be alternatively decomposed as

$$X_I(k) = \sum_{n=0}^{N/2} y(2n) \cos \frac{\pi nk}{N/2} + \sum_{n=0}^{N/2-1} y(2n+1) \cos \frac{\pi(2n+1)k}{2(N/2)}, \quad 0 \le k \le \frac{N}{2} \quad (6.107)$$

$$X_I(N-k) = \sum_{n=0}^{N/2} y(2n) \cos \frac{\pi nk}{N/2} - \sum_{n=0}^{N/2-1} y(2n+1) \cos \frac{\pi(2n+1)k}{2(N/2)}, \quad 0 \le k \le \frac{N}{2}. \quad (6.108)$$

Both (6.107) and (6.108) contain a length $N/2$ type-I DCT and a length $N/2$ type-II DCT. If the decimation-in-frequency method is used, the length N type-I DCT can be computed by

$$X_I(2k) = y(\frac{N}{2})\cos(\pi k) + \sum_{n=0}^{N/2-1}[y(n)+y(N-n)]\cos\frac{\pi nk}{N/2}, \quad 0 \le k \le \frac{N}{2} \quad (6.109)$$

$$X_I(2k+1) = \sum_{n=0}^{N/2-1}[y(n)-y(N-n)]\cos\frac{\pi n(2k+1)}{N/2}, \quad 0 \le k < \frac{N}{2} \quad (6.110)$$

which becomes a length $N/2$ type-III DCT. It can be easily proven that a similar decomposition also exists for the sine sequence.

In the following subsections, the type-I DCT is used as a basic computational block for the other types of DCTs. Therefore, the algorithm described in (6.107) and (6.108) may be used for the type-II DCT, and the computation in (6.109) and (6.110) may be used for the type-III DCT. The decomposition presented in (6.101), (6.102), (6.103) and (6.104) can be used for either the type-II or -III DCT.

6.5.3   Algorithm for type-II DCT

The type-II 1D DCT defined in (6.2) can be decomposed into

$$X_{II}(k) = \sum_{n=0}^{N/2}\left[x(2n-\frac{q+1}{2})\cos\frac{\pi(4n-q)k}{2N} + x(2n+\frac{q-1}{2})\cos\frac{\pi(4n+q)k}{2N}\right]$$

$$= \sum_{n=0}^{N/2}\left[x(2n-\frac{q+1}{2}) + x(2n+\frac{q-1}{2})\right]\cos\frac{\pi qk}{2N}\cos\frac{\pi nk}{N/2}$$

$$+ \sum_{n=1}^{N/2-1}\left[x(2n-\frac{q+1}{2}) - x(2n+\frac{q-1}{2})\right]\sin\frac{\pi qk}{2N}\sin\frac{\pi nk}{N/2}$$

$$= \cos\frac{\pi qk}{2N}A(k) + \sin\frac{\pi qk}{2N}B(k), \quad 0 \le k \le \frac{N}{2} \quad (6.111)$$

and

$$X_{II}(N-k) = (-1)^{(q-1)/2}\left[\sin\frac{\pi qk}{2N}A(k) - \cos\frac{\pi qk}{2N}B(k)\right], \quad 1 \le k < \frac{N}{2}, \quad (6.112)$$

where for (6.111) and (6.112)

$$A(k) = \sum_{n=0}^{N/2}\left[x(2n-\frac{q+1}{2}) + x(2n+\frac{q-1}{2})\right]\cos\frac{\pi nk}{N/2}, \quad 0 \le k \le \frac{N}{2} \quad (6.113)$$

$$B(k) = \sum_{n=1}^{N/2-1}\left[x(2n-\frac{q+1}{2}) - x(2n+\frac{q-1}{2})\right]\sin\frac{\pi nk}{N/2}, \quad 1 \le k < \frac{N}{2}. \quad (6.114)$$

Therefore, the computation of the type-II DCT relies on the computation of the type-I DCT defined in (6.1) and the type-I DST in (6.92).

In (6.113) and (6.114), some indices of the input sequence are either negative or larger than $(N-1)$. Therefore, a mapping process to rearrange the input sequence $x(n)$ is necessary. The validity of the mapping process is proven in Appendix F. When $q = 1$ and 3, the mapping process is defined by

$$x(n) = \begin{cases} 0, & n < 0 \\ 0, & n > N-1 \end{cases}.$$ (6.115)

The mapping process for $q > 3$ is as follows. When $(q-1) \bmod 2$ is even, the mapping procedure is defined by

$$x(n) = \begin{cases} 0, & n = N + 2t \\ 0, & n = -2t - 1 \\ x(2i), & n = -2i - 1 \\ x(N - 2i - 1), & n = N + 2i, \end{cases}$$ (6.116)

where t is the integer part of $q/4$, and $0 \le i < t$. When $(q-1) \bmod 2$ is odd,

$$x(n) = \begin{cases} 0, & n = N + 2t + 1 \\ 0, & n = -2t - 2 \\ x(2i + 1), & n = -2i - 2 \\ x(N - 2i - 2), & n = N + 2i + 1, \end{cases}$$ (6.117)

where in both (6.116) and (6.117), $x(n)$, where $n < 0$ or $n > N-1$, is replaced by either 0

Table 6.5   Index mapping for $q = 9$ and $N = 18$.

$y(-1) = x(0)$	$y(18) = x(17)$
$y(-3) = x(2)$	$y(20) = x(15)$
$y(-5) = 0$	$y(22) = 0$
$y(1) = x(1)$	$y(6) = x(6)$
$y(3) = x(3)$	$y(8) = x(8)$
$y(5) = x(5)$	$y(10) = x(10)$
$y(7) = x(7)$	$y(12) = x(12)$
$y(9) = x(9)$	$y(14) = x(14)$
$y(11) = x(11)$	$y(16) = x(16)$

or the valid data. Table 6.5 shows the mapping process for $q = 9$ and $N = 18$. In general, the number of data indices to be mapped for a DCT computation is $2(t + 1)$, which is only a small number. In contrast, the mapping process required by most other non-radix type fast algorithms has to deal with almost the entire input sequence.

Based on the computational complexity of the fast algorithm for the type-I DCT and DST, the numbers of additions and multiplications for the type-II length N DCT are

$$A_{II}^c(N) = A_I^c(\frac{N}{2}) + A_I^s(\frac{N}{2}) + 2N - q - 3, \tag{6.118}$$

$$M_{II}^c(N) = M_I^c(\frac{N}{2}) + M_I^s(\frac{N}{2}) + 2N - 3q, \tag{6.119}$$

where the terms with the negative signs in (6.118) and (6.119) are the numbers of additions and multiplications saved from the trivial twiddle factors in (6.111) and (6.112).

6.5.4 Algorithm for type-III DCT

The type-III DCT of input sequence $X(k)$ defined in (6.3) can be expressed as

$$x(2n + \frac{q-1}{2}) = \sum_{k=0}^{N-1} X(k) \cos \frac{\pi(4n + q)k}{2N}, \quad 0 \le n \le \frac{N}{2} - 1 \tag{6.120}$$

$$x(2n - \frac{q+1}{2}) = \sum_{k=0}^{N-1} X(k) \cos \frac{\pi(4n - q)k}{2N}, \quad 0 \le n \le \frac{N}{2} - 1. \tag{6.121}$$

When index n is negative or larger than $N - 1$, the mapping process described in the last subsection can be used to achieve the valid transform outputs. By combining (6.120) and (6.121), we form new sequences

$$\begin{aligned}
f(n) &= \frac{x(2n - \frac{q+1}{2}) + x(2n + \frac{q-1}{2})}{2} \\
&= \sum_{k=0}^{N} X(k) \cos \frac{\pi qk}{2N} \cos \frac{\pi nk}{N/2} = \sum_{k=0}^{N/2} F(k) \cos \frac{\pi nk}{N/2}, \quad 0 \le n \le \frac{N}{2},
\end{aligned} \tag{6.122}$$

where

$$F(k) = \begin{cases} X(0), & k = 0 \\ X(k) \cos \frac{\pi qk}{2N} + (-1)^{\frac{q-1}{2}} X(N - k) \sin \frac{\pi qk}{2N}, & 1 \le k < \frac{N}{2} \\ X(\frac{N}{2}) \cos \frac{\pi q}{4}, & k = \frac{N}{2} \end{cases} . \tag{6.123}$$

Similarly, we have

$$\begin{aligned}
g(n) &= \frac{x(2n - \frac{q+1}{2}) - x(2n + \frac{q-1}{2})}{2} \\
&= \sum_{k=1}^{N-1} X(k) \sin \frac{\pi qk}{2N} \sin \frac{\pi nk}{N/2} = \sum_{k=0}^{N/2} G(k) \sin \frac{\pi nk}{N/2}, \quad 1 \le n \le \frac{N}{2},
\end{aligned} \tag{6.124}$$

where

$$G(k) = X(k) \sin \frac{\pi qk}{2N} - (-1)^{\frac{q-1}{2}} X(N - k) \cos \frac{\pi qk}{2N}, \quad 1 \le k < \frac{N}{2}. \tag{6.125}$$

Therefore, the computation of the type-III DCT is converted into a length N type-I DCT and DST. The final transform outputs are calculated by

$$
x(n) = \begin{cases}
f(0), & n = \frac{q-1}{2} \\
f(N/2), & n = N - \frac{q+1}{2} \\
f(i) + g(i), & n = 2i - \frac{q+1}{2} \\
f(i) - g(i), & n = 2i + \frac{q-1}{2}
\end{cases} ,
\tag{6.126}
$$

where $1 \le i < N/2$. The computational complexity of the algorithm is the same as that used by the algorithm for the type-II DCT, as shown as in (6.118) and (6.119).

6.5.5   Algorithm for type-IV DCT

Based on the definition of the type-IV DCT in (6.4), we can have

$$
\begin{aligned}
F(k) &= \frac{X_{IV}(2k) + X_{IV}(2k-1)}{2} \\
&= \sum_{n=0}^{N-1} x(n) \cos \frac{\pi(2n+1)q}{4N} \cos \frac{\pi(2n+1)k}{2N}
\end{aligned}
\tag{6.127}
$$

which is a length N type-II DCT, and the final output is obtained by

$$
X_{IV}(0) = F(0), \quad X_{IV}(k) = F(k) - X_{IV}(2k-1), \quad 1 \le k < N.
\tag{6.128}
$$

It can be easily seen that the decomposition needs N multiplications in (6.107) and $N-1$ additions in (6.108). Therefore, the computational complexity needed by the fast algorithm for the length N type-IV DCT is

$$
\begin{aligned}
A_{IV}^c(N) &= A_{II}^c(N) + N - 1 \\
M_{IV}^c(N) &= M_{II}^c(N) + N.
\end{aligned}
\tag{6.129}
\tag{6.130}
$$

Similarly, we can also convert the type-IV DCT into the type-III DCT by

$$
X_{IV}(k) = \cos \frac{\pi(2k+1)}{4N} \sum_{n=0}^{N-1} f(n) \cos \frac{\pi n(2k+1)}{2N}, \quad 0 \le k < N,
\tag{6.131}
$$

where

$$
f(n) = \begin{cases}
x(0), & n = 0 \\
x(n) + x(n-1), & 1 \le n < N
\end{cases} .
\tag{6.132}
$$

The described algorithms are particularly suited to the DCT whose sequence length contains odd factors. Table 6.6 lists the number of operations that are needed by the algorithm for the type-I DCT and DST and type-II DCTs whose sequence lengths contain odd factor $q = 1, 3, 5,$ and 9.

Table 6.6   Computational complexity for $q = 1, 3, 5$ and 9.

N	Type-I DCT M_I^c	A_I^c	Type-I DST M_I^s	A_I^s	Proposed algorithm M_{II}^c	A_{II}^c	Total
			$q = 1$				
2	0	4	0	0	1	2	3
4	1	9	1	4	4	9	13
8	6	25	6	18	15	25	40
16	21	66	21	57	41	69	110
32	62	169	62	158	103	183	286
64	165	416	165	403	249	451	700
128	414	995	414	980	583	1071	1654
256	997	2312	997	2305	1337	2483	3820
512	2334	5317	2334	5298	3015	5647	8662
1024	5349	11988	5349	11967	6713	12659	19372
			$q = 3$				
3	0	8	1	2	1	4	5
6	1	17	3	10	4	16	20
12	5	45	6	36	19	45	64
24	24	114	25	103	50	123	173
48	74	285	75	272	136	307	443
96	210	688	211	673	332	743	1075
192	542	1623	543	1606	796	1739	2535
384	1338	3744	1339	3727	1844	3991	5835
768	3227	8503	3228	8484	4184	9001	13185
			$q = 5$				
5	4	18	8	8	4	13	17
10	8	40	12	29	17	38	55
20	25	98	29	85	45	101	146
40	70	239	74	224	119	255	374
80	189	574	193	557	289	615	904
160	478	1349	482	1330	687	1443	2130
320	1165	3112	1169	3091	1585	3311	4896
640	2750	7063	2754	7040	3599	7475	11074
1280	6349	15818	6353	15793	8049	16655	24704
			$q = 9$				
9	8	38	12	28	8	34	42
18	16	89	20	78	29	90	119
36	45	215	49	202	81	227	308
72	126	514	130	499	211	549	760
144	337	1207	341	1190	517	1289	1806
288	854	2784	858	2765	1227	2961	4188
576	2081	6321	2085	6300	2837	6689	9526
1152	4918	14162	4922	14139	6443	14913	21356

6.6 CHAPTER SUMMARY

In this chapter, we have considered the radix-2 algorithm, radix-q algorithm, the PFA and the algorithms based on the type-I DCT. Some of these fast algorithms have to be combined to support computation of composite sequence lengths. For example, the radix-q algorithm and the PFA are particularly useful for decomposing the whole computation task into smaller ones. The radix-2 and the algorithms based on the type-I DCT are particularly useful for dealing with DCTs whose sequence length is even.

Among these algorithms, the radix-q algorithms generally require more operations than the others. Therefore, the radix-q algorithm can be used only to decompose short sequence lengths that contain odd factors. For composite sequence lengths, we can combine either the radix-2 and radix-q algorithms, or the algorithm based on the type-I DCT and the radix-q algorithm. Both approaches need about the same computational complexity in terms of the total number of arithmetic operations, which can be seen by comparing Tables 6.4 and 6.6. The PFA can also be used to replace the radix-q algorithm if the factors are mutually prime. Substantial reduction of the computational complexity can be achieved if the PFA can be used.

All these algorithms have regular computational structures in terms of data indexing and recursive computation. The radix-2 algorithm appears to be the simplest and yet provides a good computational efficiency. The radix-q algorithm has the most complex structure and requires the most computational complexity. By using the index mapping method and the capability of in-place computation, the described PFA is very competitive in terms of the computational structure and complexity. The computation based on the type-I DCT is also regular except that mapping a few input indices is needed. In addition, the main computation kernel of the algorithms relies on the type-I DCT and DST. Some of the concepts used in this chapter can be extended to MD DCT computation, as discussed in the next chapter.

Appendix A. Subroutine for index mapping

```
mapping(q, q, index)
{    int i, j, k1, k2, s1, s2, n, NN;
     i= -1; j= -1; k1=1; k2=q; s1=1; s2=q; NN=qq;
     for (n=0; n < NN; n++)
     { i=i+k1; j=j+s1;
         if (i==k2) /*i and j are the new array indices */
         { if (k2==q) { i=i-1; k2=-1; k1=-1;}
             else if (k2==-1) { i=i+1; k2=q;  k1=1; }
         }
         if (j==s2)
         { if (s2==q) { j=j-1; s2=-1; s1=-1;}
             else if (s2==-1) { j=j+1; s2= q; s1= 1;}
         }
         index[i][j]=n;  /*index contains the relation among i, j and n*/
     } return(0);
}
```

Appendix B. Subroutine for length 9 DCT

```
type data = array[0..9] of rea;
   procedure DCT9(var x data);              (*x is the input sequence *)
   var real s, t1, s1; data u, v, t; integer i1, i2, i3;
   begin for i1=0 to 3 do
         begin u[i1]=x[i1]+x[9-i1];       v[i1]=x[i1]-x[8-i1]; end;
         t[0]=u[0]+u[2]+u[3];             t[2]=u[0]-u[2];
         t[3]=u[2]-u[3];                  t[1]=(t[2]+t[3])*0.766044443;
         t[2]=t[2]*0.173648178;           t[3]=-t[3]*0.9396926;
         t[2]=t[1]+t[2];                  t[3]=t[1]+t[3];
         t[1]=a[1]+x[4];                  t[4]=0.5*u[1]-x[4];
         x[0]=t[0]+t[1];                  x[2]=t[2]+t[4];
         x[4]=t[3]-t[4];                  x[6]=0.5*t[0]-t[1];
         x[8]=t[2]-t[3]-t[4];             t[0]= v[1]*0.866025404;
         t[1]=(v[0]+v[3])*0.984807753;    t[2]=(v[2]-v[3])*0.64278761;
         t[3]=-(v[0]+v[2])*0.342020143;   t[2]=t[1]+t[2];
         t[3]=t[1]+t[3];                  x[1]=t[2]+t[0];
         x[3]=(v[0]-v[2]-v[3])*0.866025;  x[5]=t[3]-t[0];
         x[7]=t[2]-t[3]-t[0];
   end;
```

Appendix C. Subroutine for length 5 DCT

```
type data = array[0..5] of rea;
   procedure DCT5(var x data);  (*x is the input sequence *)
   var data u, v, t;
   begin
         u[0]=x[0]+x[4];     u[1]=x[1]+x[3];      v[0]=x[0]-x[4];
         v[1]=x[1]-x[3];     t[1]=u[0]+u[1];      t[2]=u[0]-u[1];
         t[3]=t[1]*0.25;     t[4]=t[2]*0.55901699; t[3]=t[3]-t[2];
         x[0]=t[1]+x[2];     x[2]=t[3]+t[4];      x[4]=t[4]-t[3];
         t[1]=(v[0]+v[1])*0.951056515;      t[2]= v[1]*0.363271264;
         t[3]=v[0]*1.538841; x[1]=t[1]-t[2];      x[3]=t[3]-t[1];

   end;
```

Appendix D. Radix-q algorithm when $q = 9$

When $q = 9$, the algorithm can be derived from (6.49) – (6.51). However, further reduction of arithmetic operations can be made by using some properties of twiddle factors. Let us assume that $A(k), C_m(k)$ and $S_m(k)$ are the length $N/9$ DCTs defined in (6.43) – (6.45). According to (6.49) and (6.50), we further define

$$B_m(k) = C_m(k) \cos \frac{(4-m)\pi k}{N} + S_m(k) \sin \frac{(4-m)\pi k}{N}, \qquad m = 0, 1, \ldots, 3 \qquad (D.1)$$

$$E_m(k) = C_m(\frac{N}{9} - k) \sin \frac{(4-m)\pi k}{N} + S_m(\frac{N}{9} - k) \cos \frac{(4-m)\pi k}{N}, \qquad m = 0, 2, 3 \qquad (D.2)$$

and

$$E_1(k) = 0.866 \cos \frac{3\pi k}{N} \left\{ C_1(\frac{N}{9} - k) \tan \frac{3\pi k}{N} + S_1(\frac{N}{9} - k) \right\}. \qquad (D.3)$$

The radix-9 DCT can be expressed by

$$X(k) = [A(k) + B_2(k)] + [B_1(k) + B_3(k) + B_4(k)] \qquad (D.4)$$
$$F_6(k) = -[A(k) + B_2(k)] + 0.5[B_1(k) + B_3(k) + B_4(k)] \qquad (D.5)$$
$$F_2(k) = -C(k) - \{0.766D(k) - 0.174[B_3(k) - B_4(k)]\} \qquad (D.6)$$
$$F_4(k) = C(k) + \{0.766D(k) + 0.939[B_3(k) - B_4(k)]\} \qquad (D.7)$$
$$F_8(k) = C(k) - \{0.939D(k) + 0.766[B_3(k) - B_4(k)]\} \qquad (D.8)$$
$$C(k) = A(k) - 0.5B_2(k) \quad D(k) = B_1(k) - B_3(k). \qquad (D.9)$$

In the above equations, we have used the property $\cos(2\pi/9) + \cos(4\pi/9) = -\cos(8\pi/9)$. Furthermore, the last term of (D.8) can be obtained from the difference of the last terms of (D.6) and (D.7). Similarly, we have

$$F_3(k) = -0.866[E_0(k) - E_2(k) + E_3(k)] \qquad (D.10)$$
$$F_1(k) = E_1(k) + \{0.342[E_0(k) - E_3(k)] + 0.984[E_2(k) + E_3(k)]\} \qquad (D.11)$$
$$F_5(k) = -E_1(k) + \{0.984[E_0(k) - E_3(k)] + 0.642[E_2(k) + E_3(k)]\} \qquad (D.12)$$
$$F_7(k) = -E_1(k) - \{0.642[E_0(k) - E_3(k)]0.342[E_2(k) + E_3(k)]\}. \qquad (D.13)$$

The property $\sin(\pi/9) + \sin(2\pi/9) = \sin(4\pi/9)$ is used when (D.11) – (D.13) are derived. The last term of (D.13) can be obtained from the difference of the last terms of (D.11) and (D.12). It can be proven that to compute a rotation operation such as $ax + by$ and $bx + cy$, three additions and three multiplications are needed. Therefore, the above equations need 34 additions and 23 multiplications. Furthermore, 16 additions are used by (6.43), (6.44) and (6.51) for $q = 9$. Compared to the computational complexity required by (6.65) and (6.66) for a general radix-q algorithm, 21 multiplications and 6 additions can be saved by using the properties of the rotation factors.

Appendix E. Growth rate of computational complexity

In general, the number of additions or multiplications required by the radix-q algorithm can be recursively expressed by

$$f(N) = qf(\frac{N}{q}) + C_1 N - C_2, \qquad (E.1)$$

where C_1 is the coefficient associated with the number of operations needed by twiddle factors and C_2 is the number of operations saved from trivial twiddle factors. Equation (E.1) can be also expressed in a closed form

$$f(N) = A_1 N \log_q N + A_2 N + A_3, \qquad (E.2)$$

where $A_i (i = 1, 2, 3)$ are functions of C_1 and C_2. It can be proved that

$$A_1 = C_1 \qquad (E.3)$$

$$A_2 = \frac{f(q)}{q} - C_1 + \frac{C_2}{1 - q} \qquad (E.4)$$

$$A_3 = \frac{C_2}{(q-1)q}. \qquad (E.5)$$

By putting (E.3) – (E.5) into (E.2), the growth rate of $f(N)$ with N is

$$\frac{df(N)}{dN} = C_1 \left\{ \frac{1}{\ln q} - 1 + \log_2 N \right\} - \frac{C_2}{q(q-1)} + \frac{f(q)}{q}, \qquad (E.6)$$

where $f(q)$ is the number of operations needed by a length q DCT. Equation (E.6) shows that the parameter that mainly decides the growth rate is C_1 which is the number of operations needed by the twiddle factors. It is also useful to increase C_2 and reduce $f(q)$. However, the influence of the second and third terms of (E.6) decreases when q increases. Equations (E.1) – (E.6) are also valid for the radix-2 algorithm. It can be verified that the radix-2 algorithm has a smaller growth rate than that of the radix-q algorithm.

Appendix F. Mapping process

The mapping process replaces the invalid indices with the valid ones according to the definition of the type-II DCT. For easy understanding and simplicity of presentation, let us consider the 1D DCT that is defined by

$$X(k) = \sum_{n=0}^{N-1} x(n) \cos \frac{\pi(2n+1)k}{2N}, \qquad (F.1)$$

where $x(n)$ $(n = 0, 1, \ldots, N-1)$ is the input sequence, and $N = q \cdot 2^m$ is the sequence length. Parameter q is an odd positive integer. Based on (F.1), we can write

$$\begin{aligned} X(k) &= \sum_{n=0}^{N/2} \left[x(2n - \frac{q+1}{2}) \cos \frac{\pi(4n-q)k}{2N} + x(2n + \frac{q-1}{2}) \cos \frac{\pi(4n+q)k}{2N} \right] \\ &= \cos \frac{\pi qk}{2N} A(k) + \sin \frac{\pi qk}{2N} B(k) \end{aligned} \qquad (F.2)$$

for $0 \le k < N/2$ and

$$X(N-k) = (-1)^{(q-1)/2} \left[\sin \frac{\pi qk}{2N} A(k) - \cos \frac{\pi qk}{2N} B(k) \right] \qquad (F.3)$$

for $1 \le k < N/2$. The sequences $A(k)$ and $B(k)$ in (F.2) and (F.3) are defined by

$$A(k) = \sum_{n=0}^{N/2} \left[x(2n - \frac{q+1}{2}) + x(2n + \frac{q-1}{2}) \right] \cos \frac{\pi nk}{N/2} \tag{F.4}$$

$$B(k) = \sum_{n=1}^{N/2-1} \left[x(2n - \frac{q+1}{2}) - x(2n + \frac{q-1}{2}) \right] \sin \frac{\pi nk}{N/2}. \tag{F.5}$$

In (F.4) and (F.5), the indices of input sequence $x(\ )$ are invalid when $n < (q+1)/2$ and $n > N/2 - (q-1)/4$. Now, let us consider the following relations:

$$x(N + 2t) \cos \frac{\pi(2N + 4t + 1)k}{2N} \Leftrightarrow x(N - 2t - 1) \cos \frac{\pi(2N - 4t - 1)k}{2N}$$

$$x(-2t - 1) \cos \frac{-\pi(4t + 1)k}{2N} \Leftrightarrow x(2t) \cos \frac{\pi(4t + 1)k}{2N}$$

$$x(-2i - 1) \cos \frac{-\pi(4i + 1)k}{2N} \Leftrightarrow x(2i) \cos \frac{\pi(4i + 1)k}{2N} \tag{F.6}$$

$$x(N + 2i) \cos \frac{\pi(2N + 4i + 1)k}{2N} \Leftrightarrow x(N - 2i - 1) \cos \frac{\pi(2N - 4i - 1)k}{2N},$$

where t is the integer part of $q/4$ and $0 \le i < t$. By observing (F.6) we have

- the data on the left side are invalid and those on the right side are valid,

- the two cosine functions in each line can be proven to be the same.

Furthermore, by examining (F.4) and (F.5), we find that

- $x(N - 2t - 1)$ and $x(2t)$ are not needed in (F.4) and (F.5),

- $x(2i)$ and $x(N - 2i - 1)$ are needed in (F.4) and (F.5).

Therefore, it is necessary to have $x(N + 2t) = x(-2t - 1) = 0$, and $x(-2i - 1) = x(2i)$ and $x(N + 2i) = x(N - 2i - 1)$ so that the DCT can be computed by the algorithm given by (F.2) – (F.5). For $(q-1) \bmod 2$ is even, the mapping process can be summarized by

$$\begin{aligned} x(N + 2t) &= 0 \\ x(-2t - 1) &= 0 \\ x(-2i - 1) &= x(2i) \\ x(N + 2i) &= x(N - 2i - 1), \end{aligned} \tag{F.7}$$

where $0 \le i < (q-1)/4$. Similarly, for $(q-1) \bmod 2$ is odd, the mapping process is defined by

$$\begin{aligned} x(N + 2t + 1) &= 0 \\ x(-2t - 2) &= 0 \\ x(-2i - 2) &= x(2i + 1) \\ x(N + 2i + 1) &= x(N - 2i - 2), \end{aligned} \tag{F.8}$$

where $0 \le i < (q-3)/4$.

Example: $q = 7$ and $N = 14$, and $q = 9$ and $N = 18$.

$n = 0$	$q = 7$ and $N = 14$		$q = 9$ and $N = 18$	
	$x(2n - 4)$	$x(2n + 3)$	$x(2n - 5)$	$x(2n + 4)$
$n = 0$	$\mathbf{x(-4)} = 0$	$x(3)$	$\mathbf{x(-5)} = 0$	$x(4)$
1	$\mathbf{x(-2)} = x(1)$	$x(5)$	$\mathbf{x(-3)} = x(2)$	$x(6)$
2	$x(0)$	$x(7)$	$\mathbf{x(-1)} = x(0)$	$x(8)$
3	$x(2)$	$x(9)$	$x(1)$	$x(10)$
4	$x(4)$	$x(11)$	$x(3)$	$x(12)$
5	$x(6)$	$x(13)$	$x(5)$	$x(14)$
6	$x(8)$	$\mathbf{x(15)} = x(12)$	$x(7)$	$x(16)$
7	$x(10)$	$\mathbf{x(17)} = 0$	$x(9)$	$\mathbf{x(18)} = x(15)$
8	-	-	$x(11)$	$\mathbf{x(20)} = x(17)$
9	-	-	$x(13)$	$\mathbf{x(22)} = 0$

In the preceeding table, the items in boldface indicate that the invalid data indices are replaced by the valid ones. It can also be observed that the number of invalid data is only related to the value of q and independent of sequence length N. Therefore, the overhead of the mapping process is trivial when q is small and N is large. The same concept can be directly applied to the computation of MD DCTs.

REFERENCES

1. G. Bi and L. W. Yu, DCT algorithms for composite sequence lengths, *IEEE Trans. Signal Process.*, vol. 46, no. 3, 554–562, 1998.

2. G. Bi, Fast algorithms for type-III DCT of composite sequence lengths, *IEEE Trans. Signal Process.*, vol. 47, no. 7, 2053–2059, 1999.

3. G. Bi, Index mapping for prime factor algorithm of discrete cosine transform, *Electron. Lett.*, vol. 35, no. 3, 198–200,1999.

4. S. C. Chan and K. L. Ho, Efficient index mapping for computing discrete cosine transform, *Electron. Lett.*, vol. 25, no. 22, 1499–1500, 1989.

5. Y. H. Chan and W. C. Siu, Mixed-radix discrete cosine transform, *IEEE Trans. Signal process.*, vol. 41, no. 11, 3157–3161, 1993.

6. D. C. Kar and V. V. B. Tao, On the prime factor decomposition algorithm for the discrete sine transform, *IEEE Trans. Signal Process.*, vol. 42, no. 11, 3258–3260, 1994.

7. B. G. Lee, A new algorithm to compute the discrete cosine transform, *IEEE Trans. Acoust., Speech, Signal Processing*, vol. ASSP-32. no. 12, 1243-1245, 1983.

8. B. G. Lee, Input and output index mapping for a prime-factor-decomposed computation of discrete cosine transform, *IEEE Trans. Acoust., Speech, Signal Process.*, vol. 37, no. 2, 237–244, 1989.

9. P. Lee and Fang-Yu Huang, An efficient prime-factor algorithm for the discrete cosine transform and its hardware implementations, *IEEE Trans. Acoust., Speech, Signal Process.*, vol. 42, no. 8, 1996–2005, 1994.

10. H. S. Malvar, Fast computation of the discrete cosine transform and the discrete Hartley transform, *IEEE Trans. Acoust., Speech. Signal Process.*, vol. ASSP-35, no. 10, 1484–1485, 1987.

11. H. S. Malvar, *Signal Processing with Lapped Transforms*, Norwood, MA, Artech House, 1991.

12. A. D. Poularikas, *The Transforms and Applications Handbook*, Boca Raton, Fla. : CRC Press, 2000.

13. K. R. Rao and P. Yip, *Discrete Cosine Transform. Algorithms, Advantages and Applications*, Chap. 7, Academic Press, New York, 1990.

14. N. R. Murthy and M. N. S. Swampy, On the on-line computation of DCT-IV and DOST-IV transforms, *IEEE Trans. Signal Process.*, vol. 43, no. 5, 1249–1251, 1995.

15. T. C. Tan, G. Bi, Y. Zeng and H. N. Tan, DCT hardware structure for sequentially presented data, *Signal Processing*, vol. 81, 2333–2342, 2001.

16. A. Tatsaki, C. Dre, T. Storaities and C. Goutis, Prime-factor DCT algorithms, *IEEE Trans. Signal Process.*, vol. 43, no. 3, 772–776, 1995.

17. Z. Wang, Interpolation using the discrete cosine transform: Reconsideration, *Electron. Lett.*, vol. 29, no. 2, 198–200, 1993.

18. Z. Wang, On computing the discrete Fourier and cosine transforms, *IEEE Trans. Acoust., Speech, Signal Process*, vol. ASSP-33, 1985.

19. L. N. Wu, Comment on "On the shift property of DCTs and DSTs," *IEEE Trans. Acoust., Speech, Signal Process.*, vol. ASSP-38, 186–190, 1990.

20. P. N. Yang and M. J. Narashimha, Prime factor decomposition of the discrete cosine transform and its hardware realization, *Proc. IEEE ICASSP*, 642–644, 1986.

21. P. Yip and K. R. Rao, On the shift property of DCTs and DSTs, *IEEE Trans. Acoust., Speech, Signal Process.*, vol. ASSP-35, no. 3, 404–406, 1987.

7

Fast Algorithms for MD
Discrete Cosine Transform

This chapter presents fast algorithms that have recently been developed for the computation of the type-I, -II and -III multidimensional (MD) discrete cosine transforms (DCTs). Some of the concepts used in Chapter 6 are extended for the computation of MD DCTs.

- Section 7.2 derives fast algorithms for two-dimensional (2D) type-I, -II and -III DCTs based on the relations among various 2D discrete transforms. The decomposition approach is an extension of the concepts used for one-dimensional (1D) DCTs in Section 6.5.

- Section 7.3 presents prime factor algorithms for MD type-II and -III DCTs by extending the decomposition approaches used in Section 6.3.

- Section 7.4 derives radix-2 polynomial transform (PT)-based algorithms for MD type-II DCTs whose dimensional sizes are 2^m.

- Section 7.5 presents radix-2 PT-based algorithms for MD type-III DCTs whose dimensional sizes are 2^m.

- Section 7.6 presents radix-q PT-based algorithms for MD type-II DCTs whose dimensional sizes are q^m.

- Section 7.7 presents radix-q PT-based algorithms for MD type-III DCTs whose dimensional sizes are q^m.

Details of the mathematical derivations and comparisons in terms of the computational complexity and structures are provided.

7.1   INTRODUCTION

Computational complexity and structural regularity have been the major concerns for many practical applications of MD DCTs. Although there exist many fast algorithms in the literature, the row-column algorithms are widely used in practice due to their simple concept and regular computational structure. As both dimensional sizes and the number of dimensions increase, the required computational complexity quickly reaches the limits that can be handled by high speed processors. It becomes difficult to use the row-column methods to economically provide the prohibitive processing throughput needed by real time applications. Parallel processing techniques are often used to achieve the required processing throughput, which definitely increases the computation costs and hardware complexity. The other approach is to find new fast algorithms that one hopes will minimize the computational complexity. At the same time, it is desirable for these algorithm to achieve some desirable properties such as the flexibility of using the same algorithms for various dimensional sizes and/or different dimensions.

Similar to 1D DCTs, there are also four types of MD DCTs whose definitions can be easily extended from those given in the last chapter for 1D DCTs. In this chapter, we mainly consider the computation of MD type-I, -II and -III DCTs. As in the previous chapter, the constant scaling factors in the definition are ignored for simplicity of presentation. For example, the 2D type-I, -II and -III DCTs are defined by

Type-I:

$$X(k_1, k_2) = \sum_{n_1=0}^{N_1} \sum_{n_2=0}^{N_2} x(n_1, n_2) \cos \frac{\pi n_1 k_1}{N_1} \cos \frac{\pi n_2 k_2}{N_2} \qquad (7.1)$$

$$0 \leq k_1 \leq N_1, \qquad 0 \leq k_2 \leq N_2$$

Type-II:

$$X(k_1, k_2) = \sum_{n_1=0}^{N_1-1} \sum_{n_2=0}^{N_2-1} x(n_1, n_2) \cos \frac{\pi(2n_1 + 1)k_1}{2N_1} \cos \frac{\pi(2n_2 + 1)k_2}{2N_2} \qquad (7.2)$$

$$0 \leq k_1 \leq N_1 - 1, \qquad 0 \leq k_2 \leq N_2 - 1$$

Type-III:

$$X(k_1, k_2) = \sum_{n_1=0}^{N_1-1} \sum_{n_2=0}^{N_2-1} x(n_1, n_2) \cos \frac{\pi(2k_1 + 1)n_1}{2N_1} \cos \frac{\pi(2k_2 + 1)n_1}{2N_2} \qquad (7.3)$$

$$0 \leq k_1 \leq N_1 - 1, \qquad 0 \leq k_2 \leq N_2 - 1,$$

where $x(n_1, n_2)$ is the input array. The array size for a type-I DCT is $(N_1+1) \times (N_2+1)$ and that for type-II or -III DCTs is $N_1 \times N_2$. The type-I DCT is also termed the cosine-cosine transform. In general, the type-III DCT is considered to be the inverse of the type-II DCT and therefore, the type-III DCT is often defined as

$$x(n, n) = \sum_{k_1=0}^{N_1-1} \sum_{k_2=0}^{N_2-1} X(k_1, k_2) \cos \frac{\pi(2n_1 + 1)k_1}{2N_1} \cos \frac{\pi(2n_2 + 1)k_1}{2N_2} \qquad (7.4)$$

$$0 \leq n_1 \leq N_1 - 1, \quad 0 \leq n_2 \leq N_2 - 1$$

in which the input array becomes $X(k_1, k_2)$. The definitions of these types of DCTs can be easily extended for higher dimensions.

The row-column algorithms have been widely used to support some image processing applications which do not need very high processing throughput. Such an algorithm can also be easily extended for DCTs of higher dimensions. We shall use the computational complexity of this algorithm as a reference to compare the merits of fast algorithms that will be discussed in this chapter.

7.2 ALGORITHMS FOR 2D TYPE-I -II AND -III DCTs

This section extends the concepts used in Section 6.5 for 2D type-I , -II and -III DCTs. We assume the array sizes $N_1 = q_1 \cdot 2^{m_1}$ and $N_2 = q_2 \cdot 2^{m_2}$, where q_1 and q_2 are odd integers which are generally different. It is shown that 2D DCTs can be decomposed into a set of 2D arrays which can be further decomposed into smaller ones in a similar way.

7.2.1 Algorithm for 2D type-II DCT

Based on the concepts of the algorithms used for the 1D DCT, the 2D type-II DCT defined in (7.2) can be expressed by

$$
\begin{aligned}
X_{DCT}^{II}(k_1, k_2) &= \sum_{n_1=0}^{N_1-1} \sum_{n_2=0}^{N_2/2-1} \left[x(n_1, 2n_2 + \frac{q_2-1}{2}) \cos \frac{\pi(4n_2+q_2)k_2}{2N_2} \right. \\
&\quad \left. + x(n_1, 2n_2 - \frac{q_2+1}{2}) \cos \frac{\pi(4n_2-q_2)k_2}{2N_2} \right] \cos \frac{\pi(2n_1+1)k_1}{2N_1} \\
&= \sum_{n_1=0}^{N_1-1} \sum_{n_2=0}^{N_2/2-1} \left\{ \left[x(n_1, 2n_2 - \frac{q_2+1}{2}) + x(n_1, 2n_2 + \frac{q_2-1}{2}) \right] \right. \\
&\quad \cdot \cos \frac{\pi q_2 k_2}{2N_2} \cos \frac{\pi n_2 k_2}{N_2/2} + \left[x(n_1, 2n_2 - \frac{q_2+1}{2}) - x(n_1, 2n_2 + \frac{q_2-1}{2}) \right] \\
&\quad \left. \cdot \sin \frac{\pi q_2 k_2}{2N_2} \sin \frac{\pi n_2 k_2}{N_2/2} \right\} \cos \frac{\pi(2n_1+1)k_1}{2N_1},
\end{aligned}
\tag{7.5}
$$

where the subscript indicates the transform to be performed and the superscript represents the type of the transform. For example, $X_{DCT}^{II}(,)$ means the transform is the type-II DCT. The inner summation is divided into two terms and the outer summation can be similarly decomposed. Without presenting the details of the decomposition process, the final results are given in the following equations.

$$
\begin{aligned}
X_{DCT}^{II}(k_1, k_2) &= \cos(\alpha_2, k_2)[\cos(\alpha_1, k_1)X_{cc}^{I}(k_1, k_2) + \sin(\alpha_1, k_1)X_{sc}^{I}(k_1, k_2)] \\
&\quad + \sin(\alpha_2, k_2)[\cos(\alpha_1, k_1)X_{cs}^{I}(k_1, k_2) + \sin(\alpha_1, k_1)X_{ss}^{I}(k_1, k_2)] \\
&\quad\quad 0 \le k_1 \le \frac{N_1}{2}, \quad 0 \le k_2 \le \frac{N_2}{2}
\end{aligned}
\tag{7.6}
$$

$$
\begin{aligned}
&X_{DCT}^{II}(k_1, N_2 - k_2) \\
&= (-1)^{\frac{(q_2-1)}{2}} \left\{ \sin(\alpha_2, k_2)[\cos(\alpha_1, k_1)X_{cc}^{I}(k_1, k_2) + \sin(\alpha_1, k_1)X_{sc}^{I}(k_1, k_2)] \right. \\
&\quad \left. - \cos(\alpha_2, k_2) [\cos(\alpha_1, k_1)X_{cs}^{I}(k_1, k_2) + \sin(\alpha_1, k_1)X_{ss}^{I}(k_1, k_2)] \right\} \\
&\quad\quad 0 \le k_1 \le \frac{N_1}{2}, \quad 1 \le k_2 < \frac{N_2}{2}
\end{aligned}
\tag{7.7}
$$

$$
\begin{aligned}
&X_{DCT}^{II}(N_1 - k_1, k_2) \\
&= (-1)^{\frac{(q_1-1)}{2}} \left\{ \cos(\alpha_2, k_2)[\sin(\alpha_1, k_1)X_{cc}^{I}(k_1, k_2) - \cos(\alpha_1, k_1)X_{sc}^{I}(k_1, k_2)] \right. \\
&\quad \left. + \sin(\alpha_2, k_2) [\sin(\alpha_1, k_1)X_{cs}^{I}(k_1, k_2) - \cos(\alpha_1, k_1)X_{ss}^{I}(k_1, k_2)] \right\} \\
&\quad\quad 1 \le k_1 < \frac{N_1}{2}, \quad 0 \le k_2 \le \frac{N_2}{2}
\end{aligned}
\tag{7.8}
$$

$$X_{DCT}^{II}(N_1 - k_1, N_2 - k_2)$$
$$= -(-1)^{\frac{(q_1+q_2)}{2}} \left\{ \sin(\alpha_2, k_2)[\sin(\alpha_1, k_1)X_{cc}^I(k_1, k_2) - \cos(\alpha_1, k_1)X_{sc}^I(k_1, k_2)] \right.$$
$$\left. - \cos(\alpha_2, k_2) \left[\sin(\alpha_1, k_1)X_{cs}^I(k_1, k_2) - \cos(\alpha_1, k_1)X_{ss}^I(k_1, k_2)\right] \right\} \qquad (7.9)$$
$$1 \le k_1 < \frac{N_1}{2}, \quad 1 \le k_2 < \frac{N_2}{2},$$

where for $(7.6) - (7.9)$, $\alpha_i = \pi q_i / (2N_i)$ $(i = 1, 2)$ and

$$X_{cc}^I(k_1, k_2) = X_{cc}^I(k_1, N_2 - k_2) = X_{cc}^I(N_1 - k_1, k_2) = X_{cc}^I(N_1 - k_1, N_2 - k_2)$$
$$= \sum_{n_1=0}^{N_1/2} \sum_{n_2=0}^{N_2/2} y_0(n_1, n_2) \cos \frac{\pi n_1 k_1}{N_1/2} \cos \frac{\pi n_2 k_2}{N_2/2} \qquad (7.10)$$
$$0 \le k_1 \le \frac{N_1}{2}, \quad 0 \le k_2 \le \frac{N_2}{2}$$

$$X_{sc}^I(k_1, k_2) = X_{sc}^I(k_1, N_2 - k_2) = -X_{sc}^I(N_1 - k_1, k_2) = -X_{sc}^I(N_1 - k_1, N_2 - k_2)$$
$$= \sum_{n_1=1}^{N_1/2-1} \sum_{n_2=0}^{N_2/2} y_1(n_1, n_2) \sin \frac{\pi n_1 k_1}{N_1/2} \cos \frac{\pi n_2 k_2}{N_2/2} \qquad (7.11)$$
$$1 \le k_1 < \frac{N_1}{2}, \quad 0 \le k_2 \le \frac{N_2}{2}$$

$$X_{cs}^I(k_1, k_2) = -X_{cs}^I(k_1, N_2 - k_2) = X_{cs}^I(N_1 - k_1, k_2) = -X_{cs}^I(N_1 - k_1, N_2 - k_2)$$
$$= \sum_{n_1=0}^{N_1/2} \sum_{n_2=1}^{N_2/2-1} y_2(n_1, n_2) \cos \frac{\pi n_1 k_1}{N_1/2} \sin \frac{\pi n_2 k_2}{N_2/2} \qquad (7.12)$$
$$0 \le k_1 \le \frac{N_1}{2}, \quad 1 \le k_2 < \frac{N_2}{2}$$

$$X_{ss}^I(k_1, k_2) = -X_{ss}^I(k_1, N_2 - k_2) = -X_{ss}^I(N_1 - k_1, k_2) = X_{ss}^I(N_1 - k_1, N_2 - k_2)$$
$$= \sum_{n_1=1}^{N_1/2-1} \sum_{n_2=1}^{N_2/2-1} y_3(n_1, n_2) \sin \frac{\pi n_1 k_1}{N_1/2} \sin \frac{\pi n_2 k_2}{N_2/2} \qquad (7.13)$$
$$1 \le k_1 < \frac{N_1}{2}, \quad 1 \le k_2 < \frac{N_2}{2},$$

where

$$\begin{bmatrix} y_0(n_1, n_2) \\ y_1(n_1, n_2) \\ y_2(n_1, n_2) \\ y_3(n_1, n_2) \end{bmatrix} = \begin{bmatrix} 1 & 1 & 1 & 1 \\ 1 & 1 & -1 & -1 \\ 1 & -1 & 1 & -1 \\ 1 & -1 & -1 & 1 \end{bmatrix} \begin{bmatrix} x\left(2n_1 - \frac{q_1+1}{2}, 2n_2 - \frac{q_2+1}{2}\right) \\ x\left(2n_1 - \frac{q_1+1}{2}, 2n_2 + \frac{q_2-1}{2}\right) \\ x\left(2n_1 + \frac{q_1-1}{2}, 2n_2 - \frac{q_2+1}{2}\right) \\ x\left(2n_1 + \frac{q_1-1}{2}, 2n_2 + \frac{q_2+1}{2}\right) \end{bmatrix}. \qquad (7.14)$$

In $(7.10) - (7.13)$, $X_{cc}^I(k_1, k_2)$ is the 2D type-I DCT which is known as the type-I $(N_1/2 + 1) \times (N_2/2 + 1)$ cosine-cosine array. Similarly, $X_{sc}^I(k_1, k_2)$, $X_{cs}^I(k_1, k_2)$ and $X_{ss}^I(k_1, k_2)$ are known as the type-I $(N_1/2 - 1) \times (N_2/2 + 1)$ sine-cosine array, the type-I $(N_1/2 + 1) \times (N_2/2 - 1)$ cosine-sine array and the type-I $(N_1/2 - 1) \times (N_2/2 - 1)$ sine-sine array. The

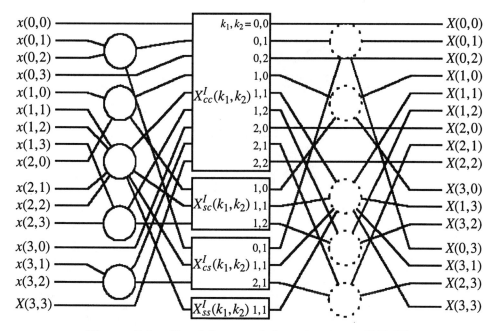

Figure 7.1   Signal flow graph for a 4 × 4 type-II DCT.

use of the subscripts c and s in these equations indicates cosine and sine functions used in the definition of these arrays, respectively. Similar to the definition of the types of DCT [1], the superscripts represent the types of cosine-cosine, cosine-sine, sine-cosine and sine-sine arrays. Therefore the type-II DCT computation can be obtained from the combination of these four arrays, as defined in (7.6) – (7.9). Figure 7.1 shows the signal flow graph of a 4×4 type-II DCT in which the solid circles on the input side perform the computation defined in (7.14) and the dashed circles on the output side combine $X_{cc}^I(k_1, k_2)$, $X_{sc}^I(k_1, k_2)$, $X_{cs}^I(k_1, k_2)$ and $X_{ss}(k_1, k_2)$ according to (7.6) – (7.9).

Some data indices in (7.14) are either negative or larger than $N_1 - 1$ (or $N_2 - 1$). A mapping process, similar to that used in Section 6.5, has to be performed to assign $x(n_1, n_2)$ with valid data in the input array. In general, the mapping process is used to change indices of certain columns and rows of the input array. As shown in Figure 7.2, the array $x(\ ,\ )$ is divided into the zero area, mapping area and unchanged area. Nothing is to be done for the unchanged area because the data in this area have valid indices. The values of the data in the zero area are zero. This happens only when $n_1 = 0$ or $n_2 = 0$ and $n_1 = N_1/2$ or $n_2 = N_2/2$. In the mapping area, the row indices above the array in the figure are replaced by those below the array. Similarly, the column indices on the left side of the array are replaced by those on the right side of the array. For simplicity of presentation, let us assume $q_1 = q_2 = q$, and $N_1 = N_2 = N$. When $(q - 1) \bmod 2$ is even, the mapping process is defined by

$$\begin{array}{ll} x(N + 2t) = 0, & x(-t - 1) = 0 \\ x(-2i - 1) = (2i), & x(N + 2i) = x(N - 2i - 1) \end{array} . \tag{7.15}$$

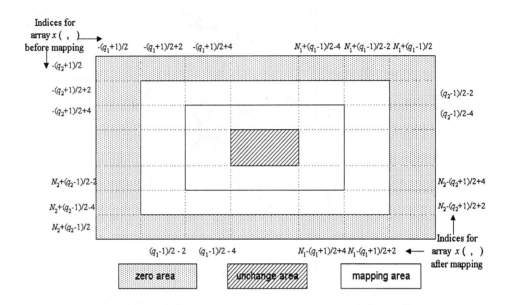

Figure 7.2   Illustration of the mapping process.

Similarly, for $(q-1)$ mod 2 odd

$$
\begin{aligned}
&x(N+2t+1)=0, &&x(-t-2)=0 \\
&x(-2i-2)=(2i+1), &&x(N+2i+1)=x(N-2i-2)
\end{aligned}, \tag{7.16}
$$

where in (7.15) and (7.16), $0 \le i < t$, and t is the integer part of $(q-1)/4$. The mapping process defined in (7.15) and (7.16) is essentially the same as that used in Section 6.5 and valid for both n_1 and n_2. It should be noted that the mapping process involves only a few rows and columns of the input array.

To achieve the above decomposition, arithmetic operations are performed to combine the input data according to (7.10) – (7.13) and to obtain the transformed outputs according to (7.6) – (7.9). The decomposition costs are

- $T = 2(N_1 N_2 - N_1 - N_2)$ additions for (7.14),

- $T_m = 3(N_1 - 2q_1)(N_2 - 2q_2) + 4(N_1 - 2q_1)(q_2 - 1) + 4(N_2 - 2q_2)(q_1 - 1) + 2(q_1 - 1)(q_2 - 1) + 4(N_1 + N_2) - 6(q_1 + q_2) - 2$ multiplications for (7.6) – (7.9),

- $T_a = 2(N_1 - 2)(N_2 - 2) - (N_2 - 2)(q_1 - 1) - (N_1 - 2)(q_2 - 1) + 2(N_2 - q_2) + 2(N_1 - q_1) - 4$ additions for (7.6) – (7.9).

The detailed calculations for the last two items are given in Appendix A. The total number of multiplications needed by the algorithm for the 2D type-II DCT is

$$
\begin{aligned}
M_{DCT}^I(N_1, N_2) =\ & M_{cc}^I\left(\frac{N_1}{2} + 1, \frac{N_2}{2} + 1\right) + M_{cs}^I\left(\frac{N_1}{2} + 1, \frac{N_2}{2} - 1\right) \\
& + M_{sc}^I\left(\frac{N_1}{2} - 1, \frac{N_2}{2} + 1\right) + M_{ss}^I\left(\frac{N_1}{2} - 1, \frac{N_2}{2} - 1\right) + T_m
\end{aligned} \tag{7.17}
$$

and the number of additions is

$$
\begin{aligned}
A_{DCT}^I(N_1, N_2) = & \ A_{cc}^I(\frac{N_1}{2}+1, \frac{N_2}{2}+1) + A_{cs}^I(\frac{N_1}{2}+1, \frac{N_2}{2}-1) \\
& + A_{sc}^I(\frac{N_1}{2}-1, \frac{N_2}{2}+1) + A_{ss}^I(\frac{N_1}{2}-1, \frac{N_2}{2}-1) + T + Ta. \quad (7.18)
\end{aligned}
$$

In both (7.17) and (7.18), the subscripts and superscripts are used in accordance with those used in (7.10) – (7.13). The computation of the type-I cosine-cosine, sine-cosine, cosine-sine and sine-sine arrays will be considered in latter subsections.

7.2.2   Algorithm for type-III DCT

The type-III 2D DCT (or inverse of the type-II DCT) defined in (7.4) can be decomposed into cosine-cosine, sine-cosine, cosine-sine and sine-sine arrays of smaller sizes. The computation can be partitioned into

$$
\begin{bmatrix} x_{11}(n_1,n_2) \\ x_{12}(n_1,n_2) \\ x_{21}(n_1,n_2) \\ x_{22}(n_1,n_2) \end{bmatrix} =
\begin{bmatrix} x(2n_1 + \frac{q_1-1}{2}, 2n_2 + \frac{q_2-1}{2}) \\ x(2n_1 + \frac{q_1-1}{2}, 2n_2 - \frac{q_2+1}{2}) \\ x(2n_1 - \frac{q_1+1}{2}, 2n_2 + \frac{q_2-1}{2}) \\ x(2n_1 - \frac{q_1+1}{2}, 2n_2 - \frac{q_2+1}{2}) \end{bmatrix}. \quad (7.19)
$$

It is possible that some indices in (7.19) are either greater than or equal to N_1 (or N_2), or less than zero. The mapping process is defined by

$$
x(-t-1) \Leftrightarrow x(t), \qquad x(N+t) \Leftrightarrow x(N-t-1) \quad (7.20)
$$
$$
0 \le t \le \frac{q-1}{2},
$$

where the data on the left side of symbol $\Leftrightarrow$ have invalid indices and therefore are replaced by the data on the right side, and

$$
x(-\frac{q+1}{2}) = 0, \qquad x(N+\frac{q-1}{2}) = 0. \quad (7.21)
$$

By using the properties $\cos(A \pm B) = \cos A \cos B \mp \sin A \sin B$, (7.4) is decomposed into

$$
\begin{bmatrix} x_{11}(n_1,n_2) \\ x_{12}(n_1,n_2) \\ x_{21}(n_1,n_2) \\ x_{22}(n_1,n_2) \end{bmatrix} =
\begin{bmatrix} 1 & -1 & -1 & 1 \\ 1 & -1 & 1 & -1 \\ 1 & 1 & -1 & -1 \\ 1 & 1 & 1 & 1 \end{bmatrix}
\begin{bmatrix} y_{cc}^I(n_1,n_2) \\ y_{sc}^I(n_1,n_2) \\ y_{cs}^I(n_1,n_2) \\ y_{ss}^I(n_1,n_2) \end{bmatrix}, \quad (7.22)
$$

where for $\alpha_1 = \frac{\pi q_1}{2N_1}$ and $\alpha_2 = \frac{\pi q_2}{2N_2}$,

$$y_{cc}^I(n_1, n_2) = \sum_{k_1=0}^{N_1-1} \sum_{k_2=0}^{N_2-1} X(k_1, k_2) \cos(\alpha_1 k_1) \cos(\alpha_2 k_2) \cos \frac{\pi n_1 k_1}{N_1/2} \cos \frac{\pi n_2 k_2}{N_2/2} \quad (7.23)$$

$$0 \le n_1 \le \frac{N_1}{2}, \quad 0 \le n_2 \le \frac{N_2}{2}$$

$$y_{sc}^I(n_1, n_2) = \sum_{k_1=0}^{N_1-1} \sum_{k_2=0}^{N_2-1} X(k_1, k_2) \sin(\alpha_1 k_1) \cos(\alpha_2 k_2) \sin \frac{\pi n_1 k_1}{N_1/2} \cos \frac{\pi n_2 k_2}{N_2/2} \quad (7.24)$$

$$0 < n_1 < \frac{N_1}{2}, \quad 0 \le n_2 \le \frac{N_2}{2}$$

$$y_{cs}^I(n_1, n_2) = \sum_{k_1=0}^{N_1-1} \sum_{k_2=0}^{N_2-1} X(k_1, k_2) \cos(\alpha_1 k_1) \sin(\alpha_2 k_2) \cos \frac{\pi n_1 k_1}{N_1/2} \sin \frac{\pi n_2 k_2}{N_2/2} \quad (7.25)$$

$$0 \le n_1 \le \frac{N_1}{2}, \quad 0 < n_2 < \frac{N_2}{2}$$

$$y_{ss}^I(n_1, n_2) = \sum_{k_1=0}^{N_1-1} \sum_{k_2=0}^{N_2-1} X(k_1, k_2) \sin(\alpha_1 k_1) \sin(\alpha_2 k_2) \sin \frac{\pi n_1 k_1}{N_1/2} \sin \frac{\pi n_2 k_2}{N_2/2} \quad (7.26)$$

$$0 < n_1 < \frac{N_1}{2}, \quad 0 < n_2 < \frac{N_2}{2}.$$

By using the property

$$\cos \frac{\pi n(N-k)}{N/2} = \cos \frac{\pi n k}{N/2}, \quad \sin \frac{\pi n(N-k)}{N/2} = -\sin \frac{\pi n k}{N/2}$$

we can obtain (7.27) – (7.30).

$$y_{cc}^I(n_1, n_2) = \sum_{k_1=0}^{N_1/2} \sum_{k_2=0}^{N_2/2} \left[\begin{array}{l} \cos(\alpha_1 k_1)[X(k_1, k_2) \cos(\alpha_2 k_2) \\ + (-1)^{(q_2-1)/2} r_2 X(k_1, N_2 - k_2) \sin(\alpha_2 k_2)] \\ + (-1)^{(q_1-1)/2} r_1 \cos(\alpha_1 k_1)[X(N_1 - k_1, k_2) \cos(\alpha_2 k_2) \\ + (-1)^{(q_2-1)/2} r_2 X(N_1 - k_1, N_2 - k_2) \sin(\alpha_2 k_2)] \end{array} \right]$$
$$\cdot \cos \frac{\pi n_1 k_1}{N_1/2} \cos \frac{\pi n_2 k_2}{N_2/2} \quad (7.27)$$

$$y_{sc}^I(n_1, n_2) = \sum_{k_1=0}^{N_1/2} \sum_{k_2=0}^{N_2/2} \left[\begin{array}{l} \sin(\alpha_1 k_1)[X(k_1, k_2) \cos(\alpha_2 k_2) \\ + (-1)^{(q_2-1)/2} r_2 X(k_1, N_2 - k_2) \sin(\alpha_2 k_2)] \\ - (-1)^{(q_1-1)/2} \cos(\alpha_1 k_1)[X(N_1 - k_1, k_2) \cos(\alpha_2 k_2) \\ + (-1)^{(q_2-1)/2} r_2 X(N_1 - k_1, N_2 - k_2) \sin(\alpha_2 k_2)] \end{array} \right]$$
$$\cdot \sin \frac{\pi n_1 k_1}{N_1/2} \cos \frac{\pi n_2 k_2}{N_2/2} \quad (7.28)$$

$$y_{cs}^I(n_1, n_2) = \sum_{k_1=0}^{N_1/2} \sum_{k_2=0}^{N_2/2} \left[\begin{array}{l} \cos(\alpha_1 k_1)[X(k_1, k_2) \sin(\alpha_2 k_2) \\ - (-1)^{(q_2-1)/2} X(k_1, N_2 - k_2) \cos(\alpha_2 k_2)] \\ + (-1)^{(q_1-1)/2} r_1 \sin(\alpha_1 k_1)[X(N_1 - k_1, k_2) \sin(\alpha_2 k_2) \\ + (-1)^{(q_2-1)/2} X(N_1 - k_1, N_2 - k_2) \cos(\alpha_2 k_2)] \end{array} \right]$$
$$\cdot \cos \frac{\pi n_1 k_1}{N_1/2} \sin \frac{\pi n_2 k_2}{N_2/2} \quad (7.29)$$

$$y_{ss}^I(n_1, n_2) = \sum_{k_1=0}^{N_1/2} \sum_{k_2=0}^{N_2/2} \begin{bmatrix} \sin(\alpha_1 k_1)[X(k_1, k_2)\sin(\alpha_2 k_2) \\ -(-1)^{(q_2-1)/2}X(k_1, N_2 - k_2)\cos(\alpha_2 k_2)] \\ -(-1)^{(q_1-1)/2}\cos(\alpha_1 k_1)[X(N_1 - k_1, k_2)\sin(\alpha_2 k_2) \\ +(-1)^{(q_2-1)/2}X(N_1 - k_1, N_2 - k_2)\cos(\alpha_2 k_2)] \end{bmatrix}$$
$$\cdot \sin\frac{\pi n_1 k_1}{N_1/2} \sin\frac{\pi n_2 k_2}{N_2/2}, \tag{7.30}$$

where

$$r_1 = \begin{cases} 0, & \text{if } k_1 = 0, \frac{N_1}{2} \\ 1, & \text{otherwise} \end{cases} \quad r_2 = \begin{cases} 0, & \text{if } k_2 = 0, \frac{N_2}{2} \\ 1, & \text{otherwise} \end{cases}.$$

It is noted that the ranges of indices k_1 and k_2 in (7.27) – (7.30) are defined according to the type of trigonometric identity. If the cosine function is involved, for example, the related index is valid from 0 to $N_i/2$, where $i = 1$ or 2. If the sine function is involved, the index is valid from 1 to $N_i/2 - 1$. Such an arrangement includes all the valid indices. It is different from the index range used in other decomposition processes in which the index is valid from 0 to $N_i/2 - 1$.

It seems that r_1 and r_2 are redundant. In fact, they are introduced for compact mathematical expressions especially when $k_i = N_i/2$. However, they do not need any extra arithmetic operations. Once $y_{cc}^I, y_{cs}^I, y_{sc}^I$ and y_{ss}^I are computed according to (7.27) – (7.30), the final transformed outputs can be combined according to (7.22). Input data indexing is straightforward, although a mapping process dealing with a few invalid output indices is needed, as defined in (7.20) and (7.21). Without providing the details, we simply state here that the decomposition costs in terms of the number of arithmetic operations are

- $T_A = 2N_1 N_2 - N_1(q_2 + 1) - N_2(q_1 + 1)$ additions for (7.27) – (7.30),

- $T_B = 2(N_1 N_2 - N_1 - N_2)$ additions for (7.22),

- $T_M = 3(N_1 - 2q_1)(N_2 - 2q_2) + 4p_2(N_1 - 2q_1) + 4q_1(N_2 - 2q_2) + 2q_1 q_2$ multiplications for (7.27) – (7.30).

The total number of multiplications is

$$M_{DCT}^{III}(N_1, N_2) = M_{DCT}^{III}(\frac{N_1}{2}, \frac{N_2}{2}) + M_{sc}^I(\frac{N_1}{2}, \frac{N_2}{2}) + M_{cs}^I(\frac{N_1}{2}, \frac{N_2}{2})$$
$$+ M_{ss}^I(\frac{N_1}{2}, \frac{N_2}{2}) + T_M \tag{7.31}$$

and the number of additions is

$$A_{DCT}^{III}(N_1, N_2) = A_{DCT}^{III}(\frac{N_1}{2}, \frac{N_2}{2}) + A_{sc}^I(\frac{N_1}{2}, \frac{N_2}{2}) + A_{cs}^I(\frac{N_1}{2}, \frac{N_2}{2})$$
$$+ A_{ss}^I(\frac{N_1}{2}, \frac{N_2}{2}) + T_A + T_B. \tag{7.32}$$

As shown above, the computation for both 2D type-II and -III DCTs depends on the computation of various subarrays that are obtained from the decomposition process. Fast algorithms for these subarrays can also be derived as seen in the following subsections.

7.2.3   Algorithm for 2D type-I DCT

Based on the decimation-in-time decomposition, (7.1) can be expressed by

$$
\begin{bmatrix}
X_{cc}^I(k_1, k_2) \\
X_{cc}^I(N_1 - k_1, k_2) \\
X_{cc}^I(k_1, N_2 - k_2) \\
X_{cc}^I(N_1 - k_1, N_2 - k_2)
\end{bmatrix}
=
\begin{bmatrix}
1 & 1 & 1 & 1 \\
1 & 1 & -1 & -1 \\
1 & -1 & 1 & -1 \\
1 & -1 & -1 & 1
\end{bmatrix}
\begin{bmatrix}
B_{cc}^I(k_1, k_2) \\
B_{cc}^{I-II}(k_1, k_2) \\
B_{cc}^{II-I}(k_1, k_2) \\
B_{cc}^{II}(k_1, k_2)
\end{bmatrix},
\tag{7.33}
$$

where

$$
\begin{aligned}
B_{cc}^I(k_1, k_2) &= B_{cc}^I(N_1 - k_1, k_2) = B_{cc}^I(k_1, N_2 - k_2) = B_{cc}^I(N_1 - k_1, N_2 - k_2) \\
&= \sum_{n_1=0}^{N_1/2} \sum_{n_2=0}^{N_2/2} u(2n_1, 2n_2) \cos\frac{\pi n_1 k_1}{N_1/2} \cos\frac{\pi n_2 k_2}{N_2/2}
\end{aligned}
\tag{7.34}
$$

$$
0 \le k_1 \le \frac{N_1}{2}, \quad 0 \le k_2 \le \frac{N_2}{2}
$$

$$
\begin{aligned}
B_{cc}^{I-II}(k_1, k_2) &= B_{cc}^{I-II}(N_1 - k_1, k_2) = -B_{cc}^{I-II}(k_1, N_2 - k_2) = -B_{cc}^{I-II}(N_1 - k_1, N_2 - k_2) \\
&= \sum_{n_1=0}^{N_1} \sum_{n_2=0}^{\frac{N_2}{2}-1} u(2n_1, 2n_2 + 1) \cos\frac{\pi n_1 k_1}{N_1/2} \cos\frac{\pi (2n_2 + 1)k_2}{2(N_2/2)}
\end{aligned}
\tag{7.35}
$$

$$
0 \le k_1 \le \frac{N_1}{2}, \quad 0 \le k_2 < \frac{N_2}{2}
$$

$$
\begin{aligned}
B_{cc}^{II-I}(k_1, k_2) &= -B_{cc}^{II-I}(N_1 - k_1, k_2) = B_{cc}^{II-I}(k_1, N_2 - k_2) = -B_{cc}^{II-I}(N_1 - k_1, N_2 - k_2) \\
&= \sum_{n_1=0}^{N_1/2-1} \sum_{n_2=1}^{N_2/2} u(2n_1 + 1, 2n_2) \cos\frac{\pi (2n_1 + 1)k_1}{2(N_1/2)} \cos\frac{\pi n_2 k_2}{N_2/2}
\end{aligned}
\tag{7.36}
$$

$$
0 \le k_1 < \frac{N_1}{2}, \quad 0 \le k_2 \le \frac{N_2}{2}
$$

$$
\begin{aligned}
B_{cc}^{II}(k_1, k_2) &= -B_{cc}^{II}(N_1 - k_1, k_2) = -B_{cc}^{II}(k_1, N_2 - k_2) = B_{cc}^{II}(N_1 - k_1, N_2 - k_2) \\
&= \sum_{n_1=0}^{N_1/2-1} \sum_{n_2=0}^{N_2/2-1} u(2n_1 + 1, 2n_2 + 1) \cos\frac{\pi (2n_1 + 1)k_1}{2(N_1/2)} \cos\frac{\pi (2n_2 + 1)k_2}{2(N_2/2)}
\end{aligned}
$$

$$
0 \le k_1 < \frac{N_1}{2}, \quad 0 \le k_2 < \frac{N_2}{2}.
\tag{7.37}
$$

It is noted that the decomposition presented in the above equations is very simple and no twiddle factor is used in the entire computation. The indexing process for input/output is also straightforward so that no mapping process is needed. All these desirable properties can effectively reduce implementation complexity. The total number of multiplications needed by the algorithm for the type-I $(N_1 + 1) \times (N_2 + 1)$ cosine-cosine array is

$$
\begin{aligned}
M_{cc}^I(N_1 + 1, N_2 + 1) &= M_{cc}^I(\frac{N_1}{2} + 1, \frac{N_2}{2} + 1) + M_{cc}^{I-II}(\frac{N_1}{2} + 1, \frac{N_2}{2}) \\
&+ M_{cc}^{II-I}(\frac{N_1}{2}, \frac{N_2}{2} + 1) + M_{DCT}^{II}(\frac{N_1}{2}, \frac{N_2}{2}).
\end{aligned}
\tag{7.38}
$$

Taking the $2N_1N_2 + N_1 + N_2$ additions needed by (7.33) into account, the total number of additions for the type-I $(N_1 + 1) \times (N_2 + 1)$ cosine-cosine array is

$$A_{cc}^I(N_1 + 1, N_2 + 1) = A_{cc}^I(\frac{N_1}{2} + 1, \frac{N_2}{2} + 1) + A_{cc}^{I-II}(\frac{N_1}{2} + 1, \frac{N_2}{2}) + A_{cc}^{II-I}(\frac{N_1}{2}, \frac{N_2}{2} + 1)$$

$$+ A_{DCT}^{II}(\frac{N_1}{2}, \frac{N_2}{2}) + 2N_1N_2 + N_1 + N_2. \tag{7.39}$$

7.2.4 Algorithm for type-I cosine-sine array

The type-I $(N_1 + 1) \times (N_2 - 1)$ cosine-sine array of input $u(n_1, n_2)$ is defined by

$$X_{cs}^I(k_1, k_2) = \sum_{n_1=0}^{N_1} \sum_{n_2=1}^{N_2-1} u(N_1, n_2) \cos \frac{\pi n_1 k_1}{N_1} \sin \frac{\pi n_2 k_2}{N_2} \tag{7.40}$$

$$0 \le k_1 \le N_1, \quad 1 \le k_2 < N_2$$

which can be decomposed into

$$\begin{bmatrix} X_{cs}^I(k_1, k_2) \\ X_{cs}^I(k_1, N_2 - k_2) \\ X_{cs}^I(N_1 - k_1, k_2) \\ X_{cs}^I(N_1 - k_1, N_2 - k_2) \end{bmatrix} = \begin{bmatrix} 1 & 1 & 1 & 1 \\ -1 & 1 & -1 & 1 \\ 1 & 1 & -1 & -1 \\ -1 & 1 & 1 & -1 \end{bmatrix} \begin{bmatrix} B_{cs}^I(k_1, k_2) \\ B_{cs}^{I-II}(k_1, k_2) \\ B_{cs}^{II-I}(k_1, k_2) \\ B_{cs}^{II}(k_1, k_2) \end{bmatrix}, \tag{7.41}$$

where

$$B_{cs}^I(k_1, k_2) = B_{cs}^I(N_1 - k_1, k_2) = -B_{cs}^I(k_1, N_2 - k_2) = -B_{cs}^I(N_1 - k_1, N_2 - k_2)$$

$$= \sum_{n_1=0}^{\frac{N_1}{2}} \sum_{n_2=1}^{\frac{N_2}{2}-1} u(2n_1, 2n_2) \cos \frac{\pi n_1 k_1}{N_1/2} \sin \frac{\pi n_2 k_2}{N_2/2} \tag{7.42}$$

$$0 \le k_1 \le \frac{N_1}{2}, \quad 1 \le k_2 < \frac{N_2}{2}$$

$$B_{cs}^{I-II}(k_1, k_2) = B_{cs}^{I-II}(N_1 - k_1, k_2) = B_{cs}^{I-II}(k_1, N_2 - k_2) = B_{cs}^{I-II}(N_1 - k_1, N_2 - k_2)$$

$$= \sum_{n_1=0}^{\frac{N_1}{2}} \sum_{n_2=0}^{\frac{N_2}{2}-1} u(2n_1, 2n_2 + 1) \cos \frac{\pi n_1 k_1}{N_1/2} \sin \frac{\pi (2n_2 + 1)k_2}{2(N_2/2)} \tag{7.43}$$

$$0 \le k_1 \le \frac{N_1}{2}, \quad 1 \le k_2 < \frac{N_2}{2}$$

$$B_{cs}^{II-I}(k_1, k_2) = -B_{cs}^{II-I}(N_1 - k_1, k_2) = -B_{cs}^{II-I}(k_1, N_2 - k_2) = B_{cs}^{II-I}(N_1 - k_1, N_2 - k_2)$$

$$= \sum_{n_1=0}^{\frac{N_1}{2}-1} \sum_{n_2=1}^{\frac{N_2}{2}-1} u(2n_1 + 1, 2n_2) \cos \frac{\pi (2n_1 + 1)k_1}{2(N_1/2)} \sin \frac{\pi n_2 k_2}{N_2/2} \tag{7.44}$$

$$0 \le k_1 < \frac{N_1}{2}, \quad 1 \le k_2 < \frac{N_2}{2}$$

$$B_{cs}^{II}(k_1, k_2) = -B_{cs}^{II}(N_1 - k_1, k_2) = B_{cs}^{II}(k_1, N_2 - k_2) = -B_{cs}^{II}(N_1 - k_1, N_2 - k_2)$$

$$= \sum_{n_1=0}^{\frac{N_1}{2}-1} \sum_{n_2=0}^{\frac{N_2}{2}-1} u(2n_1 + 1, 2n_2 + 1) \cos \frac{\pi (2n_1 + 1)k_1}{2(N_1/2)} \sin \frac{\pi (2n_2 + 1)k_2}{2(N_2/2)} \tag{7.45}$$

$$0 \le k_1 < \frac{N_1}{2}, \quad 1 \le k_2 < \frac{N_2}{2}.$$

The total number of multiplications needed by the computation of the type-I cosine-sine array is

$$
\begin{aligned}
M_{cs}^I(N_1+1, N_2-1) &= M_{cs}^I(\frac{N_1}{2}+1, \frac{N_2}{2}-1) + M_{cs}^{I-II}(\frac{N_1}{2}+1, \frac{N_2}{2}) \\
&+ M_{cs}^{II-I}(\frac{N_1}{2}, \frac{N_2}{2}-1) + M_{DCT}(\frac{N_1}{2}, \frac{N_2}{2}).
\end{aligned} \tag{7.46}
$$

For decomposition, (7.42) – (7.45) require $2N_1N_2 - 3N_1 + N_2 - 2$ additions. Therefore, the number of additions is

$$
\begin{aligned}
A_{cs}^I(N_1+1, N_2-1) &= A_{cs}^I(\frac{N_1}{2}+1, \frac{N_2}{2}-1) + A_{cs}^{I-II}(\frac{N_1}{2}+1, \frac{N_2}{2}) \\
&+ A_{cs}^{II-I}(\frac{N_1}{2}, \frac{N_2}{2}-1) + A_{DCT}(\frac{N_1}{2}, \frac{N_2}{2}) \\
&+ 2N_1N_2 - 3N_1 + N_1 - 2.
\end{aligned} \tag{7.47}
$$

In (7.42) – (7.45), B_{cs}^I is the type-I $(N_1/2 + 1) \times (N_2/2 - 1)$ cosine-sine array and B_{cs}^{II} is the type-II $(N_1/2) \times (N_2/2)$ cosine-sine array and can be converted into the type-II $(N_1/2) \times (N_2/2)$ DCT if k_2 is replaced by $(N_2/2 - k_2)$. Similarly, B_{cs}^{I-II} can be converted into the type-I-II $(N_1/2 + 1) \times (N_2/2)$ cosine-cosine array.

The type-I cosine-sine array is symmetric to the type-I sine-cosine array. For example, the type-I sine-cosine array can be obtained by swapping the functions of sine and cosine, the input indices n_1 and n_2, the output indices k_1 and k_2 and the dimension sizes N_1 and N_2. Therefore, the decomposition procedures used for the type-I cosine-sine array can be similarly applied to decompose the type-I sine-cosine array. Based on the symmetric property, the computational complexity of the type-I sine-cosine array can be derived. The number of multiplications for the type-I sine-cosine array is

$$
\begin{aligned}
M_{sc}^I(N_1-1, N_2+1) &= M_{sc}^I(\frac{N_1}{2}-1, \frac{N_2}{2}+1) + M_{sc}^{I-II}(\frac{N_1}{2}, \frac{N_2}{2}+1) \\
&+ M_{sc}^{II-I}(\frac{N_1}{2}, \frac{N_2}{2}+1) + M_{sc}^{II}(\frac{N_1}{2}, \frac{N_2}{2}).
\end{aligned} \tag{7.48}
$$

The number of additions is

$$
\begin{aligned}
A_{sc}^I(N_1-1, N_2+1) &= A_{sc}^I(\frac{N_1}{2}-1, \frac{N_2}{2}+1) + A_{sc}^{I-II}(\frac{N_1}{2}, \frac{N_2}{2}+1) + A_{sc}^{II-I}(\frac{N_1}{2}, \frac{N_2}{2}+1) \\
&+ A_{sc}^{II}(\frac{N_1}{2}, \frac{N_2}{2}) + 2N_1N_2 + N_1 - 3N_2 - 2.
\end{aligned} \tag{7.49}
$$

7.2.5   Algorithm for type-I-II cosine-cosine array

The type-I-II $(N_1 + 1) \times N_2$ cosine-cosine array of input $u(n_1, n_2)$ is defined by

$$
X_{cc}^{I-II}(k_1, k_2) = \sum_{n_1=0}^{N_1} \sum_{n_2=0}^{N_2-1} u(N_1, n_2) \cos\frac{\pi n_1 k_1}{N_1} \cos\frac{\pi(2n_2+1)k_2}{2N_2}
$$

$$
0 \le k_1 \le N_1, \quad 0 \le k_2 < N_2 \tag{7.50}
$$

which can be decomposed into

$$
\begin{aligned}
X_{cc}^{I-II}(k_1, k_2) &= \cos(\alpha_2 k_2)[B_{cc}^I(k_1, k_2) + B_{cc}^{II-I}(k_1, k_2)] \\
&\quad + \sin(\alpha_2 k_2)[B_{cs}^I(k_1, k_2) + B_{cs}^{II-I}(k_1, k_2)], \\
&\qquad 0 \le k_1 < \frac{N_1}{2}, \quad 0 \le k_2 \le \frac{N_2}{2}
\end{aligned}
\tag{7.51}
$$

$$
\begin{aligned}
X_{cc}^{I-II}(N_1 - k_1, k_2) &= \cos(\alpha_2 k_2)[B_{cc}^I(k_1, k_2) - B_{cc}^{II-I}(k_1, k_2)] \\
&\quad + \alpha_2 k_2)[B_{cs}^I(k_1, k_2) - B_{cs}^{II-I}(k_1, k_2)] \\
&\qquad 0 \le k_1 < \frac{N_1}{2}, \quad 0 \le k_2 < \frac{N_2}{2}
\end{aligned}
\tag{7.52}
$$

$$
\begin{aligned}
X_{cc}^{I-II}(k_1, N_2 - k_2) &= (-1)^{\frac{q_2-1}{2}} \left\{ \sin(\alpha_2 k_2)[B_{cc}^I(k_1, k_2) + B_{cc}^{II-I}(k_1, k_2)] \right. \\
&\quad \left. - \cos(\alpha_2 k_2)[B_{cs}^I(k_1, k_2) + B_{cs}^{II-I}(k_1, k_2)] \right\} \\
&\qquad 0 \le k_1 < \frac{N_1}{2}, \quad 0 \le k_2 < \frac{N_2}{2}
\end{aligned}
\tag{7.53}
$$

$$
\begin{aligned}
X_{cc}^{I-II}(N_1 - k_1, N_2 - k_2) &= (-1)^{\frac{q_2-1}{2}} \left\{ \sin(\alpha_2 k_2)[B_{cc}^I(k_1, k_2) - B_{cc}^{II-I}(k_1, k_2)] \right. \\
&\quad \left. - \cos(\alpha_2 k_2)[B_{cs}^I(k_1, k_2) - B_{cs}^{II-I}(k_1, k_2)] \right\} \\
&\qquad 0 \le k_1 < \frac{N_1}{2}, \quad 0 \le k_2 < \frac{N_2}{2},
\end{aligned}
\tag{7.54}
$$

where $\alpha_2 = \pi q_2/(2N_2)$ and for (7.51) – (7.54)

$$
\begin{aligned}
B_{cc}^I(k_1, k_2) &= \sum_{n_1=0}^{N_1/2} \sum_{n_2=0}^{N_2/2} \left[u(2n_1, 2n_2 - \frac{q_2+1}{2}) + u(2n_1, 2n_2 + \frac{q_2-1}{2}) \right] \\
&\quad \cdot \cos\frac{\pi n_1 k_1}{N_1/2} \cos\frac{\pi n_2 k_2}{N_2/2}, \qquad 0 \le k_1 \le \frac{N_1}{2}, \; 0 \le k_2 \le \frac{N_2}{2}
\end{aligned}
\tag{7.55}
$$

$$
\begin{aligned}
B_{cc}^{II-I}(k_1, k_2) &= \sum_{n_1=0}^{N_1/2-1} \sum_{n_2=0}^{N_2/2} \left[u(2n_1+1, 2n_2 - \frac{q_2+1}{2}) + u(2n_1+1, 2n_2 + \frac{q_2-1}{2}) \right] \\
&\quad \cdot \cos\frac{\pi(2n_1+1)k_1}{2(N_1/2)} \cos\frac{\pi n_2 k_2}{N_2/2}, \qquad 0 \le k_1 < \frac{N_1}{2}, \; 0 \le k_2 \le \frac{N_2}{2}
\end{aligned}
\tag{7.56}
$$

$$
\begin{aligned}
B_{cs}^I(k_1, k_2) &= \sum_{n_1=0}^{N_1/2} \sum_{n_2=1}^{N_2/2-1} \left[u(2n_1, 2n_2 - \frac{q_2+1}{2}) - u(2n_1, 2n_2 + \frac{q_2-1}{2}) \right] \\
&\quad \cdot \cos\frac{\pi n_1 k_1}{N_1/2} \sin\frac{\pi n_2 k_2}{N_2/2}, \qquad 0 \le k_1 \le \frac{N_1}{2}, \; 1 \le k_2 < \frac{N_2}{2}
\end{aligned}
\tag{7.57}
$$

$$
\begin{aligned}
B_{cs}^{II-I}(k_1, k_2) &= \sum_{n_1=0}^{N_1/2-1} \sum_{n_2=1}^{N_2/2-1} \left[u(2n_1+1, 2n_2 - \frac{q_2+1}{2}) - u(2n_1+1, 2n_2 + \frac{q_2-1}{2}) \right] \\
&\quad \cdot \cos\frac{\pi(2n_1+1)k_1}{2(N_1/2)} \sin\frac{\pi n_2 k_2}{N_2/2}, \qquad 0 \le k_1 < \frac{N_1}{2}, \; 1 \le k_2 < \frac{N_2}{2}.
\end{aligned}
\tag{7.58}
$$

In (7.55) – (7.58), the data indices $2n_2 - (q_2+1)/2$ and $2n_2 + (q_2-1)/2$ become invalid when $n_2 < (q_2+1)/4$ and $n_2 > N_2/2 - (q_2-1)/4$, respectively. Similar to the case of decomposing the 1D DCT, a mapping process is needed to change the invalid indices into valid ones. The mapping process is defined by

$$u\left(2n_1+1, -\frac{q_2+1}{2}\right) = 0, \quad u\left(2n_1+1, N_2+\frac{q_2+1}{2}\right) = 0 \tag{7.59}$$

and for $1 \leq n_2 \leq (q_2+1)/4$,

$$u\left(2n_1+1, 2n_2-\frac{q_2+1}{2}\right) = u\left(2n_1+1, \frac{q_2+1}{2}-2n_2\right) \tag{7.60}$$

$$u\left(2n_1+1, N_2-2n_2+\frac{q_2+1}{2}\right) = u\left(2n_1+1, N_2+2n_2+\frac{q_2+1}{2}\right). \tag{7.61}$$

In (7.55) – (7.58), $N_1 N_2 - 2N_1 + N_2 - 2$ additions are needed for processing of input data inside the square brackets. Furthermore, $N_1 N_2 - (N_1/2+1)(q_2+1) + N_2$ additions are needed by (7.51) and (7.53), and $N_1 N_2 - (q_2+1)N_1/2$ additions by (7.52) – (7.54). The number of multiplications needed by the twiddle factors in (7.51) – (7.54) is $(N_1/2+1)[N_2 - (3q_2-1)/2]$, $(N_1/2)[N_2 - (3q_2-1)/2]$, $(N_1/2+1)[N_2 - (3q_2+1)/2]$, and $(N_1/2)[N_2 - (3q_2+1)/2]$, respectively. The total number of additions for the type-I-II $(N_1/2+1) \times (N_2/2)$ cosine-cosine array is

$$\begin{aligned}
A_{cc}^{I-II}(N_1+1, N_2) &= A_{cc}^{I}(\frac{N_1}{2}+1, \frac{N_2}{2}+1) + A_{cc}^{II-I}(\frac{N_1}{2}, \frac{N_2}{2}+1) \\
&\quad + A_{cs}^{I}(\frac{N_1}{2}+1, \frac{N_2}{2}-1) + A_{cs}^{II-I}(\frac{N_1}{2}, \frac{N_2}{2}-1) \\
&\quad + 3N_1 N_2 - (q_2+3)(N_1+1) + 2N_2
\end{aligned} \tag{7.62}$$

and the total number of multiplications is

$$\begin{aligned}
M_{cc}^{I-II}(N_1+1, N_2) &= M_{cc}^{I}(\frac{N_1}{2}+1, \frac{N_2}{2}+1) + M_{cc}^{II-I}(\frac{N_1}{2}, \frac{N_2}{2}+1) \\
&\quad + M_{cs}^{I}(\frac{N_1}{2}+1, \frac{N_2}{2}-1) + M_{cs}^{II-I}(\frac{N_1}{2}, \frac{N_2}{2}-1) \\
&\quad + (N_1+1)(2N_2-3q_2).
\end{aligned} \tag{7.63}$$

By exchanging n_1 and n_2, k_1 and k_2, and N_1 and N_2, the same procedures can be used for the computation of the type-II-I cosine-cosine sequence.

7.2.6   Algorithm for type-I-II sine-cosine array

The type-I-II $(N_1-1) \times N_2$ sine-sine array of input $u(n_1, n_2)$ is

$$X_{sc}^{I-II}(k_1, k_2) = \sum_{n_1=1}^{N_1-1} \sum_{n_2=0}^{N_2-1} u(n_1, n_2) \sin \frac{\pi n_1 k_1}{N_1} \cos \frac{\pi(2n_2+1)k_2}{2N_2} \tag{7.64}$$

$$1 \leq k_1 < N_1, \quad 0 \leq k_2 < N_2$$

which can be decomposed into

$$X_{sc}^{I-II}(k_1, k_2) = \cos(\alpha_2 k_2)[B_{sc}^I(k_1, k_2) + B_{sc}^{II-I}(k_1, k_2)]$$
$$+ \sin(\alpha_2 k_2)[B_{ss}^I(k_1, k_2) + B_{ss}^{II-I}(k_1, k_2)] \tag{7.65}$$
$$1 \leq k_1 \leq \frac{N_1}{2}, \quad 0 \leq k_2 \leq \frac{N_2}{2}$$

$$X_{sc}^{I-II}(N_1 - k_1, k_2) = -\cos(\alpha_2 k_2)[B_{scc}^I(k_1, k_2) - B_{sc}^{II-I}(k_1, k_2)]$$
$$- \sin(\alpha_2 k_2)[B_{ss}^I(k_1, k_2) - B_{ss}^{II-I}(k_1, k_2)] \tag{7.66}$$
$$1 \leq k_1 < \frac{N_1}{2}, \quad 0 \leq k_2 \leq \frac{N_2}{2}$$

$$X_{sc}^{I-II}(k_1, N_2 - k_2) = (-1)^{\frac{q_2-1}{2}} \left\{ \sin(\alpha_2 k_2)[B_{sc}^I(k_1, k_2) + B_{sc}^{II-I}(k_1, k_2)] \right.$$
$$\left. - \cos(\alpha_2 k_2)[B_{ss}^I(k_1, k_2) + B_{ss}^{II-I}(k_1, k_2)] \right\} \tag{7.67}$$
$$1 \leq k_1 \leq \frac{N_1}{2}, \quad 1 \leq k_2 < \frac{N_2}{2}$$

$$X_{sc}^{I-II}(N_1 - k_1, N_2 - k_2) = (-1)^{\frac{q_2+1}{2}} \left\{ \sin(\alpha_2 k_2)[B_{sc}^I(k_1, k_2) - B_{sc}^{II-I}(k_1, k_2)] \right.$$
$$\left. - \cos(\alpha_2 k_2)[B_{ss}^I(k_1, k_2) - B_{ss}^{II-I}(k_1, k_2)] \right\} \tag{7.68}$$
$$1 \leq k_1 < \frac{N_1}{2}, \quad 1 \leq k_2 < \frac{N_2}{2},$$

where for (7.64) – (7.68)

$$B_{sc}^I(k_1, k_2) = \sum_{n_1=1}^{N_1/2-1} \sum_{n_2=0}^{N_2/2} \left[u(2n_1, 2n_2 - \frac{q_2+1}{2}) + u(2n_1, 2n_2 + \frac{q_2-1}{2}) \right]$$
$$\cdot \sin\frac{\pi n_1 k_1}{N_1/2} \cos\frac{\pi n_2 k_2}{N_2/2}, \qquad 1 < k_1 \leq \frac{N_1}{2}, \; 0 \leq k_2 \leq \frac{N_2}{2} \tag{7.69}$$

$$B_{sc}^{II-I}(k_1, k_2) = \sum_{n_1=0}^{N_1/2-1} \sum_{n_2=0}^{N_2/2} \left[u(2n_1+1, 2n_2 - \frac{q_2+1}{2}) + u(2n_1+1, 2n_2 + \frac{q_2-1}{2}) \right]$$
$$\cdot \sin\frac{\pi(2n_1+1)k_1}{2(N_1/2)} \cos\frac{\pi n_2 k_2}{N_2/2}, \qquad 1 \leq k_1 \leq \frac{N_1}{2}, \; 0 \leq k_2 \leq \frac{N_2}{2} \tag{7.70}$$

$$B_{ss}^I(k_1, k_2) = \sum_{n_1=1}^{N_1/2-1} \sum_{n_2=1}^{N_2/2-1} \left[u(2n_1, 2n_2 - \frac{q_2+1}{2}) - u(2n_1, 2n_2 + \frac{q_2-1}{2}) \right]$$
$$\cdot \sin\frac{\pi n_1 k_1}{N_1/2} \sin\frac{\pi n_2 k_2}{N_2/2}, \qquad 1 \leq k_1 < \frac{N_1}{2}, \; 1 \leq k_2 < \frac{N_2}{2} \tag{7.71}$$

$$B_{ss}^{II-I}(k_1, k_2) = \sum_{n_1=0}^{N_1/2-1} \sum_{n_2=1}^{N_2/2-1} \left[u(2n_1+1, 2n_2 - \frac{q_2+1}{2}) - u(2n_1+1, 2n_2 + \frac{q_2-1}{2}) \right]$$
$$\cdot \sin\frac{\pi(2n_1+1)k_1}{2(N_1/2)} \sin\frac{\pi n_2 k_2}{N_2/2}, \qquad 1 \leq k_1 \leq \frac{N_1}{2}, \; 1 \leq k_2 < \frac{N_2}{2}. \tag{7.72}$$

The mapping process defined in (7.59) – (7.61) is needed in (7.69) – (7.72). The number of multiplications for the twiddle factors in (7.65) – (7.68) is $(N_1/2)[N_2 - (3q_2-1)/2]$, $(N_1/2 -$

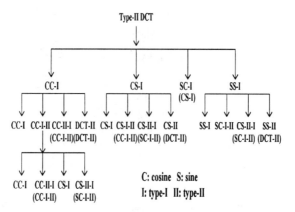

Figure 7.3   Computation flow graph of the type-II 2D DCT.

$1)[N_2 - (3q_2 - 1)/2], (N_1/2)[N_2 - (3q_2 + 1)/2]$, and $(N_1/2 - 1)[N_2 - (3q_2 + 1)/2]$, respectively. The number of additions needed by (7.65) and (7.67) is $N_1 N_2 - (q_2 + 1)N_1/2 - N_2$. For (7.66) and (7.68), $(N_1/2 - 1)(2N_2 - q_2 - 1)$ additions are needed. Furthermore, $N_1 N_2 - 2N_1 - N_2 + 2$ additions are needed to combine the input data inside the square brackets. Therefore, the number of additions for the type-I-II sine-cosine sequence is

$$
\begin{aligned}
A_{sc}^{I-II}(N_1 - 1, N_2 + 1) &= A_{sc}^I(\frac{N_1}{2} - 1, \frac{N_2}{2} + 1) + A_{sc}^{II-I}(\frac{N_1}{2}, \frac{N_2}{2} + 1) \\
&+ A_{ss}^I(\frac{N_1}{2} - 1, \frac{N_2}{2} - 1) + A_{ss}^{II-I}(\frac{N_1}{2}, \frac{N_2}{2} - 1) \\
&+ 3N_1 N_2 - (q_2 + 3)(N_1 - 1) - 4N_2
\end{aligned}
\tag{7.73}
$$

and the number of multiplications is

$$
\begin{aligned}
M_{sc}^{I-II}(N_1 - 1, N_2 + 1) &= M_{sc}^I(\frac{N_1}{2} - 1, \frac{N_2}{2} + 1) + M_{sc}^{II-I}(\frac{N_1}{2}, \frac{N_2}{2} + 1) \\
&+ M_{ss}^I(\frac{N_1}{2} - 1, \frac{N_2}{2} - 1) + A_{ss}^{II-I}(\frac{N_1}{2}, \frac{N_2}{2} - 1) \\
&+ (N_1 - 1)(2N_2 - 3q_2).
\end{aligned}
\tag{7.74}
$$

7.2.7   Implementation issues

The 2D type-II or -III DCT has been decomposed into various types of arrays. The computational complexity for each decomposed array is also given. Figure 7.3 illustrates the decomposition process and various arrays needed for the computation of the 2D type-II DCT. In general, it is not desirable to implement too many subroutines for all these arrays. The number of different arrays to be implemented can be reduced by using two properties. The symmetric property allows us to convert one array into another by swapping the input indices n_1 and n_2, the outputs indices k_1 and k_2, the dimension sizes N_1 and N_2 and the trigonometric functions, respectively. For example, X_{sc}^I and X_{sc}^{II-I} can be converted into X_{cs}^I and X_{cc}^{II-I}, respectively. When two arrays are related by the symmetric properties, the subroutine written for the computation of one array can be used for the other. For example, the same subroutine can be used for the computation of X_{sc}^I and X_{cs}^I. However, the facility for swapping the input/output indices, the dimension sizes and the trigonometric functions has to be implemented in the subroutine. The other property is to change the cosine into the sine function or vice versa. By substituting k_1 with $N_1 - k_1$, for example, the type-II sine-cosine array X_{sc}^{II} and the type-II-I sine-cosine array X_{sc}^{II-I} are converted into the type-II DCT and the type-II-I cosine-cosine array X_{cc}^{II-I}, respectively. It is also possible to apply both properties together to convert one array into another. Therefore, the entire computation of the type-II DCT can be computed by implementing six subroutines. Figure 7.3 shows that the arrays inside the parentheses are used to compute those above them.

This algorithm has a regular computational structure and can be used for various transform sizes. It is important to note that this algorithm differs from other recursive algorithms [5, 12, 14] in several aspects. It does not need a global indexing process, which can be illustrated by comparing the indexing process required by Chan's and the presented algorithms. For simplicity, let us assume $N = N_1 = N_2$ and $q_1 = q_2 = 1$. In Chan's algorithm [5], the input data, $x(2n_1, 2n_2), x(N - 2n_1 - 1, 2n_2), x(2n_1, N - 2n_2 - 1), x(N - 2n_1 - 1, N - 2n_2 - 1)$, are needed in the computation. Therefore, the entire input matrix has to be available in a memory space of N^2 data stores before computation starts. For the presented algorithm, $x(2n_1 - 1, 2n_2 - 1), x(2n_1 - 1, 2n_2), x(2n_1, 2n_2 - 1), x(2n_1, 2n_2)$ are required, as shown in (7.74). The computation can start once the first two rows of the input matrix are available.

Decimated outputs, $X(2k_1, 2k_2), X(2k_1 + 1, 2k_2), X(2k_1, 2k_2 + 1)$ and $X(2k_1 + 1, 2k_2 + 1)$, are obtained by Chan's algorithm after each stage of the computation. A bit-reversal process has to be used to achieve a natural index order. The presented algorithm directly achieves the transform outputs $X(k_1, k_2), X(k_1, N - k_2), X(N - k_1, k_2)$ and $X(N - k_1, N - k_2)$. The natural order of the transform outputs can be easily achieved by an up/down counting parameter used for loop control in software implementation. Therefore, the simple indexing process for the inputs/outputs can potentially save computation time, memory space and implementation complexity. Such desirable structural properties are particularly important if the algorithm is implemented by pipelined hardware for real time processing. Because the duration of data collection before computation is substantially reduced, the overall processing latency can be accordingly minimized.

The existence of parameters q_1 and q_2 in the angles of the twiddle factors creates more trivial ones which do not need multiplications and additions. Therefore, the presented algorithm can achieve more savings in the number of multiplications and additions from the trivial twiddle factors than other reported algorithms. In general, the number of trivial twiddle factors is proportional to the values of q_1 and q_2, which can be seen, for example, from T_a and T_m in (7.62), (7.63), (7.73) and (7.74).

7.2.8   Discussion

In general, this algorithm uses more multiplications and fewer additions than those needed by the algorithms in [5, 8]. In the past, this was considered to be a disadvantage because the computation time for a multiplication is much longer than that for an addition. However, it is not true at present since current technologies allow many processors to perform multiplication and addition at the same speed. Therefore, it is important to compare the computational complexity in terms of the total number of arithmetic operations (additions plus multiplications). Figure 7.4 compares the computational complexity needed by this algorithm and those in [5, 8] in terms of the total number of arithmetic operations per transform point. It shows that for $q_1 = q_2 = 1$, the presented one achieves savings on the number of arithmetic operations. For $q_1 > 1$ and $q_2 > 1$, a reduction of the total number of arithmetic operations is also made.

Compared to the algorithms reported in [9, 12], this algorithm needs more arithmetic operations. However, the algorithms in [9, 12] have an irregular computational structure, which substantially increases the computation time and the implementation cost when the transform sizes are larger than 32. For example, Cho reported a fast algorithm in [10] to improve the computational structure of the previously reported one [9] at the expense of using many extra additions. It was shown that the improved algorithm for a 16×16 DCT needed 510 extra additions compared to that in [9]. Therefore, the improved algorithm used 3552 (3042 + 510) operations for the 16×16 DCT, which is 360 operations more than that used by the presented algorithm.

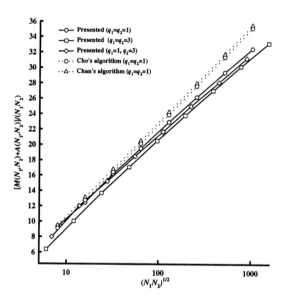

Figure 7.4 Number of arithmetic operations needed by various algorithms for 2D type-II DCT.

The presented algorithm is simple in concept and has the desirable properties for the input/output indexing process, which effectively minimizes the overhead associated with the indexing process. However, a mapping process is needed for the type-II DCT, type-I-II cosine-cosine and the type-I-II sine-cosine arrays. Because the presented algorithm does not need the entire input matrix before computation starts, the amount of memory and processing latency can also be minimized.

7.3   PRIME FACTOR ALGORITHM FOR MD DCT

It is shown in Section 6.3 that the presented prime factor algorithm (PFA) has made some desirable improvements on both computational and structural issues for the 1D DCT. It is natural to extend the concept of this algorithm to MD DCT computation [3]. Because the MD DCT has a separable kernel, the extension of the PFAs becomes straightforward. For easy understanding and description, we first consider the computation of 2D type-II and type-III DCTs. Extensions of related concepts to DCTs of higher dimensions are then outlined.

7.3.1 Algorithm for type-II DCT

The type-II 2D DCT of the input array $x(n_1, n_2)$, $0 \leq n_1 < N_1$, $0 \leq n_2 < N_2$, is defined by (7.2). We assume that the dimensional sizes are $N_1 = p_1 \cdot q_1$, and $N_2 = p_2 \cdot q_2$, where p_1 and q_1, and p_2 and q_2, are mutually prime, respectively. We further define for $i = 1, 2$

$$\mu_i = \begin{cases} 1, & \text{if } p_i k_i + m_i q_i < N_i \\ -1, & \text{if } p_i k_i + m_i q_i \geq N_i \end{cases}. \tag{7.75}$$

By using $\cos(\alpha + \beta) = \cos(\alpha)\cos(\beta) - \sin(\alpha)\sin(\beta)$, (7.2) can be decomposed into several terms, as expressed by

$$X(|\,p_1\mu_1 k_1 + q_1 m_1\,|, |\,p_2\mu_2 k_2 + q_2 m_2\,|)$$
$$= \sum_{n_2=0}^{N_2-1} \left[\sum_{n_1=0}^{N_1-1} x(n_1, n_2) \cos \frac{\pi(2n_1 + 1)(p_1\mu_1 k_1 + q_1 m_1)}{2N_1} \right]$$
$$\cdot \cos \frac{\pi(2n_2 + 1)(p_2\mu_2 k_2 + q_2 m_2)}{2N_2}$$
$$= \begin{cases} Y(k_1, m_1, k_2, m_2) - \mu_1 Y(p_1 - k_1, q_1 - m_1, k_2, m_2), & k_1 m_1 \neq 0 \\ Y(k_1, m_1, k_2, m_2), & k_1 m_1 = 0 \end{cases}, \tag{7.76}$$

where

$$Y(k_1, m_1, k_2, m_2) = \begin{cases} A(k_1, m_1, k_2, m_2) - \mu_2 A(k_1, m_1, p_2 - k_2, q_2 - m_2), & k_2 m_2 \neq 0 \\ A(k_1, m_1, k_2, m_2), & k_2 m_2 = 0 \end{cases} \tag{7.77}$$

and

$$A(k_1, m_1, k_2, m_2) = \sum_{n_2=0}^{N_2-1} \sum_{n_1=0}^{N_1-1} x(n_1, n_2) \cos \frac{\pi(2n_1 + 1)m_1}{2p_1} \cos \frac{\pi(2n_1 + 1)k_1}{2q_1}$$
$$\cdot \cos \frac{\pi(2n_2 + 1)m_2}{2p_2} \cos \frac{\pi(2n_2 + 1)k_2}{2q_2}, \tag{7.78}$$

where $0 \leq k_1 < q_1, 0 \leq k_2 < q_2, 0 \leq m_1 < p_1$ and $0 \leq m_2 < p_2$. According to (7.76) and (7.77), the 2D type-II DCT can be computed if $A(k_1, m_1, k_2, m_2)$ in (7.78) is available.

Now let us consider the computation of $A(k_1, m_1, k_2, m_2)$ which is expressed by a double summation. For convenience of derivation, the inner summation of (7.78) associated with n_1, k_1 and m_1 can be rewritten as

$$\sum_{n_1=0}^{N_1-1} x(n_1, n_2) \cos \frac{\pi(2n_1 + 1)m_1}{2p_1} \cos \frac{\pi(2n_1 + 1)k_1}{2q_1}$$
$$= \sum_{n_1=0}^{q_1-1} \left\{ \sum_{j=0}^{(p_1-1)/2} x(2jq_1 + n_1, n_2) \cos \frac{\pi(4jq_1 + 2n_1 + 1)m_1}{2p_1} \right.$$
$$\left. + \sum_{j=1}^{(p_1-1)/2} x(2jq_1 - n_1 - 1, n_2) \cos \frac{\pi(4jq_1 - 2n_1 - 1)m_1}{2p_1} \right\} \cos \frac{\pi(2n_1 + 1)k_1}{2q_1} \tag{7.79}$$

in which the outer summation is a length q_1 1D type-II DCT. The two terms in the braces can be expressed as

$$\sum_{j=-(p_1-1)/2}^{(p_1-1)/2} y(j, n_1, n_2) \cos \frac{\pi(4jq_1 + 2n_1 + 1)m_1}{2p_1}, \tag{7.80}$$

where

$$y(j, n_1, n_2) = \begin{cases} x(2jq_1 + n_1, n_2), & 0 \le j \le \frac{p_1-1}{2} \\ x(-2jq_1 - n_1 - 1, n_2), & \frac{p_1-1}{2} \le j \le 0 \end{cases}. \tag{7.81}$$

In fact, (7.80) is a 1D type-II DCT if index mapping is performed. For any given combination of p_1, q_1 and n_1, the index mapping is to find each value of $l_1, 0 \le l_1 \le p_1 - 1$, for each j, $-(p_1 - 1)/2 \le j \le (p_1 - 1)/2$, so that

$$\cos \frac{\pi(2l_1 + 1)m_1}{2p_1} \equiv \cos \frac{\pi(4jq_1 + 2n_1 + 1)m_1}{2p_1}$$

is valid. The mapping can be expressed by $l_1 \Leftrightarrow j(l_1)$, where $j(l_1)$ represents the value of each j that is related to the value of each l_1 by the above equality. By substituting $4jq_1 + 2n_1 + 1$ with $2l_1 + 1$, (7.80) can be converted into a 1D type-II DCT

$$\sum_{l_1=0}^{p_1-1} y'(j(l_1), n_1, n_2) \cos \frac{\pi(2l_1 + 1)m_1}{2p_1}, \tag{7.82}$$

where $y'(j(l_1), n_1, n_2)$ is a 3D array converted from the original input array $x(n_l, n_2)$. As shown in the previous chapter, the conversion can be performed according to a mapping relation expressed by

$$2l_1 + 1 = \begin{cases} ||4jq_1 + 2n_1 + 1| - 4p_1|, & |4jq_1 + 2n_1 + 1| > 2p_1 \\ |4jq_1 + 2n_1 + 1|, & |4jq_1 + 2n_1 + 1| < 2p_1 \end{cases}, \tag{7.83}$$

where $0 \le n_1 \le q_1 - 1$, $-(p_1 - 1)/2 \le j \le (p_1 - 1)/2$ and $0 \le l_1 \le p_1 - 1$. Thus, (7.79) or the inner summation of (7.78) can be expressed by a 2D type-II DCT:

$$\sum_{n_1=0}^{q_1-1} \sum_{l_1=0}^{p_1-1} y'(j(l_1), n_1, n_2) \cos \frac{\pi(2l_1 + 1)m_1}{2p_1} \cos \frac{\pi(2n_1 + 1)k_1}{2q_1}. \tag{7.84}$$

Similarly, the outer summation of (7.78) can also be converted into a 2D type-II DCT so that $A(k_1, m_1, k_2, m_2)$ becomes a 4D type-II DCT defined by

$$A(k_l, m_l, k_2, m_2) = \sum_{n_1=0}^{q_1-1} \sum_{l_1=0}^{p_1-1} \sum_{n_2=0}^{q_2-1} \sum_{l_2=0}^{p_2-1} y'(j(l_1), n_1, j(l_2), n_2) \cos \frac{\pi(2l_1 + 1)m_1}{2p_1}$$

$$\cdot \cos \frac{\pi(2n_1 + 1)k_1}{2q_1} \cos \frac{\pi(2l_2 + 1)m_1}{2p_2} \cos \frac{\pi(2n_2 + 1)k_2}{2q_2}, \tag{7.85}$$

where $0 \le m_1 < p_1, 0 \le k_1 < q_1, 0 \le m_2 < p_2$, and $0 \le k_2 < q_2$. Once $A(k_1, m_1, k_2, m_2)$ is available, the final transform outputs can be calculated according to (7.76) and (7.77).

The number of multiplications required by the proposed decomposition is

$$M(N_1, N_2) = M(p_1, q_1, p_2, q_2) \tag{7.86}$$

and the number of additions is

$$A(N_1, N_2) = A(p_1, q_1, p_2, q_2) + N_1 N_2 \left[(1 - \frac{1}{p_1})(1 - \frac{1}{q_1}) + (1 - \frac{1}{p_2})(1 - \frac{1}{q_2}) \right], \tag{7.87}$$

where $M(p_1, q_1, p_2, q_2)$ and $A(p_1, q_1, p_2, q_2)$ are the numbers of multiplications and additions needed by the 4D DCT, respectively. Equation (7.86) shows that no twiddle factors are needed in the computation and the second term in (7.87) is the number of additions needed by the post-processing stage defined by (7.76) and (7.77). The 4D DCT can be calculated in many ways. For example, it can be calculated by the row-column approach using 1D or 2D DCTs. If a 2D DCT is used in the row-column approach, the required numbers of multiplications and additions are

$$M(N_1, N_2) = N_1 M(p_2, q_2) + N_2 M(p_1, q_1) \tag{7.88}$$

$$A(N_1, N_2) = N_1 A(p_2, q_2) + N_2 A(p_1, q_1) + N_2 (p_1 - 1)(q_1 - 1)$$
$$+ N_1 (p_2 - 1)(q_2 - 1). \tag{7.89}$$

7.3.2 Algorithm for type-III DCT

The type-III DCT of 2D array $X(k_1, k_2), 0 \leq k_1 < N_1, 0 \leq k_2 < N_2$, is defined by

$$x(n_1, n_2) = \sum_{k_2=0}^{N_2-1} \sum_{k_1=0}^{N_1-1} X(k_1, k_2) \cos \frac{\pi(2n_1+1)k_1}{2N_1} \cos \frac{\pi(2n_2+1)k_2}{2N_2}$$
$$0 \leq n_1 < N_1, 0 \leq n_2 < N_2, \tag{7.90}$$

If p_i and q_i, where $N_i = p_i \cdot q_i, i = 1, 2$, are mutually prime, (7.90) can be rewritten as

$$x(n_1, n_2) = \sum_{k_2=0}^{q_2-1} \sum_{k_1=0}^{q_1-1} \sum_{m_2=0}^{p_2-1} \sum_{m_1=0}^{p_1-1} X(|\mu_1 p_1 k_1 + q_1 m_1|, |\mu_2 p_2 k_2 + q_2 m_2|)$$
$$\cdot \left[\cos \frac{\pi(2n_1+1)k_1}{2q_1} \cos \frac{\pi(2n_1+1)m_1}{2p_1} \cos \frac{\pi(2n_2+1)k_2}{2q_2} \cos \frac{\pi(2n_2+1)m_2}{2p_2} \right.$$
$$- \mu_2 \cos \frac{\pi(2n_1+1)k_1}{2q_1} \cos \frac{\pi(2n_1+1)m_1}{2p_1} \sin \frac{\pi(2n_2+1)k_2}{2q_2} \sin \frac{\pi(2n_2+1)m_2}{2p_2}$$
$$- \mu_1 \sin \frac{\pi(2n_1+1)k_1}{2q_1} \sin \frac{\pi(2n_1+1)m_1}{2p_1} \cos \frac{\pi(2n_2+1)k_2}{2q_2} \cos \frac{\pi(2n_2+1)m_2}{2p_2}$$
$$+ \mu_1 \mu_2 \sin \frac{\pi(2n_1+1)k_1}{2q_1} \sin \frac{\pi(2n_1+1)m_1}{2p_1}$$
$$\left. \cdot \sin \frac{\pi(2n_2+1)k_2}{2q_2} \sin \frac{\pi(2n_2+1)m_2}{2p_2} \right], \tag{7.91}$$

where $\mu_i, i = 1, 2$, is defined in (7.75). In (7.91), the sine terms can be changed into cosine terms by substituting k_i with $q_i - k_i$ and m_i with $p_i - m_i$. Therefore, (7.91) can be rewritten as

$$x(n_1, n_2) = \sum_{k_2=0}^{q_2-1} \sum_{k_1=0}^{q_1-1} \sum_{m_1=0}^{p_1-1} \sum_{m_2=0}^{p_2-1} Y(k_1, m_1, k_2, m_2) \cos \frac{\pi(2n_1+1)m_1}{2p_1}$$
$$\cdot \cos \frac{\pi(2n_2+1)m_2}{2p_2} \cos \frac{\pi(2n_1+1)k_1}{2q_1} \cos \frac{\pi(2n_2+1)k_2}{2q_2}, \tag{7.92}$$

where

$$Y(k_1, m_1, k_2, m_2)$$
$$= \begin{cases} A(k_1, m_1, k_2, m_2) - \mu_1 A(q_1 - k_1, p_1 - m_1, k_2, m_2), & k_2 m_2 \neq 0 \\ A(k_1, m_1, k_2, m_2), & k_2 m_2 = 0 \end{cases} \tag{7.93}$$

and

$$A(k_1, m_1, k_2, m_2) =$$
$$\begin{cases} X(|\mu_1 p_1 k_1 + q_1 m_1|, |\mu_2 p_2 k_2 + q_2 m_2|), & k_2 m_2 = 0 \\ -\mu_2 X(|\mu_1 p_1 k_1 + q_1 m_1|, |\mu_2 p_2 (q_2 - k_2) + q_2 (p_2 - m_2)|), & k_2 m_2 \neq 0 \end{cases} \cdot \tag{7.94}$$

From (7.92), we have

$$x(2j_1 q_1 + n_1, 2j_2 q_2 + n_2)$$
$$= \sum_{k_2=0}^{p_2-1} \sum_{k_1=0}^{p_1-1} \sum_{m_1=0}^{q_1-1} \sum_{m_2=0}^{q_2-1} Y(k_1, m_1, k_2, m_2) \cos \frac{\pi(4j_1 q_1 + 2n_1 + 1)m_1}{2p_1}$$
$$\cdot \cos \frac{\pi(4j_2 q_2 + 2n_2 + 1)m_2}{2p_2} \cos \frac{\pi(2n_1+1)k_1}{2q_1} \cos \frac{\pi(2n_2+1)k_2}{2q_2} \tag{7.95}$$
$$0 \le j_1 \le \frac{p_1-1}{2}, \quad 0 \le j_2 \le \frac{p_2-1}{2}$$

and

$$x(2j_1 q_1 - n_1 - 1, 2j_2 q_2 - n_2 - 1)$$
$$= \sum_{k_2=0}^{q_2-1} \sum_{k_1=0}^{q_1-1} \sum_{m_2=0}^{p_2-1} \sum_{m_1=0}^{p_1-1} Y(k_1, m_1, k_2, m_2) \cos \frac{\pi(4j_1 q_1 - 2n_1 - 1)m_1}{2p_1}$$
$$\cdot \cos \frac{\pi(4j_2 q_2 - 2n_2 - 1)m_2}{2p_2} \cos \frac{\pi(2n_1+1)k_1}{2q_1} \cos \frac{\pi(2n_2+1)k_2}{2q_2} \tag{7.96}$$
$$1 \le j_1 \le \frac{p_1-1}{2}, \quad 1 \le j_2 \le \frac{p_2-1}{2}.$$

If the innermost summations related to m_1 in (7.95) and (7.96) are combined, we have

$$\sum_{m_1=0}^{p_1-1} Y(k_1, m_1, k_2, m_2) \cos \frac{\pi(4j_1 q_1 + 2n_1 + 1)m_1}{2p_1}$$
$$-\frac{p_1-1}{2} \le j_1 \le \frac{p_1-1}{2}, \tag{7.97}$$

To convert (7.97) into a length p_1 type-III DCT, it is necessary to find p_1 values of l_1 for each n_1 to satisfy (7.82). Therefore, (7.97) becomes

$$\sum_{m_1=0}^{p_1-1} Y(k_1, m_1, k_2, m_2) \cos \frac{\pi(2l_1+1)m_1}{2p_1}, \quad l_1 = 0, 1, \ldots, p_1 - 1 \tag{7.98}$$

which is a 1D length p_1 DCT. The summation related to m_2 in (7.95) and (7.96) can be similarly converted into a 1D length p_2 DCT. Based on the above discussion, (7.92) can be formally expressed by a 4D type-III DCT,

$$
\begin{aligned}
x'[n_1, j(l_1), n_2, j(l_2)] = \sum_{k_2=0}^{p_2-1} \sum_{k_1=0}^{p_1-1} \sum_{m_2=0}^{q_2-1} \sum_{m_1=0}^{q_1-1} & Y(k_1, m_1, k_2, m_2) \cos \frac{\pi(2l_1+1)m_1}{2q_1} \\
& \cdot \cos \frac{\pi(2l_2+1)m_2}{2q_2} \cos \frac{\pi(2n_1+1)k_1}{2p_1} \cos \frac{\pi(2n_2+1)k_2}{2p_2},
\end{aligned}
$$

(7.99)

where $0 \le l_1 < q_1$, $0 \le l_2 < q_2$, $0 \le n_1 < p_1$ and $0 \le n_2 < p_2$. The decomposition procedure used here is basically the same as that used for the type-II 2D DCT. In comparison, a post-processing stage is used for the type-II 2D DCT and a pre-processing stage, as shown in (7.93) and (7.94), is needed for the type-III 2D DCT. It can be easily shown that the computational complexity for a type-III DCT is the same as that for a type-II DCT.

7.3.3   Input/output index mapping

We have seen that the 2D DCT is converted into a 4D DCT with a pre-processing stage for the type-III DCT (or post-processing for the type-II DCT). Therefore, the original 2D input array has to be converted into a 4D array, which is known as index mapping. An important issue of the algorithm is the computational complexity of the mapping process so that the associated overhead does not offset the savings achieved by eliminating twiddle factors. It is observed that the two input indices, n_1 and n_2, for example, are completely independent due to the fact that the kernel of the MD DCT is separable. It means that the mapping process for one dimension is completely independent of other dimensions. For example, the index mapping process for the 2D DCT can be achieved by performing twice the mapping process for the 1D DCT. If the sizes of all dimensions are the same, i.e., $N = N_1 = N_2$, the mapping information achieved for a particular dimension can be used for the mapping of the other dimensions. Therefore, the complexity of the mapping process is significantly reduced to the mapping process of N integers only. The mapping technique used by the PFA for the 1D DCT can be used for a MD DCT without any change.

The transform outputs are also required to be arranged into a natural order. The expression $|p\mu k + qm|$ is used to relate each transform output of the type-II DCT to the indices $k = 0, 1, \dots, q-1$ and $m = 0, 1, \dots, p-1$. This process can be easily implemented with operations of comparison, addition and subtraction. In Appendix B, the output index mapping and formation of the final transformed outputs from $A(k_l, m_1, k_2, m_2)$ are combined. Similar to the input mapping process, the output mapping process for the 2D type-II DCT can be performed by the mapping process of the 1D DCT. When $N_1 = N_2 = N$, for example, Appendix B uses the index (stored in array index1[ ][ ]) for the indices of both dimensions, i.e., $kk1$ and $kk2$.

7.3.4   Discussions

Compared to other reported PFAs of the 2D DCT [11, 17, 19], the presented one is similar to that in [17], but differs in the output index by introducing a factor of μ. One main achievement is that a simple and general approach is derived to minimize the computational overhead associated with the index mapping process. Due to the independence among different dimensional indices, our input/output mapping complexity for an rD DCT is in the order of $O(N)$ rather than $O(N^r)$ needed by other algorithms ([19], for example). Furthermore, the mapping procedure is simple and does not require any modulo operation. It is also general and can be used for any pair of factors that are mutually prime. In comparison, the presented mapping method removes the limitation of the tabulation method [16] which is difficult to use for large values of prime factors.

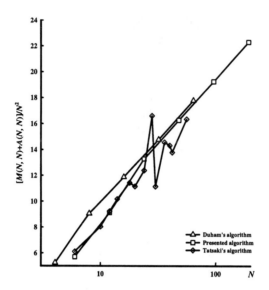

Figure 7.5   Comparison of number of arithmetic operations.

The 4D DCT can be implemented in various ways. If $N = N_1 = N_2 = p \cdot q$, for example, the row-column method can be used to perform p^2 $q \times q$ and q^2 $p \times p$ 2D DCTs. For example, with $M(3,3) = 5$ and $A(3,3) = 24$, Figure 7.5 shows the required number of arithmetic operations (multiplications plus additions) needed by the algorithm for a $3 \cdot 2^m \times 3 \cdot 2^m$ 2D type-II DCT. Based on the number of operations needed by three algorithms, as listed in Table 7.1, Figure 7.5 shows that the presented algorithm requires a smaller number of arithmetic operations than the algorithm given in [12], and a larger number of arithmetic operations than the algorithm in [21]. In general, an rD DCT can be converted into a $2r$D DCT with a post-processing stage of 2^r terms. If the dimension sizes, $N_1, N_2, \ldots, N_r$, are different, the input mapping process can be performed by repeating the mapping procedure for r times. When all the dimension sizes, $N_1, N_2, \ldots, N_r$, are same, it is sufficient that the subroutine in Appendix B is used only once because the index mapping is the same for all dimensions. In general, the number of multiplications for an rD DCT is

$$M(N_1, \ldots, N_r) = M(p_1, q_1, \ldots, p_r, q_r) \qquad (7.100)$$

and the number of additions is

$$A(N_1, \ldots, N_r) = A(p_1, q_1, \ldots, p_r, q_r) + N_1 N_2 \cdots N_r \sum_{i=1}^{r}(1 - \frac{1}{p_i})(1 - \frac{1}{q_i}), \qquad (7.101)$$

where $M(p_1, q_1, \ldots, p_r, q_r,)$ and $A(p_1, q_1, \ldots, p_r, q_r)$ are the numbers of multiplications and additions needed by the $2r$D DCT, respectively, and the second term is the number of additions used to combine the outputs of the $2r$D DCT in the post-processing stage.

Table 7.1 Computational complexity required by three algorithms.

Algorithm in [12]				Algorithm in [21]			
$N \times N$	Mul	Add	Total	$N \times N$	Mul	Add	Total
4×4	16	68	84	6×6	34	186	220
8×8	96	484	580	10×10	146	658	804
16×16	512	2531	3043	12×12	208	1122	1330
32×32	2560	12578	15138	14×14	354	1642	1996
64×64	12288	60578	72866	18×18	546	3162	3708
Presented algorithm				24×24	1192	5946	7138
6×6	38	168	206	28×28	1808	11222	13030
12×12	224	1092	1316	30×30	1714	8322	10036
24×24	1184	6468	7652	36×36	2832	16050	18882
48×48	5888	31611	37499	40×40	4136	18778	22914
96×96	28160	149296	177458	56×56	9192	42106	51298
192×192	131072	691122	822194	60×60	8656	42738	51394

7.4 PT-BASED RADIX-2 ALGORITHM FOR MD TYPE-II DCT

In the rest of this chapter, we consider fast algorithms based on the polynomial transform (PT)[23, 24, 25, 26]. This section and Section 7.5 are devoted to the type-II and -III MD DCTs whose dimensional sizes are 2^t. Section 7.6 and 7.7 describe the fast algorithms that support MD DCTs with dimensional sizes of q^t, where q is an odd prime. For easy understanding and simplicity of presentation, we first consider the 2D DCT to illustrate some basic concepts which are needed for the understanding of the PT-based algorithm for MD DCTs. Then we will show that the symmetric properties in the PT can be used to reduce the computational complexity. Finally, the algorithm for the 2D type-II DCT is generalized and extended for the rD DCT, where $r \geq 3$. The computational complexity required by the algorithm is also considered. Some basic knowledge of the polynomial transform can be reviewed in Chapter 2.

7.4.1 PT-based radix-2 algorithm for 2D type-II DCT

The 2D type-II DCT of the array $x(n, m)$ ($n = 0, 1, \ldots, N - 1$, $m = 0, 1, \ldots, M - 1$) is defined by

$$X(k, l) = \sum_{n=0}^{N-1} \sum_{m=0}^{M-1} x(n, m) \cos \frac{\pi(2n + 1)k}{2N} \cos \frac{\pi(2m + 1)l}{2M}, \qquad (7.102)$$

where $k = 0, 1, \ldots, N - 1$, $l = 0, 1, \ldots, M - 1$ and the constant scaling factors in the DCT definition are ignored for simplicity. We assume that M and N are powers of 2 and $M \geq N$. We can write $N = 2^t$ and $M = 2^J N$, where $t > 0$ and $J \geq 0$, respectively. When $M < N$, M and N can be swapped to satisfy the above assumption. By reordering the input array

$x(n, m)$ into

$$
\begin{aligned}
y(n, m) &= x(2n, 2m) \\
y(N - 1 - n, m) &= x(2n + 1, 2m) \\
y(n, M - 1 - m) &= x(2n, 2m + 1) \\
y(N - 1 - n, M - 1 - m) &= x(2n + 1, 2m + 1),
\end{aligned} \tag{7.103}
$$

where $n = 0, 1, \ldots, N/2 - 1$, $m = 0, 1, \ldots, M/2 - 1$, (7.102) can be expressed by

$$
X(k, l) = \sum_{n=0}^{N-1} \sum_{m=0}^{M-1} y(n, m) \cos \frac{\pi(4n + 1)k}{2N} \cos \frac{\pi(4m + 1)l}{2M}. \tag{7.104}
$$

For simplicity, we use the notation $p(m)$ to represent $((4p + 1)m + p) \bmod N$ in our presentation.

Lemma 9 *Let* $A = \{(n, m) \mid 0 \le n \le N - 1, \ 0 \le m \le M - 1\}$ *and* $B = \{(p(m), m) \mid 0 \le p \le N - 1, \ 0 \le m \le M - 1\}$. *Then* $A = B$.

Proof. It is obvious that $B \subseteq A$ and the number of elements in A is the same as that in B. Therefore, it is sufficient to prove that the elements in B are different from each other. Let $(p(m), m)$ and $(p'(m'), m')$ be two elements in B. If they are equal, then

$$
p(m) = p'(m'), \ m = m'.
$$

From the definition of $p(m)$, we have

$$
(4p + 1)m + p \equiv (4p' + 1)m + p' \bmod N.
$$

Hence

$$
(4m + 1)(p - p') \equiv 0 \bmod N.
$$

Since $4m + 1$ and N are relatively prime with each other, we have $p \equiv p' \bmod N$. Therefore, $p = p'$, which concludes the lemma.

Using the properties of trigonometric identities and Lemma 9, (7.104) can be computed by

$$
X(k, l) = \frac{A(k, l) + B(k, l)}{2}, \tag{7.105}
$$

where

$$
A(k, l) = \sum_{p=0}^{N-1} \sum_{m=0}^{M-1} y(p(m), m) \cos \left(\frac{\pi(4p(m) + 1)k}{2N} + \frac{\pi(4m + 1)l}{2M} \right) \tag{7.106}
$$

$$
B(k, l) = \sum_{p=0}^{N-1} \sum_{m=0}^{M-1} y(p(m), m) \cos \left(\frac{\pi(4p(m) + 1)k}{2N} - \frac{\pi(4m + 1)l}{2M} \right), \tag{7.107}
$$

where for (7.106) and (7.107), $k = 0, 1, \ldots, N - 1$ and $l = 0, 1, \ldots, M - 1$. From the definition of $p(m)$, it can easily be proven that

$$
4p(m) + 1 \equiv (4m + 1)(4p + 1) \bmod 4N.
$$

By applying this equation in (7.106) and (7.107), we obtain

$$A(k,l) = \sum_{p=0}^{N-1} \sum_{m=0}^{M-1} y(p(m),m) \cos \left(\frac{\pi(4m+1)(4p+1)k}{2N} + \frac{\pi(4m+1)l}{2M} \right)$$

$$= \sum_{p=0}^{N-1} \sum_{m=0}^{M-1} y(p(m),m) \cos \frac{\pi(4m+1)(2^J(4p+1)k+l)}{2M}, \tag{7.108}$$

where $2^J = M/N$. Similarly

$$B(k,l) = \sum_{p=0}^{N-1} \sum_{m=0}^{M-1} y(p(m),m) \cos \frac{\pi(4m+1)(2^J(4p+1)k-l)}{2M}. \tag{7.109}$$

By using the definition

$$V_p(j) = \sum_{m=0}^{M-1} y(p(m),m) \cos \frac{\pi(4m+1)j}{2M}, \tag{7.110}$$

where $p = 0, 1, \ldots, N-1$ and $j = 0, 1, \ldots, M-1$, (7.106) and (7.107) can be expressed by

$$A(k,l) = \sum_{p=0}^{N-1} V_p[2^J(4p+1)k+l] \tag{7.111}$$

$$B(k,l) = \sum_{p=0}^{N-1} V_p[2^J(4p+1)k-l]. \tag{7.112}$$

If the input $y(p(m),m)$ is reordered in a similar way to that used in (7.103), we can express (7.110) into 1D type-II DCTs, as shown in (7.113) by defining an array $\hat{y}_p(2m) = y(p(m),m)$ and $\hat{y}_p(2m+1) = y(p(M-1-m), M-1-m)$, where $m = 0, 1, \ldots, M/2 - 1$:

$$V_p(j) = \sum_{m=0}^{M-1} \hat{y}_p(m) \cos \frac{\pi(2m+1)j}{2M}, \tag{7.113}$$

where $p = 0, 1, \ldots, N-1$ and $j = 0, 1, \ldots, M-1$. Both (7.111) and (7.112) are summations of $V_p(j)$, which have the properties

$$V_p(j+u2M) = (-1)^u V_p(j), \quad V_p(2M-j) = -V_p(j), \quad V_p(M) = 0.$$

Therefore, $A(k,l)$ and $B(k,l)$ can be computed from $V_p(j)$ $(j = 0, 1, \ldots, M-1)$. In general, a direct computation of (7.111) and (7.112) needs $2NM(N-1)$ additions. However, they can be equivalently computed by using the PT, which substantially reduces the computational complexity.

Based on the properties of $V_p(j)$, it can be proven that

$$A(k, 2M-l) = \sum_{p=0}^{N-1} V_p[2^J(4p+1)k+2M-l]$$

$$= -\sum_{p=0}^{N-1} V_p[2^J(4p+1)k-l] = -B(k,l) \tag{7.114}$$

and

$$A(k,0) = B(k,0), \quad A(0,l) = B(0,l). \tag{7.115}$$

Now let us generate a polynomial

$$B_k(z) = \sum_{l=0}^{M-1} B(k,l)z^l - \sum_{l=M}^{2M-1} A(k,2M-l)z^l. \tag{7.116}$$

By substituting (7.112) and (7.114) in (7.116), we have

$$B_k(z) = \sum_{l=0}^{2M-1} \sum_{p=0}^{N-1} V_p(2^J(4p+1)k - l)z^l$$

from which $A(k,l)$ and $B(k,l)$ can be derived. In order to use the PT, we select the module $z^{2M} + 1$ such that the index l in $V_p(l - 2^J(4p+1)k)z^l \bmod z^{2M} + 1$ has a period of $2M$. Then $B_k(z)$ is expressed as

$$
\begin{aligned}
B_k(z) &\equiv \sum_{p=0}^{N-1} \sum_{l=0}^{2M-1} V_p(2^J(4p+1)k - l)z^l \bmod z^{2M} + 1 \\
&\equiv \sum_{p=0}^{N-1} \sum_{l=0}^{2M-1} V_p(l - 2^J(4p+1)k)z^l \bmod z^{2M} + 1 \\
&\equiv \sum_{p=0}^{N-1} \sum_{l=0}^{2M-1} V_p(l)z^{l + 2^J(4p+1)k} \bmod z^{2M} + 1 \\
&\equiv \left(\sum_{p=0}^{N-1} U_p(z)\hat{z}^{pk} \right) z^{2^J k} \bmod z^{2M} + 1 \\
&\equiv C_k(z)z^{2^J k} \bmod z^{2M} + 1, \tag{7.117}
\end{aligned}
$$

where $k = 0, 1, \ldots, N-1$, $\hat{z} \equiv z^{2^{J+2}} \bmod z^{2M} + 1$, and

$$U_p(z) \equiv \sum_{j=0}^{2M-1} V_p(j)z^j \bmod z^{2M} + 1 \tag{7.118}$$

$$C_k(z) \equiv \sum_{p=0}^{N-1} U_p(z)\hat{z}^{pk} \bmod z^{2M} + 1. \tag{7.119}$$

Since $\hat{z}^N \equiv 1 \bmod z^{2M} + 1$ and $\hat{z}^{N/2} \equiv -1 \bmod z^{2M} + 1$, (7.119) is a PT which can be computed by a fast algorithm to be discussed in the next subsection.

The main steps of the fast algorithm for a 2D type-II DCT are summarized as follows.

Algorithm 14 *PT-based radix-2 algorithm for 2D type-II DCT*

Step 1.   *Compute N 1D type-II DCTs of length M according to (7.113).*

Step 2.   *Compute the PT according to (7.119) and then $B_k(z)$ according to (7.117). The fast algorithm for (7.119) will be discussed in the next subsection.*

Step 3.   *Add $A(k,l)$ and $B(k,l)$, which can be obtained from the definition of $B_k(z)$ in (7.116), to form $X(k,l)$ according to (7.105).*

7.4.2   Fast polynomial transform

The polynomial transform in (7.119) can be computed by the fast polynomial transform (FPT) algorithm discussed in Chapter 2, which generally requires $2MN \log_2 N$ additions. By using the symmetric property of the input polynomial sequence $U_p(z)$, however, the number of additions can be further reduced to be smaller than $MN \log_2 N$. It is noted that the coefficients of $U_p(z)$ are $V_p(j)$ $(j = 0, 1, \ldots, 2M - 1)$ which satisfy

$$V_p(2M - j) = -V_p(j), \ V_p(M) = 0.$$

This property indicates that only one-half of the coefficients are needed to express $U_p(z)$. It can be expressed as

$$U_p(z) \equiv U_p(z^{-1}) \bmod z^{2M} + 1.$$

Based on this property, it can be proven that the polynomial sequence $C_k(z)$ in (7.119) also has a symmetric property expressed as

$$C_{N-k}(z) \equiv C_k(z^{-1}) \bmod z^{2M} + 1$$

which indicates that about one-half of the coefficients $C_k(z)$ are necessarily computed. The FPT algorithm computes a PT by $\log_2 N$ stages. In the following, we will prove that the output polynomials of each stage also have a symmetric property that provides further savings in the number of additions.

We use the decimation-in-time radix-2 FPT discussed in Chapter 2 to compute (7.119). Let $C_n^0(z)$ be the binary inverse of $U_n(z)$. Then the t stages of the algorithm are

$$C_{\hat{n}2^j+\bar{k}}^j(z) \equiv C_{\hat{n}2^j+\bar{k}}^{j-1}(z) + C_{\hat{n}2^j+\bar{k}+2^{j-1}}^{j-1}(z)\hat{z}^{2^{t-j}\bar{k}} \bmod z^{2M} + 1 \qquad (7.120)$$

$$C_{\hat{n}2^j+\bar{k}+2^{j-1}}^j(z) \equiv C_{\hat{n}2^j+\bar{k}}^{j-1}(z) - C_{\hat{n}2^j+\bar{k}+2^{j-1}}^{j-1}(z)\hat{z}^{2^{t-j}\bar{k}} \bmod z^{2M} + 1, \qquad (7.121)$$

where $\bar{k} = 0, 1, \ldots, 2^{j-1} - 1$, $\hat{n} = 0, 1, \ldots, 2^{t-j} - 1$, and $j = 1, 2, \ldots, t$. The polynomials $C_{\hat{n}2^j+\bar{k}}^j(z)$ have an important symmetric property as shown in the following lemma, which reduces the computational complexity for (7.120) and (7.121) by one-half.

Lemma 10 *We have*

$$C_{\hat{n}2^j+2^j-\bar{k}}^j(z^{-1}) \equiv C_{\hat{n}2^j+\bar{k}}^j(z) \bmod z^{2M} + 1, \qquad (7.122)$$

where $\bar{k} = 1, 2, \ldots, 2^{j-1} - 1$ *and*

$$C_{\hat{n}2^j}^j(z^{-1}) \equiv C_{\hat{n}2^j}^j(z) \bmod z^{2M} + 1 \qquad (7.123)$$

$$C_{\hat{n}2^j+2^{j-1}}^j(z^{-1}) \equiv C_{\hat{n}2^j+2^{j-1}}^j(z) \bmod z^{2M} + 1. \qquad (7.124)$$

Proof. In Chapter 2, an expression is given for the output polynomial $C_{\hat{n}2^j+\bar{k}}^j(z)$ of stage j, that is,

$$C_{\hat{n}2^j+\bar{k}}^j(z) \equiv \sum_{\bar{n}=0}^{2^j-1} U_{\bar{n}2^{t-j}+n'}(z)\hat{z}^{2^{t-j}\bar{n}\bar{k}} \bmod z^{2M} + 1.$$

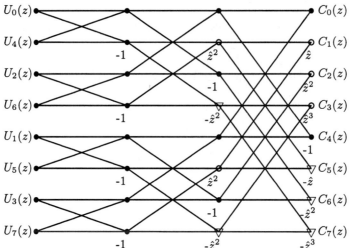

The node marked by filled circle • represents that the polynomial has symmetric property within itself. The node marked by $\triangledown$ represents that the polynomial can be derived from one of the polynomials marked by empty circles ○.

Figure 7.6   Signal flow graph of PT-based radix-2 algorithm for 2D type-II DCT.

So, we have

$$
\begin{aligned}
C^j_{\hat{n}2^j+2^j-\bar{k}}(z^{-1}) &\equiv \sum_{\bar{n}=0}^{2^j-1} U_{\hat{n}2^{t-j}+n'}(z)\hat{z}^{-2^{t-j}\bar{n}(2^j-\bar{k})} \bmod z^{2M}+1 \\
&\equiv \sum_{\bar{n}=0}^{2^j-1} U_{\bar{n}2^{t-j}+n'}(z)\hat{z}^{2^{t-j}\bar{n}\bar{k}} \bmod z^{2M}+1 \\
&\equiv C^j_{\hat{n}2^j+\bar{k}}(z) \bmod z^{2M}+1.
\end{aligned}
$$

Similarly, (7.123) and (7.124) can be easily proven.

The symmetric properties enable us to obtain $C^j_{\bar{n}2^j+\bar{k}}(z)$, $2^{j-1}+1 \le \bar{k} \le 2^j-1$, from $C^j_{\bar{n}2^j+\bar{k}}(z)$, $1 \le \bar{k} < 2^{j-1}$. Furthermore, the properties in (7.123) and (7.124) show that it is sufficient to compute only one-half of the polynomial coefficients for polynomial $C^j_{\hat{n}2^j}(z)$ and $C^j_{\hat{n}2^j+2^{j-1}}(z)$. The rest of the coefficients can be obtained from the properties without requiring any computation. Therefore, (7.120) and (7.121) need NM additions only. It is noted from (7.116) that the Mth coefficient of each output polynomial is $B(k, M)$ which is not required by our computation. Therefore, a further savings of $N/2-1$ additions is available. In total, the number of additions required by (7.119) for the fast polynomial transform is $NM \log_2 N - N/2+1$. When $N=8$, Figure 7.6 presents the signal flow graph of the FPT.

7.4.3   Computational complexity

Based on the assumption that the 1D type-II DCT of length M is computed by the radix-2 algorithm discussed in Chapter 6, which requires $\frac{1}{2}M \log_2 M$ multiplications and $\frac{3}{2}M \log_2 M - M + 1$ additions, Step 1 needs $\frac{1}{2}NM \log_2 M$ multiplications and $\frac{3}{2}NM \log_2 M - NM + N$

additions. Step 2 requires $NM \log_2 N - N/2 + 1$ additions and no multiplications, and Step 3 uses $NM - N - M + 1$ additions when the properties in (7.115) are exploited. If shifting operations are not taken into account, the presented algorithm requires

$$M_u = \frac{NM}{2} \log_2 M$$

multiplications and

$$A_d = \frac{3NM}{2} \log_2 M + NM \log_2 N - M - \frac{N}{2} + 2$$

additions.

It is known that the row-column algorithm needs

$$\frac{3NM}{2} \log_2(NM) - 2NM + N + M$$

additions and

$$\frac{NM}{2} \log_2(NM)$$

multiplications. In comparison, the presented algorithm achieves savings of about one-half of the number of multiplications and $\frac{1}{2}NM \log_2 N$ additions when $N = M$. Table 7.2 compares the number of operations for the 2D type-II DCT when $N = M = 2^t$, where M_u, A_d and $\overline{M}_u$, $\overline{A}_d$ represent the number of multiplications and number of additions of the PT-based and the row-column algorithms, respectively. For $N = M$, several algorithms

Table 7.2   Comparison of number of operations for 2D type-II DCT.

N	PT-based algorithm		Row-column based algorithm	
	M_u	A_d	$\overline{M}_u$	$\overline{A}_d$
4	16	76	32	72
8	96	470	192	464
16	512	2538	1024	2592
32	2560	12754	5120	13376
64	12288	61346	24576	65664
128	57344	286530	114688	311552
256	262144	1310338	524288	1442304
512	1179648	5897474	2359296	6554624
1024	5242880	26212866	10485760	29362176

were reported in [9, 12, 19, 22] which need nearly the same computational complexity as the presented one. One major problem of the algorithms reported in [9, 22] is that their post-processing stage is too complicated, which leads to many difficulties in implementation when the size of each dimension is larger than 16. Our PT-based algorithm utilizes the concept of the PT to concisely express the computation process, which removes the irregularities of the post-processing stage and provides simplicity of implementation. In general, most PT-based algorithms for DCT computation use the relation between the DCT and the discrete Fourier transform (DFT). The PT is used as a polynomial evaluation to derive the DCT. Such utilization of the PT requires operations on complex data [12, 18, 19]. Our algorithm directly applies the PT in the DCT computation so that operations on complex data are not necessary, which further reduces implementation complexity. The algorithms in [9, 12, 18, 19, 22] deal with the transform whose sizes of all dimensions are the same. In contrast, our PT-based algorithm has a unified process that is capable of dealing with different dimensional sizes. Such a feature is necessary for some applications (see [20], for example).

7.4.4   PT-based radix-2 algorithm for MD type-II DCT

We generalize the method used in the previous subsection for the computation of MD type-II DCTs. Let us consider the rD type-II DCT

$$
X(k_1, k_2, \ldots, k_r) = \sum_{n_1=0}^{N_1-1} \sum_{n_2=0}^{N_2-1} \cdots \sum_{n_r=0}^{N_r-1} x(n_1, n_2, \ldots, n_r)
$$
$$
\cdot \cos \frac{\pi(2n_1 + 1)k_1}{2N_1} \cdots \cos \frac{\pi(2n_r + 1)k_r}{2N_r}, \qquad (7.125)
$$

where $k_i = 0, 1, \ldots, N_i - 1$ and $i = 1, 2, \ldots, r$. We assume that N_i are powers of 2 and $N_r > N_i$ for $i = 1, 2, \ldots, r-1$. We can write that $N_r = 2^t$ and $N_r/N_i = 2^{l_i}$ for $l_i \geq 0$. If an MD array $y(n_1, \ldots, n_r)$ is defined by

$$
y(n_1, n_2, \ldots, n_r) = x(2n_1, 2n_2, \ldots, 2n_r)
$$
$$
y(N_1 - 1 - n_1, n_2, \ldots, n_r) = x(2n_1 + 1, 2n_2, \ldots, 2n_r)
$$
$$
\cdots \qquad \cdots
$$
$$
y(N_1 - 1 - n_1, \ldots, N_r - 1 - n_r) = x(2n_1 + 1, \ldots, 2n_r + 1),
$$

where $n_i = 0, 1, \ldots, N_i/2 - 1$ and $i = 1, 2, \ldots, r$, equation (7.125) becomes

$$
X(k_1, k_2, \ldots, k_r) = \sum_{n_1=0}^{N_1-1} \sum_{n_2=0}^{N_2-1} \cdots \sum_{n_r=0}^{N_r-1} y(n_1, n_2, \ldots, n_r)
$$
$$
\cdot \cos \frac{\pi(4n_1 + 1)k_1}{2N_1} \cdots \cos \frac{\pi(4n_r + 1)k_r}{2N_r}. \qquad (7.126)
$$

Now, we consider the fast algorithm for the computation of (7.126). Let us define

$$
A(k_1, k_2, \ldots, k_r) = \sum_{n_1=0}^{N_1-1} \sum_{n_2=0}^{N_2-1} \cdots \sum_{n_r=0}^{N_r-1} y(n_1, n_2, \ldots, n_r)
$$
$$
\cdot \cos \frac{\pi(4n_1 + 1)k_1}{2N_1} \cdots \cos \frac{\pi(4n_{r-2} + 1)k_{r-2}}{2N_{r-2}}
$$
$$
\cdot \cos \left(\frac{\pi(4n_{r-1} + 1)k_{r-1}}{2N_{r-1}} + \frac{\pi(4n_r + 1)k_r}{2N_r} \right)
$$

and

$$B(k_1, k_2, \ldots, k_r) = \sum_{n_1=0}^{N_1-1} \sum_{n_2=0}^{N_2-1} \cdots \sum_{n_r=0}^{N_r-1} y(n_1, n_2, \ldots, n_r)$$
$$\cdot \cos \frac{\pi(4n_1+1)k_1}{2N_1} \cdots \cos \frac{\pi(4n_{r-2}+1)k_{r-2}}{2N_{r-2}}$$
$$\cdot \cos \left(\frac{\pi(4n_{r-1}+1)k_{r-1}}{2N_{r-1}} - \frac{\pi(4n_r+1)k_r}{2N_r} \right).$$

Then, (7.126) can be computed by

$$X(k_1, k_2, \ldots, k_r) = \frac{A(k_1, k_2, \ldots, k_r) + B(k_1, k_2, \ldots, k_r)}{2}. \tag{7.127}$$

Let $p_{r-1}(n_r)$ be the least non negative remainder of $((4p_{r-1}+1)n_r + p_{r-1}) \bmod N_{r-1}$. Similar to the 2D case, we have

$$A(k_1, k_2, \ldots, k_r) = \sum_{n_1=0}^{N_1-1} \cdots \sum_{n_{r-2}=0}^{N_{r-2}-1} \sum_{p_{r-1}=0}^{N_{r-1}-1} \sum_{n_r=0}^{N_r-1} y(n_1, \ldots, n_{r-2}, p_{r-1}(n_r), n_r)$$
$$\cdot \cos \frac{\pi(4n_1+1)k_1}{2N_1} \cdots \cos \frac{\pi(4n_{r-2}+1)k_{r-2}}{2N_{r-2}}$$
$$\cdot \cos \frac{\pi(4n_r+1)(2^{l_{r-1}}(4p_{r-1}+1)k_{r-1} + k_r)}{2N_r} \tag{7.128}$$

and

$$B(k_1, k_2, \ldots, k_r) = \sum_{n_1=0}^{N_1-1} \cdots \sum_{n_{r-2}=0}^{N_{r-2}-1} \sum_{p_{r-1}=0}^{N_{r-1}-1} \sum_{n_r=0}^{N_r-1} y(n_1, \ldots, n_{r-2}, p_{r-1}(n_r), n_r)$$
$$\cdot \cos \frac{\pi(4n_1+1)k_1}{2N_1} \cdots \cos \frac{\pi(4n_{r-2}+1)k_{r-2}}{2N_{r-2}}$$
$$\cdot \cos \frac{\pi(4n_r+1)(2^{l_{r-1}}(4p_{r-1}+1)k_{r-1} - k_r)}{2N_r}. \tag{7.129}$$

If we define

$$V_{k_1,\ldots,k_{r-2},p_{r-1}}(l) = \sum_{n_1=0}^{N_1-1} \cdots \sum_{n_{r-2}=0}^{N_{r-2}-1} \sum_{n_r=0}^{N_r-1} y(n_1, \ldots, n_{r-2}, p_{r-1}(n_r), n_r)$$
$$\cdot \cos \frac{\pi(4n_1+1)k_1}{2N_1} \cdots \cos \frac{\pi(4n_{r-2}+1)k_{r-2}}{2N_{r-2}} \cos \frac{\pi(4n_r+1)l}{2N_r} \tag{7.130}$$

to be the $(r-1)$D modified type-II DCT with size $N_1 \times \cdots \times N_{r-2} \times N_r$ as defined in (7.126), (7.128) and (7.129) become

$$A(k_1, k_2, \ldots, k_r) = \sum_{p_{r-1}=0}^{N_{r-1}-1} V_{k_1,\ldots,k_{r-2},p_{r-1}}(2^{l_{r-1}}(4p_{r-1}+1)k_{r-1} + k_r)$$

$$B(k_1, k_2, \ldots, k_r) = \sum_{p_{r-1}=0}^{N_{r-1}-1} V_{k_1,\ldots,k_{r-2},p_{r-1}}(2^{l_{r-1}}(4p_{r-1}+1)k_{r-1} - k_r).$$

We form a polynomial

$$
\begin{aligned}
B_{k_1,k_2,\ldots,k_{r-1}}(z) &= \sum_{k_r=0}^{N_r-1} B(k_1, k_2, \ldots, k_r) z^{k_r} - \sum_{k_r=1}^{N_r} A(k_1, k_2, \ldots, 2N_r - k_r) z^{k_r} \\
&= \sum_{k_r=0}^{2N_r-1} B(k_1, k_2, \ldots, k_r) z^{k_r}
\end{aligned}
\tag{7.131}
$$

which can be expressed as

$$B_{k_1,k_2,\ldots,k_{r-1}}(z) \equiv C_{k_1,k_2,\ldots,k_{r-1}}(z) z^{2^{l_{r-1}}k_{r-1}} \bmod z^{2N_r} + 1, \tag{7.132}$$

where $k_i = 0, 1, \ldots, N_i - 1$, $i = 1, 2, \ldots, r-1$, $\hat{z} \equiv z^{2^{l_{r-1}+2}} \bmod z^{2N_r} + 1$,

$$C_{k_1,k_2,\ldots,k_{r-1}}(z) \equiv \sum_{p_{r-1}=0}^{N_{r-1}-1} U_{k_1,\ldots,k_{r-2},p_{r-1}}(z) \hat{z}^{p_{r-1}k_{r-1}} \bmod z^{2N_r} + 1 \tag{7.133}$$

and

$$U_{k_1,\ldots,k_{r-2},p_{r-1}}(z) = \sum_{l=0}^{2N_r-1} V_{k_1,\ldots,k_{r-2},p_{r-1}}(l) z^l,$$

which is the generating polynomial of $V_{k_1,\ldots,k_{r-2},p_{r-1}}(l)$. The fast algorithm can be summarized as follows.

Algorithm 15 *PT-based radix-2 algorithm for rD type-II DCT*

Step 1.   *Compute N_{r-1} $(r-1)D$ modified type-II DCTs with size $N_1 \times \cdots \times N_{r-2} \times N_r$ according to (7.130). If $(r-1) > 1$, a similar procedure as described above should be used to further decompose the $(r-1)D$ modified type-II DCTs into smaller sizes. If $(r-1) = 1$, the 1D type-II DCT algorithm can be used for the computation as described in the previous section.*

Step 2.   *Compute $N_1 N_2 \cdots N_{r-2}$ 1D PTs according to (7.133) and then compute $B_{k_1,k_2,\ldots,k_{r-1}}(z)$ according to (7.132).*

Step 3.   *Based on the definition of $B_{k_1,k_2,\ldots,k_{r-1}}(z)$ in (7.131), $A(k_1, k_2, \ldots, k_r)$ and $B(k_1, k_2, \ldots, k_r)$ are derived to form $X(k_1, k_2, \ldots, k_r)$ according to (7.127), which needs $N - N/N_r - N/N_{r-1} + 1$ additions, where $N = N_1 N_2 \cdots N_r$.*

Now let us consider the computational complexity needed by the above algorithm. By taking into account all operations in the three steps, the computational complexity of the rD type-II DCT with size $N_1 \times \cdots \times N_r$ is

$$
\begin{aligned}
M_u(N_1, \ldots, N_r) &= N_{r-1} M_u(N_1, \ldots, N_{r-2}, N_r) \\
A_d(N_1, \ldots, N_r) &= N_{r-1} A_d(N_1, \ldots, N_{r-2}, N_r) + N \log_2 N_{r-1} + N \\
&\quad - \frac{3N}{2N_r} - \frac{N}{N_{r-1}} + \frac{N}{N_{r-1}N_r} + 1,
\end{aligned}
$$

where $N = N_1 N_2 \cdots N_r$, $M_u(N_1, \ldots, N_r)$ is the number of multiplications and $A_d(N_1, \ldots, N_r)$ is the number of additions.

If we assume that the radix-2 algorithm for the 1D type-II DCT given in Chapter 6 is used, the total numbers of multiplications and additions needed by the rD type-II DCT can be expressed by

$$M_u(N_1, \ldots, N_r) = \frac{N}{2} \log_2 N_r$$

$$A_d(N_1, \ldots, N_r) = \frac{3N}{2} \log_2 N_r + N \log_2 \frac{N}{N_r} + (r-2)N - \frac{3r-5}{2} \frac{N}{N_r}$$

$$- N(1 - \frac{1}{N_r})(\frac{1}{N_1} + \cdots + \frac{1}{N_{r-1}}) + (r-1).$$

In general, the numbers of multiplications and additions required by the row-column method for rD type-II DCT are

$$\overline{M_u} = \frac{N}{2} \log_2 N$$

$$\overline{A_d} = \frac{3N}{2} \log_2 N - rN + N(\frac{1}{N_1} + \cdots + \frac{1}{N_r}).$$

It can be easily seen that when $N_1 = N_2 = \cdots = N_r$, the computational complexities needed by the presented and the row-column algorithms have the following relationship:

$$M_u = \frac{1}{r} \overline{M_u}, \ A_d \approx \frac{2r+1}{3r} \overline{A_d}.$$

Table 7.3 compares the number of operations for the 3D type-II DCT when $N_1 = N_2 = N_3$, where M_u, A_d and $\overline{M_u}$, $\overline{A_d}$ represent the numbers of multiplications and additions of the presented PT algorithm and the row-column method, respectively.

Table 7.3 Comparison of the number of operations for 3D type-II DCT.

N_1	FPT algorithm		Row-column algorithm	
	M_u	A_d	$\overline{M_u}$	$\overline{A_d}$
4	64	456	192	432
8	768	5648	2304	5568
16	8192	60448	24576	62208
32	81920	602176	245760	642048
64	786432	5750912	2359296	6303744
128	7340032	53412096	22020096	59817984

When $N_1 = N_2 = \cdots = N_r$, the PT-based algorithm needs about the same number of operations compared with the fast algorithm for the rD type-II DCT reported in [22]. However, this algorithm clearly defines the post-processing stage by the PT, which allows an easy implementation.

7.5   PT-BASED RADIX-2 ALGORITHM FOR MD TYPE-III DCT

The computation of the type-III DCT is considered to be the reverse process of that used for the type-II DCT. It is generally true that the fast algorithm for the 1D type-III DCT can easily be obtained from the fast algorithm for the 1D type-II DCT. However, such a conversion is certainly not a trivial matter for transforms of more than two dimensions. If an attempt is made to convert the fast algorithm for the type-II MD DCT, for example, it is a major challenge to derive a pre-processing stage (corresponding to the post-processing stage for the type-II MD DCT) that requires a reasonable number of additions and has a simple computational structure. In this section, we consider a PT-based algorithm for the rD type-III DCT whose size is $N_1 \times N_2 \times \cdots \times N_r$, where N_i $(i = 1, 2, \ldots, r)$ are powers of 2 and not necessarily the same [24].

For simplicity and easy understanding, we first describe a 2D type-III DCT to illustrate some basic concepts. In particular, symmetric properties in the PT are shown to be useful for the reduction of the computational complexity. Then, the algorithm for the 2D type-III DCT is generalized and extended for the rD type-III DCT, where $r \geq 3$. The computational complexity required by the algorithm is also considered.

7.5.1   PT-based radix-2 algorithm for 2D type-III DCT

The 2D type-III DCT of the sequence $x(n, m)$ $(n = 0, 1, \ldots, N - 1, \ m = 0, 1, \ldots, M - 1)$ is defined by

$$X(k, l) = \sum_{n=0}^{N-1} \sum_{m=0}^{M-1} x(n, m) \cos \frac{\pi(2k + 1)n}{2N} \cos \frac{\pi(2l + 1)m}{2M}, \qquad (7.134)$$

where $k = 0, 1, \ldots, N - 1$ and $l = 0, 1, \ldots, M - 1$. It is assumed that the sizes of the two dimensions are $N = 2^t$ and $M = 2^J N$, where $t > 0$ and $J \geq 0$, respectively. We convert the output sequence $X(k, l)$ into

$$
\begin{aligned}
Y(k, l) &= X(2k, 2l) \\
Y(N - 1 - k, l) &= X(2k + 1, 2l) \\
Y(k, M - 1 - l) &= X(2k, 2l + 1) \\
Y(N - 1 - k, M - 1 - l) &= X(2k + 1, 2l + 1),
\end{aligned}
$$

where $k = 0, 1, \ldots, N/2 - 1$ and $l = 0, 1, \ldots, M/2 - 1$, so that (7.134) can be expressed by

$$Y(k, l) = \sum_{n=0}^{N-1} \sum_{m=0}^{M-1} x(n, m) \cos \frac{\pi(4k + 1)n}{2N} \cos \frac{\pi(4l + 1)m}{2M}. \qquad (7.135)$$

In the following, we show that the 2D type-III DCT can be decomposed into a number of 1D DCTs whose inputs can be further processed by a PT.

Let $p(l) \equiv (4p + 1)l + p \mod N$. The following lemma shows that $Y(p(l), l)$ $(p = 0, 1, \ldots, N - 1, \ l = 0, 1, \ldots, M - 1)$ is simply a reordering of $Y(k, l)$ $(k = 0, 1, \ldots, N - 1, \ l = 0, 1, \ldots, M - 1)$.

Lemma 11 *Let* $U = \{(k, l) \mid 0 \leq k \leq N - 1, \ 0 \leq l \leq M - 1\}$, $V = \{(p(l), l) \mid 0 \leq p \leq N - 1, \ 0 \leq l \leq M - 1\}$. *Then* $U = V$.

Proof. It is proven in Lemma 9.

Therefore, it is sufficient to compute $Y(p(l), l)$ $(p = 0, 1, \ldots, N-1, \ l = 0, 1, \ldots, M-1)$. Based on the properties of trigonometric identities, (7.135) can be computed by

$$Y(p(l), l) = \frac{A(p(l), l) + B(p(l), l)}{2}, \tag{7.136}$$

where

$$A(p(l), l) = \sum_{n=0}^{N-1} \sum_{m=0}^{M-1} x(n, m) \cos\left(\frac{\pi(4p(l) + 1)n}{2N} + \frac{\pi(4l + 1)m}{2M}\right) \tag{7.137}$$

$$B(p(l), l) = \sum_{n=0}^{N-1} \sum_{m=0}^{M-1} x(n, m) \cos\left(\frac{\pi(4p(l) + 1)n}{2N} - \frac{\pi(4l + 1)m}{2M}\right). \tag{7.138}$$

From the definition of $p(l)$, it can easily be proven that

$$4p(l) + 1 \equiv (4l + 1)(4p + 1) \bmod 4N.$$

By applying this equation in (7.137) and (7.138), we obtain

$$
\begin{aligned}
A(p(l), l) &= \sum_{n=0}^{N-1} \sum_{m=0}^{M-1} x(n, m) \cos\left(\frac{\pi(4l+1)(4p+1)n}{2N} + \frac{\pi(4l+1)m}{2M}\right) \\
&= \sum_{n=0}^{N-1} \sum_{m=0}^{M-1} x(n, m) \cos\frac{\pi(4l+1)(2^J(4p+1)n + m)}{2M}.
\end{aligned}
\tag{7.139}
$$

Similarly

$$B(p(l), l) = \sum_{n=0}^{N-1} \sum_{m=0}^{M-1} x(n, m) \cos\frac{\pi(4l+1)(2^J(4p+1)n - m)}{2M}. \tag{7.140}$$

If we construct a sequence $y(n, m)$ as

$$
y(n, m) = \begin{cases}
2x(n, 0), & m = 0 \\
0, & m = M \\
x(n, m), & 1 \le m \le M - 1 \\
-x(n, 2M - m), & M + 1 \le m \le 2M - 1
\end{cases}
\tag{7.141}
$$

based on (7.139), (7.140) and the properties of trigonometric identities, (7.136) can be expressed as

$$Y(p(l), l) = \frac{1}{2} \sum_{n=0}^{N-1} \sum_{m=0}^{2M-1} y(n, m) \cos\frac{\pi(4l+1)(2^J(4p+1)n + m)}{2M}. \tag{7.142}$$

Now let us define

$$Y_p(z) \equiv \sum_{n=0}^{N-1} \sum_{m=0}^{2M-1} y(n, m) z^{2^J(4p+1)n+m} \bmod z^{2M} + 1 \tag{7.143}$$

$$\bar{Y}_p(z) = (Y_p(z) + Y_p(z^{-1})) \bmod z^{2M} + 1 = \sum_{m=0}^{2M-1} h(p,m) z^m, \qquad (7.144)$$

where $h(p,m)$ are the coefficients of the polynomials $\bar{Y}_p(z)$ and have the property

$$h(p,m) = -h(p, 2M - m), \quad m = 1, 2, \ldots, 2M - 1.$$

Since $(e^{i\frac{\pi(4l+1)}{2M}})^{2M} = -1$, we know that (7.142) can be related to (7.144) by

$$Y(p(l), l) = \frac{1}{4} \bar{Y}_p(z) \Big|_{z = e^{i\frac{\pi(4l+1)}{2M}}}$$

which can be further elaborated into

$$Y(p(l), l) = \sum_{m=0}^{M-1} \bar{h}(p,m) \cos \frac{\pi(4l+1)m}{2M}, \qquad (7.145)$$

where $p = 0, 1, \ldots, N - 1$, $l = 0, 1, \ldots, M - 1$ and

$$\bar{h}(p,0) = h(p,0)/4, \quad \bar{h}(p,m) = h(p,m)/2, \quad m = 1, 2, \ldots, M - 1.$$

Therefore, (7.145) shows that the 2D type-III DCT is converted into N 1D type-III DCTs of length M if its output sequence is reordered in a similar way to that in (7.135). However, $\bar{h}(p,m)$ (or $h(p,m)$ in (7.144)) have to be available for (7.145).

The computation of this pre-processing stage can be defined by a PT. We rewrite $Y_p(z)$ defined in (7.143) as

$$Y_p(z) \equiv \sum_{n=0}^{N-1} y_n(z) z^{2^J n} \hat{z}^{pn} \bmod z^{2M} + 1,$$

where

$$y_n(z) = \sum_{m=0}^{2M-1} y(n,m) z^m, \quad \hat{z} \equiv z^{2^{J+2}} \bmod z^{2M} + 1, \qquad (7.146)$$

where $y(n,m)$ in (7.146) is defined in (7.141). Based on the symmetric property $y(n,m) = -y(n, 2M - m)$, (7.146) can be shown to have the property

$$y_n(z) \equiv y_n(z^{-1}) \bmod z^{2M} + 1. \qquad (7.147)$$

Therefore, (7.144) can be equivalently obtained by

$$\bar{Y}_p(z) \equiv \sum_{n=0}^{N-1} \bar{y}_n(z) \hat{z}^{pn} \bmod z^{2M} + 1, \qquad (7.148)$$

where for $n = 1, 2, \ldots, N - 1$,

$$\bar{y}_0(z) \equiv 2y_0(z) \bmod z^{2M} + 1 \qquad (7.149)$$

$$\bar{y}_{N/2}(z) \equiv (y_{N/2}(z) + y_{N/2}(z) z^M) z^{-M/2} \bmod z^{2M} + 1 \qquad (7.150)$$

$$\bar{y}_n(z) \equiv (y_{N-n}(z) + y_n(z) z^M) z^{2^J n} z^{-M} \bmod z^{2M} + 1. \qquad (7.151)$$

It is obvious that for $n = 1, 2, \ldots, N/2 - 1$,

$$\bar{y}_0(z) \equiv \bar{y}_0(z^{-1}) \bmod z^{2M} + 1 \tag{7.152}$$

$$\bar{y}_{N/2}(z) \equiv \bar{y}_{N/2}(z^{-1}) \bmod z^{2M} + 1 \tag{7.153}$$

$$\bar{y}_n(z) \equiv \bar{y}_{N-n}(z^{-1}) \bmod z^{2M} + 1. \tag{7.154}$$

Therefore, we need to compute (7.149), (7.150) and (7.151) and the PT (7.148) to get $\bar{Y}_p(z)$. The PT (7.148) can be computed by a fast algorithm to be discussed in the next subsection. The main steps of the algorithm are summarized as follows.

Algorithm 16 *PT-based radix-2 algorithm for 2D type-III DCT*

Step 1. *Compute the input polynomials $\bar{y}_n(z)$ according to (7.149), (7.150) and (7.151).*

Step 2. *Compute a PT according to (7.148) to obtain $h(p, m)$ or $\bar{h}(p.m)$.*

Step 3. *Compute the N 1D type-III DCTs of length M according to (7.145).*

7.5.2 Fast polynomial transform

As we know, the PT in (7.148) can be computed by the FPT algorithm which generally requires $2MN \log_2 N$ additions (see Chapter 2). However, similar to the 2D type-II DCT algorithm discussed before, the number of additions can be further reduced by using the symmetric property of the input polynomial sequence $\bar{y}_n(z)$ stated in (7.152)–(7.154). The output polynomial $\bar{Y}_p(z)$ also has a symmetric property. In fact, from the definition (7.144) we know that

$$\bar{Y}_p(z) \equiv \bar{Y}_p(z^{-1}) \bmod z^{2M} + 1.$$

Therefore, only one-half of the coefficients of $\bar{Y}_p(z)$ are to be computed. We will prove that some kind of symmetric property also exists in the output polynomials of each stage. By using the symmetric properties, a large number of additions can be saved.

Let $\bar{Y}_n^0(z)$ be the binary inverse of $\bar{y}_n(z)$. If we use the decimation-in-time radix-2 FPT for computing (7.148) (see Chapter 2), the computational procedure is as follows:

$$\bar{Y}_{\hat{n}2^j + \bar{p}}^j(z) \equiv \bar{Y}_{\hat{n}2^j + \bar{p}}^{j-1}(z) + \bar{Y}_{\hat{n}2^j + (\bar{p} + 2^{j-1})}^{j-1}(z)\hat{z}^{2^{t-j}\bar{p}} \bmod z^{2M} + 1 \tag{7.155}$$

$$\bar{Y}_{\hat{n}2^j + (\bar{p} + 2^{j-1})}^j(z) \equiv \bar{Y}_{\hat{n}2^j + \bar{p}}^{j-1}(z) - \bar{Y}_{\hat{n}2^j + (\bar{p} + 2^{j-1})}^{j-1}(z)\hat{z}^{2^{t-j}\bar{p}} \bmod z^{2M} + 1, \tag{7.156}$$

where $\bar{p} = 0, 1, \ldots, 2^{j-1} - 1$, $\hat{n} = 0, 1, \ldots, 2^{t-j} - 1$, $j = 1, 2, \ldots, t$ and

$$\bar{Y}_{\hat{n}2^j + \bar{p}}^j(z) \equiv \sum_{\bar{n}=0}^{2^j - 1} \bar{y}_{\bar{n}2^{t-j} + n'}(z)\hat{z}^{2^{t-j}\bar{n}\bar{p}} \bmod z^{2M} + 1$$

for $\bar{p} = 0, 1, \ldots, 2^j - 1$, $\hat{n} = 0, 1, \ldots, 2^{t-j} - 1$ and n' the binary inverse of $\hat{n}$. The last stage produces the outputs, that is, $\bar{Y}_p^t(z) = \bar{Y}_p(z)$.

Lemma 12 *Let $\tilde{n}$ be the binary inverse of $2^{t-j} - n'$. We have*

$$\bar{Y}_{\tilde{n}2^j + \bar{p}}^j(z^{-1}) \equiv \bar{Y}_{\tilde{n}2^j + \bar{p}}^j(z)\hat{z}^{2^{t-j}\bar{p}} \bmod z^{2M} + 1, \tag{7.157}$$

where $\hat{n} = 1, 2, \ldots, 2^{t-j} - 1$, $\bar{p} = 0, 1, \ldots, 2^j - 1$ and

$$\bar{Y}_{\bar{p}}^j(z) \equiv \bar{Y}_{\bar{p}}^j(z^{-1}) \bmod z^{2M} + 1. \tag{7.158}$$

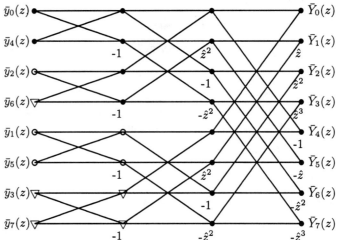

The node marked by filled circle • represents that the polynomial has symmetric property within itself. The node marked by ▽ represents that the polynomial can be derived from one of the polynomials marked by empty circles ○.

Figure 7.7   Signal flow graph of PT-based radix-2 algorithm for 2D type-III DCT.

Proof. If $\hat{n} \neq 0$, we have

$$\bar{Y}^j_{\hat{n}2^j+\bar{p}}(z^{-1}) \equiv \sum_{\bar{n}=0}^{2^j-1} \bar{y}_{\bar{n}2^{t-j}+n'}(z^{-1})\hat{z}^{-2^{t-j}\bar{n}\bar{p}} \bmod z^{2M}+1$$

$$\equiv \sum_{\bar{n}=0}^{2^j-1} \bar{y}_{2^t-\bar{n}2^{t-j}-n'}(z)\hat{z}^{-2^{t-j}\bar{n}\bar{p}} \bmod z^{2M}+1$$

$$\equiv \sum_{\bar{n}=0}^{2^j-1} \bar{y}_{2^{-j}(2^j-1-\bar{n})+2^{t-j}-n'}(z)\hat{z}^{-2^{t-j}\bar{n}\bar{p}} \bmod z^{2M}+1$$

$$\equiv \sum_{\bar{n}=0}^{2^j-1} \bar{y}_{\bar{n}2^{t-j}+2^{t-j}-n'}(z)\hat{z}^{-2^{t-j}(2^j-1-\bar{n})\bar{p}} \bmod z^{2M}+1$$

$$\equiv \left(\sum_{\bar{n}=0}^{2^j-1} \bar{y}_{\bar{n}2^{t-j}+2^{t-j}-n'}(z)\hat{z}^{2^{t-j}\bar{n}\bar{p}}\right)\hat{z}^{2^{t-j}\bar{p}} \bmod z^{2M}+1$$

$$\equiv \bar{Y}^j_{\hat{n}2^j+\bar{p}}(z)\hat{z}^{2^{t-j}\bar{p}} \bmod z^{2M}+1.$$

Similarly, (7.158) can be verified.

Both (7.157) and (7.158) show that at every stage of the FPT, only one-half of the coefficients need to be computed. The rest of the coefficients can be achieved from the above properties without requiring any computation. So, each stage requires NM additions. Furthermore, based on the symmetric property we know that at the $(t-1)$th stage, every polynomial $\bar{Y}^{t-1}_p(z)$ has a zero coefficient. Thus, $N/2-1$ additions can be saved at the tth stage. Therefore, the total number of additions for the PT (7.148) is $NM\log_2 N - N/2 + 1$. When $N = M = 8$, the signal flow graph of the FPT algorithm is given in Figure 7.7.

7.5.3   Computational complexity

Based on the symmetric property of $y_n(z)$ as shown in (7.147), it is easy to verify that Step 1 needs $NM - N - M$ additions. Step 2 needs only $NM \log_2 N - N/2 + 1$ additions and no multiplications. Step 3 needs $\frac{1}{2} NM \log_2 M$ multiplications and $\frac{3}{2} NM \log_2 M - NM + N$ additions. If shifting operations are not taken into account, the total numbers of multiplications and additions are, respectively,

$$M_u = \frac{NM}{2} \log_2 M$$

$$A_d = \frac{3NM}{2} \log_2 M + NM \log_2 N - M - \frac{N}{2} + 1.$$

It is known that the row-column algorithm needs $\frac{3}{2} NM \log_2(NM) - 2NM + N + M$ additions and $\frac{1}{2} NM \log_2(NM)$ multiplications. In comparison, the PT-based algorithm achieves savings of about one-half of the number of multiplications and $\frac{1}{2} NM \log_2 N$ additions.

7.5.4   PT-based radix-2 algorithm for MD type-III DCT

In this subsection, we generalize the method used in the previous subsection for the computation of MD type-III DCTs. Let us consider

$$X(k_1, k_2, \ldots, k_r) = \sum_{n_1=0}^{N_1-1} \sum_{n_2=0}^{N_2-1} \cdots \sum_{n_r=0}^{N_r-1} x(n_1, n_2, \ldots, n_r)$$
$$\cdot \cos \frac{\pi(2k_1 + 1)n_1}{2N_1} \cdots \cos \frac{\pi(2k_r + 1)n_r}{2N_r}, \qquad (7.159)$$

where $k_i = 0, 1, \ldots, N_i - 1$, $i = 1, 2, \ldots, r,$, $N_r = 2^t$ and $N_r/N_i = 2^{l_i}$, $l_i \geq 0$ for $i = 1, 2, \ldots, r - 1$. If we define a sequence $Y(k_1, \ldots, k_r)$ by

$$Y(k_1, k_2, \ldots, k_r) = X(2k_1, 2k_2, \ldots, 2k_r)$$
$$Y(N_1 - 1 - k_1, k_2, \ldots, k_r) = X(2k_1 + 1, 2k_2, \ldots, 2k_r)$$
$$\cdots \qquad \cdots$$
$$Y(N_1 - 1 - k_1, \ldots, N_r - 1 - k_r) = X(2k_1 + 1, \ldots, 2k_r + 1),$$

where $k_i = 0, 1, \ldots, N_i/2 - 1$ and $i = 1, 2, \ldots, r$, (7.159) becomes

$$Y(k_1, k_2, \ldots, k_r) = \sum_{n_1=0}^{N_1-1} \sum_{n_2=0}^{N_2-1} \cdots \sum_{n_r=0}^{N_r-1} y(n_1, n_2, \ldots, n_r)$$
$$\cdot \cos \frac{\pi(4k_1 + 1)n_1}{2N_1} \cdots \cos \frac{\pi(4k_r + 1)n_r}{2N_r}.$$

Let $p(k_r)$ be the least non-negative residue of $(4p + 1)k_r + p$ module N_{r-1}. Similar to the 2D case, we have

$$Y(k_1, \ldots, k_{r-2}, p(k_r), k_r) = \frac{A(k_1, \ldots, k_{r-2}, p(k_r), k_r) + B(k_1, \ldots, k_{r-2}, p(k_r), k_r)}{2}, \quad (7.160)$$

where

$$A(k_1, \ldots, k_{r-2}, p(k_r), k_r) = \sum_{n_1=0}^{N_1-1} \sum_{n_2=0}^{N_2-1} \cdots \sum_{n_r=0}^{N_r-1} x(n_1, n_2, \ldots, n_r)$$
$$\cdot \cos \frac{\pi(4k_1 + 1)n_1}{2N_1} \cdots \cos \frac{\pi(4k_{r-2} + 1)n_{r-2}}{2N_{r-2}}$$
$$\cdot \cos \frac{\pi(4k_r + 1)(2^{l_{r-1}}(4p + 1)n_{r-1} + n_r)}{2N_r}$$

and

$$B(k_1, \ldots, k_{r-2}, p(k_r), k_r) = \sum_{n_1=0}^{N_1-1} \sum_{n_2=0}^{N_2-1} \cdots \sum_{n_r=0}^{N_r-1} x(n_1, n_2, \ldots, n_r)$$
$$\cdot \cos \frac{\pi(4k_1 + 1)n_1}{2N_1} \cdots \cos \frac{\pi(4k_{r-2} + 1)n_{r-2}}{2N_{r-2}}$$
$$\cdot \cos \frac{\pi(4k_r + 1)(2^{l_{r-1}}(4p + 1)n_{r-1} - n_r)}{2N_r}.$$

If we construct a sequence $y(n_1, \ldots, n_{r-1}, n_r)$

$$y(n_1, \ldots, n_{r-1}, n_r) = \begin{cases} 2x(n_1, \ldots, n_{r-1}, 0), & n_r = 0 \\ 0, & n_r = N_r \\ x(n_1, \ldots, n_{r-1}, n_r), & 1 \leq n_r \leq N_r - 1 \\ -x(n_1, \ldots, n_{r-1}, 2N_r - n_r), & N_r + 1 \leq n_r \leq 2N_r - 1 \end{cases} \quad (7.161)$$

then we have

$$Y(k_1, \ldots, k_{r-2}, p(k_r), k_r) = \sum_{n_1=0}^{N_1-1} \cdots \sum_{n_{r-2}=0}^{N_{r-2}-1} \cos \frac{\pi(4k_1 + 1)n_1}{2N_1} \cdots \cos \frac{\pi(4k_{r-2} + 1)n_{r-2}}{2N_{r-2}}$$
$$\cdot \left\{ \frac{1}{2} \sum_{n_{r-1}=0}^{N_{r-1}-1} \sum_{n_r=0}^{2N_r-1} y(n_1, \ldots, n_{r-2}, n_{r-1}, n_r) \right.$$
$$\left. \cdot \cos \frac{\pi(4k_r + 1)(2^{l_{r-1}}(4p + 1)n_{r-1} + n_r)}{2N_r} \right\}. \quad (7.162)$$

We denote the inner summation by $\hat{y}(n_1, \ldots, n_{r-2}, p, k_r)$, that is,

$$\hat{y}(n_1, \ldots, n_{r-2}, p, k_r) = \frac{1}{2} \sum_{n_{r-1}=0}^{N_{r-1}-1} \sum_{n_r=0}^{2N_r-1} y(n_1, \ldots, n_{r-2}, n_{r-1}, n_r)$$
$$\cdot \cos \frac{\pi(4k_r + 1)(2^{l_{r-1}}(4p + 1)n_{r-1} + n_r)}{2N_r} \quad (7.163)$$

which is very similar to (7.142). Based on this, (7.163) can be converted into 1D type-III DCTs as

$$\hat{y}(n_1, \ldots, n_{r-2}, p, k_r) = \sum_{n_r=0}^{N_r-1} \bar{h}(n_1, \ldots, n_{r-2}, p, n_r) \cos \frac{\pi(4k_r + 1)n_r}{2N_r}, \quad (7.164)$$

where $\bar{h}(n_1, \ldots, n_{r-2}, p, n_r)$ are coefficients of polynomials which can be computed by the FPT algorithm as described in the first two subsections of this section. Substituting (7.164) into (7.162), we get

$$
\begin{aligned}
Y(k_1, \ldots, k_{r-2}, p(k_r), k_r) = \sum_{n_1=0}^{N_1-1} \cdots \sum_{n_{r-2}=0}^{N_{r-2}-1} \sum_{n_r=0}^{N_r-1} \bar{h}(n_1, \ldots, n_{r-2}, p, n_r) \\
\cdot \cos \frac{\pi(4k_1+1)n_1}{2N_1} \cdots \cos \frac{\pi(4k_{r-2}+1)n_{r-2}}{2N_{r-2}} \\
\cdot \cos \frac{\pi(4k_r+1)n_r}{2N_r} \qquad\qquad (7.165)
\end{aligned}
$$

which is N_{r-1} $(r-1)$D type-III DCTs. The method described above can be used to further reduce the number of dimensions of the transform. The algorithm is summarized as follows.

Algorithm 17 *PT-based radix-2 algorithm for rD type-III DCT*

Step 1. *Use the PT algorithm to get $\bar{h}(n_1, \ldots, n_{r-2}, p, n_r)$.*

Step 2. *Compute N_{r-1} $(r-1)$D type-III DCTs in (7.165).*

Now let us consider the computational complexity needed by the above algorithm and define $N = N_1 N_2 \cdots N_r$. Step 1 needs $\frac{N}{N_{r-1}N_r}(N_{r-1}N_r \log_2 N_{r-1} + N_{r-1}N_r - N_r - \frac{3}{2}N_{r-1}+1)$ additions. Hence, the computational complexity of the rD type-III DCT with size $N_1 \times \cdots \times N_r$ is the same as that of the rD type-II DCT. As previously discussed, the algorithm substantially reduces the number of arithmetic operations compared to the row-column method.

7.6 PT-BASED RADIX-q ALGORITHM FOR MD TYPE-II DCT

The PT-based algorithms previously discussed can only deal with MD DCTs whose dimensional sizes are powers of two. The requirements on dimensional sizes may not be able to support some special applications. Polynomial algorithms for rD DCT with size $q^{l_1} \times q^{l_2} \times \cdots \times q^{l_r}$, where $r > 1$ and q is an odd prime number, are reported in [25, 26]. In this section and Section 7.7, we present two PT-based algorithms for rD type-II and type-III DCTs, respectively. There are other algorithms such as [13] which can also deal with the above mentioned rD DCTs. The fast algorithm in [13] uses a recursive technique to decompose 2D DCTs of size $q^l \times q^l$, where q is a prime number, into cyclic convolution (CC), skew-cyclic convolution (SCC) and matrix multiplications. This algorithm greatly reduces the number of multiplications compared to that needed by the row-column method. However, the computation of the CC and SCC generally requires a large number of additions and an irregular computational structure. Furthermore, such fast algorithms are impractical when q is small and l is large.

7.6.1   PT-based radix-q algorithm for 2D type-II DCT

The 2D type-II DCT of the sequence $x(n, m)$ $(n = 0, 1, \ldots, N - 1,\ m = 0, 1, \ldots, M - 1)$ is defined by

$$X(k, l) = \sum_{n=0}^{N-1} \sum_{m=0}^{M-1} x(n, m) \cos \frac{\pi(2n+1)k}{2N} \cos \frac{\pi(2m+1)l}{2M}, \tag{7.166}$$

where $k = 0, 1, \ldots, N - 1$ and $l = 0, 1, \ldots, M - 1$. It is assumed that the dimensional sizes are $M = q^t$ and $N = M/q^J$, where $t > 0$, $J \geq 0$, and q is an odd prime number. The input sequence $x(n, m)$ is converted into

$$
\begin{aligned}
y(n, m) &= x(2n, 2m), & 0 \leq n \leq \frac{N-1}{2},\ 0 \leq m \leq \frac{M-1}{2} \\
y(N-1-n, m) &= x(2n+1, 2m), & 0 \leq n \leq \frac{N-3}{2},\ 0 \leq m \leq \frac{M-1}{2} \\
y(n, M-1-m) &= x(2n, 2m+1), & 0 \leq n \leq \frac{N-1}{2},\ 0 \leq m \leq \frac{M-3}{2} \\
y(N-1-n, M-1-m) &= x(2n+1, 2m+1), & 0 \leq n \leq \frac{N-3}{2},\ 0 \leq m \leq \frac{M-3}{2}
\end{aligned}
\tag{7.167}
$$

so that (7.166) can be expressed by

$$X(k, l) = \sum_{n=0}^{N-1} \sum_{m=0}^{M-1} y(n, m) \cos \frac{\pi(4n+1)k}{2N} \cos \frac{\pi(4m+1)l}{2M}, \tag{7.168}$$

where $k = 0, 1, \ldots, N - 1$ and $l = 0, 1, \ldots, M - 1$. Let $\lambda(q)$ represent the unique non-negative integer which is smaller than q and satisfies $4\lambda(q) + 1 \equiv 0 \bmod q$. The 2D type-II DCT defined in (7.168) can be decomposed into

$$X(k, l) = X^0(k, l) + X^1(k, l), \tag{7.169}$$

where

$$X_0(k, l) = \sum_{n=0}^{N-1} \sum_{m \equiv \lambda(q) \bmod q} y(n, m) \cos \frac{\pi(4n+1)k}{2N} \cos \frac{\pi(4m+1)l}{2M} \tag{7.170}$$

$$X_1(k, l) = \sum_{n=0}^{N-1} \sum_{m \not\equiv \lambda(q) \bmod q} y(n, m) \cos \frac{\pi(4n+1)k}{2N} \cos \frac{\pi(4m+1)l}{2M}, \tag{7.171}$$

where for (7.170) and (7.171), $k = 0, 1, \ldots, N - 1$ and $l = 0, 1, \ldots, M - 1$. The computation defined in (7.170) can be expressed into a 2D type-II DCT with size $N \times M/q$. In fact, let $4\lambda(q) + 1 = \delta q$, then δ must be 1 or 3. If $\delta = 1$, we have

$$X_0(k, l) = \sum_{n=0}^{N-1} \sum_{m=0}^{M/q-1} y(n, qm + \lambda(q)) \cos \frac{\pi(4n+1)k}{2N} \cos \frac{\pi(4m+1)l}{2M/q}, \tag{7.172}$$

where $k = 0, 1, \ldots, N - 1$ and $l = 0, 1, \ldots, M/q - 1$. If $\delta = 3$, we get

$$X_0(k, l) = \sum_{n=0}^{N-1} \sum_{m=0}^{M/q-1} y(n, qm + \lambda(q)) \cos \frac{\pi(4n+1)k}{2N} \cos \frac{\pi(4m+3)l}{2M/q}, \tag{7.173}$$

which can be reordered as

$$X_0(k,l) = \sum_{n=0}^{N-1} \sum_{m=0}^{M/q-1} y(n, q(2M/q-1-m) + \lambda(q))$$
$$\cdot \cos \frac{\pi(4n+1)k}{2N} \cos \frac{\pi(4m+1)l}{2M/q}. \tag{7.174}$$

It can be easily proven that $X_0(k, l+u\frac{2M}{q}) = (-1)^u X_0(k,l)$ and $X_0(k, \frac{2M}{q}-l) = -X_0(k,l)$ for an arbitrary integer u. If M/q is a multiple of q, (7.172) can be further decomposed in the same way.

We now consider the computation of (7.171). If we define $p(m) \equiv (4p+1)m+p \bmod N$, then the following lemma holds.

Lemma 13 Let $S_1 = \{(n,m) \mid 0 \le n \le N-1,\ 0 \le m \le M-1, m \not\equiv \lambda(q) \bmod q\}$, $S_2 = \{(p(m), m) \mid 0 \le p \le N-1,\ 0 \le m \le M-1, m \not\equiv \lambda(q) \bmod q\}$, Then $S_1 = S_2$.

Proof. It is obvious that $S_2 \subseteq S_1$ and the number of elements in S_1 is the same as that in S_2. Therefore, it is sufficient to prove that the elements in S_2 are different from each other. Let $(p(m), m)$ and $(p'(m'), m')$ be two elements in S_2. If they are equal, then $p(m) = p'(m'), m = m'$. From the definition of $p(m)$, we have $(4p+1)m+p \equiv (4p'+1)m+p' \bmod N$. Hence $(4m+1)(p-p') \equiv 0 \bmod N$. Since $m \not\equiv \lambda(q) \bmod q$, that is, $4m+1$ and N are relatively prime with each other, we have $p \equiv p' \bmod N$. Therefore, $p = p'$.

Based on the lemma and the properties of trigonometric identities, (7.171) can be computed by

$$X_1(k,l) = \frac{A(k,l) + B(k,l)}{2}, \tag{7.175}$$

where

$$A(k,l) = \sum_{p=0}^{N-1} \sum_{m \not\equiv \lambda(q) \bmod q}^{M-1} y(p(m), m) \cos \left(\frac{\pi(4p(m)+1)k}{2N} + \frac{\pi(4m+1)l}{2M} \right) \tag{7.176}$$

$$B(k,l) = \sum_{p=0}^{N-1} \sum_{m \not\equiv \lambda(q) \bmod q}^{M-1} y(p(m), m) \cos \left(\frac{\pi(4p(m)+1)k}{2N} - \frac{\pi(4m+1)l}{2M} \right). \tag{7.177}$$

From the definition of $p(m)$, it can easily be proven that $4p(m)+1 \equiv (4m+1)(4p+1) \bmod 4N$. After applying this equation in (7.176) and (7.177), we obtain

$$A(k,l) = \sum_{p=0}^{N-1} \sum_{m \not\equiv \lambda(q) \bmod q}^{M-1} y(p(m), m) \cos \left(\frac{\pi(4m+1)(4p+1)k}{2N} + \frac{\pi(4m+1)l}{2M} \right)$$
$$= \sum_{p=0}^{N-1} \sum_{m \not\equiv \lambda(q)}^{M-1} y(p(m), m) \cos \frac{\pi(4m+1)(q^J(4p+1)k+l)}{2M} \tag{7.178}$$

$$B(k,l) = \sum_{p=0}^{N-1} \sum_{m \not\equiv \lambda(q) \bmod q}^{M-1} y(p(m), m) \cos \frac{\pi(4m+1)(q^J(4p+1)k-l)}{2M}. \tag{7.179}$$

By defining

$$V_p(l) = \sum_{\substack{m \not\equiv \lambda(q) \bmod q}}^{M-1} y(p(m), m) \cos \frac{\pi(4m+1)l}{2M}, \tag{7.180}$$

where $p = 0, 1, \ldots, N-1$ and $l = 0, 1, \ldots, M-1$, we can express (7.178) and (7.179) into

$$A(k, l) = \sum_{p=0}^{N-1} V_p[q^J(4p+1)k + l] \tag{7.181}$$

$$B(k, l) = \sum_{p=0}^{N-1} V_p[q^J(4p+1)k - l], \tag{7.182}$$

where $k = 0, 1, \ldots, N-1$ and $l = 0, 1, \ldots, M-1$. For convenience of presentation, we define a new sequence

$$W(l) = \sum_{\substack{m \not\equiv \lambda(q) \bmod q}}^{M-1} E(m) \cos \frac{\pi(4m+1)l}{2M} \tag{7.183}$$

or

$$W(l) = \sum_{\substack{m \not\equiv \frac{q-1}{2} \bmod q}}^{M-1} F(m) \cos \frac{\pi(2m+1)l}{2M}, \tag{7.184}$$

where $l = 0, 1, \ldots, M-1$. Between (7.183) and (7.184), there exist relations $F(2m) = E(m)$ $(m = 0, 1, \ldots, \frac{M-1}{2})$ and $F(2m+1) = E(M-1-m)$ $(m = 0, 1, \ldots, \frac{M-1}{2} - 1)$. (7.184) is the same as the 1D type-II DCT except that those terms satisfying $m \equiv \frac{q-1}{2} \bmod q$ are ignored in the summation. Accordingly, (7.184) is known as a reduced 1D type-II DCT. For different p values, (7.180) specifies N reduced 1D type-II DCTs of length M. The computation of the reduced 1D type-II DCT will be discussed in a later subsection.

Both (7.181) and (7.182) are summations of $V_p(l)$ which has the properties

$$V_p(l + u2M) = (-1)^u V_p(l), \quad V_p(2M - l) = -V_p(l), \quad V_p(M) = 0.$$

By using these properties, $A(k, l)$ and $B(k, l)$ can be computed from $V_p(l)$ for $l = 0, 1, \ldots, M - 1$. In general, a direct computation of (7.181) and (7.182) needs $2NM(N-1)$ additions. However, (7.181) and (7.182) can be equivalently computed by using the PT, which substantially reduces the computational complexity.

Based on the properties of $V_p(l)$, it can be proven that

$$\begin{aligned} A(k, 2M - l) &= \sum_{p=0}^{N-1} V_p[q^J(4p+1)k + 2M - l] \\ &= -\sum_{p=0}^{N-1} V_p[q^J(4p+1)k - l] \\ &= -B(k, l) \end{aligned} \tag{7.185}$$

and

$$A(k, 0) = B(k, 0), \quad A(0, l) = B(0, l). \tag{7.186}$$

Now let us generate polynomials

$$B_k(z) = \sum_{l=0}^{M-1} B(k,l)z^l - \sum_{l=M}^{2M-1} A(k, 2M - l)z^l. \tag{7.187}$$

By using the property in (7.185), (7.187) becomes

$$B_k(z) = \sum_{l=0}^{M-1} B(k,l)z^l + \sum_{l=M}^{2M-1} B(k,l)z^l = \sum_{l=0}^{2M-1} B(k,l)z^l$$

from which $A(k,l)$ and $B(k,l)$ can be derived. $B_k(z)$ can be expressed by

$$B_k(z) \equiv \sum_{p=0}^{N-1} \sum_{l=0}^{2M-1} V_p(q^J(4p+1)k - l)z^l \bmod z^{2M} + 1$$

$$\equiv \sum_{p=0}^{N-1} \sum_{l=0}^{2M-1} V_p(l - q^J(4p+1)k)z^l \bmod z^{2M} + 1$$

$$\equiv \sum_{p=0}^{N-1} \sum_{l=0}^{2M-1} V_p(l)z^{l + q^J(4p+1)k} \bmod z^{2M} + 1$$

$$\equiv \left(\sum_{p=0}^{N-1} U_p(z)\hat{z}^{pk} \right) z^{q^J k} \bmod z^{2M} + 1$$

$$\equiv C_k(z)z^{q^J k} \bmod z^{2M} + 1, \tag{7.188}$$

where

$$U_p(z) \equiv \sum_{l=0}^{2M-1} V_p(l)z^l \bmod z^{2M} + 1 \tag{7.189}$$

$$C_k(z) \equiv \sum_{p=0}^{N-1} U_p(z)\hat{z}^{pk} \bmod z^{2M} + 1, \tag{7.190}$$

where $k = 0, 1, \ldots, N-1$ and $\hat{z} \equiv z^{4q^J} \bmod z^{2M} + 1$.

Since $\hat{z}^N \equiv 1 \bmod z^{2M} + 1$, (7.190) is a pseudo PT which can be computed by a fast algorithm to be discussed in the next subsection. The main steps of the algorithm are given below.

Algorithm 18 *PT-based radix-q algorithm for 2D type-II DCT*

Step 1. *Compute N 1D reduced type-II DCTs of length M according to (7.180).*

Step 2. *Compute a pseudo PT according to (7.190) and then $B_k(z)$ according to (7.188).*

Step 3. *Add $A(k,l)$ and $B(k,l)$, which can be obtained from the definition of $B_k(z)$ in (7.187), to form $X^1(k,l)$ according to (7.175).*

Step 4. *Compute a 2D type-II DCT with size $N \times M/q$ in (7.172) by recursively using Step 1 through Step 3.*

Step 5. *Compute $X(k,l)$ according to (7.169).*

7.6.2   Fast polynomial transform

The pseudo PT in (7.190) can be computed by the FPT algorithm which generally requires $2(q-1)MN \log_q N$ additions (see Chapter 2). However, the number of additions can be further reduced by using the symmetric property of the input polynomial sequence $U_p(z)$. It is noted that the coefficients of $U_p(z)$ are $V_p(l)$ ($l = 0, 1, \ldots, 2M-1$), which satisfies $V_p(2M-l) = -V_p(l)$ and $V_p(M) = 0$. It indicates that only one-half of the coefficients are independent. This property can be expressed as

$$U_p(z) \equiv U_p(z^{-1}) \bmod z^{2M} + 1.$$

Based on this property, it can be proven that the polynomial sequence $C_k(z)$ in (7.190) also has a symmetric property expressed as

$$C_{N-k}(z) \equiv C_k(z^{-1}) \bmod z^{2M} + 1.$$

Therefore, only one-half of the $C_k(z)$ need to be computed. The FPT algorithm computes a PT by $s = \log_q N$ stages. The computation of each stage produces temporary outputs. We will prove that these temporary output polynomials also have symmetric properties that provide further savings in the number of additions.

We use the radix-q decimation-in-time FPT, which is discussed in Chapter 2, to compute (7.190). Let $C_n^0(z)$ be the q-ary inverse of $U_n(z)$. Then, the s stages of the algorithm are as follows:

$$C^j_{\hat{n}q^j + \hat{k} + k_{j-1}q^{j-1}}(z) \equiv \sum_{n_{s-j}=0}^{q-1} \left[C^{j-1}_{\hat{n}q^j + \hat{k} + n_{s-j}q^{j-1}}(z) \hat{z}^{q^{s-j}n_{s-j}\hat{k}} \right]$$

$$\cdot \hat{z}^{q^{s-1}n_{s-j}k_{j-1}} \bmod z^{2M} + 1, \qquad (7.191)$$

where $\hat{n} = 0, 1, \ldots, q^{s-j}-1$, $\hat{k} = 0, 1, \ldots, q^{j-1}-1$, $k_{j-1} = 0, 1, \ldots, q-1$, $j = 1, 2, \ldots, s$ and

$$C^j_{\hat{n}q^j + \hat{k}}(z) \equiv \sum_{\bar{n}=0}^{q^j-1} U_{\bar{n}q^{s-j}+n'}(z) \hat{z}^{q^{s-j}\bar{n}\bar{k}} \bmod z^{2M} + 1, \qquad (7.192)$$

where $\bar{k} = 0, 1, \ldots, q^j-1$, $\hat{n} = 0, 1, \ldots, q^{s-j}-1$ and n' is the q-ary inverse of $\hat{n}$. The final stage produces the output, that is, $C_k(z) = C_k^s(z)$.

In general, (7.191) needs $2(q-1)MN$ additions. However, the number of additions can be reduced by one-half if the symmetric property is fully used.

Lemma 14 *The output polynomials of stage j have the following symmetric properties:*

$$C^j_{\hat{n}q^j}(z^{-1}) \equiv C^j_{\hat{n}q^j}(z) \bmod z^{2M} + 1 \qquad (7.193)$$

$$C^j_{\hat{n}q^j + q^j - \bar{k}}(z^{-1}) \equiv C^j_{\hat{n}q^j + \bar{k}}(z) \bmod z^{2M} + 1, \qquad (7.194)$$

where $\bar{k} = 1, 2, \ldots, q^j - 1$.

Proof. From (7.192) we have

$$
\begin{aligned}
C^j_{\hat{n}q^j+q^j-\bar{k}}(z^{-1}) &\equiv \sum_{\bar{n}=0}^{q^j-1} H_{\bar{n}q^{s-j}+n'}(z)\hat{z}^{-q^{s-j}\bar{n}(q^j-\bar{k})} \bmod z^{2M}+1 \\
&\equiv \sum_{\bar{n}=0}^{q^j-1} H_{\bar{n}q^{s-j}+n'}(z)\hat{z}^{q^{s-j}\bar{n}\bar{k}} \bmod z^{2M}+1 \\
&\equiv C^j_{\hat{n}q^j+\bar{k}}(z) \bmod z^{2M}+1.
\end{aligned}
$$

Similarly, (7.193) can be proven.

The symmetric properties mean that we need to compute $C^j_{\hat{n}q^j+\bar{k}}(z)$ for $\bar{k}=1$ to $\frac{q^j-1}{2}$ only, and $C^j_{\hat{n}q^j+\bar{k}}(z)$ for $\bar{k}=\frac{q^j+1}{2}$ to q^j-1 can be obtained by using the symmetric property. Furthermore, the property in (7.193) shows that it is enough to compute only one-half of the total polynomial coefficients for the polynomial $C^j_{\hat{n}q^j}(z)$. The rest of the coefficients can be achieved from the properties without requiring any computation. Therefore, (7.191) needs $(q-1)NM$ additions only. It is noted from (7.187) that the Mth coefficient of each output polynomial is $B(k,M)$ which is not required by our computation. Therefore, a further savings of $(N-1)/2$ additions can be made. In total, the number of additions required by (7.190) is $(q-1)NM\log_q N - (N-1)/2$.

7.6.3 Radix-q algorithm for reduced type-II DCT

In general, many fast algorithms for the 1D type-II DCT, for example, the algorithms in Chapter 6 or in [2, 7, 6, 15, 17, 28] can be amended for the reduced 1D type-II DCT. This subsection presents a fast algorithm for the reduced 1D type-II DCT based on the fast algorithm reported in [28, 27].

We divide $W(l)$ ($l=0,1,\ldots,M-1$), as defined in (7.184), into $C_j(l)=W(ql+j)$ ($j=0,1,\ldots,q-1$, $l=0,1,\ldots,M/q-1$) and show that $C_j(l)$ can be expressed as the reduced 1D type-II DCT of length M/q. For example, we have

$$
\begin{aligned}
C_0(l) = W(ql) &= \sum_{\substack{m=0 \\ m\not\equiv(q-1)/2}}^{M-1} F(m)\cos\frac{\pi(2m+1)l}{2M} \\
&= \sum_{\substack{m=0 \\ m\not\equiv(q-1)/2}}^{M/q-1} d_0(m)\cos\frac{\pi(2m+1)l}{2M/q}
\end{aligned}
\tag{7.195}
$$

which is a reduced 1D type-II DCT of length M/q whose input sequence is

$$
d_0(m) = F(m) + \sum_{i=0}^{(q-1)/2}\left[F(\frac{2iM}{q}+m) + F(\frac{2iM}{q}-1-m)\right],
\tag{7.196}
$$

where $m = 0, 1, \ldots, M/q - 1$ and $m \not\equiv \frac{q-1}{2} \bmod q$. Now let us consider $D_j(l) = C_j(l) + C_{q-j}(l-1)$ $(l = 0, 1, \ldots, M/q - 1)$ which can be expressed as

$$
\begin{aligned}
D_j(l) &= W(ql + j) + W(ql - j) \\
&= \sum_{\substack{m=0 \\ m \not\equiv (q-1)/2}}^{M-1} 2F(m) \cos \frac{\pi(2m+1)j}{2M} \cos \frac{\pi(2m+1)l}{2M/q} \\
&= \sum_{\substack{m=0 \\ m \not\equiv (q-1)/2}}^{M/q-1} d_j(m) \cos \frac{\pi(2m+1)l}{2M/q}.
\end{aligned}
\tag{7.197}
$$

Equation (7.197) is again a reduced 1D type-II DCT of input sequence

$$
\begin{aligned}
d_j(m) &= 2F(m) \cos \frac{\pi(2m+1)j}{2M} \\
&+ \sum_{i=0}^{(q-1)/2} \left\{ 2F(\frac{2iM}{q} + m) \cos \frac{\pi(4iM/q + 2m + 1)j}{2M} \right. \\
&\left. + 2F(\frac{2iM}{q} - 1 - m) \cos \frac{\pi(4iM/q - 2m - 1)j}{2M} \right\}.
\end{aligned}
\tag{7.198}
$$

By using the trigonometric properties, the computation of $d_j(m)$ can be simplified. If we define

$$
\begin{aligned}
f_0(m) &= F(m) \\
f_i(m) &= F(\frac{2iM}{q} - 1 - m) + F(\frac{2iM}{q} + m) \\
g_i(m) &= F(\frac{2iM}{q} - 1 - m) - F(\frac{2iM}{q} + m),
\end{aligned}
\tag{7.199}
$$

where $i = 1, 2, \ldots, (q-1)/2$, $0 \le m \le M/q - 1$ and $m \not\equiv (q-1)/2 \bmod q$, then (7.196) and (7.198) can be expressed by

$$
\begin{aligned}
d_0(m) &= \overline{C}_0(m) \\
d_j(m) &= \overline{C}_j(m) 2 \cos \frac{\pi(2m+1)j}{2M} + \overline{S}_j(m) 2 \sin \frac{\pi(2m+1)j}{2M},
\end{aligned}
\tag{7.200}
$$

where $j = 1, 2, \ldots, q - 1$, $0 \le m \le M/q - 1$, $m \not\equiv \frac{q-1}{2} \bmod q$ and

$$
\begin{aligned}
\overline{C}_j(m) &= \sum_{i=0}^{(q-1)/2} f_i(m) \cos \frac{2\pi ij}{q} \\
\overline{S}_j(m) &= \sum_{i=1}^{(q-1)/2} g_i(m) \sin \frac{2\pi ij}{q}.
\end{aligned}
\tag{7.201}
$$

The reduced 1D type-II DCT $W(l)$ can be obtained from $C_0(l)$ and $D_j(l)$ according to

$$
\begin{aligned}
C_0(l) &= W(ql), \quad C_j(0) = \frac{D_j(0)}{2} \\
C_j(l) &= D_j(l) - C_{q-j}(l-1)
\end{aligned}
\tag{7.202}
$$

for $l = 0, 1, \ldots, M/q - 1$ and $j = 1, 2, \ldots, q - 1$. Therefore, the reduced 1D type-II DCT of length M is decomposed into q reduced 1D type-II DCTs of length M/q at the decomposition cost required by (7.199)–(7.202). The main steps of the decomposition are listed below.

Step 1.   Compute (7.199) to get $f_i(m)$ and $g_i(m)$.

Step 2.   Compute (7.201) to get $\overline{C}_j(m)$ and $\overline{S}_j(m)$.

Step 3.   Compute (7.200) to get $d_j(m)$.

Step 4.   Compute the reduced type-II DCTs of $d_j(m)$ with length M/q to get $C_0(l)$ and $D_j(l)$.

Step 5.   Get $C_j(l)$ according to (7.202).

With the assumption that the computation in Step 2 for a fixed m needs $\alpha(q)$ multiplications and $\beta(q)$ additions, the total number of operations for Steps 1 through 3 and Step 5 are $(\alpha(q) + 2q - 2)(\frac{M}{q} - \frac{M}{q^2})$ multiplications and $(\beta(q) + 2q - 2)(\frac{M}{q} - \frac{M}{q^2}) + (q - 1)(\frac{M}{q} - 1)$ additions. We can use the same technique to further decompose the reduced 1D type-II in Step 4 if M/q, which is the length of the subsequence, is divisible by q. Let $\varphi(M)$ and $\psi(M)$ represent the numbers of multiplications and additions, respectively, for the reduced 1D type-II of length M. If the recursive decomposition process stops until the sequence length becomes q^l, the number of arithmetic operations becomes

$$\varphi(q^t) = (\alpha(q) + 2q - 2)(\frac{1}{q} - \frac{1}{q^2})(t - l)q^t + q^{t-l}\varphi(q^l)$$

$$\psi(q^t) = (\beta(q) + 2q - 2)(\frac{1}{q} - \frac{1}{q^2})(t - l)q^t + \frac{q - 1}{q}(t - l)q^t$$
$$- q^{t-l} + 1 + q^{t-l}\psi(q^l).$$

The above described decomposition procedures are valid for arbitrary odd number q. However, if q is small such as $q = 3$ and $q = 5$, some arithmetic operations can be saved by an optimized implementation. The fast algorithms for these two cases are given as examples.

Algorithm for $q = 3$

Step 1.   For $m = 0, 1, \ldots, 3^{t-1} - 1$ and $m \not\equiv 1 \bmod 3$, compute

$$
\begin{aligned}
T_1(m) &= F(2 \cdot 3^{t-1} - 1 - m) + F(2 \cdot 3^{t-1} + m) \\
T_2(m) &= F(2 \cdot 3^{t-1} - 1 - m) - F(2 \cdot 3^{t-1} + m) \\
T_3(m) &= 2F(m) - T_1(m) \\
d_0(m) &= F(m) + T_1(m) \\
d_1(m) &= T_3(m) \cos \frac{(2m + 1)\pi}{2M} + T_2(m)\sqrt{3} \sin \frac{(2m + 1)\pi}{2M} \\
d_2(m) &= \left[T_3(m) \cos \frac{(2m + 1)\pi}{2M} - T_2(m)\sqrt{3} \sin \frac{(2m + 1)\pi}{2M} \right] \\
&\quad \cdot 2\cos \frac{(2m + 1)\pi}{2M} - T_3(m).
\end{aligned}
$$

Step 2.   Compute the reduced 1D type-II DCT of $d_j(m)$ to get $D_j(l)$ ($l = 0, 1, \ldots, 3^{t-1} - 1$, $j = 0, 1, 2$).

Step 3.   Compute $C_j(l) = W(3l + j)$ by

$$C_0(l) = D_0(l)$$

$$C_1(0) = \frac{D_1(0)}{2}, \qquad\qquad C_2(0) = \frac{D_2(0)}{2}$$

$$C_1(l) = D_1(l) - C_2(l - 1), \quad C_2(l) = D_1(l) - C_1(l - 1),$$

where $l = 0, 1, \ldots, 3^{t-1} - 1$.

If the decomposition process terminates until the subsequence length is 3 and the length 3 reduced 1D type-II DCT needs 1 multiplication and 2 additions plus 1 shift, we have

$$\varphi(3^t) = (2t - 1)3^{t-1}, \quad \psi(3^t) = (20t - 17)3^{t-2}.$$

Algorithm for $q = 5$

The computation defined in Step 2 (or (7.201)) for $q = 5$ can be expressed explicitly by

$$\overline{C}_0(m) = f_0(m) + f_1(m) + f_2(m)$$

$$\overline{C}_1(m) = [f_1(m) - f_2(m)] \cos \frac{2\pi}{5} - \frac{f_2(m)}{2} + f_0(m)$$

$$\overline{C}_2(m) = -[f_1(m) - f_2(m)] \cos \frac{2\pi}{5} - \frac{f_1(m)}{2} + f_0(m)$$

$$\overline{S}_1(m) = [g_1(m) + 2g_2(m) \cos \frac{2\pi}{5}] \sin \frac{2\pi}{5}$$

$$\overline{S}_2(m) = [2g_1(m) \cos \frac{2\pi}{5} - g_2(m)] \sin \frac{2\pi}{5}.$$

It is shown in [2] that the reduced length 5 type-II DCT needs 4 multiplications and 11 additions. Therefore, the computational complexity of the reduced 1D type-II DCT of length 5^t is

$$\varphi(5^t) = (44t - 24)5^{t-2}, \quad \psi(5^t) = (88t - 38)5^{t-2} + 1.$$

7.6.4   PT-based radix-q algorithm for rD type-II DCT

This subsection generalizes the method previously used for the computation of the rD type-II DCT defined by

$$X(k_1, k_2, \ldots, k_r) = \sum_{n_1=0}^{N_1-1} \sum_{n_2=0}^{N_2-1} \cdots \sum_{n_r=0}^{N_r-1} x(n_1, n_2, \ldots, n_r)$$

$$\cdot \cos \frac{\pi(2n_1 + 1)k_1}{2N_1} \cdots \cos \frac{\pi(2n_r + 1)k_r}{2N_r},$$

where $k_i = 0, 1, \ldots, N_i - 1$ for $i = 1, 2, \ldots, r$, $N_r = q^t$ and $N_r/N_i = q^{l_i}$ for $l_i \geq 0$ and $i = 1, 2, \ldots, r - 1$. We rearrange the input array $x(n_1, \ldots, n_r)$ by

$$y(n_1, n_2, \ldots, n_r) = x(2n_1, 2n_2, \ldots, 2n_r)$$

$$y(N_1 - 1 - n_1, n_2, \ldots, n_r) = x(2n_1 + 1, 2n_2, \ldots, 2n_r)$$

$$\cdots$$

$$y(N_1 - 1 - n_1, \ldots, N_r - 1 - n_r) = x(2n_1 + 1, \ldots, 2n_r + 1),$$

where $0 \le 2n_i \le (N_i - 1)$ or $0 \le 2n_i + 1 \le (N_i - 1)$ $(i = 1, 2, \ldots, r)$, then the rD type-II DCT becomes

$$X(k_1, k_2, \ldots, k_r) = \sum_{n_1=0}^{N_1-1} \sum_{n_2=0}^{N_2-1} \cdots \sum_{n_r=0}^{N_r-1} y(n_1, n_2, \ldots, n_r)$$
$$\cdot \cos \frac{\pi(4n_1 + 1)k_1}{2N_1} \cdots \cos \frac{\pi(4n_r + 1)k_r}{2N_r}, \tag{7.203}$$

where $k_i = 0, 1, \ldots, N_i - 1$ and $i = 1, 2, \ldots, r$.

Let $\lambda(q)$ represent the unique non-negative integer smaller than q and satisfying $4\lambda(q) + 1 \equiv 0 \bmod q$. Then, (7.203) can be decomposed into two parts as

$$X(k_1, k_2, \ldots, k_r) = X_0(k_1, k_2, \ldots, k_r) + X_1(k_1, k_2, \ldots, k_r), \tag{7.204}$$

where the first part can be expressed as (or similar to the 2D case in (7.174))

$$X_0(k_1, k_2, \ldots, k_r) = \sum_{n_1=0}^{N_1-1} \cdots \sum_{n_{r-1}=0}^{N_{r-1}-1} \sum_{n_r=0}^{N_r/q-1} y(n_1, \ldots, n_{r-1}, qn_r + \lambda(q))$$
$$\cdot \cos \frac{\pi(4n_1 + 1)k_1}{2N_1} \cdots \cos \frac{\pi(4n_{r-1} + 1)k_{r-1}}{2N_{r-1}} \cos \frac{\pi(4n_r + 1)k_r}{2N_r/q} \tag{7.205}$$

and

$$X_1(k_1, k_2, \ldots, k_r) = \sum_{n_1=0}^{N_1-1} \cdots \sum_{n_{r-1}=0}^{N_{r-1}-1} \sum_{n_r \not\equiv \lambda(q)\bmod q}^{N_r-1} y(n_1, \ldots, n_{r-1}, n_r)$$
$$\cdot \cos \frac{\pi(4n_1 + 1)k_1}{2N_1} \cdots \cos \frac{\pi(4n_r + 1)k_r}{2N_r}. \tag{7.206}$$

Noting that $X_0(k_1, \ldots, k_{r-1}, k_r + u\frac{2N_r}{q}) = (-1)^u X_0(k_1, \ldots, k_{r-1}, k_r)$ and $X_0(k_1, \ldots, k_{r-1}, \frac{2N_r}{q} - k_r) = -X_0(k_1, \ldots, k_{r-1}, k_r)$, we know that (7.205) can be computed by an rD type-II DCT with size $N_1 \times \cdots \times N_{r-1} \times N_r/q$.

Now we focus our attention on (7.206). If we define

$$A(k_1, k_2, \ldots, k_r) = \sum_{n_1=0}^{N_1-1} \cdots \sum_{n_{r-1}=0}^{N_{r-1}-1} \sum_{n_r \not\equiv \lambda(q)\bmod q}^{N_r-1} y(n_1, n_2, \ldots, n_r)$$
$$\cdot \cos \frac{\pi(4n_1 + 1)k_1}{2N_1} \cdots \cos \frac{\pi(4n_{r-2} + 1)k_{r-2}}{2N_{r-2}}$$
$$\cdot \cos \left(\frac{\pi(4n_{r-1} + 1)k_{r-1}}{2N_{r-1}} + \frac{\pi(4n_r + 1)k_r}{2N_r} \right)$$

and

$$B(k_1, k_2, \ldots, k_r) = \sum_{n_1=0}^{N_1-1} \cdots \sum_{n_{r-1}=0}^{N_{r-1}-1} \sum_{n_r \not\equiv \lambda(q)\bmod q}^{N_r-1} y(n_1, n_2, \ldots, n_r)$$
$$\cdot \cos \frac{\pi(4n_1 + 1)k_1}{2N_1} \cdots \cos \frac{\pi(4n_{r-2} + 1)k_{r-2}}{2N_{r-2}}$$
$$\cdot \cos \left(\frac{\pi(4n_{r-1} + 1)k_{r-1}}{2N_{r-1}} - \frac{\pi(4n_r + 1)k_r}{2N_r} \right),$$

then (7.206) can be computed by

$$X_1(k_1, k_2, \ldots, k_r) = \frac{A(k_1, k_2, \ldots, k_r) + B(k_1, k_2, \ldots, k_r)}{2}. \tag{7.207}$$

Let $p_{r-1}(n_r)$ be the least non-negative remainder of $(4p_{r-1} + 1)n_r + p_{r-1}$ modulo N_{r-1}. Similar to the 2D case, we have

$$
\begin{aligned}
A(k_1, k_2, \ldots, k_r) = & \sum_{n_1=0}^{N_1-1} \cdots \sum_{n_{r-2}=0}^{N_{r-2}-1} \sum_{p_{r-1}=0}^{N_{r-1}-1} \sum_{\substack{n_r=0 \\ n_r \not\equiv \lambda(q)\bmod q}}^{N_r-1} y(n_1, \ldots, n_{r-2}, p_{r-1}(n_r), n_r) \\
& \cdot \cos\frac{\pi(4n_1+1)k_1}{2N_1} \cdots \cos\frac{\pi(4n_{r-2}+1)k_{r-2}}{2N_{r-2}} \\
& \cdot \cos\frac{\pi(4n_r+1)(q^{l_{r-1}}(4p_{r-1}+1)k_{r-1}+k_r)}{2N_r}
\end{aligned} \tag{7.208}
$$

and

$$
\begin{aligned}
B(k_1, k_2, \ldots, k_r) = & \sum_{n_1=0}^{N_1-1} \cdots \sum_{n_{r-2}=0}^{N_{r-2}-1} \sum_{p_{r-1}=0}^{N_{r-1}-1} \sum_{\substack{n_r=0 \\ n_r \not\equiv \lambda(q)\bmod q}}^{N_r-1} y(n_1, \ldots, n_{r-2}, p_{r-1}(n_r), n_r) \\
& \cdot \cos\frac{\pi(4n_1+1)k_1}{2N_1} \cdots \cos\frac{\pi(4n_{r-2}+1)k_{r-2}}{2N_{r-2}} \\
& \cdot \cos\frac{\pi(4n_r+1)(q^{l_{r-1}}(4p_{r-1}+1)k_{r-1}-k_r)}{2N_r}.
\end{aligned} \tag{7.209}
$$

If we define

$$
\begin{aligned}
V_{k_1,\ldots,k_{r-2},p_{r-1}}(l) = & \sum_{n_1=0}^{N_1-1} \cdots \sum_{n_{r-2}=0}^{N_{r-2}-1} \sum_{\substack{n_r=0 \\ n_r \not\equiv \lambda(q)\bmod q}}^{N_r-1} y(n_1, \ldots, n_{r-2}, p_{r-1}(n_r), n_r) \\
& \cdot \cos\frac{\pi(4n_1+1)k_1}{2N_1} \cdots \cos\frac{\pi(4n_{r-2}+1)k_{r-2}}{2N_{r-2}} \cos\frac{\pi(4n_r+1)l}{2N_r}
\end{aligned} \tag{7.210}
$$

to be the $(r-1)$D reduced type-II DCT with size $N_1 \times \cdots \times N_{r-2} \times N_r$, then (7.208) and (7.209) become

$$A(k_1, k_2, \ldots, k_r) = \sum_{p_{r-1}=0}^{N_{r-1}-1} V_{k_1,\ldots,k_{r-2},p_{r-1}}(q^{l_{r-1}}(4p_{r-1}+1)k_{r-1}+k_r)$$

$$B(k_1, k_2, \ldots, k_r) = \sum_{p_{r-1}=0}^{N_{r-1}-1} V_{k_1,\ldots,k_{r-2},p_{r-1}}(q^{l_{r-1}}(4p_{r-1}+1)k_{r-1}-k_r).$$

We form a polynomial

$$
\begin{aligned}
B_{k_1,k_2,\ldots,k_{r-1}}(z) = & \sum_{k_r=0}^{N_r-1} B(k_1, k_2, \ldots, k_r)z^{k_r} - \sum_{k_r=1}^{N_r} A(k_1, k_2, \ldots, 2N_r - k_r)z^{k_r} \\
= & \sum_{k_r=0}^{2N_r-1} B(k_1, k_2, \ldots, k_r)z^{k_r},
\end{aligned} \tag{7.211}
$$

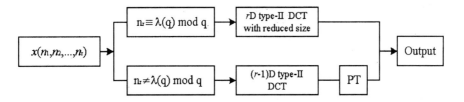

Figure 7.8   Flow chart of PT-based radix-q algorithm for rD type-II DCT.

which can be expressed as

$$B_{k_1,k_2,\ldots,k_{r-1}}(z) \equiv C_{k_1,k_2,\ldots,k_{r-1}}(z)z^{q^{l_{r-1}k_{r-1}}} \bmod z^{2N_r} + 1, \qquad (7.212)$$

where $k_i = 0, 1, \ldots, N_i - 1$, $i = 1, 2, \ldots, r-1$, $\hat{z} \equiv z^{4q^{l_{r-1}}} \bmod z^{2N_r} + 1$,

$$C_{k_1,k_2,\ldots,k_{r-1}}(z) \equiv \sum_{p_{r-1}=0}^{N_{r-1}-1} U_{k_1,\ldots,k_{r-2},p_{r-1}}(z)\hat{z}^{p_{r-1}k_{r-1}} \bmod z^{2N_r} + 1 \qquad (7.213)$$

and

$$U_{k_1,\ldots,k_{r-2},p_{r-1}}(z) = \sum_{l=0}^{2N_r-1} V_{k_1,\ldots,k_{r-2},p_{r-1}}(l)z^l,$$

which is the generating polynomial of $V_{k_1,\ldots,k_{r-2},p_{r-1}}(l)$.

The flow chart of the algorithm is given in Figure 7.8 and the main steps are summarized below.

Algorithm 19 *PT-based radix-q algorithm for rD type-II DCT*

Step 1.   Compute N_{r-1} $(r-1)D$ reduced type-II DCTs with size $N_1 \times \cdots \times N_{r-2} \times N_r$ according to (7.210).

Step 2.   Compute $N_1 N_2 \cdots N_{r-2}$ PTs according to (7.213) and then compute $B_{k_1,k_2,\ldots,k_{r-1}}(z)$ according to (7.212).

Step 3.   Based on $B_{k_1,k_2,\ldots,k_{r-1}}(z)$ defined in (7.211), $A(k_1,k_2,\ldots,k_r)$ and $B(k_1,k_2,\ldots,k_r)$ are derived to form $X^1(k_1,k_2,\ldots,k_r)$ according to (7.207).

Step 4.   Compute an rD type-II DCT with size $N_1 \times \cdots \times N_{r-1} \times N_r/q$ according to (7.205).

Step 5.   Compute $X(k_1,k_2,\ldots,k_r)$ according to (7.204).

7.6.5   Computational complexity

The described fast algorithm needs to compute (7.205), an rD type-II DCT with size $N_1 \times \cdots \times N_{r-1} \times N_r/q$, and (7.206), an rD reduced type-II DCT with size $N_1 \times \cdots \times N_{r-1} \times N_r$. Let $\varphi(N_1,\ldots,N_{r-1},N_r)$ and $\psi(N_1,\ldots,N_{r-1},N_r)$ represent the numbers of multiplications and additions, respectively, for the rD reduced type-II DCT with size $N_1 \times \cdots \times N_{r-1} \times N_r$, then we have

$$\varphi(N_1, \ldots, N_{r-1}, N_r) = N_{r-1}\varphi(N_1, \ldots, N_{r-2}, N_r) \tag{7.214}$$

$$\psi(N_1, \ldots, N_{r-1}, N_r) = N_{r-1}\psi(N_1, \ldots, N_{r-2}, N_r) + N$$
$$+(q-1)N\log_q N_{r-1}, \tag{7.215}$$

where $N = N_1 N_2 \cdots N_r$. Closed form expressions for (7.214) and (7.215) can be derived as

$$\varphi(N_1, \ldots, N_{r-1}, N_r) = \frac{N}{N_r}\varphi(N_r) \tag{7.216}$$

$$\psi(N_1, \ldots, N_{r-1}, N_r) = \frac{N}{N_r}\psi(N_r) + (r-1)N + (q-1)N\log_q \frac{N}{N_r}. \tag{7.217}$$

Therefore, the number of multiplications and additions for the rD type-II DCT with size $N_1 \times \cdots \times N_{r-1} \times N_r$ are as follows:

$$M_u(N_1, \ldots, N_{r-1}, N_r) = \frac{N}{N_r}\varphi(N_r) + M_u(N_1, \ldots, N_{r-1}, \frac{N_r}{q}) \tag{7.218}$$

$$A_d(N_1, \ldots, N_{r-1}, N_r) = \frac{N}{N_r}\psi(N_r) + rN + (q-1)N\log_q \frac{N}{N_r}$$
$$+A_d(N_1, \ldots, N_{r-1}, \frac{N_r}{q}). \tag{7.219}$$

It is difficult to derive a closed form expression of the number of operations for arbitrary N_i. When $N_1 = N_2 = \cdots = N_r = q^t$, however, the closed form expression can be obtained. For this case, let us define $M_u(q^t) = M_u(q^t, \ldots, q^t, q^t)$ and $A_d(q^t) = A_d(q^t, \ldots, q^t, q^t)$ for simplicity, and without loss of generality, assume that $\varphi(q^t) = a_1 t q^t + a_2 q^t + a_3$ and $\psi(q^t) = b_1 t q^t + b_2 q^t + b_3$. From (7.218) and (7.219) we can deduce that

$$M_u(q^t) = \frac{a_1 q}{q-1}tq^{rt} + (a_2 - \frac{a_1}{q^r - 1})\frac{q}{q-1}(q^{rt} - 1)$$
$$+\frac{a_3(q^r - 1)}{(q-1)(q^{r-1} - 1)}(q^{(r-1)t} - 1) \tag{7.220}$$

$$A_d(q^t) = (\frac{b_1 q}{q-1} + (r-1)q)tq^{rt} + (b_2 + 1 - \frac{b_1 + q^r - q}{q^r - 1})\frac{q}{q-1}(q^{rt} - 1)$$
$$+\frac{b_3(q^r - 1)}{(q-1)(q^{r-1} - 1)}(q^{(r-1)t} - 1). \tag{7.221}$$

Let $\tau(q^t)$ and $\rho(q^t)$ represent the numbers of multiplications and additions, respectively, required by a 1D type-II DCT of length q^t. Without loss of generality, we assume that $\tau(q^t) = c_1 t q^t + c_2 q^t + c_3$ and $\rho(q^t) = d_1 t q^t + d_2 q^t + d_3$. When $N_1 = N_2 = \cdots = N_r = q^t$, the computational complexity of the row-column method is

$$\overline{M}_u(q^t) = c_1 r t q^{rt} + c_2 r q^{rt} + c_3 r q^{(r-1)t}$$
$$\overline{A}_d(q^t) = d_1 r t q^{rt} + d_2 r q^{rt} + d_3 r q^{(r-1)t}.$$

Compared to the above computational complexity, our fast algorithm achieves a remarkable reduction in the number of multiplications. Based on the relationship between the type-II DCT and the reduced type-II DCT, we have $a_1 = \frac{q-1}{q}c_1$ and $a_2 \approx \frac{q-1}{q}c_2$. Therefore,

$$M_u \approx \frac{1}{r}\overline{M}_u.$$

It is rather difficult to make a comparison on the number of additions needed by the two algorithms. Let us consider two special cases for $q = 3$ and $q = 5$. When $q = 3$, it has been shown earlier that $a_1 = \frac{2}{3}$, $a_2 = -\frac{1}{3}$, $a_3 = 0$, and $b_1 = \frac{20}{9}$, $b_2 = -\frac{17}{9}$, $b_3 = 1$. Therefore, we get

$$M_u(3^t) = t3^{rt} - \frac{3^r + 1}{2(3^r - 1)}(3^{rt} - 1)$$

$$A_d(3^t) = (3r + \frac{1}{3})t3^{rt} - \frac{17 \cdot 3^r - 15}{6(3^r - 1)}(3^{rt} - 1)$$

$$+ \frac{3^r - 1}{2(3^{r-1} - 1)}(3^{(r-1)t} - 1). \tag{7.222}$$

When $q = 5$, the algorithm discussed earlier shows that $a_1 = \frac{44}{25}$, $a_2 = -\frac{24}{25}$, $a_3 = 0$, and $b_1 = \frac{88}{25}$, $b_2 = -\frac{38}{25}$, $b_3 = 1$. We have

$$M_u(5^t) = \frac{11}{5}t5^{rt} - (\frac{6}{5} + \frac{11}{5(5^r - 1)})(5^{rt} - 1)$$

$$A_d(5^t) = (5r - \frac{3}{5})t5^{rt} - (\frac{38}{25} - \frac{12}{25(5^r - 1)})(5^{rt} - 1)$$

$$+ \frac{5^r - 1}{4(5^{r-1} - 1)}(5^{(r-1)t} - 1). \tag{7.223}$$

If the algorithm for the type-II DCT discussed in Chapter 6 is used, we have

$$\overline{M}_u(3^t) = \frac{4}{3}rt3^{rt} - \frac{35}{18}r3^{rt} + 1.5r3^{(r-1)t}$$

$$\overline{A}_d(3^t) = \frac{8}{3}rt3^{rt} - \frac{16}{9}r3^{rt} + 2r3^{(r-1)t} \tag{7.224}$$

and

$$\overline{M}_u(5^t) = \frac{11}{5}rt5^{rt} - \frac{17}{10}r5^{rt} + 1.5r5^{(r-1)t}$$

$$\overline{A}_d(5^t) = \frac{21}{5}rt5^{rt} - 2r5^{rt} + 2r5^{(r-1)t}. \tag{7.225}$$

As seen from these two cases, the number of multiplications needed by the presented algorithm is approximately $\frac{1}{r}$ times that needed by the row-column method. However, the number of additions required by the algorithm is more than that needed by the row-column method. Finally, the total number of arithmetic operations (multiplications plus additions) can be reduced when t is larger, as seen from (7.222)–(7.225). The structure of the algorithm is also simple because it uses only the reduced 1D type-II DCT and the PT, which can be easily implemented.

7.7 PT-BASED RADIX-q ALGORITHM FOR MD TYPE-III DCT

By combining the PT and radix-q decomposition, this section presents an algorithm for the rD type-III DCT whose size is $q^{l_1} \times q^{l_2} \times \cdots \times q^{l_r}$, where q is an odd prime number. The rD type-III DCT is converted into a series of 1D reduced type-III DCTs. For simplicity and easy understanding, we first consider the 2D DCT to illustrate the basic concepts. Some symmetric properties in the PT are shown to be useful for the reduction of the required computational complexity. Then, the concepts used for the 2D type-III DCT are generalized for the rD type-III DCT, where $r \geq 3$. The computational complexity required by the presented and the row-column algorithms will also be discussed.

7.7.1   PT-based radix-q algorithm for 2D type-III DCT

The 2D type-III DCT of the sequence $x(n, m)$ $(n = 0, 1, \ldots, N - 1,\ m = 0, 1, \ldots, M - 1)$ is defined by

$$X(k, l) = \sum_{n=0}^{N-1} \sum_{m=0}^{M-1} x(n, m) \cos \frac{\pi(2k + 1)n}{2N} \cos \frac{\pi(2l + 1)m}{2M}, \qquad (7.226)$$

where $k = 0, 1, \ldots, N - 1$ and $l = 0, 1, \ldots, M - 1$. It is assumed that N and M are powers of an odd prime number q. Without loss of generality, we can assume that $M \geq N$. Therefore, we can write that $N = q^t$ and $M = q^J N$, where $t > 0$, $J \geq 0$, and q is an odd prime number. We convert the output sequence $Y(k, l)$ into

$$
\begin{aligned}
Y(k, l) &= X(2k, 2l), \\
&\quad k = 0, 1, \ldots, \frac{N - 1}{2},\ l = 0, 1, \ldots, \frac{M - 1}{2} \\
Y(N - 1 - k, l) &= X(2k + 1, 2l), \\
&\quad k = 0, 1, \ldots, \frac{N - 3}{2},\ l = 0, 1, \ldots, \frac{M - 1}{2} \\
Y(k, M - 1 - l) &= X(2k, 2l + 1), \\
&\quad k = 0, 1, \ldots, \frac{N - 1}{2},\ l = 0, 1, \ldots, \frac{M - 3}{2} \\
Y(N - 1 - k, M - 1 - l) &= X(2k + 1, 2l + 1), \\
&\quad k = 0, 1, \ldots, \frac{N - 3}{2};\ l = 0, 1, \ldots, \frac{M - 3}{2}
\end{aligned}
\tag{7.227}
$$

so that (7.226) can be expressed by

$$Y(k, l) = \sum_{n=0}^{N-1} \sum_{m=0}^{M-1} x(n, m) \cos \frac{\pi(4k + 1)n}{2N} \cos \frac{\pi(4l + 1)m}{2M}. \qquad (7.228)$$

Let $\lambda(q)$ represent the unique non-negative integer smaller than q and satisfying $4\lambda(q) + 1 \equiv 0 \bmod q$. Therefore, $4\lambda(q) + 1 = \delta q$, where $\delta = 1$ or 3. Then the output of (7.228) can be grouped into two parts, that is, $Y(k, ql + \lambda(q))$ $(l = 0, 1, \ldots, M/q - 1)$ and $Y(k, l)$ $(l = 0, 1, \ldots, M - 1,\ l \not\equiv \lambda(q) \bmod q)$. The first part can be expressed into a 2D type-III DCT with size $N \times M/q$,

$$Y(k, ql + \lambda(q)) = \sum_{n=0}^{N-1} \sum_{m=0}^{M/q-1} \hat{x}(n, m) \cos \frac{\pi(4k + 1)n}{2N} \cos \frac{\pi(4l + \delta)m}{2M/q}, \qquad (7.229)$$

where $k = 0, 1, \ldots, N - 1$ and $l = 0, 1, \ldots, M/q - 1$, and

$$
\hat{x}(n,m) = \begin{cases} x(n,m) + \displaystyle\sum_{i=1}^{(q-1)/2} (-1)^i \left[x(n, \tfrac{2iM}{q} + m) + x(n, \tfrac{2iM}{q} - m) \right], \\ \qquad\qquad m = 1, 2, \ldots, M/q - 1. \\ \displaystyle\sum_{i=0}^{(q-1)/2} (-1)^i x(n, \tfrac{2iM}{q}), \quad m = 0 \end{cases} \tag{7.230}
$$

If $\delta = 1$, (7.229) is a 2D type-III DCT which can be decomposed further in the same way. If $\delta = 3$, by simply reversing the output index l, we can turn (7.229) into a 2D type-III DCT defined by

$$
Y(k, q(\frac{M}{q} - 1 - l) + \lambda(q)) = \sum_{n=0}^{N-1} \sum_{m=0}^{M/q-1} \hat{x}(n,m) \cos \frac{\pi(4k+1)n}{2N} \cos \frac{\pi(4l+1)m}{2M/q}, \tag{7.231}
$$

where $k = 0, 1, \ldots, N - 1$ and $l = 0, 1, \ldots, M/q - 1$.

Now let us consider decomposition of the second term $Y(k, l)$. If we define $p(l)$ to be $((4p + 1)l + p) \bmod N$, the following lemma shows that $Y(p(l), l)$ $(p = 0, 1, \ldots, N - 1, l = 0, 1, \ldots, M - 1$ and $l \not\equiv \lambda(q) \bmod q)$ is simply a reordering of $Y(k, l)$ $(k = 0, 1, \ldots, N - 1, l = 0, 1, \ldots, M - 1$ and $l \not\equiv \lambda(q) \bmod q)$.

Lemma 15 Let $U = \{(k, l) \mid 0 \le k \le N - 1,\ 0 \le l \le M - 1$ and $l \not\equiv \lambda(q) \bmod q\}$, $V = \{(p(l), l) \mid 0 \le p \le N - 1,\ 0 \le l \le M - 1$ and $l \not\equiv \lambda(q) \bmod q\}$. Then $U = V$.

Proof. It is proven in Section 7.6 (see Lemma 13).

Based on the lemma, it is sufficient to compute $Y(p(l), l)$ $(p = 0, 1, \ldots, N - 1, l = 0, 1, \ldots, M - 1$ and $l \not\equiv \lambda(q) \bmod q)$. According to the properties of trigonometric identities, we have

$$
Y(p(l), l) = \frac{A(p(l), l) + B(p(l), l)}{2}, \tag{7.232}
$$

where for $p = 0, 1, \ldots, N - 1, l = 0, 1, \ldots, M - 1$ and $l \not\equiv \lambda(q) \bmod q$,

$$
A(p(l), l) = \sum_{n=0}^{N-1} \sum_{m=0}^{M-1} x(n,m) \cos \left(\frac{\pi(4p(l)+1)n}{2N} + \frac{\pi(4l+1)m}{2M} \right) \tag{7.233}
$$

$$
B(p(l), l) = \sum_{n=0}^{N-1} \sum_{m=0}^{M-1} x(n,m) \cos \left(\frac{\pi(4p(l)+1)n}{2N} - \frac{\pi(4l+1)m}{2M} \right). \tag{7.234}
$$

By applying the relation $4p(l) + 1 \equiv (4l + 1)(4p + 1) \bmod 4N$ in (7.233) and (7.234), we obtain

$$
\begin{aligned}
A(p(l), l) &= \sum_{n=0}^{N-1} \sum_{m=0}^{M-1} x(n,m) \cos \left(\frac{\pi(4l+1)(4p+1)n}{2N} + \frac{\pi(4l+1)m}{2M} \right) \\
&= \sum_{n=0}^{N-1} \sum_{m=0}^{M-1} x(n,m) \cos \frac{\pi(4l+1)(q^J(4p+1)n + m)}{2M}
\end{aligned} \tag{7.235}
$$

and

$$
B(p(l), l) = \sum_{n=0}^{N-1} \sum_{m=0}^{M-1} x(n,m) \cos \frac{\pi(4l+1)(q^J(4p+1)n - m)}{2M}. \tag{7.236}
$$

If we construct a sequence $y(n, m)$ as

$$y(n, m) = \begin{cases} 2x(n, 0), & m = 0 \\ 0, & m = M \\ x(n, m), & 1 \leq m \leq M - 1 \\ -x(n, 2M - m), & M + 1 \leq m \leq 2M - 1 \end{cases} \quad (7.237)$$

and apply (7.235), (7.236) and the properties of trigonometric identities, (7.232) can be expressed as

$$Y(p(l), l) = \frac{1}{2} \sum_{n=0}^{N-1} \sum_{m=0}^{2M-1} y(n, m) \cos \frac{\pi(4l + 1)(q^J(4p + 1)n + m)}{2M}. \quad (7.238)$$

Now let us define

$$Y_p(z) \equiv \sum_{n=0}^{N-1} \sum_{m=0}^{2M-1} y(n, m) z^{q^J(4p+1)n+m} \bmod z^{2M} + 1 \quad (7.239)$$

and

$$\bar{Y}_p(z) = (Y_p(z) + Y_p(z^{-1})) \bmod z^{2M} + 1 = \sum_{m=0}^{2M-1} h(p, m) z^m, \quad (7.240)$$

where $h(p, m)$ are the coefficients of the polynomials $\bar{Y}_p(z)$ and have the property

$$h(p, m) = -h(p, 2M - m), \quad m = 1, 2, \ldots, 2M - 1.$$

Since $(e^{i\frac{\pi(4l+1)}{2M}})^{2M} = -1$, we know that (7.238) can be related to (7.240) by

$$Y(p(l), l) = \frac{1}{4} \bar{Y}_p(z) \Big|_{z=e^{i\frac{\pi(4l+1)}{2M}}}$$

which can be further elaborated into

$$Y(p(l), l) = \sum_{m=0}^{M-1} \bar{h}(p, m) \cos \frac{\pi(4l + 1)m}{2M}, \quad (7.241)$$

where $p = 0, 1, \ldots, N - 1$, $l = 0, 1, \ldots, M - 1$ and $l \not\equiv \lambda(q) \bmod q$. It can be proven that $\bar{h}(p, 0) = h(p, 0)/4$, $\bar{h}(p, m) = h(p, m)/2$ $(m = 1, 2, \ldots, M - 1)$. If we define a discrete transform by

$$E(l) = \sum_{m=0}^{M-1} F(m) \cos \frac{\pi(4l + 1)m}{2M}, \quad l = 0, 1, \ldots, M - 1 \text{ and } l \not\equiv \lambda(q) \bmod q, \quad (7.242)$$

then (7.241) is just N such transforms. For easy computation, we write (7.242) as

$$W(l) = \sum_{m=0}^{M-1} F(m) \cos \frac{\pi(2l + 1)m}{2M}, \quad (7.243)$$

where $W(2l) = E(l)$ $(l = 0, 1, \ldots, \frac{M-1}{2})$ and $W(2l+1) = E(M-1-l)$ $(l = 0, 1, \ldots, \frac{M-1}{2}-1)$. Equation (7.243) is the same as the 1D type-III DCT except that the output terms satisfying

$l \equiv \frac{q-1}{2} \bmod q$ are ignored, which is the reason why (7.243) is known as the reduced 1D type-III DCT. Therefore, (7.241) can be obtained by computing N reduced 1D type-III DCTs of length M.

In order to compute (7.241), $\bar{h}(p, m)$ (or $h(p, m)$ in (7.240)) have to be computed first. The computation of this pre-processing stage can be defined by a pseudo PT, as discussed below. We rewrite $Y_p(z)$ defined in (7.239) as

$$Y_p(z) \equiv \sum_{n=0}^{N-1} y_n(z) z^{q^J n} \hat{z}^{pn} \bmod z^{2M} + 1,$$

where

$$y_n(z) = \sum_{m=0}^{2M-1} y(n, m) z^m = 2x(n, 0) + \sum_{m=1}^{M-1} x(n, m)(z^m - z^{2M-m}) \tag{7.244}$$

$$\hat{z} \equiv z^{4q^J} \bmod z^{2M} + 1,$$

where $y(n, m)$ is defined in (7.237). It is obvious that

$$y_n(z) \equiv y_n(z^{-1}) \bmod z^{2M} + 1. \tag{7.245}$$

Therefore, (7.240) can be equivalently obtained by

$$\bar{Y}_p(z) \equiv \sum_{n=0}^{N-1} \bar{y}_n(z) \hat{z}^{pn} \bmod z^{2M} + 1, \tag{7.246}$$

where for $n = 1, 2, \ldots, N-1$,

$$\bar{y}_0(z) \equiv 2y_0(z) \bmod z^{2M} + 1$$
$$\bar{y}_n(z) \equiv (y_{N-n}(z) + y_n(z) z^M) z^{q^J n} z^{-M} \bmod z^{2M} + 1. \tag{7.247}$$

From (7.245), it is easy to show that

$$\bar{y}_0(z) \equiv \bar{y}_0(z^{-1}) \bmod z^{2M} + 1$$
$$\bar{y}_n(z) \equiv \bar{y}_{N-n}(z^{-1}) \bmod z^{2M} + 1. \tag{7.248}$$

Therefore, we need to compute (7.247) and the pseudo PT (7.246) to get $\bar{Y}_p(z)$. The PT (7.246) can be computed by a fast algorithm to be discussed in the next subsection.

In summary, the main steps of the algorithm are listed below and also shown in Figure 7.9.

Algorithm 20 *PT-based radix-q algorithm for 2D type-III DCT*

Step 1.   *Compute the input polynomials $\bar{y}_n(z)$ according to (7.247).*

Step 2.   *Compute a pseudo PT according to (7.246) to obtain $h(p, m)$ or $\bar{h}(p, m)$ as defined in (7.240).*

Step 3.   *Compute N 1D reduced type-III DCTs of length M according to (7.241).*

Step 4.   *Compute $\hat{x}(n, m)$ according to (7.230).*

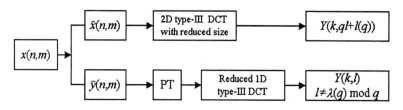

Figure 7.9   Flow chart of PT-based radix-q algorithm for 2D type-III DCT.

Step 5.   *Compute a 2D type-III DCT with size $N \times M/q$ in (7.229) or (7.231) by recursively using Step 1 through 3.*

For easy understanding, let us consider the example presented below.

Example 13 *Computation of 3×9 type-III DCT*

Step 1.   *Let $y_n(z) = 2x(n,0) + \sum_{m=1}^{8} x(n,m)(z^m - z^{18-m})$. Compute*

$$\bar{y}_0(z) \equiv 2y_0(z) \bmod z^{18} + 1$$
$$\bar{y}_1(z) \equiv y_1(z)z^3 + y_2(z)z^{-6} \bmod z^{18} + 1$$
$$\bar{y}_2(z) \equiv y_1(z^{-1}) \bmod z^{18} + 1.$$

Step 2.   *Compute a PT*

$$\bar{Y}_p(z) \equiv \sum_{n=0}^{2} \bar{y}_n(z)z^{12pn} \bmod z^{18} + 1.$$

We have

$$\bar{Y}_0(z) \equiv \bar{y}_0(z) + \bar{y}_1(z) + \bar{y}_2(z) \bmod z^{18} + 1$$
$$\bar{Y}_1(z) \equiv \bar{y}_0(z) + \bar{y}_1(z)z^{12} + \bar{y}_2(z)z^{24} \bmod z^{18} + 1$$
$$\bar{Y}_2(z) \equiv \bar{y}_0(z) + \bar{y}_1(z)z^{24} + \bar{y}_2(z)z^{48} \bmod z^{18} + 1.$$

We can use the symmetries in the polynomials to save operations (this will be discussed in detail in the next subsection). In fact, since $\bar{y}_0(z) \equiv \bar{y}_0(z^{-1}) \bmod z^{18} + 1$ and $\bar{y}_1(z) \equiv \bar{y}_2(z^{-1}) \bmod z^{18} + 1$, we can easily deduce that

$$\bar{Y}_0(z^{-1}) \equiv \bar{y}_0(z^{-1}) + \bar{y}_1(z^{-1}) + \bar{y}_2(z^{-1})$$
$$\equiv \bar{y}_0(z) + \bar{y}_2(z) + \bar{y}_1(z) \equiv \bar{Y}_0(z) \bmod z^{18} + 1$$
$$\bar{Y}_1(z^{-1}) \equiv \bar{y}_0(z^{-1}) + \bar{y}_1(z^{-1})z^{-12} + \bar{y}_2(z^{-1})z^{-24}$$
$$\equiv \bar{y}_0(z) + \bar{y}_2(z)z^{24} + \bar{y}_1(z)z^{12} \equiv \bar{Y}_1(z) \bmod z^{18} + 1$$

and $\bar{Y}_2(z^{-1}) \equiv \bar{Y}_2(z) \bmod z^{18} + 1$ for the same reason. This means that we need only to compute and store one-half of the polynomial coefficients for $\bar{Y}_p(z)$.

Let $\bar{Y}_p(z) = \sum_{m=0}^{17} h(p,m)z^m$ and $\bar{h}(p,0) = h(p,0)/4$, $\bar{h}(p,m) = h(p,m)/2$ ($m = 1, 2, \ldots, 8$).

Step 3.   *Compute 3 1D reduced type-III DCTs of length 9*

$$Y((4p+1)l + p \bmod 3, l) = \sum_{m=0}^{8} \bar{h}(p,m) \cos \frac{\pi(4l+1)m}{18},$$

where $p = 0, 1, 2$, $l = 0, 1, 3, 4, 6, 7$ and $l \not\equiv 2 \bmod 3$.

Step 4.   *Compute $\hat{x}(n, 0) = x(n, 0) - x(n, 6)$, $\hat{x}(n, m) = x(n, m) - x(n, 6+m) - x(n, 6-m)$, where $n = 0, 1, 2$ and $m = 1, 2$.*

Step 5.   *Compute a 3×3 type-III DCT:*

$$Y(k, 3l + 2) = \sum_{n=0}^{2} \sum_{m=0}^{2} \hat{x}(n, m) \cos \frac{\pi(4k+1)n}{6} \cos \frac{\pi(4l+3)m}{6}.$$

7.7.2   Fast polynomial transform

The PT in (7.246) can be computed by the FPT algorithm which generally requires $2(q - 1)MN \log_q N$ additions. However, the number of additions can be further reduced by using the symmetric property of the input polynomial sequence $\bar{y}_n(z)$ stated in (7.248). The output polynomial $\bar{Y}_p(z)$ also has a symmetric property. In fact, from the definition (7.240) we know that

$$\bar{Y}_p(z) \equiv \bar{Y}_p(z^{-1}) \bmod z^{2M} + 1.$$

Therefore, only one-half of the coefficients of $\bar{Y}_p(z)$ need to be computed. The FPT algorithm computes a PT by $\log_2 N$ stages. The computation of each stage produces temporary outputs which also have a symmetric property that provides further savings in the number of additions.

Let $\bar{Y}_n^0(z)$ be the q-ary inverse of $\bar{y}_n(z)$. The stages of the decimation-in-time radix-q FPT discussed in Chapter 2 for (7.246) is as follows:

$$\bar{Y}_{\hat{n}q^j + \hat{p} + p_{j-1}q^{j-1}}^j(z) \equiv \sum_{n_{t-j}=0}^{q-1} \left[\bar{Y}_{\hat{n}q^j + \hat{p} + n_{t-j}q^{j-1}}^{j-1}(z) \hat{z}^{q^{t-j} n_{t-j} \hat{p}} \right]$$
$$\cdot \hat{z}^{q^{t-1} n_{t-j} p_{j-1}} \bmod z^{2M} + 1, \qquad (7.249)$$

where $\hat{p} = 0, 1, \ldots, q^{j-1} - 1$, $\hat{n} = 0, 1, \ldots, q^{t-j} - 1$, and for $\bar{p} = 0, 1, \ldots, q^j - 1$, $\hat{n} = 0, 1, \ldots, q^{t-j} - 1$,

$$\bar{Y}_{\hat{n}q^j + \bar{p}}^j(z) \equiv \sum_{\bar{n}=0}^{q^j-1} \bar{y}_{\bar{n}q^{t-j} + n'}(z) \hat{z}^{q^{t-j} \bar{n} \bar{p}} \bmod z^{2M} + 1, \qquad (7.250)$$

where n' is the q-ary inverse of $\hat{n}$. The last stage $\bar{Y}_p^t(z)$ produces the outputs $\bar{Y}_p(z)$. In general, (7.249) needs $2(q - 1)MN$ additions. However, the number of addition can be reduced by one-half if the symmetric properties are fully used. Now we show that symmetries exist among the polynomials $\bar{Y}_p^j(z)$.

Lemma 16 *We have*

$$\bar{Y}_{\hat{n}q^j + \bar{p}}^j(z^{-1}) \equiv \bar{Y}_{\hat{n}q^j + \bar{p}}^j(z) \hat{z}^{q^{t-j} \bar{p}} \bmod z^{2M} + 1, \qquad (7.251)$$

where $\tilde{n}$ is the q-ary inverse of $q^{t-j} - n'$ and $\hat{n} \neq 0$, and

$$\bar{Y}_{\bar{p}}^j(z) \equiv \bar{Y}_{\bar{p}}^j(z^{-1}) \bmod z^{2M} + 1. \qquad (7.252)$$

Proof. If $\hat{n} \neq 0$, from (7.250) we have

$$
\bar{Y}^j_{\hat{n}q^j+\bar{p}}(z^{-1}) \equiv \sum_{\bar{n}=0}^{q^j-1} \bar{y}_{\bar{n}q^t-j+n'}(z^{-1})\hat{z}^{-q^{t-j}\bar{n}\bar{p}} \bmod z^{2M}+1
$$

$$
\equiv \sum_{\bar{n}=0}^{q^j-1} \bar{y}_{q^t-\bar{n}q^t-j-n'}(z)\hat{z}^{-q^{t-j}\bar{n}\bar{p}} \bmod z^{2M}+1
$$

$$
\equiv \sum_{\bar{n}=0}^{q^j-1} \bar{y}_{q^{t-j}(q^j-1-\bar{n})+q^{t-j}-n'}(z)\hat{z}^{-q^{t-j}\bar{n}\bar{p}} \bmod z^{2M}+1
$$

$$
\equiv \sum_{\bar{n}=0}^{q^j-1} \bar{y}_{\bar{n}q^t-j+q^t-j-n'}(z)\hat{z}^{-q^{t-j}(q^j-1-\bar{n})\bar{p}} \bmod z^{2M}+1
$$

$$
\equiv \left[\sum_{\bar{n}=0}^{q^j-1} \bar{y}_{\bar{n}q^t-j+q^t-j-n'}(z)\hat{z}^{q^{t-j}\bar{n}\bar{p}} \right] \hat{z}^{q^{t-j}\bar{p}} \bmod z^{2M}+1
$$

$$
\equiv \bar{Y}^j_{\hat{n}q^j+\bar{p}}(z)\hat{z}^{q^{t-j}\bar{p}} \bmod z^{2M}+1.
$$

Similarly, (7.252) can be proven.

Both (7.251) and (7.252) show that at every stage, only one-half of the coefficients need to be computed. The rest of the coefficients can be achieved from the preceding properties without requiring any computation. Therefore, each stage requires only $(q-1)NM$ additions. Furthermore, based on the symmetric property we know that at the $(t-1)$th stage, every polynomial $\bar{Y}^{t-1}_p(z)$ has a zero coefficient. It means that $(N-1)/2$ additions can be saved at the tth stage. Therefore, the total number of additions for the PT defined in (7.246) is $(q-1)NM \log_q N - (N-1)/2$.

7.7.3   PT-based radix-q algorithm for rD type-III DCT

In this subsection, we generalize the method used in the previous section for the computation of the MD type-III DCT defined by

$$
X(k_1, k_2, \ldots, k_r) = \sum_{n_1=0}^{N_1-1} \sum_{n_2=0}^{N_2-1} \cdots \sum_{n_r=0}^{N_r-1} x(n_1, n_2, \ldots, n_r)
$$
$$
\cdot \cos\frac{\pi(2k_1+1)n_1}{2N_1} \cdots \cos\frac{\pi(2k_r+1)n_r}{2N_r}, \qquad (7.253)
$$

where $k_i = 0, 1, \ldots, N_i - 1$ and $i = 1, 2, \ldots, r$. We also have $N_r = q^t$, $N_r/N_i = q^{l_i}$, where $l_i \geq 0$ for $i = 1, 2, \ldots, r-1$ and q is an odd prime number. If we define an array $Y(k_1, \ldots, k_r)$ by

$$
Y(k_1, k_2, \ldots, k_r) = X(2k_1, 2k_2, \ldots, 2k_r)
$$
$$
Y(N_1-1-k_1, k_2, \ldots, k_r) = X(2k_1+1, 2k_2, \ldots, 2k_r)
$$
$$
\cdots
$$
$$
Y(N_1-1-k_1, \ldots, N_r-1-k_r) = X(2k_1+1, \ldots, 2k_r+1)
$$

for $0 \leq 2k_i \leq N_i - 1$ or $0 \leq 2k_i + 1 \leq N_i - 1$ for $i = 1, 2, \ldots, r$, then (7.253) becomes

$$
\begin{aligned}
Y(k_1, k_2, \ldots, k_r) = &\sum_{n_1=0}^{N_1-1} \sum_{n_2=0}^{N_2-1} \cdots \sum_{n_r=0}^{N_r-1} x(n_1, n_2, \ldots, n_r) \\
&\cdot \cos \frac{\pi(4k_1+1)n_1}{2N_1} \cdots \cos \frac{\pi(4k_r+1)n_r}{2N_r}.
\end{aligned} \tag{7.254}
$$

Similar to the 2D case, let $\lambda(q)$ represent the unique non-negative integer smaller than q and satisfying $4\lambda(q) + 1 \equiv 0 \bmod q$. Therefore, $4\lambda(q) + 1 = \delta q$, where $\delta = 1$ or 3. The outputs of (7.254) can be grouped into two terms, that is, $Y(k_1, \ldots, k_{r-1}, qk_r + \lambda(q))$ ($k_r = 0, 1, \ldots, N_r/q - 1$), and $Y(k_1, \ldots, k_{r-1}, k_r)$ ($k_r = 0, 1, \ldots, N_r - 1$) and $k_r \not\equiv \lambda(q) \bmod q$. The first term can be converted into an rD type-III DCT of size $N_1 \times \cdots \times N_{r-1} \times N_r/q$ as

$$
\begin{aligned}
Y(k_1, \ldots, k_{r-1}, qk_r + \lambda(q)) = &\sum_{n_1=0}^{N_1-1} \cdots \sum_{n_{r-1}=0}^{N_{r-1}-1} \sum_{n_r=0}^{N_r-1} \hat{x}(n_1, \ldots, n_{r-1}, n_r) \\
&\cdot \cos \frac{\pi(4k_1+1)n_1}{2N_1} \cdots \cos \frac{\pi(4k_{r-1}+1)n_{r-1}}{2N_{r-1}} \\
&\cdot \cos \frac{\pi(4k_r+\delta)n_r}{2N_r/q},
\end{aligned} \tag{7.255}
$$

where $k_i = 0, 1, \ldots, N_i - 1$, $i = 1, 2, \ldots, r-1$, $k_r = 0, 1, \ldots, N_r/q - 1$ and

$$
\hat{x}(n_1, \ldots, n_{r-1}, n_r) = \begin{cases} x(n_1, \ldots, n_{r-1}, n_r) + \displaystyle\sum_{i=1}^{(q-1)/2} (-1)^i \left[x\left(n_1, \ldots, n_{r-1}, \frac{2iN_r}{q} + n_r\right) \right. \\ \left. \quad + x\left(n_1, \ldots, n_{r-1}, \frac{2iN_r}{q} - n_r\right) \right], \quad n_r = 1, 2, \ldots, \frac{N_r}{q} - 1. \\ \displaystyle\sum_{i=0}^{(q-1)/2} (-1)^i x\left(n_1, \ldots, n_{r-1}, \frac{2iN_r}{q}\right), \quad n_r = 0 \end{cases} \tag{7.256}
$$

Now we consider the computation of the second term. Let $p(k_r)$ be the least non-negative residue of $(4p+1)k_r + p$ module N_{r-1}. Similar to the 2D case, we have

$$
Y(k_1, \ldots, k_{r-2}, p(k_r), k_r) = \frac{A(k_1, \ldots, k_{r-2}, p(k_r), k_r) + B(k_1, \ldots, k_{r-2}, p(k_r), k_r)}{2}, \tag{7.257}
$$

where

$$
\begin{aligned}
A(k_1, \ldots, k_{r-2}, p(k_r), k_r) = &\sum_{n_1=0}^{N_1-1} \sum_{n_2=0}^{N_2-1} \cdots \sum_{n_r=0}^{N_r-1} x(n_1, n_2, \ldots, n_r) \\
&\cdot \cos \frac{\pi(4k_1+1)n_1}{2N_1} \cdots \cos \frac{\pi(4k_{r-2}+1)n_{r-2}}{2N_{r-2}} \\
&\cdot \cos \left(\frac{\pi(4p(k_r)+1)n_{r-1}}{2N_{r-1}} + \frac{\pi(4k_r+1)n_r}{2N_r} \right)
\end{aligned}
$$

$$
= \sum_{n_1=0}^{N_1-1} \sum_{n_2=0}^{N_2-1} \cdots \sum_{n_r=0}^{N_r-1} x(n_1, n_2, \ldots, n_r)
$$

$$
\cdot \cos \frac{\pi(4k_1+1)n_1}{2N_1} \cdots \cos \frac{\pi(4k_{r-2}+1)n_{r-2}}{2N_{r-2}}
$$

$$
\cdot \cos \frac{\pi(4k_r+1)(q^{l_{r-1}}(4p+1)n_{r-1}+n_r)}{2N_r}
$$

and

$$
B(k_1, \ldots, k_{r-2}, p(k_r), k_r) = \sum_{n_1=0}^{N_1-1} \sum_{n_2=0}^{N_2-1} \cdots \sum_{n_r=0}^{N_r-1} x(n_1, n_2, \ldots, n_r)
$$

$$
\cdot \cos \frac{\pi(4k_1+1)n_1}{2N_1} \cdots \cos \frac{\pi(4k_{r-2}+1)n_{r-2}}{2N_{r-2}}
$$

$$
\cdot \cos \left(\frac{\pi(4p(k_r)+1)n_{r-1}}{2N_{r-1}} - \frac{\pi(4k_r+1)n_r}{2N_r} \right)
$$

$$
= \sum_{n_1=0}^{N_1-1} \sum_{n_2=0}^{N_2-1} \cdots \sum_{n_r=0}^{N_r-1} x(n_1, n_2, \ldots, n_r)
$$

$$
\cdot \cos \frac{\pi(4k_1+1)n_1}{2N_1} \cdots \cos \frac{\pi(4k_{r-2}+1)n_{r-2}}{2N_{r-2}}
$$

$$
\cdot \cos \frac{\pi(4k_r+1)(q^{l_{r-1}}(4p+1)n_{r-1}-n_r)}{2N_r}.
$$

By constructing a sequence $y(n_1, \ldots, n_{r-1}, n_r)$ as

$$
y(n_1, \ldots, n_{r-1}, n_r) = \begin{cases} 2x(n_1, \ldots, n_{r-1}, 0), & n_r = 0 \\ 0, & n_r = N_r \\ x(n_1, \ldots, n_{r-1}, n_r), & 1 \le n_r \le N_r - 1 \\ -x(n_1, \ldots, n_{r-1}, 2N_r - n_r), & N_r + 1 \le n_r \le 2N_r - 1 \end{cases} \tag{7.258}
$$

we have

$$
Y(k_1, \ldots, k_{r-2}, p(k_r), k_r) = \frac{1}{2} \sum_{n_1=0}^{N_1-1} \cdots \sum_{n_{r-1}=0}^{N_{r-1}-1} \sum_{n_r=0}^{2N_r-1} y(n_1, n_2, \ldots, n_r)
$$

$$
\cdot \cos \frac{\pi(4k_1+1)n_1}{2N_1} \cdots \cos \frac{\pi(4k_{r-2}+1)n_{r-2}}{2N_{r-2}}
$$

$$
\cdot \cos \frac{\pi(4k_r+1)(q^{l_{r-1}}(4p+1)n_{r-1}+n_r)}{2N_r}
$$

$$
= \sum_{n_1=0}^{N_1-1} \cdots \sum_{n_{r-2}=0}^{N_{r-2}-1} \cos \frac{\pi(4k_1+1)n_1}{2N_1} \cdots \cos \frac{\pi(4k_{r-2}+1)n_{r-2}}{2N_{r-2}}
$$

$$
\cdot \left\{ \frac{1}{2} \sum_{n_{r-1}=0}^{N_{r-1}-1} \sum_{n_r=0}^{2N_r-1} y(n_1, \ldots, n_{r-2}, n_{r-1}, n_r) \right.
$$

$$
\left. \cdot \cos \frac{\pi(4k_r+1)(q^{l_{r-1}}(4p+1)n_{r-1}+n_r)}{2N_r} \right\}. \tag{7.259}
$$

We denote the inner summation by $\hat{y}(n_1, \ldots, n_{r-2}, p, k_r)$, that is,

$$
\begin{aligned}
\hat{y}(n_1, \ldots, n_{r-2}, p, k_r) &= \frac{1}{2} \sum_{n_{r-1}=0}^{N_{r-1}-1} \sum_{n_r=0}^{2N_r-1} y(n_1, \ldots, n_{r-2}, n_{r-1}, n_r) \\
&\quad \cdot \cos \frac{\pi(4k_r + 1)(q^{l_{r-1}}(4p + 1)n_{r-1} + n_r)}{2N_r},
\end{aligned}
\tag{7.260}
$$

which is very similar to (7.238). Based on (7.238), we know that (7.260) can be converted into reduced 1D type-III DCTs as

$$
\hat{y}(n_1, \ldots, n_{r-2}, p, k_r) = \sum_{n_r=0}^{N_r-1} \bar{h}(n_1, \ldots, n_{r-2}, p, n_r) \cos \frac{\pi(4k_r + 1)n_r}{2N_r},
\tag{7.261}
$$

where $k_r = 0, 1, \ldots, N_r - 1$, $k_r \not\equiv \lambda(q) \bmod q$, and $\bar{h}(n_1, \ldots, n_{r-2}, p, n_r)$ are coefficients of polynomials which can be computed by the FPT. By substituting (7.261) into (7.259), we achieve

$$
\begin{aligned}
Y(k_1, \ldots, k_{r-2}, p(k_r), k_r) &= \sum_{n_1=0}^{N_1-1} \cdots \sum_{n_{r-2}=0}^{N_{r-2}-1} \sum_{n_r=0}^{N_r-1} \bar{h}(n_1, \ldots, n_{r-2}, p, n_r) \\
&\quad \cdot \cos \frac{\pi(4k_1 + 1)n_1}{2N_1} \cdots \cos \frac{\pi(4k_{r-2} + 1)n_{r-2}}{2N_{r-2}} \\
&\quad \cdot \cos \frac{\pi(4k_r + 1)n_r}{2N_r},
\end{aligned}
\tag{7.262}
$$

where $k_r = 0, 1, \ldots, N_r - 1$ and $k_r \not\equiv \lambda(q) \bmod q$. Equation (7.262) is N_{r-1} $(r-1)$D reduced type-III DCTs. The method described above can be used to further reduce the number of dimensions of the transform. The fast algorithm is summarized as follows.

Algorithm 21 *PT-based radix-q algorithm for rD type-III DCT*

Step 1. *Use the improved FPT algorithm and the algorithm for the 2D DCT-III to get $\bar{h}(n_1, \ldots, n_{r-2}, p, n_r)$.*

Step 2. *Compute N_{r-1} $(r-1)$D reduced type-III DCTs in (7.262).*

Step 3. *Compute $\hat{x}(n_1, \ldots, n_{r-1}, n_r)$ in (7.256).*

Step 4. *Compute an rD type-III DCT with size $N_1 \times \cdots \times N_{r-1} \times N_r/q$ in (7.255).*

7.7.4 Computational complexity

Now we analyze the number of operations of the fast algorithm for the rD type-III DCT. The algorithm groups the outputs of an rD type-III DCT with size $N_1 \times \cdots \times N_{r-1} \times N_r$ into two terms, namely, $Y(k_1, \ldots, k_{r-1}, qk_r + \lambda(q))$ $(k_r = 0, 1, \ldots, N_r/q - 1)$ and $Y(k_1, \ldots, k_{r-1}, k_r)$ $(k_r = 0, 1, \ldots, N_r - 1)$ and $k_r \not\equiv \lambda(q) \bmod q$. The first term is computed by an rD type-III DCT with size $N_1 \times \cdots \times N_{r-1} \times N_r/q$ in (7.255) while at most N additions are needed for computing the input array described in (7.256). The second term is an rD reduced type-III DCT with size $N_1 \times \cdots \times N_{r-1} \times N_r$ which is computed by Steps 1 and 2. Step 1 needs fewer than $(q - 1)N \log_q N_{r-1}$ additions for the computation of PTs and at most

N additions for the computations of input polynomials, where $N = N_1 N_2 \cdots N_r$. Step 2 is used for computing N_{r-1} $(r-1)$D reduced type-III DCTs. Therefore, similar to the analysis of the computational complexity of the PT-based algorithm for the rD type-II DCT, we can get an expression for the computational complexity of the algorithm. When $N_1 = N_2 = \cdots = N_r = q^t$, the expression is

$$M_u(q^t) = \frac{a_1 q}{q-1} t q^{rt} + \left(a_2 - \frac{a_1}{q^r - 1} \right) \frac{q}{q-1} (q^{rt} - 1)$$

$$+ \frac{a_3(q^r - 1)}{(q-1)(q^{r-1} - 1)} (q^{(r-1)t} - 1) \tag{7.263}$$

$$A_d(q^t) = \left(\frac{b_1 q}{q-1} + (r-1)q \right) t q^{rt} + \left(b_2 + 1 - \frac{b_1 + q^r - q}{q^r - 1} \right) \frac{q(q^{rt} - 1)}{q-1}$$

$$+ \frac{b_3(q^r - 1)}{(q-1)(q^{r-1} - 1)} (q^{(r-1)t} - 1), \tag{7.264}$$

where $M_u(q^t)$ and $A_d(q^t)$ stand for the numbers of multiplications and additions, respectively, given that the numbers of multiplications and additions for a 1D reduced type-III DCT with length q^t are $\varphi(q^t) = a_1 t q^t + a_2 q^t + a_3$ and $\psi(q^t) = b_1 t q^t + b_2 q^t + b_3$, respectively.

Let $\tau(q^t) = c_1 t q^t + c_2 q^t + c_3$ and $\rho(q^t) = d_1 t q^t + d_2 q^t + d_3$ represent the numbers of multiplications and additions for computing a 1D type-III DCT of length q^t, respectively. Then, the computational complexity of the commonly used row-column method is

$$\overline{M}_u(q^t) = c_1 r t q^{rt} + c_2 r q^{rt} + c_3 r q^{(r-1)t}$$

$$\overline{A}_d(q^t) = d_1 r t q^{rt} + d_2 r q^{rt} + d_3 r q^{(r-1)t}. \tag{7.265}$$

Since the 1D reduced type-III DCT only retains $q-1$ valid outputs of the length q 1D type-III DCT, we can express the additive and multiplicative complexity to be approximately $\varphi(q^t) = \frac{q-1}{q}\tau(q^t)$ and $\psi(q^t) = \frac{q-1}{q}\rho(q^t)$. A comparison between (7.263) and (7.265) shows that

$$M_u \simeq \frac{1}{r}\overline{M}_u.$$

It is much more difficult to compare the number of additions needed by the two fast algorithms. Table 7.4 gives the comparison when $r = 3$ and $q = 3$.

Table 7.4   Comparison of the number of operations for a 3D type-III DCT.

3^t	Presented algorithm			Row-column algorithm		
	M_u	A_d	$M_u + A_d$	$\overline{M}_u$	$\overline{A}_d$	$\overline{M}_u + \overline{A}_d$
9	1300	11796	13096	2916	7776	10692
27	61321	489048	550369	150903	358668	509571
81	2364232	17810088	20174320	6141096	13856832	19997928
243	82966117	605229540	688195657	222673779	488217132	710890911

Until now, very few algorithms other than the row-column method have been proposed for the rD DCT with size $q^{l_1} \times q^{l_2} \times \cdots \times q^{l_r}$, where q is an odd prime number. In [13], a recursive technique is reported to decompose the 2D type-II DCT of size $p^l \times p^l$ into CC, SCC and matrix multiplications, where p is a prime number. The main advantage of the algorithm is the reduction in multiplications. However, it increases the number of additions greatly. The total number of operations (multiplications plus additions) is much more than that of the row-column method when p is an odd prime number. For example, for a 9×9 2D DCT, the algorithm in [13] uses 1139 operations $(80 + 1059)$, the presented algorithm uses 964 operations $(136 + 828)$ and the row-column method uses 792 operations $(216 + 576)$. The situation becomes even worse when the size p^l becomes large since the number of additions for the CC and SCC computed by the Winograd algorithm increases rapidly with the size [4]. Also, no general formula is available to express the number of operations for the CC and SCC based on the Winograd algorithm. It is difficult to obtain the number of operations for [13] when the size is large. Another problem of the algorithm reported in [13] is that the Winograd algorithm for the CC and SCC requires an irregular computational structure and is seldom used in practice, while the presented algorithm only uses the 1D type-III DCT and PT with a regular structure.

The presented algorithm for the rD type-III DCT can handle transforms whose dimensional sizes are $q^{l_1} \times q^{l_2} \times \cdots \times q^{l_r}$, where q is an odd prime number. Considerable savings in the number of multiplications are achieved. The structure of the algorithm is also simple because it uses only the 1D DCT and the PT, which can be easily implemented. Although the number of additions increases compared to the row-column method, the algorithm achieves savings in terms of the total number of arithmetic operations (additions plus multiplications). Most additions are used for PTs, for which we use the simplest FPT algorithm to minimize the additive complexity. Since the PT can be viewed as a kind of DFT in the polynomial residue rings, further improvement on the FPT algorithm is possible, which is similar to the improvements made to the FFT algorithms since the Cooley–Turkey algorithm was reported. It is worthwhile to investigate new FPT algorithms that can be used for computing transforms such as the DFT, DCT and DHT. The presented fast algorithm can also be easily extended for rD type-III DCTs whose dimensional sizes are $2^{k_1} q^{l_1} \times 2^{k_2} q^{l_2} \times \cdots \times 2^{k_r} q^{l_r}$. However, it seems that it is more difficult to extend the concepts to rD type-III DCTs whose dimensional sizes are $q_1^{l_1} \times q_2^{l_2} \times \cdots \times q_r^{l_r}$, where q_i are different prime numbers.

7.8 CHAPTER SUMMARY

This chapter is focused on the recently reported fast algorithms for MD DCTs whose transform sizes are composite. The algorithms in Section 7.2 can be used for the type-I, -II and -III DCTs whose transform sizes contain odd numbers. The concepts used for fast algorithms based on prime factor decomposition and PT can be easily extended for DCTs of any dimensions. These fast algorithms have a regular computational structure that provides the flexibility for different transform sizes and dimensions. The computational complexity is also substantially reduced.

For high computational throughput, it is likely that parallel processing is needed. Important features of the presented fast algorithms are that they are readily used for parallel processing. For example, the algorithm in Section 7.2 decomposes the processing task into several computational blocks that can be handled with multiple computing engines. Similarly, the PT-based fast algorithms require 1D DCT transforms which can also be computed by multiple processors. Further research is needed to find efficient approaches for the parallel processing.

Appendix A. Computational complexity

The algorithms can be rewritten as

$$
\begin{aligned}
X(k_1, k_2) &= \cos\frac{\pi q_2 k_2}{2N_2} \cos\frac{\pi q_1 k_1}{2N_1}\left[A(k_1, k_2) + \tan\frac{\pi q_1 k_1}{2N_1}B(k_1, k_2)\right] \\
&+ \sin\frac{\pi q_2 k_2}{2N_2} \cos\frac{\pi q_1 k_1}{2N_1}\left[C(k_1, k_2) + \tan\frac{\pi q_1 k_1}{2N_1}D(k_1, k_2)\right] \\
&\qquad 0 \leq k_1 \leq \frac{N_1}{2}, \ 0 \leq k_2 \leq \frac{N_2}{2}
\end{aligned}
\tag{A.1}
$$

$$
\begin{aligned}
X(k_1, N_2 - k_2) &= (-1)^{\frac{(q_2-1)}{2}}\left\{\sin\frac{\pi q_2 k_2}{2N_2}\cos\frac{\pi q_1 k_1}{2N_1}\left[A(k_1, k_2) + \tan\frac{\pi q_1 k_1}{2N_1}B(k_1, k_2)\right]\right. \\
&\left. - \cos\frac{\pi q_2 k_2}{2N_2}\cos\frac{\pi q_1 k_1}{2N_1}\left[C(k_1, k_2) + \tan\frac{\pi q_1 k_1}{2N_1}D(k_1, k_2)\right]\right\} \\
&\qquad 0 \leq k_1 \leq \frac{N_1}{2}, \ 0 \leq k_2 < \frac{N_2}{2}
\end{aligned}
\tag{A.2}
$$

$$
\begin{aligned}
X(N_1 - k_1, k_2) &= (-1)^{\frac{(q_1-1)}{2}}\left\{\cos\frac{\pi q_2 k_2}{2N_2}\cos\frac{\pi q_1 k_1}{2N_1}\left[\tan\frac{\pi q_1 k_1}{2N_1}A(k_1, k_2) - B(k_1, k_2)\right]\right. \\
&\left. + \sin\frac{\pi q_2 k_2}{2N_2}\cos\frac{\pi q_1 k_1}{2N_1}\left[\tan\frac{\pi q_1 k_1}{2N_1}C(k_1, k_2) - D(k_1, k_2)\right]\right\} \\
&\qquad 0 \leq k_1 < \frac{N_1}{2}, \ 0 \leq k_2 \leq \frac{N_2}{2}
\end{aligned}
\tag{A.3}
$$

$$
\begin{aligned}
X(N_1 - k_1, N_2 - k_2) &= \\
-(-1)^{\frac{(q_1+q_2)}{2}}&\left\{\sin\frac{\pi q_2 k_2}{2N_2}\cos\frac{\pi q_1 k_1}{2N_1}\left[\tan\frac{\pi q_1 k_1}{2N_1}A(k_1, k_2) - B(k_1, k_2)\right]\right. \\
&\left. - \cos\frac{\pi q_2 k_2}{2N_2}\cos\frac{\pi q_1 k_1}{2N_1}\left[\tan\frac{\pi q_1 k_1}{2N_1}C(k_1, k_2) - D(k_1, k_2)\right]\right\} \\
&\qquad 0 \leq k_1 < \frac{N_1}{2}, \ 0 \leq k_2 < \frac{N_2}{2}.
\end{aligned}
\tag{A.4}
$$

Multiplications can be substantially saved compared to that needed by the computation based on (7.6) and (7.9) – (7.14). One problem of using the above equations is that when $\cos[\pi q_1 k_1/(2N_1)] = 0$, the factor $\tan[\pi q_1 k_1/(2N_1)]$ goes to infinity. For such cases, (7.6) and (7.9) – (7.14) are used for computation. The table below lists the numbers of additions and multiplications needed by the computation associated with the indices listed on the left of the table.

k_1	k_2	Mul	Add
0	0	0	0
0	$N_2/2$	1	0
$N_1/2$	0	1	0
$N_1/2$	$N_2/2$	0	0
0	$1, 2, \ldots, N_2/2 - 1$	$2N_2 - 3q_2 - 1$	$N_2 - q_2 - 1$
$N_1/2$	$1, 2, \ldots, N_2/2 - 1$	$2N_2 - 3q_2 - 1$	$N_2 - q_2 - 1$
$1, 2, \ldots, N_1/2 - 1$	0	$2N_1 - 3q_1 - 1$	$N_1 - q_1 - 1$
$1, 2, \ldots, N_1/2 - 1$	$N_2/2$	$2N_1 - 3q_1 - 1$	$N_1 - q_1 - 1$

When the twiddle factors in (7.6) – (7.9) become 0.0, 0.5, 1.0 and 0.707, a few additions and multiplications can be saved. Therefore, the total number of multiplications for twiddle factors is

$$3(N_1 - 2q_1)(N_2 - 2q_2) + 4(N_1 - 2q_1)(q_2 - 1) + 4(N_2 - 2q_2)(q_1 - 1)$$
$$+ 2(q_1 - 1)(q_2 - 1) + 4(N_1 + N_2) - 6(q_1 + q_2) - 2 \tag{A.5}$$

The total number of additions for twiddle factors and the output processing is

$$2(N_1 - 2)(N_2 - 2) - (N_2 - 2)(q_1 - 1) - (N_1 - 2)(q_2 - 1)$$
$$+ 2(N_2 - q_2) + 2(N_1 - q_1) - 4 \tag{A.6}$$

Appendix B. Subroutine for PFA of type-II 2D DCT

```
pfadct2d(p1, q1, x)        x is a 2-D array *

{ float a[ ][ ][ ][ ], b[ ][ ][ ][ ], t};
  int l1, m1, m2, k1, k2, kk1, kk2, nn1;
  int index1[ ][ ], oindex1[ ], oindex2[ ];
    nn1=p1*q1;
    inputMapping(p1,q1,index1)    /*index mapping*/
    dct4d(p1, q1, index1, x, a) /*Computing the 4-D DCT */
    for (k1=0; k1 < p1; k1++)    /* The first step post-processing */
    {       for (m1=0; m1 < q1; m1++)
          { kk2= =p1; /* calculating output index */
              for (m2=0; m2 < q1; m2++)
              { kk2 = kk2 + p1;    oindex2[m2]=kk2;
                  index2[0][m2]= 1; oindex1[m2]=kk2;
                  index1[0][m2]=kk2;b[k1][m1][0][m2]=a[k1][m1][0][m2];
              }
              for (k2=1; k2<p1; k2++)
              {   oindex1[0]=oindex1[0]+q2; oindex1[k2][0]=oindex1[0];
                  b[k1][m1][k2][0]=a[k1][m1][k2][0];
                  for(m2=1; m2 < q1; m2++)
                  { oindex1[m2]=oindex1[m2]+q1; kk2=oindex1[m2];
                      if(kk2 > nn2)
                      { l1oindex2[m2]1;
                              if(kk2<l1) W=l1-W;  else kk2=kk2-l1;
                              index2[k2][m2]= -1;
                      } else kk2=kk2-l1; index2[k2][m2]= -1;
                      else index2[k2][m2]=1; index1[k2][m2]=kk2;
                  if (index2[k2][m2]== -1)
                     b[k1][m1][k2][m2]=a[k1][m1][k2][m2]+a[k1][m1][p1-k2][q1-m2];
                     else b[k1][m1][k2][m2]=a[k1][m1][k2][m2]-a[k1][m1][p1-k2][q1-m2];
                     }
                  }
              }
    }
    for (k2=0; k2 < p1; k2++) /* final step of post-processing */

    { for(m2=0; m2<q1; m2++)
        { kk2=index1[k2][m2];
            for (m1=0; m1<q1; m1++)
            { kk1=index1[0][m1];
                x[kk1][kk2]=b[0][m1][k2][m2];
            }
            for (k1=1; k1 < p1; k1++)
```

```
            {   kk1=index1[k1][0];  x[kk1][kk2]=b[k1][0][k2][m2];
                for(m1=1;m1<q1;m1++)
                {   kk1=index1[k1][m1];
                if (index2[k1][m1]==1)
                    x[kk1][kk2]=b[k1][m1][k2][m2]-b[p1-k1][q1-m1][k2][m2];
                else x[kk1][kk2]=b[k1][m1][k2][m2]+b[p1-k1][q1-m1][k2][m2];
                }
            }
        }
    }
}
```

REFERENCES

1. N. Ahmed and K. R. Rao, *Orthogonal Transforms for Digital Signal Processing*, Springer-Verlag, New York, 1995.

2. G. Bi, Fast algorithms for type-III DCT of composite sequence lengths, *IEEE Trans. Signal Process.*, vol. 47, no. 7, 2053 – 2059, 1999.

3. G. Bi, Y. H. Zeng and Y. Chen, Prime factor algorithm for multi-dimensional discrete cosine transform, *IEEE Trans. Signal Process.*, vol. 49, no. 9, 2156 – 2161, 2001.

4. R. E. Blahut. *Fast Algorithms for Digital Signal Processing*, Addison-Wesley, Reading, MA., 1984.

5. S. C. Chan and K. L. Ho, A new two-dimensional fast cosine transform, *IEEE Trans. Signal Process.*, vol. 39, 481 – 485, 1991.

6. S. C. Chan and K. L. Ho, Efficient index mapping for computing discrete cosine transform, *Electron. Lett.*, vol. 25, no. 22, 1499 – 1500, 1989.

7. S. C. Chan and K. L. Ho, Fast algorithm for computing the discrete cosine transform, *IEEE Trans. Circuits Syst. II*, vol. 39, no. 3, 185 – 190, 1992.

8. N. I. Cho and S. U. Lee, A fast 4×4 algorithm and implementation of 2-D discrete cosine transform, *IEEE Trans. Signal Process.,* vol. 40, no. 9, 2166 – 2173, 1992.

9. N. I. Cho and S. U. Lee, Fast algorithm and implementation for 2-D discrete cosine transform, *IEEE Trans. Circuits Syst. II*, vol. 38, no. 3, 297 – 305, 1991.

10. N. I. Cho and S. U. Lee, On the regular structure for the fast 2D DCT algorithm, *IEEE Trans. Circuits Syst. II*, vol. 40, no. 4, 259 – 266, 1993.

11. N. I. Cho and S. U. Lee, DCT algorithms for VLSI parallel implementations, *IEEE Trans. Acoustics, Speech, Signal Process.*, vol. 38, no. 1, 121 – 127, 1990.

12. P. Duhamel and C. Guillemot, Polynomial transform computation of the 2D DCT, *Proc. Int. Conf. Acoust., Speech, Signal Process*, 1515 – 1518, 1990.

13. W. H. Fang, N. C. Hu and S. K. Shih, Recursive fast computation of the two-dimensional discrete cosine transform, *IEE Proc. Vision, Image and Signal Processing*, vol. 146, no. 1, 25-33, 1999.

14. M. A. Haque, A two dimensional fast cosine transform, *IEEE Trans. Acoust., Speech, Signal Process.*, vol. 33. no. 12, 1532 – 1539, 1985.

15. D. C. Kar and V. V. B. Tao, On the prime factor decomposition algorithm for the discrete sine transform, *IEEE Trans. Signal Process.*, vol. 42, no. 11, 3258 – 3260, 1994.

16. B. G. Lee, Input and output index mapping for a prime-factor-decomposed computation of discrete cosine transform, *IEEE Trans. Acoust., Speech, Signal Process.*, vol. 37, no. 2, 237 – 244, 1989.

17. P. Lee and F. Y. Huang, An efficient prime-factor algorithm for the discrete cosine transform and its hardware implementations, *IEEE Trans. Acoust., Speech, Signal Process.*, vol. 42, no. 8, 1996 – 2005, 1994.

18. H. J. Nussbaumer and P. Quandalle, Fast polynomial transform computation of the 2-D DCT, *Proc. ICDSP*, Italy, 276 – 283, 1981.

19. J. Prado and P. Duhamel, A polynomial transform based computation of the 2-D DCT with minimum multiplicative complexity, *Proc. ICASSP*, vol. 3, 1347 – 1350, 1996.

20. Y. L. Siu and W. C. Siu, Variable temporal-length 3-D discrete cosine transform coding, *IEEE Trans. Image Processing*, vol. 6, no. 5, 758 – 763, 1997.

21. A. Tatsaki, C. Dre, T. Storaities and C. Goutis, Prime-factor DCT algorithms, *IEEE Trans. Signal process.*, vol. 43, no. 3, 772 – 776, 1995.

22. Z. S. Wang, Z. Y. He, C.R. Zou and J. D. Z. Chen, A generalized fast algorithm for n-D discrete cosine transform and its application to motion picture coding, *IEEE Trans. Circuits Syst. II*, vol. 46, no. 5, 617 – 627, 1999.

23. Y. H. Zeng, G. Bi and A. R. Leyman, New polynomial transform algorithm for multidimensional DCT, *IEEE Trans. Signal Process.*, vol. 48, no. 10, 2814 – 2821, 2000.

24. Y. H. Zeng, G. Bi and A. C. Kot, New algorithm for multidimensional type-III DCT, *IEEE Trans. Circuits Syst. II* , vol. 47, no. 12, 1523 – 1529, 2000.

25. Y. H. Zeng, G. Bi and A. R. Leyman, New algorithm for r-dimensional DCT-II, *IEE Proc., Vision, Image and Signal Processing*, vol. 148, no. 1, 1 – 8, 2000.

26. Y. H. Zeng, G. Bi and Z. P. Lin, Combined polynomial transform and radix-q algorithm for multi-dimensional DCT-III, *Multidimensional System Signal Process.*, vol. 13, no. 1, 79 - 99, 2002.

27. Y. H. Zeng, L. Z. Cheng and M. Zhou, *Parallel Algorithms for Digital Signal Processing*, National University of Defense Technology Press, Changsha, P. R. China, 1998 (in Chinese).

28. Y. H. Zeng, Fast algorithms for discrete cosine transform of arbitrary length, *Math. Numer. Sinica* (in Chinese), vol. 15, no. 3, 295 – 302, 1993.

8

Integer Transforms and Fast
Algorithms

This chapter presents integer transforms and their fast algorithms.

- Section 8.2 provides some basic concepts and methods for decomposing an arbitrary matrix into lifting matrices. Approximation of a lifting matrix by an integer matrix is also discussed.

- Section 8.3 deals with integer discrete cosine transforms (DCTs) and their fast algorithms.

- Section 8.4 presents integer discrete Hartley transforms (DHTs) and their fast algorithms.

- Section 8.5 presents multidimensional (MD) integer DCTs and their fast algorithms.

- Section 8.6 presents MD integer DHTs and their fast algorithms.

Details of mathematical derivations and comparisons in terms of the computational complexity and structures are provided. The approximation performance of integer transforms to their corresponding floating transforms is also discussed.

8.1 INTRODUCTION

Discrete sinusoidal transforms have a wide range of applications such as data compression, feature extraction, multi-frame detection and filter banks [2, 12, 17, 18, 20]. Floating-point multiplications and additions are inevitably used to implement these transforms. For some applications such as mobile computing and lossless compression, it is desired to further simplify the computation complexity to support low power consumption and high resolution of the processed outputs. The mobile devices consume the most power on arithmetic computation, especially for floating-point multiplications. It is also desired to eliminate floating-point multiplications for coding of video and graphic signals in mobile communications. Due to round-off and/or quantization errors, it is impossible to recover the input signal losslessly from the transform coefficients. Therefore it is not surprising that lossless coding schemes are hardly based on discrete sinusoidal transforms. Integer transforms possess some useful features, such as the de-correlation property and recursive structure and reversibility, and require a much lower computational complexity because they use integer arithmetic (additions and possibly multiplications). The processing speed can be accordingly increased to support real time applications. If a sufficient word length is used to represent the partially processed data, the quantization and computation errors can completely be eliminated. The integer transforms have been applied for image coding, filter banks and other areas of interest [6, 14, 16, 21, 22]. Based on lifting schemes, for example, the integer DCT has been used for image compression. The compression ratio is comparable to that using the original DCT while the computational complexity is substantially reduced [6, 16, 22]. The integer transforms can be used for lossless compression as recommended in the JPEG-2000 proposal for lossless image coding [13].

This chapter shows that, by factoring the transform into lifting steps, a sinusoidal transform can be easily converted into an integer transform which does not need floating-point multiplications. This is done by simply replacing the non-zero non-diagonal elements of the lifting matrices whose elements are of the form $\beta/2^\lambda$, where β and λ are integers. The resultant transform approximates to the original transform. Furthermore, the resultant transform is invertible and its inverse also does not need floating-point multiplications. By factoring a transform into lifting steps, we can also obtain its approximation by an invertible non-linear transform which maps integer to integer.

Lifting factorization has become a major tool for constructing integer transforms. Research on lifting schemes and integer transforms has been active in recent years [4, 6, 21, 22, 29, 30, 31]. Theoretically, any matrix with a unit determinant can be factored into products of lifting matrices [4, 21], which can be achieved with the Gaussian elimination method. However, the known methods based on Gaussian elimination generally produce $O(N^2)$ lifting matrices from a matrix of order N. If such methods are used to factor the kernel matrix of a discrete transform and then construct the integer transform, the required computational complexity will be at least in the order of $O(N^2)$, which is not acceptable for many practical applications of digital signal processing. We need methods that can factor a kernel matrix of order N into $O(N \log_2 N)$ lifting matrices and simple matrices (whose elements are 1, -1 or 0). There already exist been many fast algorithms for discrete transforms. Some of them [23] can lead to rotation matrix factorization and can therefore be used for lifting factorization. This chapter presents some simple methods to factor the discrete transform matrix of order N into $O(N \log_2 N)$ lifting steps and simple matrices. The approximation performance of the integer transforms will be shown by numerical experiments.

8.2   PRELIMINARIES

The lifting matrix is a major tool for constructing integer wavelet and integer discrete transforms. We now consider its definition and some simple properties.

Definition 18 *A lifting matrix is a matrix whose diagonal elements are 1 and in which only one non-diagonal element is nonzero. If the order of a lifting matrix is N, we use the notation $L_{i,j}(s)$ $(i \neq j)$ to denote the lifting matrix whose only non-zero element, with a non-zero value s, is at the ith row and the jth column $(0 \leq i, j \leq N - 1)$.*

For example, matrix

$$A = \begin{bmatrix} 1 & 0 & 0 & 0 \\ 0 & 1 & 0 & 0 \\ 0 & 2.3 & 1 & 0 \\ 0 & 0 & 0 & 1 \end{bmatrix} \tag{8.1}$$

is a lifting matrix, and based on Definition 18, A can be expressed by the notation $L_{2,1}(2.3)$. To multiply a lifting matrix with a vector, say, $y = L_{i,j}(s)x$, where x and y are column vectors of length N (the order of the lifting matrix), we have

$$y(i) = x(i) + sx(j), \ y(k) = x(k), \ k \neq i \tag{8.2}$$

which is illustrated in Figure 8.1. For example, if A is defined by (8.1), then

$$y(0) = x(0), \ y(1) = x(1), \ y(2) = x(2) + 2.3x(1), \ y(3) = x(3).$$

Definition 19 *A lifting step is multiplying a lifting matrix with a vector, as shown in (8.2).*

A distinguished feature of the lifting matrix is that its inverse is still a lifting matrix in a similar form. In fact, based on (8.2) we can reconstruct vector x from vector y by

$$x(i) = y(i) - sy(j), \ x(k) = y(k), \ k \neq i, \tag{8.3}$$

that is,

$$L_{i,j}^{-1}(s) = L_{i,j}(-s). \tag{8.4}$$

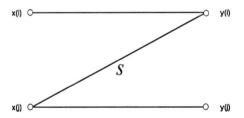

Figure 8.1   Signal flow graph of a lifting step.

In conclusion, if a matrix can be factored into a product of lifting matrices, its inverse is also a product of lifting matrices. In (8.2), floating-point multiplication is needed if s is an irrational number or even a rational number. In this case, we should approximate s by another number. For easy realization, the number should be of the form $\beta/2^\lambda$, where β and λ are integers.

Definition 20 *Let* $\mathrm{RB}(s)$ *denote a number that is of the form* $\beta/2^\lambda$ *(a rational number with its denominator being a power of two) to approximate the real number* s.

The value of $\mathrm{RB}(s)$ is not unique and depends on the required accuracy of the approximation. For example, 2.3 in (8.1) can be approximated by 2, 5/2, 9/4 or other numbers. When s is approximated by $\mathrm{RB}(s)$, matrix $L_{i,j}(s)$ is then approximated by $L_{i,j}(\mathrm{RB}(s))$ which is still invertible and whose inverse is $L_{i,j}(-\mathrm{RB}(s))$.

When lifting matrix $L_{i,j}(s)$ is approximated by $L_{i,j}(\text{RB}(s))$, the transform $y = L_{i,j}(s)x$ is approximated by $\bar{y} = L_{i,j}(\text{RB}(s))x$. We can reconstruct x from $\bar{y}$ by $x = L_{i,j}(-\text{RB}(s))\bar{y}$, that is,

$$x(i) = \bar{y}(i) - \text{RB}(s)\bar{y}(j), \ x(k) = \bar{y}(k), \ k \neq i. \tag{8.5}$$

For example, if we approximate the lifting matrix $L_{2,1}(2.3)$ defined in (8.1) by $L_{2,1}(\frac{5}{2})$ and $\bar{y} = L_{2,1}(\frac{5}{2})x$, then we can reconstruct x from $\bar{y}$ by

$$x(2) = \bar{y}(2) - \frac{5}{2}\bar{y}(1), \ x(k) = \bar{y}(k), \ k = 0, 1, 3.$$

We can also approximate a lifting step $y = L_{i,j}(s)x$ by using non-linear transforms. For example, y can be approximated by $\hat{y}$ defined as

$$\hat{y}(i) = x(i) + \lfloor sx(j) \rfloor, \ \hat{y}(k) = x(k), \ k \neq i. \tag{8.6}$$

This transform is non-linear because the floor operation $\lfloor \cdot \rfloor$ truncates a real number in a non-linear manner! A very useful property of this transform is that it maps integers into integers. This non-linear transform is invertible and its inverse is

$$x(i) = \hat{y}(i) - \lfloor s\hat{y}(j) \rfloor, \ x(k) = \hat{y}(k), \ k \neq i. \tag{8.7}$$

Example 14 *If the element s of the lifting matrix A in (8.1) is approximated by $\frac{5}{2}$ and the input is $x = (1, 3, 4, 2)'$, then y, $\bar{y}$ and $\hat{y}$ are, respectively,*

$$y = (1, 3, 10.9, 2)', \ \bar{y} = (1, 3, 11.5, 2)', \ \hat{y} = (1, 3, 10, 2)'.$$

In general, when a transform is factored into lifting steps, it can be easily approximated by another transform, which needs no floating-point multiplications, or by an integer-to-integer non-linear transform. Furthermore, the resultant transform is invertible and its inverse also needs no floating-point multiplications. This is the main reason that we factor a discrete transform into lifting steps. Having known the importance of the lifting matrix, one may ask a question: which matrix can be factored into lifting matrices and how shall we do it? We will answer this question as follows.

Firstly, it is known in [4] that an order-2 rotation matrix or diagonal matrix with a unit determinant can be factored into lifting matrices and their factorizations are

$$\begin{bmatrix} \cos\alpha & -\sin\alpha \\ \sin\alpha & \cos\alpha \end{bmatrix} = \begin{bmatrix} 1 & -\tan\frac{\alpha}{2} \\ 0 & 1 \end{bmatrix} \begin{bmatrix} 1 & 0 \\ \sin\alpha & 1 \end{bmatrix} \begin{bmatrix} 1 & -\tan\frac{\alpha}{2} \\ 0 & 1 \end{bmatrix} \tag{8.8}$$

or

$$\begin{bmatrix} \cos\alpha & -\sin\alpha \\ \sin\alpha & \cos\alpha \end{bmatrix} = \begin{bmatrix} 1 & 0 \\ \tan\frac{\alpha}{2} & 1 \end{bmatrix} \begin{bmatrix} 1 & -\sin\alpha \\ 0 & 1 \end{bmatrix} \begin{bmatrix} 1 & 0 \\ \tan\frac{\alpha}{2} & 1 \end{bmatrix} \tag{8.9}$$

$$\begin{bmatrix} c & 0 \\ 0 & \frac{1}{c} \end{bmatrix} = \begin{bmatrix} 1 & c-1 \\ 0 & 1 \end{bmatrix} \begin{bmatrix} 1 & 0 \\ 1 & 1 \end{bmatrix} \begin{bmatrix} 1 & \frac{1}{c}-1 \\ 0 & 1 \end{bmatrix} \begin{bmatrix} 1 & 0 \\ -c & 1 \end{bmatrix}. \tag{8.10}$$

For convenience, we use R_α to represent the rotation matrix,

$$R_\alpha = \begin{bmatrix} \cos\alpha & -\sin\alpha \\ \sin\alpha & \cos\alpha \end{bmatrix}. \tag{8.11}$$

Secondly, we can generalize the factorization in (8.10) to factor any diagonal matrix with a unit determinant.

Lemma 17 *Assume that D is a diagonal matrix with a unit determinant, say,*

$$D = \text{diag}(b_0, b_1, \ldots, b_{L-1}),$$

where $b_0 b_1 \cdots b_{L-1} = 1$. Let $\alpha_0 = b_0$, $\alpha_k = \alpha_{k-1} b_k$ $(k = 1, 2, \ldots, L-1)$. Then $D = B_1 B_2 = B_2 B_1$, where

$$B_1 = \text{diag}(\alpha_0, \frac{1}{\alpha_0}, \alpha_2, \frac{1}{\alpha_2}, \ldots, \alpha_{L-2}, \frac{1}{\alpha_{L-2}})$$

$$B_2 = \text{diag}(1, \alpha_1, \frac{1}{\alpha_1}, \alpha_3, \frac{1}{\alpha_3}, \ldots, 1)$$

for L being even, and

$$B_1 = \text{diag}(\alpha_0, \frac{1}{\alpha_0}, \alpha_2, \frac{1}{\alpha_2}, \ldots, 1)$$

$$B_2 = \text{diag}(1, \alpha_1, \frac{1}{\alpha_1}, \alpha_3, \frac{1}{\alpha_3}, \ldots, \alpha_{L-2}, \frac{1}{\alpha_{L-2}})$$

for L being odd. The matrices B_1 and B_2 can be factored as products of lifting matrices, respectively.

Proof. Based on the definition of D, B_1 and B_2, it is easy to verify that $D = B_1 B_2 = B_2 B_1$. To factor B_1 and B_2 into lifting matrices, we note that they can be expressed as block diagonal matrices where each block is of the form

$$\begin{bmatrix} c & 0 \\ 0 & \frac{1}{c} \end{bmatrix}$$

or an identity matrix. The matrix of order 2 can be expressed as the product of lifting matrices shown in (8.10). So, B_1 and B_2 can also be factored into products of lifting matrices.

It is known [4] that any matrix with a unit determinant can be factored into lifting matrices and permutation matrices, which is formally stated in the theorem below.

Theorem 17 *A sufficient and necessary condition that a real matrix A can be decomposed into products of lifting matrices and permutation matrices is $|A| = 1$.*

Proof. It is obvious that the necessary condition holds. Now we show the sufficiency of the condition. By means of triangular factorization of a matrix (Gaussian elimination), there exist lower triangular lifting matrices L_l $(l = 1, 2, \ldots, K_1)$, upper triangular lifting matrices U_k $(k = 1, 2, \ldots, K_2)$, a diagonal matrix $D = \text{diag}(a_0, a_1, \ldots, a_{N-1})$ with $|D| = 1$ and a permutation matrix P such that

$$A = P L_1 L_2 \cdots L_{K_1} D U_1 U_2 \cdots U_{K_2}. \tag{8.12}$$

Combining (8.12) and Lemma 17, we complete the proof.

Suppose that the determinant of a matrix A is a nonzero integer n. If we define an integer matrix $\Delta = \text{diag}(1, 1, \ldots, 1, n)$ and let $A = \Delta B$, then $|B| = 1$. Since B can be factored into lifting matrices and permutation matrices, we know that A can be factored into lifting matrices and integer matrices (permutation matrix P is a special integer matrix). Since our attention is on integer transform, an integer matrix is essential. We write this important result as a theorem below.

Theorem 18 *Any matrix with its determinant being a non-zero integer can be factored into lifting matrices and integer matrices.*

In general, any invertible matrix can be factored into products of lifting matrices, permutation matrices and a diagonal matrix whose elements may not be integers.

Theorems 17 and 18 give a general framework for lifting factorization. However, the method based on Gaussian elimination is not practical for most signal processing applications. Since Gaussian elimination generally requires $O(N^2)$ lifting matrices in the factorization of a matrix of order N, it is often unacceptable for signal processing applications. If an integer transform is based on such a factorization with $O(N^2)$ lifting matrices, its computational complexity will be at least in the order of $O(N^2)$. We know that most useful discrete transforms have fast algorithms with computational complexity in the order of $O(N \log_2 N)$. Any integer transforms should also be within this computational complexity limit. General methods are needed to factor a transform matrix of order N into $O(N \log_2 N)$ lifting matrices and simple matrices. There have already been many fast algorithms for discrete transforms. Some of them [23] can lead to rotation matrix factorization and therefore can be used for lifting factorization. In the following sections, we will discuss simple methods to factor the transform kernel matrix of order N into $O(N \log_2 N)$ lifting steps and simple matrices.

8.3   INTEGER DCT AND FAST ALGORITHMS

In this section, a general method is developed to factor the type-IV DCT into lifting steps and additions. Based on the relationships among various types of DCTs, we can also factor type-II and -III DCTs into lifting steps and additions. After approximating the lifting matrices, we will derive type-II, -III and -IV integer DCTs which do not need floating-point multiplication. Integer-to-integer transforms which approximate the DCT are also considered. Fast algorithms are given for these transforms and their computational complexities are analyzed. Numerical experiments will be used to show the approximation performance of these transforms.

8.3.1   Lifting factorization of type-IV DCT

Let $x(n)$ $(n = 0, 1, \ldots, N-1)$ be a real-valued input sequence. We assume that $N = 2^t$, where t is a positive integer. The length N type-IV DCT of $x(n)$ is defined as

$$X(k) = \sum_{n=0}^{N-1} x(n) \cos \frac{\pi(2n+1)(2k+1)}{4N}, \quad k = 0, 1, \ldots, N-1. \tag{8.13}$$

Let C_N^{IV} be the corresponding transform matrix, that is,

$$C_N^{IV} = \left(\cos \frac{\pi(2n+1)(2k+1)}{4N} \right)_{k,n=0,1,\ldots,N-1}. \tag{8.14}$$

In order to factor the kernel matrix, we first derive a simple fast algorithm. Let

$$H(k) = \frac{X(2k) + X(2k+1)}{2}, \quad G(k) = \frac{X(2k) - X(2k+1)}{2},$$

where $k = 0, 1, \ldots, N/2 - 1$. Based on the properties of trigonometric identities, we have

$$
\begin{aligned}
H(k) &= \sum_{n=0}^{N-1} x(n) \cos \frac{\pi(2n+1)}{4N} \cos \frac{\pi(2n+1)(2k+1)}{2N} \\
&= \sum_{n=0}^{N/2-1} \left[x(n) \cos \frac{\pi(2n+1)}{4N} - x(N-1-n) \sin \frac{\pi(2n+1)}{4N} \right] \\
&\quad \cdot \cos \frac{\pi(2n+1)(2k+1)}{2N}
\end{aligned}
\tag{8.15}
$$

$$
\begin{aligned}
G(k) &= \sum_{n=0}^{N-1} x(n) \sin \frac{\pi(2n+1)}{4N} \sin \frac{\pi(2n+1)(2k+1)}{2N} \\
&= \sum_{n=0}^{N/2-1} \left[x(n) \sin \frac{\pi(2n+1)}{4N} + x(N-1-n) \cos \frac{\pi(2n+1)}{4N} \right] \\
&\quad \cdot \sin \frac{\pi(2n+1)(2k+1)}{2N}.
\end{aligned}
\tag{8.16}
$$

Let

$$
h(n) = x(n) \cos \frac{\pi(2n+1)}{4N} - x(N-1-n) \sin \frac{\pi(2n+1)}{4N}
\tag{8.17}
$$

$$
g(n) = x(n) \sin \frac{\pi(2n+1)}{4N} + x(N-1-n) \cos \frac{\pi(2n+1)}{4N},
\tag{8.18}
$$

where $n = 0, 1, \ldots, N/2 - 1$. Then $H(k)$ and $G(k)$ are type-IV DCTs of $h(n)$ and $g(n)$, respectively. In fact, we have

$$
H(k) = \sum_{n=0}^{N/2-1} h(n) \cos \frac{\pi(2n+1)(2k+1)}{2N}
\tag{8.19}
$$

$$
G(k) = (-1)^k \sum_{n=0}^{N/2-1} g(N/2-1-n) \cos \frac{\pi(2n+1)(2k+1)}{2N}.
\tag{8.20}
$$

Therefore, a length N type-IV DCT is expressed into two length $N/2$ type-IV DCTs at the cost of a pre-processing stage defined in (8.17) and (8.18) and a post-processing stage defined by

$$
X(2k) = H(k) + G(k), \quad X(2k+1) = H(k) - G(k)
\tag{8.21}
$$

for $k = 0, 1, \ldots, N/2-1$. Based on the algorithm, Lemma 18 formally states the factorization process of matrix C_N^{IV}.

Lemma 18 *The transform matrix of the type-IV DCT can be factored recursively as follows:*

$$
C_2^{IV} = \begin{bmatrix} \cos \frac{\pi}{8} & \sin \frac{\pi}{8} \\ \sin \frac{\pi}{8} & -\cos \frac{\pi}{8} \end{bmatrix}
\tag{8.22}
$$

$$C_N^{IV} = P_N \begin{bmatrix} I_{N/2} & I_{N/2} \\ I_{N/2} & -I_{N/2} \end{bmatrix} \begin{bmatrix} I_{N/2} & 0 \\ 0 & B_{N/2} \end{bmatrix} \begin{bmatrix} C_{N/2}^{IV} & 0 \\ 0 & C_{N/2}^{IV} \end{bmatrix} \begin{bmatrix} I_{N/2} & 0 \\ 0 & \hat{I}_{N/2} \end{bmatrix} P_N' D_N Q_N, \quad (8.23)$$

where $I_{N/2}$ is the identity matrix of order $N/2$, $\hat{I}_{N/2}$ is the matrix obtained by reversing the rows of $I_{N/2}$, D_N is a block diagonal matrix defined by

$$D_N = \mathrm{diag}(T(0), \ T(1), \ \ldots, \ T(N/2 - 1)), \quad (8.24)$$

where $T(n)$ $(n = 0, 1, \ldots, N/2 - 1)$ are rotation matrices of order 2 defined by

$$T(n) = \begin{bmatrix} \cos \frac{\pi(2n+1)}{4N} & -\sin \frac{\pi(2n+1)}{4N} \\ \sin \frac{\pi(2n+1)}{4N} & \cos \frac{\pi(2n+1)}{4N} \end{bmatrix} \quad (8.25)$$

$$B_{N/2} = \mathrm{diag}(1, \ -1, \ \ldots, \ 1, \ -1), \quad (8.26)$$

P_N and Q_N are permutation matrices defined respectively by

$$P_N = \begin{bmatrix} 1 & 0 & \cdots & 0 & 0 & 0 & \cdots & 0 \\ 0 & 0 & \cdots & 0 & 1 & 0 & \cdots & 0 \\ 0 & 1 & \cdots & 0 & 0 & 0 & \cdots & 0 \\ 0 & 0 & \cdots & 0 & 0 & 1 & \cdots & 0 \\ & & \cdots & & & \cdots & & \\ 0 & 0 & \cdots & 1 & 0 & 0 & \cdots & 0 \\ 0 & 0 & \cdots & 0 & 0 & 0 & \cdots & 1 \end{bmatrix} \quad (8.27)$$

$$Q_N = \begin{bmatrix} 1 & 0 & \cdots & 0 & 0 & \cdots & 0 & 0 \\ 0 & 0 & \cdots & 0 & 0 & \cdots & 0 & 1 \\ 0 & 1 & \cdots & 0 & 0 & \cdots & 0 & 0 \\ 0 & 0 & \cdots & 0 & 0 & \cdots & 1 & 0 \\ & & \cdots & & & \cdots & & \\ 0 & 0 & \cdots & 1 & 0 & \cdots & 0 & 0 \\ 0 & 0 & \cdots & 0 & 1 & \cdots & 0 & 0 \end{bmatrix} \quad (8.28)$$

$C_{N/2}^{IV}$ is the transform matrix of the type-IV DCT of length $N/2$ and P_N' is the transposition of P_N.

Example 15 *Factorization of short-length type-IV DCT*

$$C_4^{IV} = \begin{bmatrix} 1 & 0 & 0 & 0 \\ 0 & 0 & 1 & 0 \\ 0 & 1 & 0 & 0 \\ 0 & 0 & 0 & 1 \end{bmatrix} \begin{bmatrix} 1 & 0 & 1 & 0 \\ 0 & 1 & 0 & 1 \\ 1 & 0 & -1 & 0 \\ 0 & 1 & 0 & -1 \end{bmatrix} \begin{bmatrix} 1 & 0 & 0 & 0 \\ 0 & 1 & 0 & 0 \\ 0 & 0 & 1 & 0 \\ 0 & 0 & 0 & -1 \end{bmatrix} \begin{bmatrix} C_2^{IV} & 0 \\ 0 & C_2^{IV} \end{bmatrix} \begin{bmatrix} 1 & 0 & 0 & 0 \\ 0 & 1 & 0 & 0 \\ 0 & 0 & 0 & 1 \\ 0 & 0 & 1 & 0 \end{bmatrix}$$

$$\cdot \begin{bmatrix} 1 & 0 & 0 & 0 \\ 0 & 0 & 1 & 0 \\ 0 & 1 & 0 & 0 \\ 0 & 0 & 0 & 1 \end{bmatrix} \begin{bmatrix} \cos \frac{\pi}{16} & -\sin \frac{\pi}{16} & 0 & 0 \\ \sin \frac{\pi}{16} & \cos \frac{\pi}{16} & 0 & 0 \\ 0 & 0 & \cos \frac{3\pi}{16} & -\sin \frac{3\pi}{16} \\ 0 & 0 & \sin \frac{3\pi}{16} & \cos \frac{3\pi}{16} \end{bmatrix} \begin{bmatrix} 1 & 0 & 0 & 0 \\ 0 & 0 & 0 & 1 \\ 0 & 1 & 0 & 0 \\ 0 & 0 & 1 & 0 \end{bmatrix} \quad (8.29)$$

$$C_8^{IV} = P_8 \begin{bmatrix} I_4 & I_4 \\ I_4 & -I_4 \end{bmatrix} \begin{bmatrix} I_4 & 0 \\ 0 & B_4 \end{bmatrix} \begin{bmatrix} C_4^{IV} & 0 \\ 0 & C_4^{IV} \end{bmatrix} \begin{bmatrix} I_4 & 0 \\ 0 & \hat{I}_4 \end{bmatrix} P_8' D_8 Q_8, \quad (8.30)$$

where

$$D_8 = \begin{bmatrix} \cos\frac{\pi}{32} & -\sin\frac{\pi}{32} \\ \sin\frac{\pi}{32} & \cos\frac{\pi}{32} \\ & & \cos\frac{3\pi}{32} & -\sin\frac{3\pi}{32} \\ & & \sin\frac{3\pi}{32} & \cos\frac{3\pi}{32} \\ & & & & \cos\frac{5\pi}{32} & -\sin\frac{5\pi}{32} \\ & & & & \sin\frac{5\pi}{32} & \cos\frac{5\pi}{32} \\ & & & & & & \cos\frac{7\pi}{32} & -\sin\frac{7\pi}{32} \\ & & & & & & \sin\frac{7\pi}{32} & \cos\frac{7\pi}{32} \end{bmatrix} \tag{8.31}$$

$$P_8 = \begin{bmatrix} 1&0&0&0&0&0&0&0 \\ 0&0&0&0&1&0&0&0 \\ 0&1&0&0&0&0&0&0 \\ 0&0&0&0&0&1&0&0 \\ 0&0&1&0&0&0&0&0 \\ 0&0&0&0&0&0&1&0 \\ 0&0&0&1&0&0&0&0 \\ 0&0&0&0&0&0&0&1 \end{bmatrix} \quad Q_8 = \begin{bmatrix} 1&0&0&0&0&0&0&0 \\ 0&0&0&0&0&0&0&1 \\ 0&1&0&0&0&0&0&0 \\ 0&0&0&0&0&0&1&0 \\ 0&0&1&0&0&0&0&0 \\ 0&0&0&0&0&1&0&0 \\ 0&0&0&1&0&0&0&0 \\ 0&0&0&0&1&0&0&0 \end{bmatrix}. \tag{8.32}$$

Floating-point multiplications are needed when matrix D_N is multiplied by a vector. In order to avoid floating-point multiplications, we convert this matrix into products of lifting matrices and then approximate the elements of the lifting matrices to be in the form of $\beta/2^\lambda$, where β and λ are integers. D_N can be easily turned into products of lifting matrices based on the factorization in (8.8). In fact, each block in D_N can be factored as

$$T(n) = \begin{bmatrix} 1 & -\tan\frac{\alpha_n}{2} \\ 0 & 1 \end{bmatrix} \begin{bmatrix} 1 & 0 \\ \sin\alpha_n & 1 \end{bmatrix} \begin{bmatrix} 1 & -\tan\frac{\alpha_n}{2} \\ 0 & 1 \end{bmatrix}, \tag{8.33}$$

where $\alpha_n = \frac{\pi(2n+1)}{4N}$. Therefore, D_N is factored into $\frac{3N}{2}$ lifting matrices. Obviously,

$$C_2^{IV} = \begin{bmatrix} 1 & -\tan\frac{\pi}{16} \\ 0 & 1 \end{bmatrix} \begin{bmatrix} 1 & 0 \\ \sin\frac{\pi}{8} & 1 \end{bmatrix} \begin{bmatrix} 1 & \tan\frac{\pi}{16} \\ 0 & -1 \end{bmatrix}. \tag{8.34}$$

If the same method is used to factor matrix $C_{N/2}^{IV}$ recursively until the matrix order is 2, we get the complete factorization of C_N^{IV}.

8.3.2 Integer type-IV DCT and fast algorithm

By approximating every non-zero non-diagonal element s of the lifting matrices in (8.33) and (8.34) by RB(s) defined in Definition 20, we can approximate matrices $T(n)$ and C_2^{IV} by $\bar{T}(n)$ and $\bar{C}_2^{IV}$, respectively, where

$$\bar{T}(n) = \begin{bmatrix} 1 & -\mathrm{RB}(\tan\frac{\alpha_n}{2}) \\ 0 & 1 \end{bmatrix} \begin{bmatrix} 1 & 0 \\ \mathrm{RB}(\sin\alpha_n) & 1 \end{bmatrix} \begin{bmatrix} 1 & -\mathrm{RB}(\tan\frac{\alpha_n}{2}) \\ 0 & 1 \end{bmatrix} \tag{8.35}$$

$$\bar{C}_2^{IV} = \begin{bmatrix} 1 & -\mathrm{RB}(\tan\frac{\pi}{16}) \\ 0 & 1 \end{bmatrix} \begin{bmatrix} 1 & 0 \\ \mathrm{RB}(\sin\frac{\pi}{8}) & 1 \end{bmatrix} \begin{bmatrix} 1 & \mathrm{RB}(\tan\frac{\pi}{16}) \\ 0 & -1 \end{bmatrix}. \tag{8.36}$$

Then D_N is approximated by $\bar{D}_N$

$$\bar{D}_N = \mathrm{diag}(\bar{T}(0),\ \bar{T}(1),\ \ldots,\ \bar{T}(N/2-1)). \tag{8.37}$$

We finally achieve an approximation matrix $\bar{C}_N^{IV}$ for C_N^{IV} to define a new transform without floating-point multiplications. Because the transform only needs integer operations and shift operations, we call it the type-IV integer DCT (IntDCT-IV). However, it does not mean that the transform IntDCT-IV maps integer to integer. An integer-to-integer transform will be introduced in the next subsection.

Definition 21 *Assume that $N = 2^t$. The transform matrix, $\bar{C}_N^{IV}$, of an IntDCT-IV is recursively defined by*

$$\bar{C}_N^{IV} = P_N \begin{bmatrix} I_{N/2} & I_{N/2} \\ I_{N/2} & -I_{N/2} \end{bmatrix} \begin{bmatrix} I_{N/2} & 0 \\ 0 & B_{N/2} \end{bmatrix} \begin{bmatrix} \bar{C}_{N/2}^{IV} & 0 \\ 0 & \bar{C}_{N/2}^{IV} \end{bmatrix} \begin{bmatrix} I_{N/2} & 0 \\ 0 & \hat{I}_{N/2} \end{bmatrix} P_N' \bar{D}_N Q_N, (8.38)$$

where the notation is the same as that in Lemma 18. The transform matrix of a length 2 IntDCT-IV is defined in (8.36).

The transform is not unique. Actually, any choice of function RB determines a transform. Based on the definition, we get a fast algorithm for the IntDCT-IV as follows.

Algorithm 22 *Fast algorithm for IntDCT-IV*

Step 1.   *Compute*

$$\begin{aligned} \bar{g}(n) &= f_n[x(n) - e_n x(N-1-n)] + x(N-1-n) \\ \bar{h}(n) &= x(n) - e_n x(N-1-n) - e_n \bar{g}(n), \end{aligned}$$

where $e_n = \mathrm{RB}(\tan\frac{\alpha_n}{2})$, $f_n = \mathrm{RB}(\sin\alpha_n)$ and $n = 0,1,\ldots,N/2-1$.

Step 2.   *Compute the length $N/2$ IntDCT-IVs $\bar{H}(k)$ and $\bar{G}(k)$ of sequences $\bar{h}(n)$ and $\bar{g}(N/2-1-n)$, respectively. If $N/2 > 2$, the length $N/2$ IntDCT-IV is decomposed into smaller ones until the length becomes 2. Based on the matrix decomposition in (8.36), the length 2 IntDCT-IV can be computed by 3 lifting steps.*

Step 3.   *Compute*

$$X(2k) = \bar{H}(k) + (-1)^k \bar{G}(k), \quad X(2k+1) = \bar{H}(k) - (-1)^k \bar{G}(k),$$

where $k = 0,1,\ldots,N/2-1$.

Now we consider the computational complexity of the algorithm. Step 1 needs $\frac{3N}{2}$ lifting steps, and Step 3 needs N additions. Let $L_{C^{IV}}(N)$ and $A_{C^{IV}}(N)$ represent the numbers of lifting steps and additions, respectively, for computing a length N IntDCT-IV. Then, we have

$$\begin{aligned} L_{C^{IV}}(N) &= 2L_{C^{IV}}(\tfrac{N}{2}) + \tfrac{3N}{2}, & L_{C^{IV}}(2) &= 3 \\ A_{C^{IV}}(N) &= 2A_{C^{IV}}(\tfrac{N}{2}) + N, & A_{C^{IV}}(2) &= 0 \end{aligned}$$

whose closed form expressions are

$$L_{C^{IV}}(N) = \frac{3N}{2}\log_2 N, \quad A_{C^{IV}}(N) = N\log_2 N - N, \quad N > 1. \tag{8.39}$$

The matrix $\bar{C}_N^{IV}$ is invertible and its inverse matrix can also be factored as products of lifting matrices and other simple matrices whose elements are 0, ± 1 or $\pm 1/2$. In fact, we have

$$(\bar{C}_2^{IV})^{-1} = \begin{bmatrix} 1 & \text{RB}(\tan\frac{\pi}{16}) \\ 0 & -1 \end{bmatrix} \begin{bmatrix} 1 & 0 \\ -\text{RB}(\sin\frac{\pi}{8}) & 1 \end{bmatrix} \begin{bmatrix} 1 & \text{RB}(\tan\frac{\pi}{16}) \\ 0 & 1 \end{bmatrix} \qquad (8.40)$$

$$(\bar{C}_N^{IV})^{-1} = \frac{1}{2} Q'_N (\bar{D}_N)^{-1} P_N \begin{bmatrix} I_{N/2} & 0 \\ 0 & \hat{I}_{N/2} \end{bmatrix} \begin{bmatrix} (\bar{C}_{N/2}^{IV})^{-1} & 0 \\ 0 & (\bar{C}_{N/2}^{IV})^{-1} \end{bmatrix}$$
$$\cdot \begin{bmatrix} I_{N/2} & 0 \\ 0 & B_{N/2} \end{bmatrix} \begin{bmatrix} I_{N/2} & I_{N/2} \\ I_{N/2} & -I_{N/2} \end{bmatrix} P'_N, \qquad (8.41)$$

where for $n = 0, 1, \ldots, N/2 - 1$,

$$(\bar{D}_N)^{-1} = \text{diag}((\bar{T}(0))^{-1}, (\bar{T}(1))^{-1}, \ldots, (\bar{T}(N/2 - 1))^{-1}) \qquad (8.42)$$

$$(\bar{T}(n))^{-1} = \begin{bmatrix} 1 & \text{RB}(\tan\frac{\alpha_n}{2}) \\ 0 & 1 \end{bmatrix} \begin{bmatrix} 1 & 0 \\ -\text{RB}(\sin\alpha_n) & 1 \end{bmatrix} \begin{bmatrix} 1 & \text{RB}(\tan\frac{\alpha_n}{2}) \\ 0 & 1 \end{bmatrix}. \qquad (8.43)$$

Based on the factorization in (8.41), we have a fast algorithm for the inverse IntDCT-IV **(reconstruction algorithm for IntDCT-IV)** as follows, where $X(k)$ is the input and $x(n)$ is the output.

Algorithm 23 *Reconstruction algorithm for IntDCT-IV*

Step 1.   *Compute*

$$\bar{H}(k) = \frac{X(2k) + X(2k+1)}{2}, \quad \bar{G}(k) = \frac{(-1)^k[X(2k) - X(2k+1)]}{2},$$

where $k = 0, 1, \ldots, N/2 - 1$.

Step 2.   *Compute the inverse IntDCT-IVs of length $N/2$ of sequences $\bar{H}(k)$ and $\bar{G}(k)$, and let the outputs be $\bar{h}(n)$ and $\bar{g}(n)$, respectively. If $N/2 > 2$, the length $N/2$ inverse IntDCT-IV is decomposed into smaller ones until the length becomes 2. Based on the matrix decomposition in (8.40), the length 2 inverse IntDCT-IV can be computed by 3 lifting steps.*

Step 3.   *Compute*

$$x(n) = \bar{h}(n) + e_n \bar{g}(N/2 - 1 - n) + e_n x(N - 1 - n)$$
$$x(N - 1 - n) = -f_n[\bar{h}(n) + e_n \bar{g}(N/2 - 1 - n)] - \bar{g}(N/2 - 1 - n),$$

where $e_n = \text{RB}(\tan\frac{\alpha_n}{2})$, $f_n = \text{RB}(\sin\alpha_n)$ and $n = 0, 1, \ldots, N/2 - 1$.

The number of operations for reconstruction is the same as that for the forward transform. It should be noted that the matrix $\bar{C}_N^{IV}$ is not orthogonal in general and we cannot use $(\bar{C}_N^{IV})'$ for reconstruction.

8.3.3   Integer-to-integer transform

Based on the preceding subsections, we know that the matrix of a type-IV DCT can be factored into products of lifting matrices and some simple matrices whose elements are either ± 1 or 0. We have also shown in (8.6) and (8.7) that a lifting step can be approximated by a non-linear integer-to-integer transform. Now, let us achieve the integer-to-integer transform to approximate the type-IV DCT by approximating all the lifting steps into integer computation. We use the notation II-DCT-IV for the transform. Based on Lemma 18, a fast algorithm of II-DCT-IV is given below.

Algorithm 24 *Fast algorithm for II-DCT-IV*

Step 1.   *For $n = 0, 1, \ldots, N/2 - 1$, compute*

$$
\begin{aligned}
h_1(n) &= x(n) - \lfloor x(N-1-n)\tan(\alpha_n/2) \rfloor, & g_1(n) &= x(N-1-n) \\
h_2(n) &= h_1(n), & g_2(n) &= \lfloor h_1(n)\sin(\alpha_n) \rfloor + g_1(n) \\
h(n) &= h_2(n) - \lfloor g_2(n)\tan(\alpha_n/2) \rfloor, & g(n) &= g_2(n).
\end{aligned}
$$

Step 2.   *Compute the length $N/2$ II-DCT-IV, $H(k)$ and $G(k)$, for the sequences $h(n)$ and $g(N/2-1-n)$, respectively. If $N/2 > 2$, the length $N/2$ II-DCT-IV is decomposed into smaller ones until the length becomes 2. Based on the matrix decomposition in (8.34), the length 2 II-DCT-IV can be computed by 3 lifting steps (every lifting step is approximated as in (8.6)).*

Step 3.   *Compute*

$$X(2k) = H(k) + (-1)^k G(k), \quad X(2k+1) = H(k) - (-1)^k G(k),$$

where $k = 0, 1, \ldots, N/2 - 1$.

The II-DCT-IV is invertible and its inverse is also an integer-to-integer transform. By inverting the steps of Algorithm 24, we can derive the reconstruction algorithm for the II-DCT-IV easily.

8.3.4   Integer type-II and -III DCTs

In this subsection, we use the relationships among various types of DCTs to factor the type-II and -III DCTs into lifting steps and additions. Based on the factorization, we can achieve the type-II and -III integer DCTs. Because all types of the discrete sine transform (DST) have simple relationships with the corresponding DCTs [10, 17], the integral versions and their corresponding fast algorithms can also be achieved easily.

Lifting factorization of type-II DCT
    Let $x(n)$ ($n = 0, 1, \ldots, N - 1$) be a real input sequence. We assume that $N = 2^t$ where $t > 0$. The type-II DCT of $x(n)$ is defined as

$$X(k) = \tau(k)\sum_{n=0}^{N-1} x(n)\cos\frac{\pi(2n+1)k}{2N}, \quad k = 0, 1, \ldots, N-1, \tag{8.44}$$

where

$$\tau(0) = \sqrt{2}/2, \ \tau(k) = 1, \ k = 1, 2, \ldots, N-1. \tag{8.45}$$

The type-II DCT can be computed by using the type-IV DCT. Based on the relationship stated in [10, 20], we get the following factorization in Lemma 19.

Lemma 19 *Let C_N^{II} be the kernel matrix of the type-II DCT,*

$$C_N^{II} = \left(\tau(k) \cos \frac{\pi k(2n+1)}{2N} \right).$$

Then

$$C_2^{II} = \begin{bmatrix} \sqrt{2}/2 & \sqrt{2}/2 \\ \sqrt{2}/2 & -\sqrt{2}/2 \end{bmatrix} \tag{8.46}$$

$$C_N^{II} = P_N \begin{bmatrix} C_{N/2}^{II} & 0 \\ 0 & C_{N/2}^{IV} \end{bmatrix} \begin{bmatrix} I_{N/2} & \hat{I}_{N/2} \\ I_{N/2} & -\hat{I}_{N/2} \end{bmatrix}, \tag{8.47}$$

where P_N, $I_{N/2}$ and $\hat{I}_{N/2}$ are defined in Lemma 18.

Example 16 *Factorization of length 4 and 8 type-II DCTs*

$$C_4^{II} = \begin{bmatrix} 1 & 0 & 0 & 0 \\ 0 & 0 & 1 & 0 \\ 0 & 1 & 0 & 0 \\ 0 & 0 & 0 & 1 \end{bmatrix} \begin{bmatrix} \sqrt{2}/2 & \sqrt{2}/2 & 0 & 0 \\ \sqrt{2}/2 & -\sqrt{2}/2 & 0 & 0 \\ 0 & 0 & \cos\frac{\pi}{8} & \sin\frac{\pi}{8} \\ 0 & 0 & \sin\frac{\pi}{8} & -\cos\frac{\pi}{8} \end{bmatrix} \begin{bmatrix} 1 & 0 & 0 & 1 \\ 0 & 1 & 1 & 0 \\ 1 & 0 & 0 & -1 \\ 0 & 1 & -1 & 0 \end{bmatrix} \tag{8.48}$$

$$C_8^{II} = P_8 \begin{bmatrix} C_4^{II} & 0 \\ 0 & C_4^{IV} \end{bmatrix} \begin{bmatrix} I_4 & \hat{I}_4 \\ I_4 & -\hat{I}_4 \end{bmatrix}, \tag{8.49}$$

where P_8 is defined in (8.32).

Definition 22 *Assume that $N = 2^t$. The kernel matrix of the type-II integer DCT (IntDCT-II) of length N is defined by*

$$\bar{C}_2^{II} = \begin{bmatrix} 1 & RB(\sqrt{2})-1 \\ 0 & -1 \end{bmatrix} \begin{bmatrix} 1 & 0 \\ -RB(\frac{\sqrt{2}}{2}) & 1 \end{bmatrix} \begin{bmatrix} 1 & RB(\sqrt{2})-1 \\ 0 & 1 \end{bmatrix} \tag{8.50}$$

$$\bar{C}_N^{II} = P_N \begin{bmatrix} \bar{C}_{N/2}^{II} & 0 \\ 0 & \bar{C}_{N/2}^{IV} \end{bmatrix} \begin{bmatrix} I_{N/2} & \hat{I}_{N/2} \\ I_{N/2} & -\hat{I}_{N/2} \end{bmatrix}, \quad N > 2. \tag{8.51}$$

Based on the definition, we derive a fast algorithm for IntDCT-II without requiring any floating-point multiplication.

Algorithm 25 *Fast algorithm for IntDCT-II*

Step 1.   *Compute*

$$h(n) = x(n) + x(N-1-n), \quad g(n) = x(n) - x(N-1-n),$$

where $n = 0, 1, \ldots, N/2 - 1$.

Step 2. *Use Algorithm 22 to compute $X(2k+1)$, which is the IntDCT-IV of sequence $g(n)$. Compute $X(2k)$, which is the IntDCT-II of sequence $h(n)$. If $N/2 > 2$, the length $N/2$ IntDCT-II is decomposed into smaller ones until the length becomes 2. Based on the matrix decomposition in (8.50), the length 2 IntDCT-II can be computed by 3 lifting steps.*

The numbers of lifting steps and additions are

$$L_{C^{II}}(N) = \frac{3N}{2}\log_2 N - 3N + 6$$
$$A_{C^{II}}(N) = N\log_2 N - N. \tag{8.52}$$

The inverse of $\bar{C}_N^{II}$ is

$$(\bar{C}_2^{II})^{-1} = \begin{bmatrix} 1 & 1 - \mathrm{RB}(\sqrt{2}) \\ 0 & 1 \end{bmatrix} \begin{bmatrix} 1 & 0 \\ \mathrm{RB}(\frac{\sqrt{2}}{2}) & 1 \end{bmatrix} \begin{bmatrix} 1 & \mathrm{RB}(\sqrt{2}) - 1 \\ 0 & -1 \end{bmatrix} \tag{8.53}$$

$$(\bar{C}_N^{II})^{-1} = \frac{1}{2} \begin{bmatrix} I_{N/2} & I_{N/2} \\ \hat{I}_{N/2} & -\hat{I}_{N/2} \end{bmatrix} \begin{bmatrix} (\bar{C}_{N/2}^{II})^{-1} & 0 \\ 0 & (\bar{C}_{N/2}^{IV})^{-1} \end{bmatrix} P_N', \quad N > 2. \tag{8.54}$$

Lifting factorization of the type-III DCT

Let $x(n)$ $(n = 0, 1, \ldots, N-1)$ be a real input sequence. We assume that $N = 2^t$ where $t > 0$. The type-III DCT of $x(n)$ is defined by

$$X(k) = \sum_{n=0}^{N-1} \tau(n)x(n)\cos\frac{\pi n(2k+1)}{2N}, \quad k = 0, 1, \ldots, N-1, \tag{8.55}$$

where $\tau(n)$ is defined in (8.45). The type-III DCT can be computed by using a type-IV DCT. Based on the relationship stated in [10, 20], we have the following factorization in Lemma 20.

Lemma 20 *Let C_N^{III} be the kernel matrix of the type-III DCT,*

$$C_N^{III} = \left(\tau(n)\cos\frac{\pi(2k+1)n}{2N}\right).$$

Then

$$C_2^{III} = \begin{bmatrix} \sqrt{2}/2 & \sqrt{2}/2 \\ \sqrt{2}/2 & -\sqrt{2}/2 \end{bmatrix} \tag{8.56}$$

$$C_N^{III} = \begin{bmatrix} I_{N/2} & I_{N/2} \\ \hat{I}_{N/2} & -\hat{I}_{N/2} \end{bmatrix} \begin{bmatrix} C_{N/2}^{III} & 0 \\ 0 & C_{N/2}^{IV} \end{bmatrix} P_N', \tag{8.57}$$

where P_N, $I_{N/2}$ and $\hat{I}_{N/2}$ are defined in Lemma 18.

Definition 23 *Assume that $N = 2^t$. The kernel matrix of a type-III integer DCT (IntDCT-III) of length N is defined by*

$$\bar{C}_2^{III} = \begin{bmatrix} 1 & \mathrm{RB}(\sqrt{2}) - 1 \\ 0 & -1 \end{bmatrix} \begin{bmatrix} 1 & 0 \\ -\mathrm{RB}(\frac{\sqrt{2}}{2}) & 1 \end{bmatrix} \begin{bmatrix} 1 & \mathrm{RB}(\sqrt{2}) - 1 \\ 0 & 1 \end{bmatrix} \tag{8.58}$$

$$\bar{C}_N^{III} = \begin{bmatrix} I_{N/2} & I_{N/2} \\ \hat{I}_{N/2} & -\hat{I}_{N/2} \end{bmatrix} \begin{bmatrix} \bar{C}_{N/2}^{III} & 0 \\ 0 & \bar{C}_{N/2}^{IV} \end{bmatrix} P_N', \quad N > 2. \tag{8.59}$$

Based on the factorization, we obtain a fast algorithm for IntDCT-III which does not need floating-point multiplications.

Algorithm 26 *Fast algorithm for IntDCT-III*

Step 1.   *Use Algorithm 22 to compute the IntDCT-IV $G(k)$ of sequence $x(2n+1)$. Compute the IntDCT-III $H(k)$ of sequence $x(2n)$. If $N/2 > 2$, the length $N/2$ IntDCT-III is further decomposed into smaller ones until the length becomes 2. Based on the matrix decomposition in (8.56), the length 2 IntDCT-III can be computed by 3 lifting steps.*

Step 2.   *For $k = 0, 1, \ldots, N/2 - 1$, compute*

$$X(k) = H(k) + G(k), \quad X(N - 1 - k) = H(k) - G(k).$$

The numbers of lifting steps and additions are

$$L_{C^{III}}(N) = \frac{3N}{2} \log_2 N - 3N + 6$$
$$A_{C^{III}}(N) = N \log_2 N - N. \tag{8.60}$$

The inverse of $\bar{C}_N^{III}$ is

$$(\bar{C}_2^{III})^{-1} = \begin{bmatrix} 1 & 1 - \mathrm{RB}(\sqrt{2}) \\ 0 & 1 \end{bmatrix} \begin{bmatrix} 1 & 0 \\ \mathrm{RB}(\frac{\sqrt{2}}{2}) & 1 \end{bmatrix} \begin{bmatrix} 1 & \mathrm{RB}(\sqrt{2}) - 1 \\ 0 & -1 \end{bmatrix}$$

$$(\bar{C}_N^{III})^{-1} = \frac{1}{2} P_N \begin{bmatrix} (\bar{C}_{N/2}^{III})^{-1} & 0 \\ 0 & (\bar{C}_{N/2}^{IV})^{-1} \end{bmatrix} \begin{bmatrix} I_{N/2} & \hat{I}_{N/2} \\ I_{N/2} & -\hat{I}_{N/2} \end{bmatrix}, \quad N > 2. \tag{8.61}$$

We have obtained three types of integer transforms. Table 8.1 lists the number of operations for IntDCTs and DCTs, where the number of operations for DCTs are from [3, 8, 11] or Chapter 6.

Table 8.1   Number of operations for IntDCT and DCT.

Transform	Lifting steps	Additions
IntDCT-II	$\frac{3}{2}N \log_2 N - 3N + 6$	$N \log_2 N - N$
IntDCT-III	$\frac{3}{2}N \log_2 N - 3N + 6$	$N \log_2 N - N$
IntDCT-IV	$\frac{3}{2}N \log_2 N$	$N \log_2 N - N$

	Multiplications	Additions
DCT-II	$\frac{1}{2}N \log_2 N$	$\frac{3}{2}N \log_2 N - N + 1$
DCT-III	$\frac{1}{2}N \log_2 N$	$\frac{3}{2}N \log_2 N - N + 1$
DCT-IV	$\frac{1}{2}N \log_2 N + N$	$\frac{3}{2}N \log_2 N$

8.3.5   Approximation performance

It is difficult to obtain a theoretical bound for the error between the DCT and IntDCT or II-DCT, which remains an open question. In general, the approximation performance of the IntDCT or II-DCT depends on the number of bits used for the lifting multipliers. In our algorithms, these lifting multipliers are $\tan(\alpha_n/2)$ and $\sin(\alpha_n)$, where $\alpha_n = \frac{\pi(2n+1)}{4M}$ ($0 \leq n \leq M/2 - 1$, $M = 2^i$, $i = 1, 2, \ldots, t$). Obviously, $0 < \tan(\alpha_n/2) < 1$. We use $\text{RB}(\tan(\alpha_n/2))$ and $\text{RB}(\sin(\alpha_n))$ (see Definition 20) to approximate $\tan(\alpha_n/2)$ and $\sin(\alpha_n)$. In general, the approximation to $\text{RB}(\tan(\alpha_n/2))$ and $\text{RB}(\sin(\alpha_n))$ with more bits provides a better performance. We have done some numerical experiments on the approximation accuracy of the presented transforms. Some of the experimental results are given in Tables 8.2 and 8.3. Table 8.2 is for the IntDCT-IV where the input signals are $x(n) = \lfloor (n+1)\cos(n+1) \rfloor$ (signal 1) and $x(n) = (n+1)\cos[(n+1) + 5(n+1)^2]$ (signal 2) ($n = 0, 1, \ldots, N-1$), respectively. The mean squared error (MSE) is defined by

Table 8.2   MSE between IntDCT and DCT.

N	MSE (signal 1)	MSE (signal 2)	N	MSE (signal 1)	MSE (signal 2)
8	0.0082	0.0088	16	0.0082	0.0071
32	0.0090	0.0093	64	0.0098	0.0104
128	0.0109	0.0103	256	0.0112	0.0109
512	0.0117	0.0122	1024	0.0121	0.0121

Table 8.3   MSE between II-DCT and DCT.

N	MSE (signal 1)	MSE (signal 2)	N	MSE (signal 1)	MSE (signal 2)
32	0.1620	0.1384	64	0.1020	0.0894
128	0.0543	0.0536	256	0.0287	0.0296
512	0.0165	0.0173	1024	0.0091	0.0091

$$\text{MSE} = \sqrt{\frac{\sum_{k=0}^{N-1}(X(k) - X_1(k))^2}{\sum_{k=0}^{N-1} X_1(k)^2}}, \tag{8.62}$$

where $X(k)$ and $X_1(k)$ are the outputs of the IntDCT-IV and the type-IV DCT, respectively. We choose the function RB to be

$$\text{RB}(s) = \frac{\lfloor 256s \rfloor}{256}. \tag{8.63}$$

Figures 8.2 (a) and (b) show the outputs of the type-IV DCT and IntDCT-IV, respectively, with $N = 8$, while Figures 8.3 (a) and (b) show those with $N = 128$ (for signal 1). Here the outputs are normalized, that is, they are divided by $\sqrt{N}$. From Table 8.2, we know that for a fixed function RB, the MSE increases with the length N. For the RB used, the errors are tolerable for many signal processing applications such as data compression even for some large sequence lengths. In many practical applications such as image processing, a very short-length transform is used. For short-length IntDCT, we can use a coarse approximation function RB. When $N = 8$, for example, using the approximation

$$\text{RB}(s) = \frac{\lfloor 64s \rfloor}{64} \tag{8.64}$$

the MSE is 0.0197. Figure 8.4 shows the outputs of the IntDCT-IV. The results in Table 8.3 are for a II-DCT-IV of the same set of input signals. For a II-DCT-IV, if the input is integral, the output is also integral (see Algorithm 24 for the details). When N is very small, this kind of approximation will cause big errors if the number itself is very small. The MSE decreases as N increases. For signal 1, Figures 8.5 (a) and (b) show the outputs of the type-IV DCT and II-DCT-IV, respectively, with $N = 32$, and Figures 8.6 (a) and (b) show those with $N = 128$, respectively. Figures 8.7 (a) and (b) are the outputs of the type-IV DCT and IntDCT-IV of signal 2, respectively, with $N = 128$.

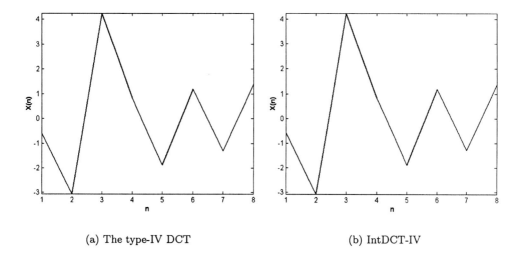

(a) The type-IV DCT   (b) IntDCT-IV

Figure 8.2   Outputs of type-IV DCT and IntDCT-IV ($N = 8$, normalized).

8.3.6   Improvements on the short-length transforms

The integer transforms are based on the lifting factorization of the DCT matrices. The lifting factorizations previously discussed are simple in structure and easy for realization, but they are not the best in terms of the required number of lifting matrices. In this subsection, we will discuss an improved method for factorization to reduce the number of lifting steps for the integer DCT at the cost of a more complex structure [7]. This method is better when the size of the transform is small.

We already have the factorization in (8.47) which turns a length N type-II DCT into a length $N/2$ type-II DCT and the length $N/2$ type-IV DCT. The factorization of the type-IV DCT is given in (8.23). It is known in [10, 23] that the type-IV DCT of length N can also be factored into two type-II DCTs of length $N/2$. Therefore, we have the following factorization:

$$C_N^{IV} = P_N^{(1)} E_N P_N^{(1)} P_N^{(2)} \begin{bmatrix} C_{\frac{N}{2}}^{II} & 0 \\ 0 & C_{\frac{N}{2}}^{II} \end{bmatrix} \begin{bmatrix} I_{\frac{N}{2}} & 0 \\ 0 & \tilde{J}_{\frac{N}{2}} \end{bmatrix} F_N \begin{bmatrix} I_{\frac{N}{2}} & 0 \\ 0 & -I_{\frac{N}{2}} \end{bmatrix}, \tag{8.65}$$

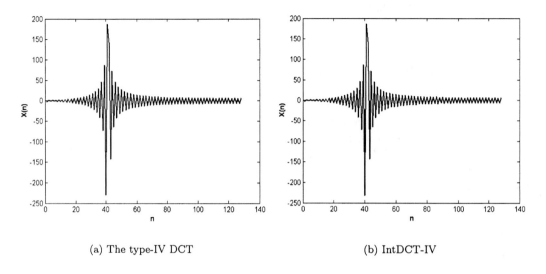

(a) The type-IV DCT   (b) IntDCT-IV

Figure 8.3   Outputs of type-IV DCT and IntDCT-IV ($N = 128$, normalized).

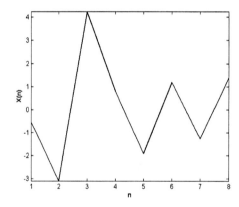

Figure 8.4   Outputs of IntDCT-IV ($N = 8$, normalized, by approximation (8.64)).

where $I_{N/2}$ is defined as before, $P_N^{(1)}$ and $P_N^{(2)}$ are permutation matrices defined by

$$
P_N^{(1)} = \begin{bmatrix} 1 & 0 & \cdots & 0 \\ 0 & 0 & \cdots & 1 \\ \vdots & \vdots & I_{N-3} & \vdots \\ 0 & 1 & \cdots & 0 \end{bmatrix}, \quad
P_N^{(2)} = \begin{bmatrix}
1 & 0 & 0 & \cdots & 0 & 0 & \cdots & 0 & 0 \\
0 & 0 & 0 & \cdots & 0 & 0 & \cdots & 0 & 1 \\
0 & 1 & 0 & \cdots & 0 & 0 & \cdots & 0 & 0 \\
0 & 0 & 0 & \cdots & 0 & 0 & \cdots & 1 & 0 \\
\vdots & \vdots & \vdots & & \vdots & \vdots & & \vdots & \vdots \\
0 & 0 & 0 & \cdots & 1 & 0 & \cdots & 0 & 0 \\
0 & 0 & 0 & \cdots & 0 & 1 & \cdots & 0 & 0
\end{bmatrix}, \quad (8.66)
$$

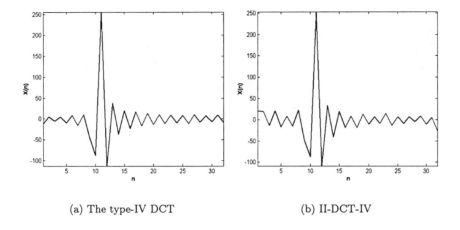

(a) The type-IV DCT   (b) II-DCT-IV

Figure 8.5   Outputs of type-IV DCT and II-DCT-IV ($N = 32$).

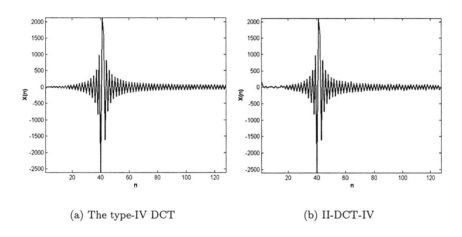

(a) The type-IV DCT   (b) II-DCT-IV

Figure 8.6   Outputs of type-IV DCT and II-DCT-IV ($N = 128$).

$$\tilde{J}_N = \begin{bmatrix} & & & & 1 \\ & & & -1 & \\ & & 1 & & \\ & & \cdot & & \\ & \cdot & & & \\ \cdot & & & & \\ -1 & & & & \end{bmatrix} \tag{8.67}$$

$$E_N = \mathrm{diag}(\sqrt{2}, \sqrt{2}, D, D, \ldots, D), \tag{8.68}$$

where

$$D = \begin{bmatrix} 1 & 1 \\ -1 & 1 \end{bmatrix}$$

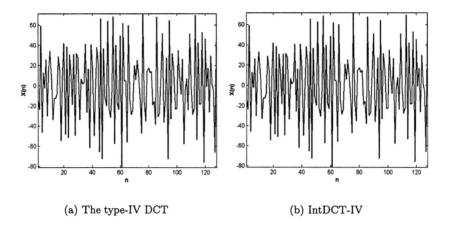

(a) The type-IV DCT   (b) IntDCT-IV

Figure 8.7   Outputs of type-IV DCT and IntDCT-IV ($N = 128$, signal 2, normalized).

and

$$
F_N = \begin{bmatrix}
\alpha_{4N}^1 & \cdots & & \cdots & & \cdots & \cdots & & -\beta_{4N}^1 \\
\vdots & \alpha_{4N}^3 & & & & & -\beta_{4N}^3 & & \vdots \\
\vdots & & \ddots & & & & & & \vdots \\
& & & \alpha_{4N}^{N-1} & -\beta_{4N}^{N-1} & & & & \\
& & & \beta_{4N}^{N-1} & \alpha_{4N}^{N-1} & & & & \vdots \\
\vdots & & \beta_{4N}^3 & & & \ddots & \alpha_{4N}^3 & & \vdots \\
\beta_{4N}^1 & \cdots & & \cdots & & \cdots & \cdots & & \alpha_{4N}^1
\end{bmatrix},
\tag{8.69}
$$

where $\alpha_{4N}^i = \cos\frac{i\pi}{4N}$ and $\beta_{4N}^i = \sin\frac{i\pi}{4N}$ are used for simplicity of presentation.

Using (8.47), (8.65) and the lifting factorizations of the rotation matrix and diagonal matrix, we can factor the DCT kernel matrices. For example, the factorization of an 8-point DCT-II is given below. The signal graph diagram of this transform is shown in Figure 8.8.

$$
C_8^{II} = P_8^{(2)} \begin{bmatrix} I_4 & 0 \\ 0 & \hat{I}_4 \end{bmatrix} \begin{bmatrix} I_5 & 0 & 0 \\ 0 & R_{-\frac{\pi}{4}} & 0 \\ 0 & 0 & 1 \end{bmatrix} \begin{bmatrix} P_4^{(2)} & 0 \\ 0 & P_4^{(2)} \end{bmatrix} \begin{bmatrix} I_2 & & & \\ & \hat{I}_2 & & \\ & & 1 & 1 \\ & & 1 & -1 \\ & & & & -1 & 1 \\ & & & & 1 & 1 \end{bmatrix}
$$

$$
\cdot \begin{bmatrix} R_{\frac{\pi}{4}} & 0 & 0 \\ 0 & R_{\frac{\pi}{8}} & 0 \\ 0 & 0 & F_4 \end{bmatrix} \begin{bmatrix} 1 & 0 & 0 & 1 \\ 0 & -1 & -1 & 0 \\ 1 & 0 & 0 & -1 \\ 0 & -1 & 1 & 0 \\ & & & & I_2 \\ & & & & & -I_2 \end{bmatrix} \begin{bmatrix} I_4 & J_4 \\ I_4 & -\hat{I}_4 \end{bmatrix}.
\tag{8.70}
$$

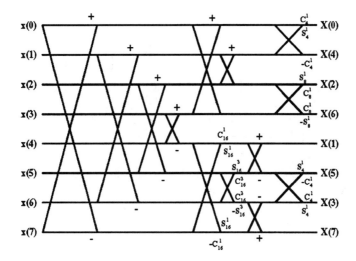

Figure 8.8   Signal flow graph of 8-point DCT-II.

8.3.7   Comparison between DCT and IntDCT/II-DCT

The IntDCT or II-DCT has a few important advantages over the conventional DCT. The IntDCT needs no floating-point multiplications. The floating-point multiplications are replaced by lifting steps which need only integer operations. If a sufficient word length is used to represent the partially processed data of the IntDCT, round-off errors can be eliminated completely. There is no information lost after the transform even if it is computed in a fixed-point computer. This is one of the main reasons that the integer discrete wavelet transform was recommended in JPEG-2000 to be used for lossless compression [13]. Although the DCT is also an invertible transform, a round-off error always exists when we approximate the transform kernel matrix. It is impossible to perfectly reconstruct the original data. Finally, the II-DCT can map integer to integer, which is useful for some applications.

The disadvantage of the IntDCT is the performance such as its de-correlation ability. Since the DCT is a nearly optimal block transform for image compression, we cannot expect the IntDCT to work better than the non-integral transform. In general, the performance of the IntDCT is related to the approximation used for the lifting multipliers or rotation matrices. Experiments in [6, 22] have shown that the performance of an 8-point IntDCT-II is close to that of a type-II DCT if all quantized lifting multipliers have 8-bit resolution. Longer word length generally provides better performance to approximate the DCT. A compromise should be made between the number of bits (or computational complexity) and the performance.

Practical applications usually require integer transforms to have some special properties. For example, the transforms are required to eliminate DC leakage in image coding, that is, constant input should be completely captured by the DC subband. The integer transforms above may not satisfy such requirements. To meet such requirements, some amendments to the integer approximations of rotation matrices or lifting matrices are needed. For example, we can approximate each rotation matrix R_α by $\bar{R}_\alpha$ in the factorization of the length 8 and the length 16 DCT-II to get a length 8 and length 16 IntDCT-II as follows. For a length 8 IntDCT-II, we use the following approximations:

$$\overline{R}_{-\frac{\pi}{4}} = \begin{bmatrix} 1 & \frac{3}{8} \\ 0 & 1 \end{bmatrix} \begin{bmatrix} -1 & 0 \\ \frac{5}{8} & 1 \end{bmatrix}, \qquad \overline{R}_{\frac{\pi}{4}} = \begin{bmatrix} 1 & 0 \\ \frac{1}{2} & -1 \end{bmatrix} \begin{bmatrix} 1 & 1 \\ 0 & 1 \end{bmatrix} \tag{8.71}$$

$$\overline{R}_{\frac{\pi}{8}} = \begin{bmatrix} 1 & -\frac{3}{16} \\ 0 & 1 \end{bmatrix} \begin{bmatrix} 1 & 0 \\ \frac{3}{8} & 1 \end{bmatrix} \begin{bmatrix} 1 & -\frac{3}{16} \\ 0 & 1 \end{bmatrix} \tag{8.72}$$

$$\overline{R}_{\frac{\pi}{16}} = \begin{bmatrix} 0 & -1 \\ 1 & 0 \end{bmatrix} \begin{bmatrix} 1 & 0 \\ -\frac{1}{8} & 1 \end{bmatrix}, \qquad \overline{R}_{\frac{3\pi}{16}} = \begin{bmatrix} 1 & 0 \\ -\frac{7}{8} & 1 \end{bmatrix} \begin{bmatrix} 1 & \frac{1}{2} \\ 0 & 1 \end{bmatrix}. \tag{8.73}$$

For a length 16 IntDCT-II, both a length 8 IntDCT-II and length 8 IntDCT-IV are required. Approximations of rotation matrices for the length 8 IntDCT-II are mentioned above. The length 8 IntDCT-IV is constructed by the following rotation matrix approximations:

$$\overline{R}_{\frac{\pi}{4}} = \begin{bmatrix} -1 & \frac{5}{8} \\ 0 & 1 \end{bmatrix} \begin{bmatrix} 1 & 0 \\ \frac{3}{8} & 1 \end{bmatrix}, \qquad \overline{R}_{\frac{\pi}{8}} = \begin{bmatrix} 1 & -\frac{3}{16} \\ 0 & 1 \end{bmatrix} \begin{bmatrix} 1 & 0 \\ \frac{3}{8} & 1 \end{bmatrix} \begin{bmatrix} 1 & -\frac{3}{16} \\ 0 & 1 \end{bmatrix} \tag{8.74}$$

$$\overline{R}_{\frac{\pi}{32}} = \begin{bmatrix} 1 & -\frac{3}{64} \\ 0 & 1 \end{bmatrix} \begin{bmatrix} 1 & 0 \\ \frac{3}{32} & 1 \end{bmatrix} \begin{bmatrix} 1 & -\frac{3}{64} \\ 0 & 1 \end{bmatrix}, \qquad \overline{R}_{\frac{3\pi}{32}} = \begin{bmatrix} 1 & -\frac{5}{32} \\ 0 & 1 \end{bmatrix} \begin{bmatrix} 1 & 0 \\ \frac{9}{32} & 1 \end{bmatrix} \begin{bmatrix} 1 & -\frac{5}{32} \\ 0 & 1 \end{bmatrix} \tag{8.75}$$

$$\overline{R}_{\frac{5\pi}{32}} = \begin{bmatrix} 1 & -\frac{1}{4} \\ 0 & 1 \end{bmatrix} \begin{bmatrix} 1 & 0 \\ \frac{29}{64} & 1 \end{bmatrix} \begin{bmatrix} 1 & -\frac{1}{4} \\ 0 & 1 \end{bmatrix}, \qquad \overline{R}_{\frac{7\pi}{32}} = \begin{bmatrix} 1 & 0 \\ \frac{11}{32} & 1 \end{bmatrix} \begin{bmatrix} 1 & -\frac{5}{8} \\ 0 & 1 \end{bmatrix} \begin{bmatrix} 1 & 0 \\ \frac{11}{32} & 1 \end{bmatrix}. \tag{8.76}$$

We have used the length 8 and length 16 integer transforms discussed above for image compression. Table 8.4 shows the results.

Table 8.4   Objective coding result comparison (peak signal-to-noise ratio (PSNR) in dB).

512×512	Comp. Ratio	IntDCT ($N = 8$)	DCT ($N = 8$)	IntDCT ($N = 16$)	DCT ($N = 16$)
			PSNR (dB)		
Lena	1:8	39.36	39.81	39.82	39.46
	1:16	35.71	36.55	36.65	36.17
	1:32	31.83	33.14	33.42	33.01
	1:64	28.18	29.82	30.28	20.12
	1:128	25.08	26.86	27.41	27.82
Goldhill	1:8	35.73	36.22	36.13	36.28
	1:16	32.12	32.79	32.82	32.90
	1:32	29.31	30.07	30.33	30.43
	1:64	26.85	27.96	28.22	28.42
	1:128	24.57	26.02	26.48	26.66
Barbra	1:8	35.78	37.14	37.40	36.78
	1:16	30.57	32.17	32.79	32.19
	1:32	26.62	28.33	29.13	28.77
	1:64	23.74	25.43	26.22	26.11
	1:128	21.64	22.93	23.79	24.14

8.4   INTEGER DHT AND FAST ALGORITHMS

This section considers a general method of factoring the kernel matrix of a generalized discrete Hartley transform (DHT) [1, 9] into O($N \log N$) lifting matrices and simple matrices, where N is the order of the transform matrix. Based on the lifting factorization, we develop four types of integer DHTs which do not need floating-point multiplications. Integer-to-integer transforms are also presented to approximate the generalized DHTs. Fast algorithms for these transforms are also provided. In this section, we first give a unified lifting factorization of the type-IV DHT. An integer transform which approximates the type-IV DHT is then constructed. Based on the relationships among the different types of DHTs, the lifting factorization of the type-I, -II and -III DHTs, and the type-I, -II and -III integer DHTs are developed. Finally, a comparison between the original DHT and the integer DHT is made.

8.4.1   Factorization of type-IV DHT

Let $x(n)$ $(n = 0, 1, \ldots, N-1)$ be a real input sequence. We assume that $N = 2^t$, where $t > 0$. The type-IV DHT of $x(n)$ is defined as [24]

$$X(k) = \sum_{n=0}^{N-1} x(n) \mathrm{cas} \frac{\pi(2n+1)(2k+1)}{2N}, \quad k = 0, 1, \ldots, N-1. \tag{8.77}$$

Let W_N^{IV} be the kernel matrix of the type-IV DHT

$$W_N^{IV} = \left(\mathrm{cas} \frac{\pi(2k+1)(2n+1)}{2N} \right)_{k,n=0,1,\ldots,N-1}. \tag{8.78}$$

We first consider a simple fast algorithm to factor the transform matrix,

$$H(k) = \frac{X(2k) + X(2k+1)}{2}, \quad G(k) = \frac{X(2k+1) - X(2k)}{2},$$

where $k = 0, 1, \ldots, N/2 - 1$. Based on the properties of the trigonometric identities, we have

$$\begin{aligned}
H(k) &= \sum_{n=0}^{N-1} x(n) \cos \frac{\pi(2n+1)}{2N} \mathrm{cas} \frac{\pi(2n+1)(2k+1)}{N} \\
&= \sum_{n=0}^{N/2-1} \left[x(n) \cos \frac{\pi(2n+1)}{2N} + x(\frac{N}{2}+n) \sin \frac{\pi(2n+1)}{2N} \right] \\
&\quad \cdot \mathrm{cas} \frac{\pi(2n+1)(2k+1)}{N} \\
G(k) &= \sum_{n=0}^{N-1} x(n) \sin \frac{\pi(2n+1)}{2N} \mathrm{cas} \frac{-\pi(2n+1)(2k+1)}{N} \\
&= \sum_{n=0}^{N/2-1} \left[x(n) \sin \frac{\pi(2n+1)}{2N} - x(\frac{N}{2}+n) \cos \frac{\pi(2n+1)}{2N} \right] \\
&\quad \cdot \mathrm{cas} \frac{-\pi(2n+1)(2k+1)}{N}.
\end{aligned}$$

Let

$$h(n) = x(n) \cos \frac{\pi(2n+1)}{2N} + x(\frac{N}{2}+n) \sin \frac{\pi(2n+1)}{2N} \tag{8.79}$$

$$g(n) = -x(n) \sin \frac{\pi(2n+1)}{2N} + x(\frac{N}{2}+n) \cos \frac{\pi(2n+1)}{2N}. \tag{8.80}$$

Then $H(k)$ is the type-IV DHT of $h(n)$ and $G(\frac{N}{2} - 1 - k)$ is the type-IV DHT of $g(n)$. In fact, we have

$$H(k) = \sum_{n=0}^{N/2-1} h(n) \mathrm{cas} \frac{\pi(2n+1)(2k+1)}{N} \tag{8.81}$$

$$G(\frac{N}{2} - 1 - k) = \sum_{n=0}^{N/2-1} g(n) \mathrm{cas} \frac{\pi(2n+1)(2k+1)}{N}, \tag{8.82}$$

where $k = 0, 1, \ldots, N/2 - 1$. Therefore, a length N type-IV DHT is converted into two type-IV DHTs of length $N/2$ in (8.81) and (8.82) at the cost of a pre-processing stage defined in (8.79) and (8.80) and a post-processing stage defined by

$$X(2k) = H(k) - G(k), \quad X(2k+1) = H(k) + G(k), \tag{8.83}$$

where $k = 0, 1, \cdots, N/2 - 1$. Based on the algorithm, we have the following lemma for the factorization of matrix W_N^{IV}.

Lemma 21 *The kernel matrix of the type-IV DHT can be factored recursively by*

$$W_1^{IV} = [1] \tag{8.84}$$

$$W_N^{IV} = P_N \begin{bmatrix} I_{N/2} & -I_{N/2} \\ I_{N/2} & I_{N/2} \end{bmatrix} \begin{bmatrix} I_{N/2} & 0 \\ 0 & \hat{I}_{N/2} \end{bmatrix} \begin{bmatrix} W_{N/2}^{IV} & 0 \\ 0 & W_{N/2}^{IV} \end{bmatrix} P'_N D_N P_N, \tag{8.85}$$

where $I_{N/2}$ is the identity matrix of order $N/2$, $\hat{I}_{N/2}$ is the matrix by reversing the rows of $I_{N/2}$, D_N is a block diagonal matrix defined by

$$D_N = \mathrm{diag}(T(0), \ T(1), \ \ldots, \ T(N/2 - 1)), \tag{8.86}$$

where $T(n)$ $(n = 0, 1, \ldots, N/2 - 1)$ are rotation matrices defined by

$$T(n) = \begin{bmatrix} \cos \frac{\pi(2n+1)}{2N} & \sin \frac{\pi(2n+1)}{2N} \\ -\sin \frac{\pi(2n+1)}{2N} & \cos \frac{\pi(2n+1)}{2N} \end{bmatrix} = R_{-\frac{\pi(2n+1)}{2N}}, \tag{8.87}$$

P_N is a permutation matrix

$$P_N = \begin{bmatrix} 1 & 0 & \cdots & 0 & 0 & 0 & \cdots & 0 \\ 0 & 0 & \cdots & 0 & 1 & 0 & \cdots & 0 \\ 0 & 1 & \cdots & 0 & 0 & 0 & \cdots & 0 \\ 0 & 0 & \cdots & 0 & 0 & 1 & \cdots & 0 \\ & & \cdots & & & \cdots & & \\ 0 & 0 & \cdots & 1 & 0 & 0 & \cdots & 0 \\ 0 & 0 & \cdots & 0 & 0 & 0 & \cdots & 1 \end{bmatrix}, \tag{8.88}$$

P'_N is the transposition of P_N and $W^{IV}_{N/2}$ is the transform matrix of the length $N/2$ type-IV DHT.

Example 17 *Factorization of short-length type-IV DHT*

$$W_2^{IV} = \begin{bmatrix} 1 & -1 \\ 1 & 1 \end{bmatrix} \begin{bmatrix} \sqrt{2}/2 & \sqrt{2}/2 \\ -\sqrt{2}/2 & \sqrt{2}/2 \end{bmatrix} \tag{8.89}$$

$$W_4^{IV} = \begin{bmatrix} 1 & 0 & 0 & 0 \\ 0 & 0 & 1 & 0 \\ 0 & 1 & 0 & 0 \\ 0 & 0 & 0 & 1 \end{bmatrix} \begin{bmatrix} 1 & 0 & -1 & 0 \\ 0 & 1 & 0 & -1 \\ 1 & 0 & 1 & 0 \\ 0 & 1 & 0 & 1 \end{bmatrix} \begin{bmatrix} 1 & 0 & 0 & 0 \\ 0 & 1 & 0 & 0 \\ 0 & 0 & 0 & 1 \\ 0 & 0 & 1 & 0 \end{bmatrix} \begin{bmatrix} W_2^{IV} & 0 \\ 0 & W_2^{IV} \end{bmatrix} \begin{bmatrix} 1 & 0 & 0 & 0 \\ 0 & 0 & 1 & 0 \\ 0 & 1 & 0 & 0 \\ 0 & 0 & 0 & 1 \end{bmatrix}$$

$$\cdot \begin{bmatrix} \cos\frac{\pi}{8} & \sin\frac{\pi}{8} & 0 & 0 \\ -\sin\frac{\pi}{8} & \cos\frac{\pi}{8} & 0 & 0 \\ 0 & 0 & \cos\frac{3\pi}{8} & \sin\frac{3\pi}{8} \\ 0 & 0 & -\sin\frac{3\pi}{8} & \cos\frac{3\pi}{8} \end{bmatrix} \begin{bmatrix} 1 & 0 & 0 & 0 \\ 0 & 0 & 1 & 0 \\ 0 & 1 & 0 & 0 \\ 0 & 0 & 0 & 1 \end{bmatrix} \tag{8.90}$$

$$W_8^{IV} = P_8 \begin{bmatrix} I_4 & -I_4 \\ I_4 & I_4 \end{bmatrix} \begin{bmatrix} I_4 & 0 \\ 0 & \hat{I}_4 \end{bmatrix} \begin{bmatrix} W_4^{IV} & 0 \\ 0 & W_4^{IV} \end{bmatrix} P'_8 D_8 P_8, \tag{8.91}$$

where

$$D_8 = \begin{bmatrix} \cos\frac{\pi}{16} & \sin\frac{\pi}{16} & & & & & & \\ -\sin\frac{\pi}{16} & \cos\frac{\pi}{16} & & & & & & \\ & & \cos\frac{3\pi}{16} & \sin\frac{3\pi}{16} & & & & \\ & & -\sin\frac{3\pi}{16} & \cos\frac{3\pi}{16} & & & & \\ & & & & \cos\frac{5\pi}{16} & \sin\frac{5\pi}{16} & & \\ & & & & -\sin\frac{5\pi}{16} & \cos\frac{5\pi}{16} & & \\ & & & & & & \cos\frac{7\pi}{16} & \sin\frac{7\pi}{16} \\ & & & & & & -\sin\frac{7\pi}{16} & \cos\frac{7\pi}{16} \end{bmatrix} \tag{8.92}$$

and

$$P_8 = \begin{bmatrix} 1 & 0 & 0 & 0 & 0 & 0 & 0 & 0 \\ 0 & 0 & 0 & 0 & 1 & 0 & 0 & 0 \\ 0 & 1 & 0 & 0 & 0 & 0 & 0 & 0 \\ 0 & 0 & 0 & 0 & 0 & 1 & 0 & 0 \\ 0 & 0 & 1 & 0 & 0 & 0 & 0 & 0 \\ 0 & 0 & 0 & 0 & 0 & 0 & 1 & 0 \\ 0 & 0 & 0 & 1 & 0 & 0 & 0 & 0 \\ 0 & 0 & 0 & 0 & 0 & 0 & 0 & 1 \end{bmatrix}. \tag{8.93}$$

Floating-point multiplications are needed when matrix D_N is multiplied by a vector. In order to avoid floating-point multiplications, this matrix is decomposed into products of lifting matrices. Then the elements of the lifting matrices are approximated to be of the form $\beta/2^\lambda$, where β and λ are integers. D_N can be easily turned into products of lifting matrices based on the factorization in (8.8). In fact, each block in D_N can be factored as

$$T(n) = \begin{bmatrix} 1 & \tan\frac{\alpha_n}{2} \\ 0 & 1 \end{bmatrix} \begin{bmatrix} 1 & 0 \\ -\sin\alpha_n & 1 \end{bmatrix} \begin{bmatrix} 1 & \tan\frac{\alpha_n}{2} \\ 0 & 1 \end{bmatrix}, \tag{8.94}$$

where $\alpha_n = \frac{\pi(2n+1)}{2N}$. Therefore, D_N is factored into $\frac{3N}{2}$ lifting matrices. If the same method is used to factor the matrix $W_{N/2}^{IV}$ recursively until the order of the submatrix is two, the complete factorization of W_N^{IV} is achieved.

8.4.2 Integer type-IV DHT and fast algorithm

By approximating every non-zero non-diagonal element s of the lifting matrices in (8.94) with RB(s) defined in Definition 20, we construct an approximation matrix $\bar{T}(n)$ for $T(n)$,

$$\bar{T}(n) = \begin{bmatrix} 1 & \mathrm{RB}(\tan\frac{\alpha_n}{2}) \\ 0 & 1 \end{bmatrix} \begin{bmatrix} 1 & 0 \\ -\mathrm{RB}(\sin\alpha_n) & 1 \end{bmatrix} \begin{bmatrix} 1 & \mathrm{RB}(\tan\frac{\alpha_n}{2}) \\ 0 & 1 \end{bmatrix}. \tag{8.95}$$

Then D_N is approximated by $\bar{D}_N$ given by

$$\bar{D}_N = \mathrm{diag}(\bar{T}(0),\ \bar{T}(1),\ \ldots,\ \bar{T}(N/2-1)). \tag{8.96}$$

We obtain an approximation matrix $\bar{W}_N^{IV}$ for W_N^{IV} which defines a new transform without floating-point multiplications. Let us call the new transform the type-IV integer DHT (IntDHT-IV). However, this does not mean that the transform IntDHT-IV maps integer to integer. An integer-to-integer transform will be introduced in the next subsection.

Definition 24 *Assume that $N = 2^t$. The kernel matrix $\bar{W}_N^{IV}$ of an IntDHT-IV is defined recursively by*

$$\bar{W}_1^{IV} = [1] \tag{8.97}$$

$$\bar{W}_N^{IV} = P_N \begin{bmatrix} I_{N/2} & -I_{N/2} \\ I_{N/2} & -I_{N/2} \end{bmatrix} \begin{bmatrix} I_{N/2} & 0 \\ 0 & \hat{I}_{N/2} \end{bmatrix} \begin{bmatrix} \bar{W}_{N/2}^{IV} & 0 \\ 0 & \bar{W}_{N/2}^{IV} \end{bmatrix} P_N' \bar{D}_N P_N, \tag{8.98}$$

where the notation used is the same as that in Lemma 21.

Because any choice of function RB determines a transform, the transform is not unique. Based on the definition, we have a fast algorithm for the IntDHT-IV as stated below.

Algorithm 27 *Fast algorithm for the IntDHT-IV*

Step 1.   *For $n = 0, 1, \cdots, N/2 - 1$, compute*

$$\bar{g}(n) = -f_n[x(n) + e_n x(\frac{N}{2} + n)] + x(\frac{N}{2} + n)$$

$$\bar{h}(n) = x(n) + e_n x(\frac{N}{2} + n) + e_n \bar{g}(n),$$

where $e_n = \mathrm{RB}(\tan\frac{\alpha_n}{2})$ and $f_n = \mathrm{RB}(\sin\alpha_n)$.

Step 2.   *Compute the length $N/2$ IntDHT-IVs, $\bar{H}(k)$ and $\bar{G}(k)$, of sequences $\bar{h}(n)$ and $\bar{g}(n)$, respectively. If $N/2 > 1$, the length $N/2$ IntDHT-IV is further decomposed into smaller ones until the length becomes one that can be calculated directly.*

Step 3.   *Compute*

$$X(2k) = \bar{H}(k) - \bar{G}(\frac{N}{2} - 1 - k), \qquad X(2k+1) = \bar{H}(k) + \bar{G}(\frac{N}{2} - 1 - k),$$

where $k = 0, 1, \ldots, N/2 - 1$.

Now we consider the computational complexity of Algorithm 27. Step 1 needs $\frac{3N}{2}$ lifting steps and Step 3 needs N additions. Let $L_{W^{IV}}(N)$ and $A_{W^{IV}}(N)$ represent the numbers of lifting steps and additions for computing a length N IntDHT-IV. Then, we have

$$L_{W^{IV}}(N) = 2L_{W^{IV}}\left(\frac{N}{2}\right) + \frac{3N}{2}, \qquad L_{W^{IV}}(1) = 0$$

$$A_{W^{IV}}(N) = 2A_{W^{IV}}\left(\frac{N}{2}\right) + N, \qquad A_{W^{IV}}(1) = 0.$$

Therefore, we finally get

$$L_{W^{IV}}(N) = \frac{3N}{2} \log_2 N, \qquad A_{W^{IV}}(N) = N \log_2 N.$$

The matrix $\bar{W}_N^{IV}$ is invertible and its inverse matrix can also be factored as products of lifting matrices and simple matrices whose elements are 0, ± 1 or $\pm 1/2$. This can be seen from the following equations:

$$(\bar{W}_1^{IV})^{-1} = [1] \tag{8.99}$$

$$(\bar{W}_N^{IV})^{-1} = \frac{1}{2} P_N' (\bar{D}_N)^{-1} P_N \begin{bmatrix} (\bar{W}_{N/2}^{IV})^{-1} & 0 \\ 0 & (\bar{W}_{N/2}^{IV})^{-1} \end{bmatrix}$$
$$\cdot \begin{bmatrix} I_{N/2} & 0 \\ 0 & \hat{I}_{N/2} \end{bmatrix} \begin{bmatrix} I_{N/2} & -I_{N/2} \\ I_{N/2} & I_{N/2} \end{bmatrix} P_N', \tag{8.100}$$

where

$$(\bar{D}_N)^{-1} = \mathrm{diag}[(\bar{T}(0))^{-1}, (\bar{T}(1))^{-1}, \ldots, (\bar{T}(N/2 - 1))^{-1}] \tag{8.101}$$

$$(\bar{T}(n))^{-1} = \begin{bmatrix} 1 & -\mathrm{RB}(\tan \frac{\alpha_n}{2}) \\ 0 & 1 \end{bmatrix} \begin{bmatrix} 1 & 0 \\ \mathrm{RB}(\sin \alpha_n) & 1 \end{bmatrix} \begin{bmatrix} 1 & -\mathrm{RB}(\tan \frac{\alpha_n}{2}) \\ 0 & 1 \end{bmatrix}. \tag{8.102}$$

Based on the factorization (8.100), we have a fast algorithm for the inverse IntDHT-IV (**reconstruction algorithm for IntDHT-IV**) as follows, where $X(k)$ is the input and $x(n)$ is the output.

Algorithm 28 *Reconstruction algorithm for IntDHT-IV*

Step 1. *Compute*

$$\bar{H}(k) = \frac{X(2k) + X(2k+1)}{2}, \qquad \bar{G}\left(\frac{N}{2} - 1 - k\right) = \frac{X(2k+1) - X(2k)}{2},$$

where $k = 0, 1, \ldots, N/2 - 1$.

Step 2. *Compute the length $N/2$ inverse IntDHT-IVs, $\bar{h}(n)$ and $\bar{g}(n)$, of input sequences $\bar{H}(k)$ and $\bar{G}(k)$, respectively. If $N/2 > 1$, the length 2 inverse IntDHT-IV is further decomposed into smaller ones until the length becomes 1.*

Step 3.   *Compute*

$$x(\frac{N}{2} + n) = f_n[\bar{h}(n) - e_n\bar{g}(n)] + \bar{g}(n), \quad x(n) = \bar{h}(n) - e_n\bar{g}(n) - e_n x(\frac{N}{2} + n),$$

where $e_n = \mathrm{RB}(\tan\frac{\alpha_n}{2})$, $f_n = \mathrm{RB}(\sin\alpha_n)$ and $n = 0, 1, \ldots, N/2 - 1$.

The number of operations for reconstruction is the same as that for the forward transform. It should be noted that the matrix $\bar{W}_N^{IV}$ is not orthogonal in general and we can not use $(\bar{W}_N^{IV})'$ for reconstruction.

8.4.3   Integer-to-integer type-IV DHT

The previous subsections show that the kernel matrix of the type-IV DHT is factored into products of lifting matrices and some simple matrices. We have also shown in (8.6) and (8.7) that a lifting step can be approximated by a non-linear integer-to-integer transform. If all the lifting steps in the factorization of the type-IV DHT are approximated by integer-to-integer transforms, we finally get an integer-to-integer transform that approximates to the type-IV DHT. We use the notation II-DHT-IV for the transform. Based on Lemma 21, a fast algorithm is given for the II-DHT-IV as follows.

Algorithm 29 *Fast algorithm for the II-DHT-IV*

Step 1.   *For $n = 0, 1, \ldots, N/2 - 1$, compute*

$$\begin{aligned}
&h_1(n) = x(n) + \lfloor x(\tfrac{N}{2} + n)\tan(\alpha_n/2)\rfloor, \quad &&g_1(n) = x(\tfrac{N}{2} + n)\\
&h_2(n) = h_1(n), \quad &&g_2(n) = -\lfloor h_1(n)\sin(\alpha_n)\rfloor + g_1(n)\\
&h(n) = h_2(n) + \lfloor g_2(n)\tan(\alpha_n/2)\rfloor, \quad &&g(n) = g_2(n).
\end{aligned}$$

Step 2.   *Compute the length $N/2$ II-DHT-IVs, $H(k)$ and $G(k)$, of sequences $h(n)$ and $g(n)$, respectively. If $N/2 > 1$, the length $N/2$ II-DHT-IV is decomposed into smaller ones until the length becomes 1.*

Step 3.   *Compute*

$$X(2k) = H(k) - G(\frac{N}{2} - 1 - k), \quad X(2k+1) = H(k) + G(\frac{N}{2} - 1 - k),$$

where $k = 0, 1, \ldots, N/2 - 1$.

The II-DHT-IV is invertible and its inverse is also an integer to integer transform. By inverting the steps of Algorithm 29, we can easily achieve the reconstruction algorithm for the II-DHT-IV.

8.4.4   Other types of integer DHTs

Based on the relationships among the four types of DHTs [10, 25], this subsection considers the factorization of type-I, -II and -III DHTs into lifting steps and additions. Based on the factorization, the integer type-I, -II and -III DHTs can be derived.

Integer type-II DHT

Let $x(n)$ $(n = 0, 1, \ldots, N - 1)$ be a real input sequence. We assume that $N = 2^t$ where $t > 0$. The type-II DHT of $x(n)$ is defined as [24]

$$X(k) = \sum_{n=0}^{N-1} x(n) \cos \frac{\pi(2n+1)k}{N}, \quad k = 0, 1, \ldots, N - 1 \tag{8.103}$$

which can be computed by using the type-IV DHT. The following factorization in Lemma 22 is based on the relationship stated in [10, 25].

Lemma 22 *Let W_N^{II} be the kernel matrix of the type-II DHT, that is,*

$$W_N^{II} = \left(\cos \frac{\pi k(2n+1)}{N} \right)_{k,n=0,1,\ldots,N-1}.$$

Then

$$W_2^{II} = \begin{bmatrix} 1 & 1 \\ 1 & -1 \end{bmatrix} \tag{8.104}$$

$$W_N^{II} = P_N \begin{bmatrix} W_{N/2}^{II} & 0 \\ 0 & W_{N/2}^{IV} \end{bmatrix} \begin{bmatrix} I_{N/2} & I_{N/2} \\ I_{N/2} & -I_{N/2} \end{bmatrix}, \tag{8.105}$$

where P_N and $I_{N/2}$ are defined in Lemma 21.

Definition 25 *Assume that $N = 2^t$. The kernel matrix of a type-II integer DHT (IntDHT-II) of length N is defined by*

$$\bar{W}_2^{II} = \begin{bmatrix} 1 & 1 \\ 1 & -1 \end{bmatrix} \tag{8.106}$$

$$\bar{W}_N^{II} = P_N \begin{bmatrix} \bar{W}_{N/2}^{II} & 0 \\ 0 & \bar{W}_{N/2}^{IV} \end{bmatrix} \begin{bmatrix} I_{N/2} & I_{N/2} \\ I_{N/2} & -I_{N/2} \end{bmatrix}, \quad N > 2. \tag{8.107}$$

Based on the factorization, floating-point multiplications are not needed for the IntDHT-II.

Algorithm 30 *Fast algorithm for the IntDHT-II*

Step 1. *For $n = 0, 1, \ldots, N/2 - 1$, compute*

$$h(n) = x(n) + x(\frac{N}{2} + n), \quad g(n) = x(n) - x(\frac{N}{2} + n).$$

Step 2. *Use Algorithm 27 to compute $X(2k + 1)$, which is the IntDHT-IV of sequence $g(n)$, and $X(2k)$, which is the IntDHT-II of sequence $h(n)$. If $N/2 > 2$, the length $N/2$ IntDHT-II is decomposed into smaller ones until the length becomes 2. Based on the matrix decomposition in (8.106), the length 2 IntDHT-II can be computed by 2 additions.*

The numbers of lifting steps and additions needed by the fast algorithm are

$$L_{W^{II}}(N) = \frac{3N}{2}\log_2 N - 3N + 3, \quad A_{W^{II}}(N) = N\log_2 N - 2N + 4.$$

The inverse of $\bar{W}_N^{II}$ is

$$(\bar{W}_2^{II})^{-1} = \frac{1}{2}\begin{bmatrix} 1 & 1 \\ 1 & -1 \end{bmatrix} \tag{8.108}$$

$$(\bar{W}_N^{II})^{-1} = \frac{1}{2}\begin{bmatrix} I_{N/2} & I_{N/2} \\ I_{N/2} & -I_{N/2} \end{bmatrix}\begin{bmatrix} (\bar{W}_{N/2}^{II})^{-1} & 0 \\ 0 & (\bar{W}_{N/2}^{IV})^{-1} \end{bmatrix} P_N', \tag{8.109}$$

where $N > 2$.

Integer type-III DHT

Let $x(n)$ $(n = 0, 1, \ldots, N-1)$ be a real input sequence. We assume that $N = 2^t$ where $t > 0$. The type-III DHT of $x(n)$ is defined as

$$X(k) = \sum_{k=0}^{N-1} x(n)\mathrm{cas}\frac{\pi n(2k+1)}{N}, \quad n = 0, 1, \ldots, N-1. \tag{8.110}$$

The kernel matrix of the type-III DHT can be obtained from the transposition of the transform matrix of the type-II DHT. By transposing the matrices in (8.105), the factorization can be achieved by the lemma below.

Lemma 23 *Let W_N^{III} be the kernel matrix of the type-III DHT,*

$$W_N^{III} = \left(\mathrm{cas}\frac{\pi(2k+1)n}{N}\right)_{k,n=0,1,\ldots,N-1}.$$

Then

$$W_2^{III} = \begin{bmatrix} 1 & 1 \\ 1 & -1 \end{bmatrix} \tag{8.111}$$

$$W_N^{III} = \begin{bmatrix} I_{N/2} & I_{N/2} \\ I_{N/2} & -I_{N/2} \end{bmatrix}\begin{bmatrix} W_{N/2}^{III} & 0 \\ 0 & W_{N/2}^{IV} \end{bmatrix} P_N', \tag{8.112}$$

where P_N and $I_{N/2}$ are defined in Lemma 21.

Example 18 *Factorization of short-length DHT-III*

$$W_4^{III} = \begin{bmatrix} 1 & 0 & 1 & 0 \\ 0 & 1 & 0 & 1 \\ 1 & 0 & -1 & 0 \\ 0 & 1 & 0 & -1 \end{bmatrix}\begin{bmatrix} 1 & 0 & 0 & 0 \\ 0 & 1 & 0 & 0 \\ 0 & 0 & 1 & -1 \\ 0 & 0 & 1 & 1 \end{bmatrix}\begin{bmatrix} 1 & 1 & 0 & 0 \\ 1 & -1 & 0 & 0 \\ 0 & 0 & \sqrt{2}/2 & \sqrt{2}/2 \\ 0 & 0 & -\sqrt{2}/2 & \sqrt{2}/2 \end{bmatrix}\begin{bmatrix} 1 & 0 & 0 & 0 \\ 0 & 0 & 1 & 0 \\ 0 & 1 & 0 & 0 \\ 0 & 0 & 0 & 1 \end{bmatrix} \tag{8.113}$$

$$W_8^{III} = \begin{bmatrix} I_4 & I_4 \\ I_4 & -I_4 \end{bmatrix}\begin{bmatrix} W_4^{III} & 0 \\ 0 & W_4^{IV} \end{bmatrix} P_8', \tag{8.114}$$

where P_8 is defined in (8.93).

Definition 26 *Assume that $N = 2^t$. The kernel matrix of the type-III Integer DHT (IntDHT-III) of length N is defined by*

$$\bar{W}_2^{III} = \begin{bmatrix} 1 & 1 \\ 1 & -1 \end{bmatrix} \tag{8.115}$$

$$\bar{W}_N^{III} = \begin{bmatrix} I_{N/2} & I_{N/2} \\ I_{N/2} & -I_{N/2} \end{bmatrix} \begin{bmatrix} \bar{W}_{N/2}^{III} & 0 \\ 0 & \bar{W}_{N/2}^{IV} \end{bmatrix} P_N', \tag{8.116}$$

where $N > 2$. Based on the factorization, we have a floating-point multiplication-free fast algorithm for IntDHT-III.

Algorithm 31 *Fast algorithm for IntDHT-III*

Step 1.   *Use Algorithm 27 to compute $G(k)$, which is the IntDHT-IV of sequence $x(2n+1)$, and $H(k)$, which is the IntDHT-III of sequence $x(2n)$. If $N/2 > 2$, the length $N/2$ IntDHT-III is further decomposed until the length becomes 2. Based on the matrix decomposition in (8.115), the length 2 IntDHT-III can be computed by 3 additions.*

Step 2.   *For $k = 0, 1, \ldots, N/2 - 1$, compute*

$$X(k) = H(k) + G(k), \quad X(\frac{N}{2} + k) = H(k) - G(k).$$

The numbers of lifting steps and additions needed by the fast algorithm are

$$L_{W^{III}}(N) = \frac{3N}{2} \log_2 N - 3N + 3, \quad A_{W^{III}}(N) = N \log_2 N - 2N + 4. \tag{8.117}$$

The inverse of $\bar{W}_N^{III}$ is

$$(\bar{W}_2^{III})^{-1} = \frac{1}{2} \begin{bmatrix} 1 & 1 \\ 1 & -1 \end{bmatrix} \tag{8.118}$$

$$(\bar{W}_N^{III})^{-1} = \frac{1}{2} P_N \begin{bmatrix} (\bar{W}_{N/2}^{III})^{-1} & 0 \\ 0 & (\bar{W}_{N/2}^{IV})^{-1} \end{bmatrix} \begin{bmatrix} I_{N/2} & I_{N/2} \\ I_{N/2} & -I_{N/2} \end{bmatrix}. \tag{8.119}$$

Integer type-I DHT

Let $x(n)$ $(n = 0, 1, \ldots, N - 1)$ be a real input sequence. We assume that $N = 2^t$ where $t > 0$. The type-I DHT of $x(n)$ is defined as [24]

$$X(k) = \sum_{n=0}^{N-1} x(n) \text{cas} \frac{2\pi nk}{N}, \quad k = 0, 1, \ldots, N - 1. \tag{8.120}$$

The type-I DHT can be computed by using the type-II DHT or type-III DHT. The following factorization in Lemma 24 can be found in [10, 25].

Lemma 24 *Let W_N^I be the kernel matrix of the type-I DHT, that is,*

$$W_N^I = \left(\text{cas} \frac{2\pi kn}{N} \right)_{k,n=0,1,\ldots,N-1}.$$

Then

$$W_4^I = \begin{bmatrix} 1 & 1 & 1 & 1 \\ 1 & 1 & -1 & -1 \\ 1 & -1 & 1 & -1 \\ 1 & -1 & -1 & 1 \end{bmatrix} \tag{8.121}$$

$$W_N^I = P_N \begin{bmatrix} W_{N/2}^I & 0 \\ 0 & W_{N/2}^{III} \end{bmatrix} \begin{bmatrix} I_{N/2} & I_{N/2} \\ I_{N/2} & -I_{N/2} \end{bmatrix}, \tag{8.122}$$

where $N > 4$, and P_N and $I_{N/2}$ are defined in Lemma 21.

Example 19 *Factorization of short-length DHT-I*
The factorization of W_4^I is shown in (8.121),

$$W_8^I = P_8 \begin{bmatrix} W_4^I & 0 \\ 0 & W_4^{III} \end{bmatrix} \begin{bmatrix} I_4 & I_4 \\ I_4 & -I_4 \end{bmatrix}, \tag{8.123}$$

where P_8 is defined in (8.93) and W_4^{III} is defined in (8.113).

Definition 27 *Assume that $N = 2^t$. The kernel matrix of the type-I integer DHT (IntDHT-I) of length N is defined by*

$$\bar{W}_4^I = \begin{bmatrix} 1 & 1 & 1 & 1 \\ 1 & 1 & -1 & -1 \\ 1 & -1 & 1 & -1 \\ 1 & -1 & -1 & 1 \end{bmatrix} \tag{8.124}$$

$$\bar{W}_N^I = P_N \begin{bmatrix} \bar{W}_{N/2}^I & 0 \\ 0 & \bar{W}_{N/2}^{III} \end{bmatrix} \begin{bmatrix} I_{N/2} & I_{N/2} \\ I_{N/2} & -I_{N/2} \end{bmatrix}, \quad N > 4. \tag{8.125}$$

Based on the factorization, we obtain a floating-point multiplication-free fast algorithm for the IntDHT-I.

Algorithm 32 *Fast algorithm for IntDHT-I*

Step 1.   *For $n = 0, 1, \ldots, N/2 - 1$, compute*

$$h(n) = x(n) + x(\frac{N}{2} + n), \quad g(n) = x(n) - x(\frac{N}{2} + n).$$

Step 2.   *Use Algorithm 31 to compute the IntDHT-III of sequence $g(n)$ to get output $X(2k + 1)$. Compute the IntDHT-I of sequence $h(n)$ to get outputs $X(2k)$. If $N/2 > 4$, the length $N/2$ IntDHT-I is further decomposed into smaller ones until* *the length becomes 4. Based on the matrix decomposition in (8.124), the length 4 IntDHT-I can be computed by 8 additions.*

The computational complexity in terms of the numbers of lifting steps and additions is

$$L_{W^I}(N) = \frac{3N}{2} \log_2 N - 6N + 3 \log_2 N + 6$$
$$A_{W^I}(N) = N \log_2 N - 4N + 4 \log_2 N + 8. \tag{8.126}$$

The inverse of $\bar{W}_N^I$ is obtained by

$$(\bar{W}_2^I)^{-1} = \frac{1}{4} \begin{bmatrix} 1 & 1 & 1 & 1 \\ 1 & 1 & -1 & -1 \\ 1 & -1 & 1 & -1 \\ 1 & -1 & -1 & 1 \end{bmatrix} \tag{8.127}$$

$$(\bar{W}_N^I)^{-1} = \frac{1}{2} \begin{bmatrix} I_{N/2} & I_{N/2} \\ I_{N/2} & -I_{N/2} \end{bmatrix} \begin{bmatrix} (\bar{W}_{N/2}^I)^{-1} & 0 \\ 0 & (\bar{W}_{N/2}^{III})^{-1} \end{bmatrix} P_N'. \tag{8.128}$$

Table 8.5   MSE between IntDHT and DHT.

N	MSE	N	MSE
8	0.0032	16	0.0056
32	0.0056	64	0.0058
128	0.0091	256	0.0105
512	0.0102	1024	0.0107

Table 8.6   MSE between II-DHT and DHT.

N	MSE	N	MSE
32	0.1472	64	0.0754
128	0.0410	256	0.0207
512	0.0114	1024	0.0058

8.4.5   Performance comparison

Because the IntDHT or II-DHT approximates the generalized DHT, assessment of their approximation performance must be made. Similar to the IntDCT or II-DCT, it is difficult to obtain a theoretical bound for the error between the generalized DHT and the corresponding IntDHT or II-DHT. In general, the approximation performance of the IntDHT or II-DHT is closely related to the number of bits used for the lifting multipliers. In our algorithms, these lifting multipliers are $\tan(\alpha_n/2)$ and $\sin(\alpha_n)$, where $\alpha_n = \frac{\pi(2n+1)}{2M}$ ($0 \le n \le M/2-1$, $M = 2^i$, $i = 1, 2, \ldots, t$). We use RB($\tan(\alpha_n/2)$) and RB($\sin(\alpha_n)$) (see Definition 20) to approximate $\tan(\alpha_n/2)$ and $\sin(\alpha_n)$. For comparison, some numerical experiments were conducted on the approximation accuracy of the proposed transforms to the generalized DHT. Some of the experimental results are given in Tables 8.5 and 8.6. Table 8.5 shows results for the IntDHT-IV where the input signal is $x(n) = \lfloor (n+1)\sin(n+1) \rfloor$ $(n = 0, 1, \ldots, N-1)$. The MSE between the output of the IntDHT or II-DHT and DHT is defined as before (see (8.62)). We choose the approximation function RB to be the one defined in (8.63). Figures 8.9 (a) and (b) show the outputs of the type-IV DHT and IntDHT-IV, respectively, for $N = 8$, while Figures 8.10 (a) and (b) show those for $N = 64$. Here the outputs are normalized by a factor $1/\sqrt{N}$. From Table 8.5, we know that for a fixed function RB, the MSE increases with the length N. For the RB used, the errors are tolerable for many signal processing applications such as data compression even if the length is large enough.

In many practical applications such as image processing, a short-length transform is generally used. For a short-length IntDHT, we can use a low precision approximation function RB. For example, when $N = 8$, by using the approximation defined in (8.64), the MSE becomes 0.0232. Figure 8.11 shows the outputs of the IntDHT-IV in this case. Table 8.6 shows results for the II-DHT-IV. The input signal is the same as that used before. For the II-DHT-IV, if the input is integral, the output is also integral (see Algorithm 29 for the details). When N is very small, this kind of approximation produces relatively large errors. The MSE decreases with increasing N. We show the outputs of the DHT-IV and II-DHT-IV for $N = 128$ in Figures 8.12 (a) and (b), respectively.

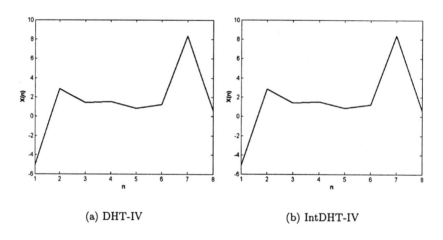

(a) DHT-IV   (b) IntDHT-IV

Figure 8.9   Outputs of DHT-IV and IntDHT-IV ($N = 8$).

Table 8.7   The number of operations for IntDHT and DHT.

Transform	Lifting steps	Additions
IntDHT-II	$\frac{3}{2}N\log_2 N - 3N + 3$	$N\log_2 N - 2N + 4$
IntDHT-III	$\frac{3}{2}N\log_2 N - 3N + 3$	$N\log_2 N - 2N + 4$
IntDHT-IV	$\frac{3}{2}N\log_2 N$	$N\log_2 N$
	Multiplications	Additions
DHT-I	$\frac{1}{2}N\log_2 N - \frac{3}{2}N + 2$	$\frac{3}{2}N\log_2 N - \frac{5}{2}N + 6$
DHT-II	$\frac{1}{2}N\log_2 N - \frac{1}{2}N$	$\frac{3}{2}N\log_2 N - \frac{3}{2}N$
DHT-III	$\frac{1}{2}N\log_2 N - \frac{1}{2}N$	$\frac{3}{2}N\log_2 N - \frac{3}{2}N$
DHT-IV	$\frac{1}{2}N\log_2 N + \frac{1}{2}N$	$\frac{3}{2}N\log_2 N - \frac{1}{2}N$

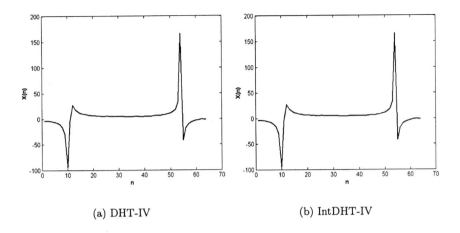

(a) DHT-IV   (b) IntDHT-IV

Figure 8.10   Outputs of DHT-IV and IntDHT-IV ($N = 64$).

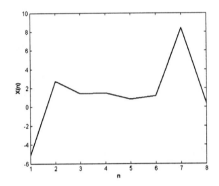

Figure 8.11   Outputs of IntDHT-IV ($N = 8$, by approximation (8.64)).

The IntDHT/II-DHT can also be treated as new transforms because they are invertible. The advantages of the IntDHT/II-DHT over the DHT are the same as those of the IntDCT/II-DCT over the DCT, that is, no floating-point multiplications, no round-off error and integer-to-integer mapping. Table 8.7 gives the number of operations for the IntDHT and DHT, where the number of operations for the DHT is based on the best known algorithms for the DHT in [10, 25] or Chapter 4. However, since the IntDHT/II-DHT are new transforms, their performance, such as the de-correlation property, remains to be questioned. Experiments in [6, 22] have shown that the performance of the 8-point IntDCT is close to that of the DCT if all quantized lifting multipliers have 8-bits resolution. It is reasonable to expect similar results for the IntDHT. Further research in this area should be conducted.

8.5   MD INTEGER DCT AND FAST ALGORITHMS

An MD transform can be generated by simply using the tensor products of the corresponding 1D transforms. For example, the MD DCT, which has a separable kernel, is constructed using the tensor-product method. To process a 2D array by a tensor product 2D transform,

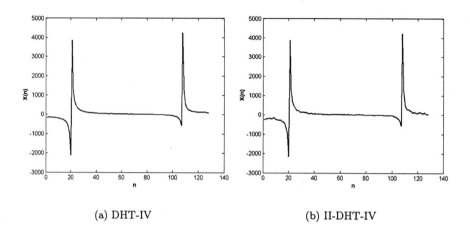

(a) DHT-IV   (b) II-DHT-IV

Figure 8.12   Outputs of DHT-IV and II-DHT-IV ($N = 128$).

we simply implement a 1D transform along its rows and columns consecutively. So, the tensor-product method is also called the row-column method. Other methods are also reported to construct an MD transform which has a non-separable kernel. Sometimes a non-separable MD transform is more useful due to the non-separable property of MD signals. In this section, we study a non-separable 2D integer transform by combining the 1D integer transform and the polynomial transform (PT).

8.5.1   Row-column MD type-II integer DCT

By implementing the 1D transform along the dimensions of an MD input array consecutively, we get a row-column MD transform.

Definition 28 *(Row-column 2D IntDCT) Let $x(n, m)$ $(n = 0, 1, \ldots, N - 1, m = 0, 1, \ldots,$ $M - 1)$ be a 2D input array. We first compute the length M 1D IntDCT of each row of $x(n, m)$ and use $y(n, l)$ $(n = 0, 1, \ldots, N - 1,$ $l = 0, 1, \ldots, M - 1)$ to denote the transform result, that is, $y(n, l)$ $(l = 0, 1, \ldots, M - 1)$ is the 1D IntDCT of $x(n, m)$ $(m = 0, 1, \ldots, M - 1)$ for every n $(n = 0, 1, \ldots, N - 1)$. Then calculate the length N 1D IntDCT of each column of $y(n, l)$ and use $X(k, l)$ $(k = 0, 1, \ldots, N - 1,$ $l = 0, 1, \ldots, M - 1)$ to denote the transform result, that is, $X(k, l)$ $(k = 0, 1, \ldots, N - 1)$ is the 1D IntDCT of $y(n, l)$ $(n = 0, 1, \ldots, N - 1)$ for every l $(0 \le l \le M - 1)$. $X(k, l)$ is called the row-column 2D IntDCT of $x(n, m)$.*

The definition can be easily extended to the MD IntDCT.

Based on the fast algorithms of the 1D IntDCT and the definition of the 2D IntDCT, we can easily get the computational complexity for the 2D IntDCT. In fact, the computational complexity of the row-column 2D IntDCT is equivalent to that of N length M 1D IntDCTs and M length N 1D IntDCTs. Using the IntDCT-II as an example, the numbers of lifting steps and additions for the row-column 2D IntDCT-II are

$$NL_{C^{II}}(M) + ML_{C^{II}}(N) \tag{8.129}$$

$$NA_{C^{II}}(M) + MA_{C^{II}}(N). \tag{8.130}$$

8.5.2   PT-based 2D Integer DCT-II

Let $x(n,m)$ $(n = 0, 1, \ldots, N-1,\ m = 0, 1, \ldots, M-1)$ be a 2D input array, and N and M be powers of two and $M \geq N$. We can write $N = 2^t$ and $M = 2^J N$, where $t > 0$ and $J \geq 0$, respectively. If $M < N$, the definition can also be used by simply interchanging N and M. The following 2D integer DCT-II is derived based on the 1D integer DCT-II and the PT.

Definition 29 *The PT-based 2D integer DCT-II (2D IntDCT-II) of $x(n,m)$ is $X(k,l)$ ($k = 0, 1, \ldots, N-1,\ l = 0, 1, \ldots, M-1$) which can be computed by the following steps.*

Step 1.   *Reorder $x(n,m)$ to get $y(p,m) = x(q(p,m), m)$, where for $p = 0, 1, \ldots, N-1$ and $m = 0, 1, \ldots, M-1$,*

$$
q(p,m) = \begin{cases}
2f(p, \frac{m}{2}), & m \text{ is even and } f(p, \frac{m}{2}) < \frac{N}{2} \\
2N - 1 - 2f(p, \frac{m}{2}), & m \text{ is even and } f(p, \frac{m}{2}) \geq \frac{N}{2} \\
2f(p, N - \frac{m+1}{2}), & m \text{ is odd and } f(p, N - \frac{m+1}{2}) < \frac{N}{2} \\
2N - 1 - 2f(p, N - \frac{m+1}{2}), & m \text{ is odd and } f(p, N - \frac{m+1}{2}) \geq \frac{N}{2}
\end{cases}
\tag{8.131}
$$

and the function $f(p,m) = ((4p+1)m + p) \bmod N$.

Step 2.   *Compute a 1D IntDCT-II for each row of array $y(p,m)$, and let the output array be $V(p,l)$.*

Step 3.   *Compute a PT*

$$
A_k(z) \equiv \sum_{p=0}^{N-1} V_p(z) \hat{z}^{pk} \bmod z^{2M} + 1, \quad k = 0, 1, \ldots, N-1,
\tag{8.132}
$$

where

$$
V_p(z) = \sum_{l=0}^{M-1} V(p,l) z^l - \sum_{l=M+1}^{2M-1} V(p, 2M-l) z^l, \quad \hat{z} \equiv z^{2^{J+2}} \bmod z^{2M} + 1
$$

and then get

$$
B_k(z) \equiv A_k(z) z^{2^J k} \bmod z^{2M} + 1.
\tag{8.133}
$$

Step 4.   *Compute*

$$
X_k(z) = \sum_{l=0}^{2M-1} X(k,l) z^l \equiv \frac{1}{2}(B_k(z) + B_k(z^{-1})) \bmod z^{2M} + 1,
\tag{8.134}
$$

where only $X(k,l)$ ($k = 0, 1, \ldots, N-1,\ l = 0, 1, \ldots, M-1$) are used as the outputs.

In Chapter 7 and [26] we proved that if the IntDCT-II in Step 2 is replaced by the conventional type-II DCT, the above defined 2D IntDCT is the conventional 2D type-II DCT. Therefore, the 2D IntDCT can also be viewed as the integer approximation to the floating-point 2D type-II DCT. However, it is different from the row-column 2D integer DCT.

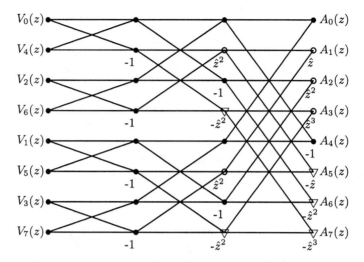

Note: The node marked by filled circle represents that the polynomial has symmetric property within itself. The node marked by ▽ represents that the polynomial can be simply derived from one of the polynomials marked by empty circles.

Figure 8.13   Signal flow graph of Step 3.

8.5.3   Comparison

If the input array is integral, the entire computation of the PT-based 2D IntDCT-II can be done by integer additions and shifting. Although Step 1 needs some integer operations, they are not related to the input, and hence can be done in advance. The PT in Step 3 has been discussed in detail in Chapter 2 and refined in Chapter 7, and requires $NM \log_2 N - N/2 + 1$ additions. Figure 8.13 gives the signal flow graph of Step 3 when $N = M = 8$. Thus, the numbers of lifting steps and additions for the PT-based 2D IntDCT-II are

$$NL_{C^{II}}(M) \tag{8.135}$$

and

$$NA_{C^{II}}(M) + NM \log_2 N + NM - M - \frac{3N}{2} + 2. \tag{8.136}$$

The PT-based 2D IntDCT generally needs fewer lifting steps and additions than that needed by the row-column 2D IntDCT-II (see (8.129) and (8.130)). The larger the size is, the more operations are reduced. Table 8.8 gives the comparisons when $N = M$.

Table 8.8   Comparison of 2D IntDCT-IIs.

Method	Lifting steps	Additions
PT-based	$\frac{3}{2}N^2 \log_2 N - 3N^2 + 6N$	$2N^2 \log_2 N - \frac{7}{2}N + 2$
Row-column	$3N^2 \log_2 N - 6N^2 + 12N$	$2N^2 \log_2 N - 2N^2$

8.5.4   Reconstruction algorithm for 2D IntDCT-II

For the row-column 2D IntDCT, we can use the row-column inverse 2D IntDCT for reconstruction. The PT-based 2D IntDCT-II is invertible because each step in Definition 29 is invertible. The following algorithm can be used for inverting the transform.

Algorithm 33 *Fast algorithm for PT-based inverse 2D IntDCT-II*

Step 1.   *Generate the polynomials*

$$X_k(z) = \sum_{l=0}^{M-1} X(k,l)z^l - \sum_{l=M+1}^{2M-1} X(k,2M-l)z^l$$

and compute

$$A_k(z) \equiv (X_k(z) + X_{N-k}(z)z^{-N})z^{k2^J} \bmod (z^{2M}+1),$$

where $k = 0,1,\ldots,N-1$.

Step 2.   *Compute a PT*

$$V_p(z) \equiv \frac{1}{N} \sum_{k=0}^{N-1} A_k(z)\hat{z}^{pk} \bmod z^{2M}+1,$$

where $p = 0,1,\ldots,N-1$, *and denote the coefficients of* $V_p(z)$ *by* $V(p,l)$.

Step 3.   *Compute the inverse 1D IntDCT for each row of array* $V(p,l)$, *and let the output array be* $y(p,m)$.

Step 4.   *Reorder the array* $y(p,m)$ *to get* $x(n,m) = y(r(n,m),m)$, *where*

$$r(n,m) = \begin{cases} g(\frac{n}{2},\frac{m}{2}), & m \text{ and } n \text{ are even} \\ g(N-\frac{n+1}{2},\frac{m}{2}), & m \text{ is even and } n \text{ is odd} \\ g(\frac{n}{2},N-\frac{m+1}{2}), & m \text{ is odd and } n \text{ is even} \\ g(N-\frac{n+1}{2},N-\frac{m+1}{2}), & m \text{ and } n \text{ are odd} \end{cases},$$

where $n = 0,1,\ldots,N-1$, $m = 0,1,\ldots,M-1$, *and the function* $g(n,m) = ((4m+1)^{-1}(n-m)) \bmod N$.

8.6   MD INTEGER DHT AND FAST ALGORITHMS

As discussed in the last section, the row-column method can be used to generate the MD integer DHT by simply implementing 1D integer DHTs along the dimensions of an MD array consecutively. The row-column MD integer DHT has a separable kernel. We can also use the PT and 1D integer DHT to generate the MD integer DHT, which produces transforms with non-separable kernels. The PT-based MD integer transform not only has a non-separable kernel, but also needs a much smaller number of operations than the row-column MD transform.

8.6.1   Row-column MD integer DHT

By implementing a 1D integer DHT along the dimensions of an MD input array consecutively, we have a row-column MD integer DHT.

Definition 30 (*Row-column rD IntDHT*) *Let* $x(n_1, n_2, \ldots, n_r)$ $(n_i = 0, 1, \ldots, N_i - 1,\ i = 1, 2, \ldots, r)$ *be an rD input sequence. We first compute a length* N_r *1D IntDHT along the rth dimension of* $x(n_1, n_2, \ldots, n_r)$ *and use* $y_1(n_1, \ldots, n_{r-1}, k_r)$ *to denote the transform result, that is,* $y_1(n_1, \ldots, n_{r-1}, k_r)$ $(k_r = 0, 1, \ldots, N_r - 1)$ *is the 1D IntDHT of* $x(n_1, \ldots, n_{r-1}, n_r)$ $(n_r = 0, 1, \ldots, N_r - 1)$ *for every* n_i $(n_i = 0, 1, \ldots, N_i - 1,\ i = 1, 2, \ldots, r - 1)$. *Then calculate a length* N_{r-1} *1D IntDHT along the* $(r - 1)$*th dimension of* $y_1(n_1, \ldots, n_{r-1}, k_r)$ *and use* $y_2(n_1, \ldots, n_{r-2}, k_{r-1}, k_r)$ *to denote the transform result. Finally, compute a length* N_1 *1D IntDHT along the first dimension of* $y_{r-1}(n_1, k_2, \ldots, k_r)$ *and use* $X(k_1, k_2, \ldots, k_r)$ *to denote the transform result.* $X(k_1, k_2, \ldots, k_r)$ *is called the row-column rD IntDHT of* $x(n_1, n_2, \ldots, n_r)$.

If N_i are powers of two, the 1D IntDHTs can be computed by the fast algorithms discussed before. Based on the fast algorithms of the 1D IntDHT and the above definition, we can easily get the computational complexity for the rD IntDHT. In fact, the computational complexity of the row-column rD IntDHT is equivalent to the total computational complexity of M_i length N_i 1D IntDHTs $(i = 1, 2, \ldots, r)$, where

$$M_i = \frac{\hat{N}}{N_i}, \quad \hat{N} = N_1 N_2 \cdots N_r.$$

Therefore, the numbers of lifting steps and additions needed by the row-column rD IntDHT-II are, respectively,

$$\sum_{i=1}^{r} [M_i L_{W^{II}}(N_i)] \tag{8.137}$$

$$\sum_{i=1}^{r} [M_i A_{W^{II}}(N_i)]. \tag{8.138}$$

8.6.2   PT-based MD type-II integer DHT

Consider a type-II rD integer DHT of size $N_1 \times N_2 \times \cdots \times N_r$. We assume that N_i $(i = 1, 2, \ldots, r)$ is a power of two. Without loss of generality, it is assumed that N_r is the greatest integer among N_i. Therefore, we can write $N_r = 2^t$ and $N_r/N_i = 2^{l_i}$ $(i = 1, 2, \ldots, r - 1)$, where $l_i \geq 0$ is an integer.

Definition 31 *The rD integer type-II DHT (rD-IntDHT-II) of* $x(n_1, n_2, \ldots, n_r)$ $(n_i = 0, 1, \ldots, N_i - 1,\ i = 1, 2, \ldots, r)$ *is defined and computed by the following steps.*

Step 1.   *Reorder the input array to get*

$$y(p_1, p_2, \ldots, p_r) = x(p_1(p_r), \ldots, p_{r-1}(p_r), p_r), \tag{8.139}$$

where $p_i(p_r)$ $(p_i = 0, 1, \ldots, N_i - 1,\ i = 1, 2, \ldots, r)$ *is defined as the least nonnegative remainder of* $(2p_i + 1)p_r + p_i$ *modulo* N_i $(i = 1, 2, \ldots, r - 1)$.

Step 2. *Compute* $N_1 N_2 \cdots N_{r-1}$ *1D IntDHT-IIs of length* N_r *along the rth dimension of the array* $y(p_1, p_2, \ldots, p_r)$ *to get* $V(p_1, \ldots, p_{r-1}, m)$. *Let*

$$V_{p_1,\ldots,p_{r-1}}(z) = \sum_{m=0}^{N_r-1} V(p_1, \ldots, p_{r-1}, m) z^m.$$

Step 3. *Compute the* $(r-1)D$ *PT of the polynomial sequence* $V_{p_1,\ldots,p_{r-1}}(z)$, *that is*

$$Y_{k_1,\ldots,k_{r-1}}(z) \equiv \sum_{p_1=0}^{N_1-1} \cdots \sum_{p_{r-1}=0}^{N_{r-1}-1} V_{p_1,\ldots,p_{r-1}}(z) z_1^{p_1 k_1} \cdots z_{r-1}^{p_{r-1} k_{r-1}} \bmod (z^{N_r}+1),$$

$$(8.140)$$

where $k_i = 0, 1, \ldots, N_i - 1$, $i = 1, 2, \ldots, r-1$ *and*

$$z_i \equiv z^{-2^{l_i+1}} \bmod (z^{N_r}+1).$$

Step 4. *Reorder the coefficients of the polynomials* $Y_{k_1,\ldots,k_{r-1}}(z)$ *to get* $X_{k_1,\ldots,k_{r-1}}(z)$, *that is,*

$$X_{k_1,\ldots,k_{r-1}}(z) \equiv Y_{k_1,\ldots,k_{r-1}}(z) z^{-k_1 2^{l_1} - \cdots - k_{r-1} 2^{l_{r-1}}} \bmod (z^{N_r}+1). \quad (8.141)$$

The coefficients of $X_{k_1,\ldots,k_{r-1}}(z)$ *are* $X(k_1, \ldots, k_r)$, *which are the outputs of the* rD *IntDHT-II.*

In Chapter 5 and [27], we proved that if the IntDHT-II in Step 2 is replaced by the conventional type-II DHT, the above defined rD IntDHT-II becomes the conventional type-II rD DHT. Therefore, the rD IntDHT-II can also be viewed as the integer approximation to the conventional type-II rD DHT. Furthermore, the transform is non-separable and different from the row-column rD integer type-II DHT.

It can be seen that the PT-based rD IntDHT-II needs $\frac{\hat{N}}{N_r}$ 1D IntDHT-IIs and $\hat{N} \log_2 \frac{\hat{N}}{N_r}$ additions (operations for the $(r-1)D$ PT in Step 3, see Chapter 2), where $\hat{N} = N_1 N_2 \cdots N_r$. Therefore, the required numbers of lifting steps and additions are

$$\frac{\hat{N}}{N_r} L_{WII}(N_r) \quad (8.142)$$

$$\frac{\hat{N}}{N_r} A_{WII}(N_r) + \hat{N} \log_2 \frac{\hat{N}}{N_r}. \quad (8.143)$$

Obviously, the PT-based rD IntDHT-II reduces the number of operations considerably compared to the row-column rD integer type-II DHT. Table 8.9 gives the comparisons when $N_1 = N_2 = \cdots = N_r = N$. The lifting steps for the PT-based transform is only $1/r$ times that of the row-column method. The number of additions is also decreased greatly.

Numerical experiments have been made to test the approximation ability of the PT-based 2D IntDHT-II. An example is shown in Figure 8.14, where the input signal is $x(n_1, n_2) = \cos[10(n_1+1) + (n_1+1)^3] + (n_2+1) \sin[(n_2+1)^2]$ and the approximation function RB is given in (8.64).

Table 8.9   Comparison of the rD integer DHT-IIs.

Method	Lifting steps	Additions
PT-based	$\frac{3}{2}N^r \log_2 N - 3N^r + 3N^{r-1}$	$(r+1)N^r \log_2 N - \frac{3}{2}N^r + N^{r-1}$
Row-column	$\frac{3}{2}rN^r \log_2 N - 3rN^r + 3rN^{r-1}$	$2rN^r \log_2 N - \frac{3}{2}rN^r + rN^{r-1}$

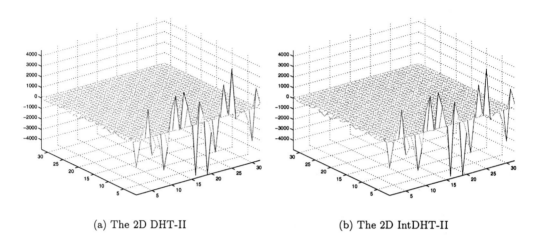

(a) The 2D DHT-II   (b) The 2D IntDHT-II

Figure 8.14   Outputs of 32×32 2D DHT-II and 2D IntDHT-II.

8.6.3   Inverse MD integer DHT

For the row-column rD IntDHT, we can use the row-column inverse rD IntDHT for reconstruction by implementing the 1D inverse IntDHT along each dimension of the transformed rD array consecutively. The PT-based rD IntDHT-II is invertible because each step in Definition 31 is invertible. The computation of its inverting transform (r**D IntDHT-III**) can be described in the following steps.

Algorithm 34 *Reconstruction algorithm for rD IntDHT-II*

Step 1.   *Generate polynomials* $X_{k_1,\ldots,k_{r-1}}(z)$ *from the input array* $X(k_1,\ldots,k_r)$ *by*

$$X_{k_1,\ldots,k_{r-1}}(z) = \sum_{k_r=0}^{N_r-1} X(k_1,\ldots,k_r)z^{k_r}.$$

Then reorder the coefficients of the polynomials $X_{k_1,\ldots,k_{r-1}}(z)$ *to get* $Y_{k_1,\ldots,k_{r-1}}(z)$, *that is,*

$$Y_{k_1,\ldots,k_{r-1}}(z) \equiv X_{k_1,\ldots,k_{r-1}}(z)z^{k_1 2^{l_1}+\cdots+k_{r-1}2^{l_{r-1}}} \mod (z^{N_r}+1), \qquad (8.144)$$

where $k_i = 0,1,\ldots,N_i-1$ *and* $i = 1,2,\ldots,r-1$.

Step 2. *Compute the $(r-1)D$ PT of the polynomial sequence $Y_{k_1,\dots,k_{r-1}}(z)$,*

$$V_{p_1,\dots,p_{r-1}}(z) \equiv \frac{1}{N_1 N_2 \cdots N_{r-1}} \sum_{k_1=0}^{N_1-1} \cdots \sum_{k_{r-1}=0}^{N_{r-1}-1} Y_{k_1,\dots,k_{r-1}}(z)$$
$$\cdot z_1^{-p_1 k_1} \cdots z_{r-1}^{-p_{r-1} k_{r-1}} \bmod (z^{N_r}+1), \qquad (8.145)$$

where $p_i = 0, 1, \dots, N_i - 1$, $i = 1, 2, \dots, r-1$ and

$$z_i \equiv z^{-2^{l_i+1}} \bmod (z^{N_r}+1), \quad i = 1, 2, \dots, r-1.$$

Step 3. *Compute $N_1 N_2 \cdots N_{r-1}$ inverse 1D IntDHT-IIs (IntDHT-IIIs) with length N_r along the rth dimension of the array $V(p_1, p_2, \dots, p_r)$, the coefficients of $V_{p_1,\dots,p_{r-1}}(z)$, to get $y(p_1, p_2, \dots, p_r)$.*

Step 4. *Compute the output*

$$x(p_1(p_r), \dots, p_{r-1}(p_r), p_r) = y(p_1, p_2, \dots, p_r) \qquad (8.146)$$

$$p_i = 0, 1, \dots, N_i - 1, \quad i = 1, 2, \dots, r,$$

where $p_i(p_r)$ is defined as the least non-negative remainder of $(2p_i + 1)p_r + p_i$ modulo N_i, $i = 1, 2, \dots, r-1$.

We can construct other types of MD integer DHTs in a similar way. For example, based on the idea in [28], we can construct a type-I PT-based MD integer DHT by combining the 1D IntDHT-I and the PT.

8.7 CHAPTER SUMMARY

Methods are presented to factor various DCTs and DHTs into lifting steps and additions. By approximating the lifting matrices, integer discrete transforms, which are free of floating-point multiplications, are obtained. Integer-to-integer transforms are also constructed. Fast algorithms are given for the transforms and their computational complexities are analyzed. Numerical experiments are presented to show their approximation performances. Known relationships among various types of discrete transforms can be used to construct other types of integer transforms. For example, based on the relationships between the DCT and DST, we can construct the integer DST. Similarly, relationships between the DHT and DFT can be used to generate the integer DFT easily. Of course, constructing an integer DFT directly is also available as described in [19]. The presented transforms can be used in mobile computing, lossless image coding, multiplier-less filter banks and other related fields.

There are other integer transforms which are not based on lifting factorizations [5, 15]. These transforms generally lack a fast algorithm and do not approximate to the original transforms. Furthermore, the methods used for constructing integer transforms in [5, 15] depend heavily on the size of the transforms. It is difficult to generate integer transforms with large sizes using these methods.

REFERENCES

1. *Proceedings of IEEE (Special issue on the Hartley transform)*, vol. 82, no. 3, 1994.

2. G. P. Apousleman, M. W. Marcellin and B. R. Hunt, Compression of hyperspectral imagery using the 3-D DCT and hybrid DPCM/DCT, *IEEE Trans. Geoscience Remote Sensing*, vol. 33, no. 1, 26–34, 1995.

3. V. Britanak and K. R. Rao, The fast generalized discrete Fourier transforms: a unified approach to the discrete sinusoidal transforms computation, *Signal Processing*, vol. 79, no. 2, 135–140, 1999.

4. R. Calderbank, I. Daubechies, W. Sweldens and B. L. Yeo, Wavelet transforms that map integers to integers, *Appl. Comput. Harmon. Anal.*, vol. 5, no. 3, 332–369, 1998.

5. W. K. Cham and P. C. Yip, Integer sinusoidal transforms for image processing, *International J. Electronics*, vol. 70, no. 6, 1015–1030, 1991.

6. Y. J. Chen, S. Oraintara and T. Nguyen, Integer discrete cosine transform (IntDCT), Invited paper, *The 2nd Intern. Conf. Inform. Comm. and Sig. Proc.*, Singapore, Dec. 1999.

7. L. Z. Cheng, Y. H. Zeng, G. Bi and Z. P. Lin, Fast multiplierless approximation of type-II and type-IV DCT, *Preprint*, 2001.

8. H. S. Hou, A fast recursive algorithm for computing discrete cosine transform, *IEEE Trans. Acoust., Speech, Signal Process.*, vol. 35, no. 10, 1455–1461, 1987.

9. N. C. Hu, H. Chang and O. K. Ersoy, Generalized discrete Hartley transform, *IEEE Trans. Signal Process.*, vol. 40, no. 12, 2951–2960, 1992.

10. Z. R. Jiang, Y. Zeng and P. N. Yu, *Fast Algorithms*, National University of Defense Technology Press, Changsha, P. R. China, 1994 (in Chinese).

11. B. G. Lee, A new algorithm to compute the discrete cosine transform, *IEEE Trans. Acoust., Speech, Signal Process.*, vol. 32, 1243–1245, 1984.

12. H. S. Malvar, *Signal Processing with Lapped Transform*, Artech House, Norwood, MA, 1991.

13. M. W. Marcellin, M. J. Gormish, A. Bilgin and M. P. Boliek, An overview of JPEG-2000, *Proceedings of Data Compression Conference*, 523–541, 2000.

14. N. Memon, X. Wu and B. L. Yeo, Improved techniques for lossless image compression with reversible integer wavelet transforms, *IEEE Int. Conf. Image Processing*, vol. 3, 891–895, 1998.

15. S. C. Pei and J. J. Ding, The integer transforms analogous to discrete trigonometric transforms, *IEEE Trans. Signal Process.*, vol. 48, no. 12, 3345 –3364, 2000.

16. W. Philps, Lossless DCT for combined lossy/lossless image coding, *IEEE Int. Conf. Image Processing*, vol. 3, 871–875, 1998.

17. K. R. Rao and P. Yip, *Discrete Cosine Transform: Algorithms, Advantages and Applications*, Academic Press, New York, 1990.

18. Y. L. Siu and W. C. Siu, Variable temporal-length 3-D discrete cosine transform coding, *IEEE Trans. Image Process.*, vol. 6, no. 5, 758–763, 1997.

19. S. Oraintara, Y. J. Chen and T. Nguyen, Integer fast Fourier transform (INTFFT), *Proc. ICASSP*, Salt Lake City, UT, 2001.

20. G. Strang and T. Nguyen, *Wavelets and Filter Banks*, Wellesley-Cambridge Press, Wellesley, 1997

21. W. Sweldens, The lifting scheme: a construction of second generation wavelets, *SIAM J. Math. Anal.*, vol. 29, no. 2, 511–546, 1998.

22. T. D. Tran, The BinDCT: fast multiplierless approximation of the DCT, *IEEE Signal Process. Lett.*, vol. 7, no. 6, 141–144, 2000.

23. Z. Wang, Fast algorithms for the discrete W transform and for the discrete Fourier transform, *IEEE Trans. Acoust., Speech, Signal Process.*, vol. 32, no. 4, 803–816, 1984.

24. Z. Wang and B. R. Hunt, The discrete W transform, *Appl. Math. Comput.*, vol. 16, 19–48, 1985.

25. Z. Wang, The fast W transform–algorithms and programs, *Science in China* (series A), vol. 32, 338–350, 1989.

26. Y. H. Zeng, G. Bi and A. R. Leyman, New polynomial transform algorithms for multi-dimensional DCT, *IEEE Trans. Signal Process.*, vol. 48, no. 10, 2814–2821, 2000.

27. Y. H. Zeng and X. M. Li, Multidimensional polynomial transform algorithms for multidimensional discrete W transform, *IEEE Trans. Signal Process.*, vol. 47, no. 7, 2050–2053, 1999.

28. Y. H. Zeng, G. Bi and A. R. Leyman, Polynomial transform algorithms for multi-dimensional discrete Hartley transform, *Proc. IEEE International Symposium on Circuits and Systems*, Geneva, Switzerland, V517–520, May 2000.

29. Y. H. Zeng, L. Z. Cheng, G. Bi and A. C. Kot, Integer discrete cosine transform and fast algorithms, *IEEE Trans. Signal Process.*, vol. 49, no. 11, 2774–2782, 2001.

30. Y. H. Zeng, G. Bi and Z. P. Lin, Lifting factorization of discrete W transform, *Circuits, Systems, Signal Process.*, vol. 21, no. 3, 277–298, 2002.

31. Y. H. Zeng, G. Bi and Z. P. Lin, Integer sinusoidal transforms based on lifting factorization, *IEEE ICASSP 2001*, Salt Lake City, UT.

9

New Methods of
Time-Frequency Analysis

This chapter presents two new transforms that are useful for the analysis of time-varying signals.

- Section 9.2 provides some basic concepts of time-frequency analysis and briefly discusses a few widely used time-frequency transforms in terms of their merits and shortcomings.

- Section 9.3 presents a harmonic transform for signals containing a fundamental and harmonics. In addition to the derivation of the transform and its inverse, application examples show the performance effectiveness and advantages in terms of the improved resolutions in the time-frequency domain.

- Section 9.4 describes the tomographic time-frequency transform based on the relationship between the fractional Fourier transform and the Radon transform of the time-frequency distribution of a signal. The theoretical development of the transform and application examples are provided.

9.1 INTRODUCTION

Signal analysis studies the characteristics of signals, which can be described in many different ways from a mathematical point of view. In addition to signal analysis in the time domain, signal representation in the frequency domain is a widely used analysis approach. The standard Fourier transform (FT) generates a spectrum that describes signals as individual frequency components with relative intensities. The frequency spectrum does not, however, provide information on the changes of these frequencies in time. Therefore, it cannot be used to process some natural and synthetic signals whose spectral contents are changing so rapidly that Fourier analysis becomes incapable of clearly showing the information in the frequency domain. For many practical applications, it is more important to characterize the time-varying signals in both the time and frequency domains simultaneously, which is known as the combined time-frequency description. The fundamental idea of time-frequency analysis is to describe how the frequency contents of a signal change in time. Suitable tools for the time-frequency analysis are needed to describe a time-varying spectrum.

Time-frequency transforms (TFTs) used for time-frequency analysis can be categorized into linear and quadratic transforms. The main linear transforms are the short-time Fourier transform (STFT) and the wavelet transform. The quadratic transforms include the spectrogram, scalogram, Wigner distribution (WD) ambiguity function, smoothed WD and various classes of quadratic TFTs. TFTs have been applied to analyze, modify and synthesize non-stationary or time-varying signals. Three-dimensional plots of TFT surfaces are used as graphical representations to show how spectral components of a signal vary with time. TFT synthesis algorithms have also been employed to recover a signal from a TFT representation, thus allowing a time-frequency implementation of signal design, time-varying filtering, noise suppression and time warping, etc. We now briefly discuss a few widely used TFTs.

Short-time Fourier transform

The STFT [22, 29] is designed for analyzing a signal in both the time and frequency domains. The spectrum at a given time instant is represented by the FT of a small signal segment around that particular time instant. A window function is used to capture the signal segment with some weighting effect. The FT of the captured segment can be interpreted as the result of passing the FT of the signal through a filter whose frequency response is the FT of the window function [22]. Thus, it is clear that a good frequency resolution of the STFT requires a narrow band filter in the frequency domain, which means a wide window function in the time domain. On the other hand it is clear that a good resolution of the transform in the time domain requires a short window function, which has a wide bandwidth in the frequency domain. Unfortunately, the uncertainty principle prohibits the existence of windows which have an arbitrarily small duration in the time domain and an arbitrarily small bandwidth in the frequency domain [24]. Hence, there exists a fundamental resolution tradeoff: improving the time resolution by using a short window results in a loss of frequency resolution, and vice versa.

Although linearity is a desirable property, the quadratic TFT can be interpreted to be a time-frequency energy distribution or the instantaneous power spectrum [10]. There also exists another interpretation in terms of correlation functions [4]. In comparison, the quadratic TFT does not have the linearity property because the quadratic TFT of the sum of two signals is not simply the sum of the quadratic TFTs of those signals. There exist some cross-terms between the individual quadratic TFTs. According to the quadratic superposition principle, the quadratic TFT of an N-component signal is comprised of N signal terms and $\frac{N(N-1)}{2}$ interference terms. The presence of the interference terms often makes visual analysis of multicomponent signals difficult.

Spectrogram and scalogram

The spectrogram and the scalogram, defined as the squared magnitudes of the STFT and the wavelet transform, respectively, may be loosely interpreted in terms of signal energy. The spectrogram has been used extensively to analyze speech signals [29] and other non-stationary signals. The interference terms of the spectrogram and scalogram have oscillatory structures, which are restricted to the regions of the time-frequency plane where the corresponding signal terms overlap. Hence, if two signal components are sufficiently separated in the time-frequency plane, their interference terms will be essentially zero [15]. On the other hand, a disadvantage of the spectrogram and scalogram is their poor time-frequency concentration (or resolution). The interference terms can be suppressed through a low pass filter [6], which also results in a poor time-frequency concentration [15]. There exists a general tradeoff between a good time-frequency concentration and small interference terms.

Wigner distribution and ambiguity function

Among all quadratic TFTs with an energetic interpretation, the WD satisfies an exceptionally large number of desirable mathematical properties [5, 7]. The WD can be loosely interpreted as a 2D distribution of signal energy over the time-frequency plane. However, the uncertainty principle prohibits the interpretation as a pointwise time-frequency energy density [20]. This restriction is also reflected by the fact that the WD may locally assume negative values [21]. Among the class of correlative TFTs, an equally important role is played by the ambiguity function (AF) [2, 39, 40]. The AF can be interpreted as a joint time-frequency correlation function. The WD and AF are dual in the sense that they are an FT pair [6].

The WD has an improved time-frequency concentration and an extensive list of desirable mathematical properties [14]. On the other hand, certain characteristics of the interference terms often cause problems in practical applications. Whereas the interference terms of the spectrogram or scalogram will be zero if the corresponding signal terms do not overlap, the interference terms of the WD will be non-zero regardless of the distance between the two signal terms. Because interference terms are oscillatory, they may be attenuated by means of a smoothing operation (i.e., low pass filtering) [11] at the cost of degraded time-frequency concentration since smoothing generally broadens the signal terms of the WD [14]. Another disadvantage of smoothing is the loss of desirable TFT properties. It is clear that there exists a fundamental compromise between interference and time-frequency concentration. A significant performance gain may often be obtained by adapting the smoothing characteristics of a smoothed WD to the signal to be analyzed [13].

The AF and its squared magnitude (the ambiguity surface) have been used extensively in the fields of radar, sonar, radio astronomy, communications and optics. The AF has been applied to design and evaluate the performance of a large variety of radar signals [19] including chirps [17] and other frequency modulated (FM) signals [33].

Cohen and affine classes

Besides the WD, there exist many other quadratic TFTs which have an energetic interpretation. Most of these TFTs satisfy the basic property of time-frequency shift invariance: if the signal $f(t)$ is delayed in time and/or shifted in frequency, its TFT will be shifted by the same time delay and/or modulation frequency. The class of all time-frequency shift-invariant quadratic TFTs is known as the quadratic Cohen class [4]. Prominent members of Cohen's class are the spectrogram and the WD. An alternative to the energetic Cohen class is provided by the affine class comprising all energetic and quadratic TFTs [34]. Any TFT which is an element of the affine class can be derived from the WD by means of an affine transformation [34]. They are especially advantageous in the case of hyperbolic FM signals [3] such as those emitted by bats [9].

In this chapter, two new TFTs, the short-time harmonic transform (STHT) and the tomographic time-frequency transform (TTFT) are presented. The STHT is designed for harmonic signals, which are composed of one fundamental and some harmonics. The TTFT is based on an intuitive assumption that the linear integral of the time-frequency distribution along the time axis is the FT of the signal. The reconstruction algorithms of computed tomography are used to obtain some interesting results.

9.2   PRELIMINARIES

9.2.1   Basic concepts

Marginal conditions

For quadratic TFTs, the summation of the energy distributions in the frequency domain at a particular time should give the instantaneous energy, or the summation of the energy distributions in the time domain at a particular frequency should give the energy density spectrum. Therefore, an ideal time-frequency transform should satisfy

$$\int_{-\infty}^{+\infty} P(t,\omega)d\omega = |s(t)|^2 \tag{9.1}$$

$$\int_{-\infty}^{+\infty} P(t,\omega)dt = |S(\omega)|^2, \tag{9.2}$$

where $s(t)$ is the signal, $S(\omega)$ is the corresponding spectrum, $|s(t)|^2$ is the intensity per unit time at time t, $|S(\omega)|^2$ is the intensity per unit frequency at ω and $P(t,\omega)$ is the intensity at time t and frequency ω. Equations (9.1) and (9.2) are known as the time and the frequency marginal conditions.

Total energy

The total energy of the distribution should be the total energy of the signal

$$E = \int_{-\infty}^{+\infty} \int_{-\infty}^{+\infty} P(t,\omega)d\omega dt = \int_{-\infty}^{+\infty} |s(t)|^2 dt = \int_{-\infty}^{+\infty} |S(\omega)|^2 d\omega. \tag{9.3}$$

If a time-frequency transform satisfies the marginal conditions, it automatically satisfies the total energy requirement. However, it is possible that a transform can satisfy the total energy requirement without satisfying the marginal conditions. One example is the spectrogram, which is the squared magnitude of the STFT. The total energy requirement is a weak condition, which is why many distributions that do not satisfy the marginal conditions may give good representations of the time-frequency structures.

Time and frequency shift invariance

For a signal $s(t)$, we have

$$\text{if}\quad s(t) \to s(t-t_0) \quad \text{then}\quad P(t,\omega) \to P(t-t_0,\omega), \tag{9.4}$$

where t_0 is a time constant. In the frequency domain, we similarly have

$$\text{if}\quad S(\omega) \to S(\omega-\omega_0) \quad \text{then}\quad P(t,\omega) \to P(t,\omega-\omega_0), \tag{9.5}$$

where ω_0 is a frequency constant. Both cases can be handled together. A signal $s(t)$ that is shifted by a constant t_0 in the time domain and ω_0 in the frequency domain is given by $e^{j\omega_0 t}s(t-t_0)$. Accordingly, we expect the distribution to be shifted accordingly in time and frequency,

$$\text{if}\quad s(t) \to e^{j\omega_0 t}s(t-t_0) \quad \text{then}\quad P(t,\omega) \to P(t-t_0,\omega-\omega_0), \tag{9.6}$$

Linear scaling

The scaled version of a signal $s(t)$ is expressed by $s_{sc}(t) = \sqrt{a}s(at)$ where a is the scaling factor. The spectrum of $s_{sc}(t)$ is

$$S_{sc}(\omega) = \frac{1}{\sqrt{|a|}}S(\frac{\omega}{a}) \quad \text{if} \quad s_{sc}(t) = \sqrt{|a|}s(at). \tag{9.7}$$

When the signal shrinks, its spectrum expands, and the signal expands while its spectrum shrinks. If these relations are to hold for the time-frequency distribution, we must have the scaled distribution satisfy the marginals of the scaled signal,

$$\int_{-\infty}^{+\infty} P_{sc}(t,\omega)d\omega = |s_{sc}(t)|^2 = a|s(at)|^2 \tag{9.8}$$

$$\int_{-\infty}^{+\infty} P_{sc}(t,\omega)dt = |S_{sc}(\omega)|^2 = \frac{1}{a}|S(\frac{\omega}{a})|^2. \tag{9.9}$$

Other conditions, such as global averages and local averages, are also very important. The uncertainty principle, for example, means that with a narrow window function, we can obtain a high resolution in the time domain and a low frequency resolution in the frequency domain, and vice versa. We cannot obtain arbitrarily high resolutions in both the time and frequency domains with one window function.

9.2.2 STFT

The STFT is based on the assumption that a segment of $f(t)$ in a time duration $\tau \in [t - \frac{\Delta t}{2}, t + \frac{\Delta t}{2}]$ has an almost constant spectrum. The spectrum of the segment can be expressed as

$$F(t,\omega) = F[f(\tau)w(t-\tau)] \tag{9.10}$$

which is the FT of the signal truncated by a window $w(t)$, where

$$w(\tau) = \begin{cases} > 0, & |\tau| \le \Delta t \\ = 0, & |\tau| > \Delta t \end{cases}. \tag{9.11}$$

The STFT of $f(t)$ is written as

$$\text{STFT}(t,\omega) = \int_{-\infty}^{+\infty} w(\tau - t)f(\tau)e^{-j\omega\tau}d\tau. \tag{9.12}$$

The energy density, or the spectrogram, is defined by

$$P_f(t,\omega) = |\text{STFT}(t,\omega)|^2 = |\int_{-\infty}^{+\infty} f(\tau)w(\tau - t)e^{-j\omega\tau}d\tau|^2. \tag{9.13}$$

The total energy of the spectrogram is defined as

$$E_f = \int_{-\infty}^{+\infty}\int_{-\infty}^{+\infty} P_f(t,\omega)dtd\omega = \int_{-\infty}^{+\infty}|f(t)|^2dt \int_{-\infty}^{+\infty}|w(t)|^2dt. \tag{9.14}$$

Therefore, if the energy of the window function is taken to be one, the energy of the spectrogram is equal to the energy of the signal.

The time marginal is obtained by

$$P(t) = \int_{-\infty}^{+\infty} |\text{STFT}(t,\omega)|^2 d\omega = \int_{-\infty}^{+\infty} |f(\tau)|^2 |w(\tau - t)|^2 d\tau. \tag{9.15}$$

Similarly, the frequency marginal is

$$P(\omega) = \int_{-\infty}^{+\infty} |\text{STFT}(t,\omega)|^2 dt = \int_{-\infty}^{+\infty} |F(\omega')|^2 |W(\omega - \omega')|^2 d\omega', \tag{9.16}$$

where $F(\omega)$ and $W(\omega)$ are the FTs of $f(t)$ and $w(t)$, respectively. The marginals of the spectrogram given in (9.15) and (9.16) do not satisfy the marginal conditions, that is,

$$P(t) \neq |f(t)|^2, \quad P(\omega) \neq |F(\omega)|^2. \tag{9.17}$$

The reason is that the spectrogram mixes the energy distribution of the signal with that of the window function, which introduces some undesirable effects unrelated to the original signal. The time marginal of the spectrogram only depends on the magnitude of the signal and the window function rather than their phases. Similarly, the frequency marginal only depends on the amplitudes of the FTs of the signal and the window function.

A narrow window $w(t)$ in the time domain gives a good time resolution, and a narrow window $W(\omega)$ in the frequency domain gives a good frequency localization. Both $w(t)$ and $W(\omega)$ cannot be made arbitrarily narrow at the same time; hence there is an inherent tradeoff between the time and the frequency localizations in the spectrogram for a particular window. Sometimes it is difficult to decide which window is the best as it is possible that one window works well with some parts of the signal while other windows work well with other parts. One improvement for the STFT is the adaptive choice of the window function for different parts of the signal, which is known as the adaptive STFT [13].

The spectrogram of a signal generally does not depend the signal solely because the STFT entangles the signal and the window. Therefore we must be cautious in interpreting the STFT to disentangle the window. In fact, because of the product relation between the window function and the signal, we have to be careful to avoid using the signal to study the window. The distinction must come only from a judicious choice of the window, which is totally under the control of the user.

Examples

For many cases, such as signals with smooth frequency variation, the signals give excellent time-frequency structures consistent with our intuitive understanding. Figure 9.1 gives the discrete Fourier transform (DFT) and STFT of a synthetic signal containing one chirp and one constant frequency component,

$$f_1(t) = e^{j2\pi 100(1+0.5t)t} + e^{j2\pi 100t}, \quad t \in [0, 2]. \tag{9.18}$$

The STFT can reveal the chirp and the constant frequency component, but the FT fails to provide enough information for the chirp component. The STFT gives a blurred frequency representation of the chirp component but a sharp one for the constant frequency, which means that the STFT is more suitable for signals with a slow frequency variation. The difference between the STFT of constant frequency and the chirp components also means that it is possible to make improvements on the concentration of the STFT for chirp signal processing.

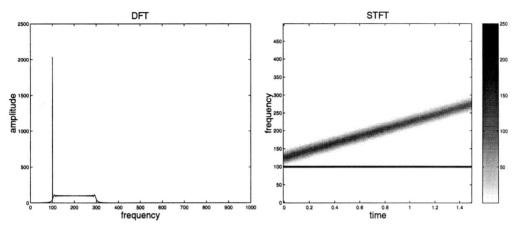

Figure 9.1   The DFT and STFT of $f_1(t)$.

9.2.3   WD

The WD of a signal $s(t)$ is

$$\begin{aligned}
\mathrm{WD}(t,\omega) &= \int_{-\infty}^{+\infty} s^*(t - \tfrac{1}{2}\tau)s(t + \tfrac{1}{2}\tau)e^{-j\tau\omega}d\tau \\
&= \int_{-\infty}^{+\infty} S^*(\omega - \tfrac{1}{2}\theta)S(\omega + \tfrac{1}{2}\theta)e^{-jt\theta}d\theta,
\end{aligned} \tag{9.19}$$

where $S(\omega)$ is the FT of $s(t)$. Because the signal appears twice in (9.19), the WD is known to be quadratic. Another important point is that the distribution is highly non-local because (9.19) has the same weight on the signal at any time instant.

The WD satisfies the time-frequency marginals

$$\int_{-\infty}^{+\infty} \mathrm{WD}(t,\omega)d\omega = |s(t)|^2 \tag{9.20}$$

$$\int_{-\infty}^{+\infty} \mathrm{WD}(t,\omega)dt = |S(\omega)|^2. \tag{9.21}$$

Therefore, the total energy condition is also automatically satisfied,

$$E = \int_{-\infty}^{+\infty}\int_{-\infty}^{+\infty} \mathrm{WD}(t,\omega)d\omega dt = \int_{-\infty}^{+\infty} |s(t)|^2 dt. \tag{9.22}$$

As a bilinear transform, the quadratic superposition principle applies to the WD, that is, for a signal as the sum of two components,

$$s(t) = s_1(t) + s_2(t), \tag{9.23}$$

we have

$$\mathrm{WD}_s(t,\omega) = \mathrm{WD}_{s1}(t,\omega) + \mathrm{WD}_{s2}(t,\omega) + 2\mathrm{Re}\{\mathrm{WD}_{s1,s2}(t,\omega)\}, \tag{9.24}$$

where

$$\mathrm{WD}_{s1,s2}(t,\omega) = \int_{-\infty}^{+\infty} s_1(t + \tau/2)s_2^*(t - \tau/2)e^{-j\omega\tau}d\tau \tag{9.25}$$

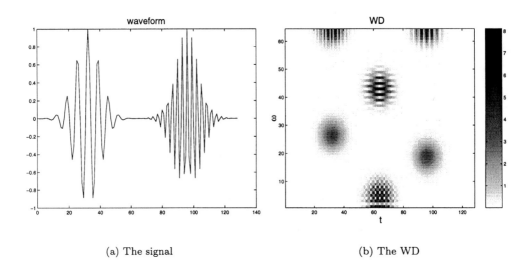

(a) The signal   (b) The WD

Figure 9.2   The signal and its WD.

is the cross-WD of $s_1(t)$ and $s_2(t)$. The WD, as given in (9.24), of the signal containing two components has an additional term which is the cross-term interference, as shown in Figure 9.2, in which the waveform shows that the signal contains two components and the cross-term interference is between the WDs of the components.

Because the cross-term interference usually oscillates and its magnitude is usually larger than that of the auto-terms, it often obscures the useful time dependent spectrum patterns. Reducing of the cross-term interference without destroying the useful properties of the WD becomes an important issue. The major motivations for the smoothing WD are that for certain types of smoothing, a positive distribution is obtained and the cross-term interference is suppressed. The fundamental idea is to smooth the WD by using a low pass 2D filter $H(t, \omega)$,

$$\text{SWD}(t, \omega) = \int_{-\infty}^{+\infty} \int_{-\infty}^{+\infty} \text{WD}(t', \omega') H(t' - t, \omega' - \omega) dt' d\omega'. \tag{9.26}$$

Usually, if the signal is simple, the low pass filter can substantially suppress the cross-terms. However, smoothing will reduce the resolution and become less effective for complex signals. A tradeoff exists between the degree of smoothing and resolution. For complex signals such as voice, smoothing is useless because of the overlap of the spectrum and the cross-term interference.

The pseudo Wigner distribution (PWD) is defined as

$$\text{PWD}(t, \omega) = \int_{-\infty}^{+\infty} s(t + \frac{\tau}{2}) s^*(t - \frac{\tau}{2}) w(\tau) e^{-j\omega\tau} d\tau, \tag{9.27}$$

which is a windowed WD to improve the time location of the WD. Because of its definition, the PWD shares some properties of the WD, such as the high concentration for chirp signals and the cross-term interference. The window function also affects the performance of the PWD. Similar to the STFT, the resolution is reduced by the window function in (9.27) and the marginal conditions are no longer satisfied. Several methods [12, 27, 28, 37, 38]

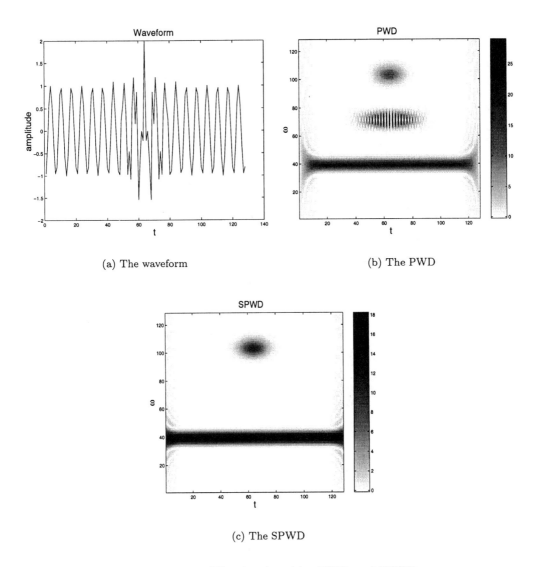

(a) The waveform

(b) The PWD

(c) The SPWD

Figure 9.3   The signal and its PWD and SPWD.

were introduced to suppress the interference. As an example, the smoothed pseudo Wigner distribution (SPWD) [26] is defined as

$$\text{SPWD}(t,\omega) = \int_{-\infty}^{+\infty} w(\tau) \int_{-\infty}^{+\infty} g(t'-t)s(t'+\tau/2)s^*(t'-\tau/2)dt'e^{-j\omega\tau}d\tau, \qquad (9.28)$$

where $g(t)$ is the filter. When $g(t) = \delta(t)$, the PWD is obtained. Figure 9.3 shows the signal and its PWD and SPWD. The concentration of the PWD is high, but sometimes the cross-term interference makes the interpretation of the results difficult. Additionally, low pass filters can suppress the cross-term interference and do not work well for complex signals.

9.2.4   Fractional Fourier transform

The fractional Fourier transform (FRFT) is selected to be a main tool for the time-frequency analysis introduced in this chapter. The FRFT was developed and treated as a rotation of the signal in the time-frequency plane for arbitrary angles [1]. For example, the FT of a signal can be interpreted as a rotation of the signal by an angle of $\pi/2$. The FRFT has been widely applied in the areas of quantum mechanics [32, 37] and signal processing [23, 25]. The transform kernel of the FRFT is defined as

$$K_\alpha(t,u) = \begin{cases} \sqrt{\dfrac{1-j\cot\alpha}{2\pi}}e^{j\frac{t^2+u^2}{2}\cot\alpha-jut\csc\alpha}, & \text{if } \alpha \text{ is not a multiple of } \pi \\ \delta(t-u), & \text{if } \alpha \text{ is a multiple of } 2\pi \\ \delta(t+u), & \text{if } \alpha+\pi \text{ is a multiple of } 2\pi \end{cases} \qquad (9.29)$$

where α is the rotation angle. The kernel has the following properties to support the rotation operation [1]:

$$K_\alpha(t,u) = K_\alpha(u,t) \qquad (9.30)$$

$$K_{-\alpha}(t,u) = K_\alpha^\star(u,t) \qquad (9.31)$$

$$K_\alpha(-t,u) = K_\alpha(t,-u) \qquad (9.32)$$

$$\int_{-\infty}^{+\infty} K_\alpha(t,u)K_\beta(u,z)du = K_{\alpha+\beta}(t,z) \qquad (9.33)$$

$$\int_{-\infty}^{+\infty} K_\alpha(t,u)K_\alpha^\star(t,u')dt = \delta(u-u'). \qquad (9.34)$$

The FRFT of $x(t)$ is

$$\begin{aligned} X_\alpha(u) &= \int_{-\infty}^{+\infty} x(t)K_\alpha(t,u)dt = \mathcal{F}_\alpha[x(t)] \\ &= \begin{cases} \sqrt{\dfrac{1-j\cot\alpha}{2\pi}}e^{j\frac{u^2}{2}\cot\alpha} \\ \quad \cdot \int_{-\infty}^{+\infty} x(t)e^{j\frac{t^2}{2}\cot\alpha}e^{-jut\csc\alpha}dt, & \text{if } \alpha \text{ is not a multiple of } \pi \\ x(t), & \text{if } \alpha \text{ is a multiple of } 2\pi \\ x(-t), & \text{if } \alpha+\pi \text{ is a multiple of } 2\pi. \end{cases} \end{aligned} \qquad (9.35)$$

where $\mathcal{F}_\alpha$ is the operator of the FRFT. Some properties and the FRFTs of a few common signals are listed in Table 9.1 [1]. When $\alpha = \frac{\pi}{2}$, the FRFT becomes the FT, which is

$$X_{\frac{\pi}{2}}(u) = \frac{1}{\sqrt{2\pi}}\int_{-\infty}^{+\infty} x(t)e^{-jut}dt. \qquad (9.36)$$

Considering (9.33), we also have

$$X_{\frac{\pi}{2}+\alpha}(u) = \frac{1}{\sqrt{2\pi}}\int_{-\infty}^{+\infty} X_\alpha(t)e^{-jut}dt. \qquad (9.37)$$

Table 9.1 The properties of the FRFT and transforms of some common signals.

	Signal	Fractional Fourier transform
1	$x(t-\tau)$	$X_\alpha(u-\tau\cos\alpha)e^{j\frac{\tau^2}{2}\sin\alpha\cos\alpha-ju\tau\sin\alpha}$
2	$x(t)e^{jvt}$	$X_\alpha(u-v\sin\alpha)e^{-j\frac{v^2}{2}\sin\alpha\cos\alpha+juv\cos\alpha}$
3	$x'(t)$	$X'_\alpha(u)\cos\alpha+juX_\alpha(u)\sin\alpha$
4*	$\int_0^t x(t')dt'$	$\sec\alpha e^{-j\frac{u^2}{2}\tan\alpha}\int_0^u X_\alpha(z)e^{j\frac{z^2}{2}\tan\alpha}dz$
5	$tx(t)$	$uX_\alpha(u)\cos\alpha+jX'_\alpha(u)\sin\alpha$
6**	$\frac{x(t)}{t}$	$-j\sec\alpha e^{j\frac{u^2}{2}\cot\alpha}\int_{-\infty}^u x(z)e^{-j\frac{z^2}{2}\cot\alpha}dz$
7	$x(-t)$	$X_\alpha(-u)$
8	$X_\beta(t)$	$X_{\alpha+\beta}(u)$
9	$ax(t)+by(t)$	$aX_\alpha(u)+bY_\alpha(u)$, a and b are constants
10	$x(ct)$	$\sqrt{\frac{1-j\cot\alpha}{c^2-j\cot\alpha}}e^{j\frac{u^2}{2}\cot\alpha\left(1-\frac{\cos^2\beta}{\cos^2\alpha}\right)}X_\beta(u\frac{\sin\beta}{c\sin\alpha})$ $\beta=\arctan(c^2\tan\alpha)$
11**	$\delta(t-\tau)$	$\sqrt{\frac{1-j\cot\alpha}{2\pi}}e^{j\frac{\tau^2+u^2}{2}\cot\alpha-ju\tau\csc\alpha}$
12*	1	$\sqrt{1+j\tan\alpha}\,e^{-j\frac{u^2}{2}\tan\alpha}$
13*	e^{jvt}	$\sqrt{1+j\tan\alpha}\,e^{-j\frac{v^2+u^2}{2}\tan\alpha+juv\sec\alpha}$
14+	$e^{j\frac{c}{2}t^2}$	$\sqrt{\frac{1+j\tan\alpha}{1+c\tan\alpha}}e^{j\frac{u^2}{2}\frac{c-\tan\alpha}{1+c\tan\alpha}}$
15	$e^{-\frac{t^2}{2}}$	$e^{-\frac{u^2}{2}}$
16	$H_n(t)e^{-\frac{t^2}{2}}$	$e^{-n\alpha}H_n(u)e^{-j\frac{u^2}{2}}$ $H_n(t)$ is Hermite polynomial.
17	$e^{-c\frac{t^2}{2}}$	$\sqrt{\frac{1-j\cot\alpha}{c-j\cot\alpha}}e^{j\frac{u^2}{2}\frac{(c^2-1)\cot\alpha}{c^2+\cot^2\alpha}}e^{-\frac{u^2}{2}\frac{c\csc^2\alpha}{c^2+\cot^2\alpha}}$

* if $\alpha-\frac{\pi}{2}$ is not a multiple of π. + if $\alpha-\arctan c-\frac{\pi}{2}$ is not a multiple of π.
** if α is not a multiple of π.

If we treat α as a variable, (9.35) can be rewritten as

$$X(\alpha,u) = \int_{-\infty}^{+\infty} x(t)K_\alpha(t,u)dt$$

$$= \begin{cases} \sqrt{\frac{1-j\cot\alpha}{2\pi}}e^{j\frac{u^2}{2}\cot\alpha} \\ \quad \cdot\int_{-\infty}^{+\infty} x(t)e^{j\frac{t^2}{2}\cot\alpha}e^{-jut\csc\alpha}dt, & \text{if } \alpha \text{ is not a multiple of } \pi \quad (9.38) \\ x(t), & \text{if } \alpha \text{ is a multiple of } 2\pi \\ x(-t), & \text{if } \alpha+\pi \text{ is a multiple of } 2\pi. \end{cases}$$

In this chapter, $X_\alpha(t)$ represents the FRFT of $x(t)$ for a particular angle α, and $X(\alpha,u)$ represents the FRFT of $x(t)$ as a function of α.

9.3 HARMONIC TRANSFORM

The STFT is the most widely used method for studying non-stationary signals. For certain situations, however, the STFT may not be the best method available. This section attempts

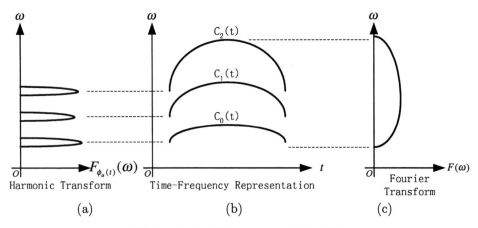

Figure 9.4   The concept of the HT.

to make use of the harmonic transform (HT) to achieve improvements for harmonic signals (e.g., voiced speech and music), which are composed of the fundamental and harmonics. We know that the FT decomposes the signals into the sum of a number of sinusoidal components and can be loosely interpreted as the integral of the signal in the time-frequency domain along the time axis. The difference between the HT and FT is the harmonic kernel $e^{j\omega\phi_u(t)}$, which can be loosely interpreted as the integration of the signal along the curves $\omega\phi'_u(t)$, where $\phi'_u(t)$ is the derivative of $\phi_u(t)$. When $\phi_u(t) = t$, the HT becomes the FT.

The HT decomposes the signal into a fundamental and harmonics. Compared with the FT, it gives a more concise representation of harmonic signals. This section considers the definition of the continuous HT and its discrete calculation. The advantages of the HT are demonstrated by a few examples of synthetic signals and speech signals. The application of the HT for time-frequency analysis is the short-time harmonic transform (STHT), which is the result of replacing the FT used in the STFT with the HT. Examples show that the STHT of a voiced speech signal has a higher concentration than the STFT.

9.3.1   Description of HT

The FT, a pervasive and versatile method, is used in many fields of science to map a problem into the frequency domain in which it can be more easily solved. The FT is effective for signals consisting of elements with fixed frequencies. The resultant representation in the frequency domain exhibits a number of sharp peaks that succinctly describe the intensities and the associated phases in terms of frequencies. The original signal can be reconstructed from these peaks. These characteristics show that the FT effectively captures the *essence* of this class of signals. The analytical power of the FT, however, drops quickly when the frequencies of signal components become time variant. The resultant Fourier spectrum will contain overlapping lumps that reveal little useful information about the signals.

The HT can make substantial improvements on the performance of the FT for signals containing variable frequencies, such as voiced speech and music. In the time-frequency plane, the HT is interpreted as the integrals of the signal along the nominal frequencies of the fundamental and harmonics. This interpretation is illustrated in Figure 9.4. The HT is also equivalent to transforming the fundamental and harmonics to constant frequency components and then calculating the FT of the processed signal. The most concise way to describe the spectrum of the signal is to follow its pattern, calculating the sum of the

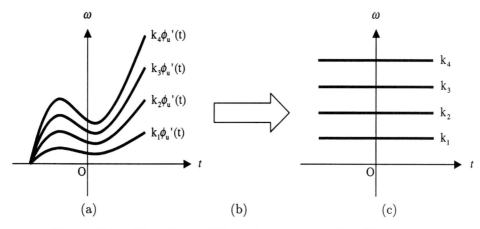

Figure 9.5   The effect of HT on the fundamental and harmonics.

fundamental and its harmonics separately. In practice, we can convert the fundamental to a constant frequency component by using the frequency function of the fundamental in the kernel of the FT. This computation should preserve the harmonical relation between components, that is, every component has constant frequency after the computation. Figure 9.5 shows this principle.

Definition
The HT of a signal $f(t)$ is defined as

$$F_{\phi_u(t)}(\omega) = \int_{-\infty}^{+\infty} f(t)\phi_u'(t)e^{-j\omega\phi_u(t)}dt, \tag{9.39}$$

where $\phi_u(t)$ is known as the unit phase function and $\phi_u'(t)$ is the derivative of $\phi_u(t)$. The unit phase function is the phase function of the fundamental of $f(t)$ divided by a constant. The constant is the instantaneous frequency of the fundamental at time $t = 0$. The constant can be other values, for it only scales the HT and has no impact on its concentration. We note that the HT becomes the FT when $\phi_u(t) = t$. The inverse HT (IHT) is

$$f(t) = \frac{1}{2\pi} \int_{-\infty}^{+\infty} F_{\phi_u(t)}(\omega)e^{j\omega\phi_u(t)}d\omega. \tag{9.40}$$

The notation for the HT and its inverse with unit phase function $\phi_u(t)$ are $\mathcal{F}_{\phi_u(t)}$ and $\mathcal{F}_{\phi_u(t)}^{-1}$, respectively. That is,

$$F_{\phi_u(t)}(\omega) = \mathcal{F}_{\phi_u(t)}[f(t)], \quad f(t) = \mathcal{F}_{\phi_u(t)}^{-1}[F_{\phi_u(t)}(\omega)]. \tag{9.41}$$

We now prove the existence of the HT. Let $\phi_u^{-1}(z)$ be the inverse function of $\phi_u(t)$; we have $z = \phi_u(t)$ and $t = \phi_u^{-1}(z)$. Substituting $t = \phi_u^{-1}(z)$ and $z = \phi_u(t)$ in (9.39) gives

$$\begin{aligned}
F_{\phi_u(t)}(\omega) &= \int_{-\infty}^{+\infty} f(t)e^{-j\omega\phi_u(t)}d\phi_u(t) \\
&= \int_{-\infty}^{+\infty} f(\phi_u^{-1}(z))e^{-j\omega z}dz.
\end{aligned} \tag{9.42}$$

Substituting $t = \phi_u^{-1}(z)$ and $z = \phi_u(t)$ in (9.40), we obtain

$$
\begin{aligned}
f(\phi_u^{-1}(z)) &= \frac{1}{2\pi} \int_{-\infty}^{+\infty} F_{\phi_u(t)}(\omega) e^{j\omega\phi_u(t)} d\omega \\
&= \frac{1}{2\pi} \int_{-\infty}^{+\infty} F_{\phi_u(t)}(\omega) e^{j\omega z} d\omega.
\end{aligned}
\tag{9.43}
$$

Comparing (9.42) and (9.43) with the definition of the FT and its inverse FT, respectively, we can see that (9.42) and (9.43) are the FT and the IFT of $f(\phi_u^{-1}(z))$, respectively. The existence of the FT of $f(\phi_u^{-1}(z))$ guarantees the existence of the HT and IHT defined by (9.39) and (9.40). In other words, $f(\phi_u^{-1}(z))$ should have finite energy. When $f(t)$ is defined over a finite range, $\phi_u(t)$ only needs to be invertible within this range instead of $(-\infty, +\infty)$. Assuming that $f_r(t)$ is a signal within a finite duration (a, b), that is,

$$
f_r(t) = 0, \quad t \notin (a, b)
\tag{9.44}
$$

and $\phi_u(t)$ is invertible in (a, b), function $\phi_{ur}(t)$ is invertible in $(-\infty, +\infty)$ and satisfies

$$
\phi_{ur}(t) = \phi_u(t), \quad \phi_{ur}'(t) = \phi_u'(t), \quad t \in (a, b).
\tag{9.45}
$$

Then the HT of $f_r(t)$ with $\phi_{ur}(t)$ is

$$
\begin{aligned}
F_{r\phi_{ur}(t)}(\omega) &= \int_{-\infty}^{+\infty} f_r(t)\phi_{ur}'(t) e^{-j\omega\phi_{ur}(t)} dt \\
&= \int_a^b f_r(t)\phi_{ur}'(t) e^{-j\omega\phi_{ur}(t)} dt \quad \text{using (9.44)} \\
&= \int_a^b f_r(t)\phi_u'(t) e^{-j\omega\phi_u(t)} dt \quad \text{using (9.45)} \\
&= \int_{-\infty}^{+\infty} f_r(t)\phi_u'(t) e^{-j\omega\phi_u(t)} dt \quad \text{using (9.44).}
\end{aligned}
\tag{9.46}
$$

The IHT with $\phi_{ur}(t)$ is

$$
\begin{aligned}
f_r(t) &= \frac{1}{2\pi} \int_{-\infty}^{+\infty} F_{r\phi_{ur}(t)}(\omega) e^{j\omega\phi_{ur}(t)} d\omega \\
&= \begin{cases} \frac{1}{2\pi} \int_{-\infty}^{+\infty} F_{r\phi_{ur}(t)}(\omega) e^{j\omega\phi_{ur}(t)} d\omega, & t \in (a, b) \\ 0, & \text{otherwise} \end{cases} \\
&= \begin{cases} \frac{1}{2\pi} \int_{-\infty}^{+\infty} F_{r\phi_{ur}(t)}(\omega) e^{j\omega\phi_u(t)} d\omega, & t \in (a, b) \\ 0, & \text{otherwise.} \end{cases}
\end{aligned}
\tag{9.47}
$$

Replacing $F_{r\phi_{ur}(t)}(\omega)$ with $F_{r\phi_u(t)}(\omega)$ gives the HT and its inverse of the finite signal $f_r(t)$,

$$
F_{r\phi_u(t)}(\omega) = \int_{-\infty}^{+\infty} f_r(t)\phi_u'(t) e^{-j\omega\phi_u(t)} dt
\tag{9.48}
$$

$$
f_r(t) = \begin{cases} \frac{1}{2\pi} \int_{-\infty}^{+\infty} F_{r\phi_u(t)}(\omega) e^{j\omega\phi_u(t)} d\omega, & t \in (a, b) \\ 0, & \text{otherwise} \end{cases}.
\tag{9.49}
$$

A comparison of (9.39) and (9.40) with (9.48) and (9.49) shows that the only difference between the HTs of infinite signals and the HTs of finite signals is that the IHTs of finite signals can only be used to recover the finite signals within their duration.

The following example gives the HT of $f_e(t)$ which is a varying frequency signal composed of one fundamental and harmonics,

$$f_e(t) = \sum_{k=1}^{M} c_k e^{jk\alpha(t)}, \tag{9.50}$$

where c_k $(k = 1 \ldots M)$ are constant coefficients and $\alpha(t)$ is the phase function of the fundamental. For such a signal, its FT is not as concise as that of constant frequency signals, since $\alpha(t)$ may not be a linear function. However, this problem can be solved by the HT. When $\phi_u(t) = \alpha(t)$, the HT of $f_e(t)$ is

$$F_{e\alpha}(\omega) = \sum_{k=1}^{M} c_k \delta(\omega - k), \tag{9.51}$$

which contains δ functions for any $\alpha(t)$. The corresponding δ functions in $F_{e\alpha}(\omega)$ also share the same coefficients with the fundamental and the harmonics. This example shows that an HT with a suitable $\phi_u(t)$ can decompose a varying frequency signal into a sum of a fundamental and harmonics to provide a concise spectrum.

Properties
Existence condition – The existence condition of the HT requires that the product of the signal $f(t)$ and $\phi'_u(t)$ should be absolutely integrable over the entire range

$$\int_{-\infty}^{+\infty} |f(t)||\phi'_u(t)| dt < M_T, \tag{9.52}$$

where M_T is a positive constant. This can be proved through the existence condition of the FT, which is

$$\int_{-\infty}^{+\infty} |f(t)| dt < M_T \tag{9.53}$$

and $f(t)$ must be well behaved. Applying this existence condition to (9.42) gives

$$\int_{-\infty}^{+\infty} |f(\phi_u^{-1}(t))| dz = \int_{-\infty}^{+\infty} |f(t) \phi'_u(t)| dt < M_T. \tag{9.54}$$

Time domain convolution – Time domain convolution implies that the HT of the convolution of two signals is the product of the HTs of the corresponding signals. Since $t = \phi_u^{-1}(z)$ and $z = \phi_u(t)$, we have the following proof using the property of the FT for the time domain convolution:

$$\int_{-\infty}^{+\infty} x(t) * y(t)\phi_u'(t)e^{-j\omega\phi_u(t)}dt$$

$$= \int_{-\infty}^{+\infty} x(\phi_u^{-1}(z)) * y(\phi_u^{-1}(z))e^{-j\omega z}dz$$

$$= \int_{-\infty}^{+\infty} x(\phi_u^{-1}(z))e^{-j\omega z}dz \int_{-\infty}^{+\infty} y(\phi_u^{-1}(z))e^{-j\omega z}dz \qquad (9.55)$$

$$= \int_{-\infty}^{+\infty} x(t)\phi_u'(t)e^{-j\omega\phi_u(t)}dt \int_{-\infty}^{+\infty} y(t)\phi_u'(t)e^{-j\omega\phi_u(t)}dt$$

$$= X_{\phi_u(t)}(\omega)Y_{\phi_u(t)}(\omega),$$

where $x(t)$, $y(t)$, $X_{\phi_u(t)}(\omega)$ and $Y_{\phi_u(t)}(\omega)$ are signals and their HTs with $\phi_u(t)$, respectively.

Energy property – It is interesting to note that unlike the FT, the HT does not preserve the energy of the signal, which can be proved with Parseval's theorem,

$$\int_{-\infty}^{+\infty} |x(t)|^2 dt = \int_{-\infty}^{+\infty} |X(\omega)|^2 d\omega, \qquad (9.56)$$

where $X(\omega)$ is the FT of $x(t)$. Using Parseval's theorem with (9.42) gives

$$\int_{-\infty}^{+\infty} |F_{\phi_u(t)}(\omega)|^2 d\omega = \int_{-\infty}^{+\infty} |f(\phi_u^{-1}(z))|^2 dz$$

$$= \int_{-\infty}^{+\infty} |f(\phi_u^{-1}(\phi_u(t)))|^2 d\phi_u(t) \qquad (9.57)$$

$$= \int_{-\infty}^{+\infty} |f^2(t)\phi_u'(t)|dt.$$

This is the energy property of the HT, in which $\phi_u(t)$ becomes a contributive factor. Apart from the properties described above, the HT also has a number of useful properties listed in Table 9.2. It is noted that most of these properties are the extensions of the corresponding properties of the FT. However, not all properties of the FT can be extended to the HT. For example, the scaling property, which is not listed in the table, depends on the function $\phi_u(t)$ and is determined by the relation between $\phi_u(t)$ and $\phi_u(\frac{t}{a})$, where a is a constant.

It is not difficult to prove these properties. The linear property can be obtained from the definitions in (9.39) and (9.40). Others can be proved via the counterparts of the FT and the relation between the FT and HT as described in (9.42) and (9.43). In summary, the HT is an extension of the FT and inherits some properties from the FT. Certain properties of the HT depend on the properties of the phase function $\phi_u(t)$.

9.3.2   Discrete calculation

Harmonic transform

A discrete calculation is needed to compute the discrete form of $F_{\phi_u(t)}(\omega)$ when $f(t)$ and

Table 9.2   The properties of the HT.

Property	Signal	Harmonic Transform	Conditions
Linearity *	$ax(t) + by(t)$	$aX_{\phi_u(t)}(\omega) + bY_{\phi_u(t)}(\omega)$	-
Decomposition	$x_e(t) + x_o(t)$, where $x_e(t) = \frac{1}{2}[x(t) + x(-t)]$ $x_o(t) = \frac{1}{2}[x(t) - x(-t)]$	$X_e(\omega) + X_o(\omega)$, where $X_e(\omega) = \mathrm{Re}[X_{\phi_u(t)}(\omega)]$ $X_o(\omega) = j\mathrm{Im}[X_{\phi_u(t)}(\omega)]$	odd
Time inverse	$x(-t)$	$X_{\phi_u(t)}(-\omega)$	odd
Conjugation	$x^*(t)$	$X^*_{\phi_u(t)}(\omega)$	-
Convolution	$x(t) * y(t)$	$X_{\phi_u(t)}(\omega)Y_{\phi_u(t)}(\omega)$	-
	$x(t)y(t)$	$X_{\phi_u(t)}(\omega) * Y_{\phi_u(t)}(\omega)$	-
Symmetry	$\frac{X_{\phi_u(t)}(t)}{\phi'_u(t)}$	$x(-\omega)$	-
Delta function	$\delta(t - t_0)$	$\phi'_u(t_0)e^{-j\omega\phi_u(t_0)}$	-
	$e^{j\omega_0\phi_u(t)}$	$\delta(\omega - \omega_0)$	-
Energy	$\int_{-\infty}^{+\infty} |x(t)|^2\phi'_u(t)dt$	$\int_{-\infty}^{+\infty} |X_{\phi_u(t)}(\omega)|^2 d\omega$	-

$X_{\phi_u(t)}(\omega) = \mathcal{F}_{\phi_u(t)}[x(t)]$, $Y_{\phi_u(t)}(\omega) = \mathcal{F}_{\phi_u(t)}[y(t)]$
* a and b are constants.

$\phi_u(t)$ are in discrete form, $f(n)$ and $\phi_u(n)$, which are the results of sampling $f(t)$ and $\phi_u(t)$:

$$f(n) = f(t)\Big|_{t=n\triangle t}, \qquad n \in [0, N-1] \tag{9.58}$$

$$\phi_u(n) = \phi_u(t)\Big|_{t=n\triangle t}, \qquad n \in [0, N-1], \tag{9.59}$$

where $\triangle t$ is the sample period. In this equation, the variable t of $f(t)$ and $\phi_u(t)$ is limited to the range $[0, T]$, where $T = N\triangle t$. A time shift is needed if the range of t is not equal to $[0, T]$. $\phi_u(n)$ should satisfy

$$\phi_u(0) = 0 \tag{9.60}$$

$$\phi_u(N-1) = (N-1)\triangle t. \tag{9.61}$$

If $\phi_{us}(n)$, the result of sampling $\phi_{us}(t)$, does not satisfy this condition, a linear transform

$$\phi_u(n) = c[\phi_{us}(n) - \phi_{us}(0)],$$

where

$$c = \frac{(N-1)\triangle t}{\phi_{us}(N-1) - \phi_{us}(0)}.$$

is applied to $\phi_{us}(n)$ and the result $\phi_u(n)$ will satisfy (9.60) and (9.61). In the continuous domain, the linear transform is

$$\phi_u(t) = c[\phi_{us}(t) - \phi_{us}(0)]. \tag{9.62}$$

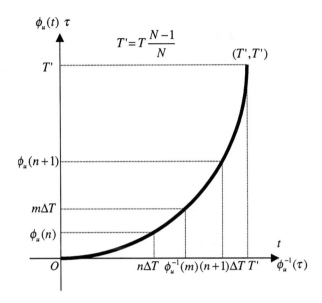

Figure 9.6   The relation between $\phi_u(t)$ and $\phi_u^{-1}(t)$.

The impact of the linear transform on the HT of $f(t)$ means a scaling and phase shifting operation because

$$
\begin{aligned}
F_{\phi_u(t)}(\omega) &= \int_{-\infty}^{+\infty} f(t)\phi_u'(t)e^{-j\omega\phi_u(t)}dt \\
&= \int_{-\infty}^{+\infty} c\, f(t)\, \phi_u'(t)e^{-j\omega c[\phi_{us}(t)-\phi_{us}(0)]}dt \\
&= ce^{jc\omega\phi_{us}(0)} \int_{-\infty}^{+\infty} \phi_{us}'(t)e^{-jc\omega\phi_{us}(t)}dt \\
&= ce^{jc\omega\phi_{us}(0)} F_{\phi_{us}(t)}(c\omega)
\end{aligned}
\tag{9.63}
$$

The sample of $\phi_u^{-1}(t)$ is

$$
\phi_u^{-1}(n) = \phi_u^{-1}(t)\Big|_{t=n\triangle t}, \qquad n \in [0, N-1].
\tag{9.64}
$$

From (9.60) and (9.61) we have

$$
\phi_u^{-1}(0) = 0
\tag{9.65}
$$
$$
\phi_u^{-1}(N-1) = (N-1)\triangle t.
\tag{9.66}
$$

The relationship between $\phi_u(t)$ and $\phi_u^{-1}(t)$ is shown in Figure 9.6. With the sampling theorem, $\phi_u^{-1}(t)$ can be recovered from $\phi_u^{-1}(n)$,

$$
\phi_u^{-1}(t) = \sum_{n=0}^{N-1} \phi_u^{-1}(n)\mathrm{sinc}\left[\frac{\pi(t-n\triangle t)}{\triangle t}\right].
\tag{9.67}
$$

Since $\phi_u^{-1}(t)$ is the inverse function of $\phi_u(t)$

$$
\phi_u^{-1}(\phi_u(t)) = t,
\tag{9.68}
$$

we have

$$k\Delta t = \sum_{n=0}^{N-1} \phi_u^{-1}(n)\mathrm{sinc}\left[\frac{\pi(\phi_u(k) - n\Delta t)}{\Delta t}\right], \qquad k \in [0, N-1]. \tag{9.69}$$

Expressing the above equation into a matrix form,

$$\mathbf{C}_{N\times 1} = \mathbf{S}_{N\times 1}\mathbf{B}_{N\times 1}, \tag{9.70}$$

where

$$\mathbf{C}_{N\times 1} = \begin{bmatrix} 0 & \Delta t & \cdots & (N-1)\Delta t \end{bmatrix}' \tag{9.71}$$

$$\mathbf{B}_{N\times 1} = \begin{bmatrix} \phi_u^{-1}(0) & \phi_u^{-1}(1) & \cdots & \phi_u^{-1}(N-1) \end{bmatrix}' \tag{9.72}$$

$$\mathbf{S}_{N\times N} = \begin{bmatrix} s_{0,0} & s_{0,1} & \cdots & s_{0,N-1} \\ s_{1,0} & s_{1,1} & \cdots & s_{1,N-1} \\ \cdots & \cdots & \cdots & \cdots \\ s_{N-1,0} & s_{N-1,1} & \cdots & s_{N-1,N-1} \end{bmatrix} \tag{9.73}$$

$$s_{n,m} = \mathrm{sinc}\left\{\frac{\pi[\phi_u(m) - n\Delta t]}{\Delta t}\right\}. \tag{9.74}$$

Since there exists one and only one $\mathbf{B}_{N\times 1}$, we have

$$\mathbf{B}_{N\times 1} = \mathbf{S}_{N\times N}^{-1}\mathbf{C}_{N\times 1} \tag{9.75}$$

with which $\phi_u^{-1}(n)$ can be obtained from $\phi_u(n)$.

Since the relationship between $f(t)$ and $f(n)$ is

$$f(t) = \sum_{m=0}^{N-1} f(m)\mathrm{sinc}\left[\frac{\pi(t - m\Delta t)}{\Delta t}\right], \tag{9.76}$$

the sampling result of $f(\phi_u^{-1}(t))$, written as $f(\phi_u^{-1}(n))$, is

$$f(\phi_u^{-1}(n)) = f(\phi_u^{-1}(t))\Big|_{t=n\Delta t}, \qquad n \in [0, N-1], \tag{9.77}$$

which can be rewritten as

$$f(\phi_u^{-1}(n)) = \sum_{m=0}^{N-1} f(m)\mathrm{sinc}\left\{\frac{\pi\left[\phi_u^{-1}(n) - m\Delta t\right]}{\Delta t}\right\}, \qquad n \in [0, N-1]. \tag{9.78}$$

Then the HT of $f(t)$ or the FT of $f(\phi_u^{-1}(t))$ is

$$F_{\phi_u(n)}(\omega) = \begin{cases} F_s(\omega), & \omega \in [0, \Omega) \\ F_s(\omega + 2\Omega), & \omega \in [-\Omega, 0) \end{cases}, \tag{9.79}$$

where for $\Delta t = \frac{T}{N} = \frac{1}{\Omega}$,

$$F_s(\omega) = \frac{\Delta t}{2\pi}\sum_{k=0}^{N-1} H(k)\mathrm{sinc}\left[\frac{\pi(\omega - k\Delta t)}{\Delta t}\right], \qquad \omega \in [0, 2\Omega) \tag{9.80}$$

$$H(k) = \mathcal{F}\left[f(\phi_u^{-1}(n))\right], \qquad k \in [0, N-1]. \tag{9.81}$$

Equations (9.78) – (9.81) give the discrete calculation of the HT when the signal and the phase function are in discrete form.

Inverse harmonic transform

Next we discuss the discrete calculation of the IHT, that is to obtain $f(t)$ from its HT $F_{\phi_u(t)}(\omega)$. From (9.79), $F_s(\omega)$ can be expressed as

$$F_s(\omega) = \begin{cases} F_{\phi_u(t)}(\omega), & \omega \in [0, \Omega) \\ F_{\phi_u(t)}(\omega - 2\Omega), & \omega \in [\Omega, 2\Omega) \end{cases}. \tag{9.82}$$

Then sampling $F_s(\omega)$ gives

$$F_s(i) = F_s(\omega)\Big|_{\omega = \triangle\omega i}, \qquad i \in [0, N-1], \tag{9.83}$$

where

$$\triangle\omega = \frac{2\Omega}{N}. \tag{9.84}$$

Substituting the above equation into (9.83), we have

$$F_s(i) = \frac{\triangle t}{2\pi} H(i) = \frac{\triangle t}{2\pi} \mathcal{F}\left[f(\phi_u^{-1}(n))\right]. \tag{9.85}$$

Thus we can get $f(\phi_u^{-1}(n))$ as

$$f(\phi_u^{-1}(n)) = \frac{2\pi}{\triangle t} \mathcal{F}^{-1}\left[F_s(i)\right], \qquad n \in [0, N-1]. \tag{9.86}$$

Since $f(\phi_u^{-1}(n))$ is the sampling result of $f(\phi_u^{-1}(t))$, it can be reconstructed by

$$f(\phi_u^{-1}(t)) = \sum_{k=0}^{N-1} f(\phi_u^{-1}(k)) \mathrm{sinc}\left[\frac{\pi(t - k\triangle t)}{\triangle t}\right]. \tag{9.87}$$

Let $t = \phi_u(n)$ and get

$$f(\phi_u^{-1}(\phi_u(n))) = \sum_{k=0}^{N-1} f(\phi_u^{-1}(k)) \mathrm{sinc}\left\{\frac{\pi[\phi_u(n) - k\triangle t]}{\triangle t}\right\}, \qquad n \in [0, N-1]. \tag{9.88}$$

As we know, $\phi_u(t)$ and $\phi_u^{-1}(t)$ are an inverse function pair, that is,

$$\phi_u(\phi_u^{-1}(t)) = \phi_u^{-1}(\phi_u(t)) = t. \tag{9.89}$$

So we have

$$f(\phi_u^{-1}(\phi_u(t))) = f(t). \tag{9.90}$$

Sampling both sides gives

$$f(\phi_u^{-1}(\phi_u(t)))\Big|_{t=n\triangle t} = f(t)\Big|_{t=n\triangle t}, \qquad n \in [0, N-1]. \tag{9.91}$$

Considering the definitions of $f(n)$ and $\phi_u(n)$ defined in (9.58) and (9.59), the above equation is rewritten as

$$f(\phi_u^{-1}(\phi_u(n))) = f(n), \qquad n \in [0, N-1]. \tag{9.92}$$

Substituting the above equation into (9.88) gives

$$f(n) = \sum_{k=0}^{N-1} f(\phi_u^{-1}(n)) \text{sinc} \left\{ \frac{\pi \left[\phi_u^{-1}(n) - k\Delta t\right]}{\Delta t} \right\}, \qquad n \in [0, N-1]. \tag{9.93}$$

Finally, using the sampling theorem we can recover $f(t)$ from $f(n)$ by

$$f(t) = \sum_{n=0}^{N-1} f(n) \text{sinc} \left[\frac{\pi(t - n\Delta t)}{\Delta t} \right]. \tag{9.94}$$

The discrete calculation of the IHT can be obtained by using (9.93), (9.94) and other related equations.

9.3.3   Discrete harmonic transform

Definition
The discrete harmonic transform (DHT) [1] is defined as the sampled version of the HT. We assume that $s(n)$ is the sampled version of signal $s(t)$ and the sampling rate is normalized to be one. The relation between the DHT and HT should be similar to the relation between the DFT and FT. When $\phi_u(n) = t$, the DHT becomes the DFT. We similarly assume that $s(t)$ has a finite duration $[0, T]$ in the time domain and a bandwidth $[-\Omega, \Omega]$ in the frequency domain. The sampling rate is $\Delta t = \frac{T}{N}$ in the time domain and $\Delta \omega = \frac{2\Omega}{N}$ in the frequency domain. Therefore, the equations in Section 9.3.2 on the sampling results are also valid for the DHT with

$$\Delta t = 1. \tag{9.95}$$

Substituting $\Delta t = 1$ in (9.60), (9.61), (9.65), and (9.66) gives the conditions for the unit phase function $\phi_u(n)$, $n \in [0, N-1]$, and its inverse function $\phi_u^{-1}(n)$, $n \in [0, N-1]$, that is,

$$\phi_u(0) = \phi_u^{-1}(0) = 0, \qquad \phi_u(N-1) = \phi_u^{-1}(N-1) = N-1. \tag{9.96}$$

With (9.75) and (9.95), the relation between $\phi_u(n)$ and $\phi_u^{-1}(n)$ can be written as

$$\mathbf{B}_{N \times 1} = \mathbf{Q}_{N \times N}^{-1} \mathbf{D}_{N \times 1}, \tag{9.97}$$

where

$$\mathbf{D}_{N \times 1} = \begin{bmatrix} 0 & 1 & \cdots & (N-1) \end{bmatrix}' \tag{9.98}$$

$$\mathbf{B}_{N \times 1} = \begin{bmatrix} \phi_u^{-1}(0) & \phi_u^{-1}(1) & \cdots & \phi_u^{-1}(N-1) \end{bmatrix}' \tag{9.99}$$

---

[1]It should not be confused with the discrete Hartley transform discussed in Chapters 4 and 5

$$\mathbf{Q}_{N \times N} = \begin{bmatrix} q_{0,0} & q_{0,1} & \cdots & q_{0,N-1} \\ q_{1,0} & q_{1,1} & \cdots & q_{1,N-1} \\ \cdots & \cdots & \cdots & \cdots \\ q_{N-1,0} & q_{N-1,1} & \cdots & q_{N-1,N-1} \end{bmatrix} \tag{9.100}$$

$$q_{n,m} = \text{sinc}\left\{\pi[\phi_u(m) - n]\right\}. \tag{9.101}$$

Sampling the HT of $f(t)$ in (9.79) gives

$$F_{\phi_u(n)}(i) = F_{\phi_u(t)}(\omega)\Big|_{\omega = i\Delta\omega}, \qquad i \in [-\frac{N}{2}, \frac{N}{2} - 1], \tag{9.102}$$

where N is even. A similar derivation can be easily obtained when N is odd, which we will not repeat. Like the DFT, $F_{\phi_u(n)}(i)$ is assumed to be periodic. We obtain

$$F_{\phi_u(n)}(i) = F_s(\omega)\Big|_{\omega = i\Delta\omega} = \frac{\Delta t}{2\pi} H(i) = \frac{\Delta t}{2\pi} \mathcal{F}[f(\phi_u^{-1}(n))], \qquad i \in [0, N-1]. \tag{9.103}$$

By moving 2π to the left side and setting $\Delta t = 1$ in the above equation, we have

$$2\pi F_{\phi_u(n)}(i) = \mathcal{F}[f(\phi_u^{-1}(n))] = \mathcal{F}[f_{\phi_u^{-1}}(n)], \qquad i \in [0, N-1], \tag{9.104}$$

where

$$f_{\phi_u^{-1}}(n) = \sum_{m=0}^{N-1} f(m)\text{sinc}\left\{\pi\left[\phi_u^{-1}(n) - m\right]\right\}. \tag{9.105}$$

In (9.104), $2\pi F_{\phi_u(n)}(i)$ is the DHT of $f(n)$. Substituting $\Delta t = 1$ in (9.93) gives

$$f(n) = \sum_{k=0}^{N-1} f(\phi_u^{-1}(k))\text{sinc}\left\{\frac{\pi}{\Delta t}[\phi_u(n) - k\Delta t]\right\}\Big|_{\Delta t = 1}$$
$$= \sum_{k=0}^{N-1} f_{\phi_u^{-1}}(k)\text{sinc}\left\{\pi[\phi_u(n) - k]\right\}, \qquad n \in [0, N-1]. \tag{9.106}$$

Replacing $f(n)$ with $s(n)$ in (9.104), (9.105), and (9.106) gives the DHT of $s(n)$ and the corresponding inverse DHT (IDHT),

$$S_{\phi_u(n)}(i) = \mathcal{F}_{\phi_u(n)}[s(n)] = \mathcal{F}[s_{\phi_u^{-1}}(n)] \tag{9.107}$$

$$s(n) = \mathcal{F}_{\phi_u(n)}^{-1}[S_{\phi_u(n)}(i)] = \sum_{k=0}^{N-1} \mathcal{F}^{-1}[S_{\phi_u(n)}(i)]\text{sinc}\left\{\pi[\phi_u(n) - k]\right\}, \tag{9.108}$$

where $i, n \in [0, N-1]$, and

$$s_{\phi_u^{-1}}(n) = \sum_{m=0}^{N-1} s(m)\text{sinc}\left\{\pi\left[\phi_u^{-1}(n) - m\right]\right\}$$
$$= \mathcal{F}^{-1}[S_{\phi_u(n)}(i)]. \tag{9.109}$$

$\mathcal{F}_{\phi_u(n)}$ and $\mathcal{F}_{\phi_u(n)}^{-1}$ are the operators of the DHT and the IDHT with $\phi_u(n)$, respectively. With the matrix notation, (9.107) can be expressed as

$$\mathbf{S}_{\phi_u(n)N \times 1} = \mathbf{F}_{N \times N}\mathbf{P}_{N \times N}\mathbf{s}_{N \times 1}, \tag{9.110}$$

where

$$\mathbf{S}_{\phi_u(n)N\times1} = \begin{bmatrix} S_{\phi_u(n)}(0) & S_{\phi_u(n)}(1) & \cdots & S_{\phi_u(n)}(N-1) \end{bmatrix}' \tag{9.111}$$

$$\mathbf{s}_{N\times1} = \begin{bmatrix} s(0) & s(1) & \cdots & s(N-1) \end{bmatrix}' \tag{9.112}$$

$$\mathbf{P}_{N\times N} = \begin{bmatrix} p_{0,0} & p_{0,1} & \cdots & p_{0,N-1} \\ p_{1,0} & p_{1,1} & \cdots & p_{1,N-1} \\ \cdots & \cdots & \cdots & \cdots \\ p_{N-1,0} & p_{N-1,1} & \cdots & p_{N-1,N-1} \end{bmatrix} \tag{9.113}$$

$$p_{n,m} = \operatorname{sinc}\left\{\pi[\phi_u^{-1}(m) - n]\right\}$$

$$\mathbf{F}_{N\times N} = \begin{bmatrix} f_{0,0} & f_{0,1} & \cdots & f_{0,N-1} \\ f_{1,0} & f_{1,1} & \cdots & f_{1,N-1} \\ \cdots & \cdots & \cdots & \cdots \\ f_{N-1,0} & f_{N-1,1} & \cdots & f_{N-1,N-1} \end{bmatrix} \tag{9.114}$$

$$f_{i,n} = W_N^{in}, \quad W_N = e^{-j\frac{2\pi}{N}}. \tag{9.115}$$

Similarly, (9.108) becomes

$$\mathbf{s}_{N\times1} = \mathbf{Q}_{N\times N}\mathbf{E}_{N\times N}\mathbf{S}_{\phi_u(n)N\times1}, \tag{9.116}$$

where

$$\mathbf{E}_{N\times N} = \begin{bmatrix} e_{0,0} & e_{0,1} & \cdots & e_{0,N-1} \\ e_{1,0} & e_{1,1} & \cdots & e_{1,N-1} \\ \cdots & \cdots & \cdots & \cdots \\ e_{N-1,0} & e_{N-1,1} & \cdots & e_{N-1,N-1} \end{bmatrix} \tag{9.117}$$

$$e_{i,n} = \frac{1}{N}W_N^{-in} \tag{9.118}$$

and $\mathbf{Q}_{N\times N}$ is defined in (9.100).

Proof of DHT
Substituting (9.110) into the right side of (9.116) gives

$$\mathbf{s}'_{N\times1} = \mathbf{Q}_{N\times N}\mathbf{E}_{N\times N}\mathbf{F}_{N\times N}\mathbf{P}_{N\times N}\mathbf{s}_{N\times1}, \tag{9.119}$$

where $\mathbf{F}_{N\times N}$ and $\mathbf{E}_{N\times N}$ represent the DFT and inverse DFT, respectively. We have

$$\mathbf{I}_{N\times N} = \mathbf{F}_{N\times N}\mathbf{E}_{N\times N} = \mathbf{E}_{N\times N}\mathbf{F}_{N\times N}, \tag{9.120}$$

where $\mathbf{I}_{N\times N}$ is

$$\mathbf{I}_{N\times N} = \begin{bmatrix} 1 & 0 & \cdots & 0 \\ 0 & 1 & \cdots & 0 \\ \cdots & \cdots & \cdots & \cdots \\ 0 & 0 & \cdots & 1 \end{bmatrix}. \tag{9.121}$$

Table 9.3   The properties of the DHT.

Property	Signal	DHT	$\phi_u(t)$
Harmonic transform	$x(n)$, $y(n)$	$X_{\phi_u(n)}(i)$, $Y_{\phi_u(n)}(i)$	-
Linearity *	$ax(n) + by(n)$	$aX_{\phi_u(n)}(i) + bY_{\phi_u(n)}(i)$	-
Decomposition	$x_e(n) + x_o(n)$, where $x_e(n) = \frac{1}{2}[x(n) + x(-n)]$ $x_o(n) = \frac{1}{2}[x(n) - x(-n)]$	$X_e(i) + X_o(i)$, where $X_e(i) = \text{Re}[X_{\phi_u(n)}(i)]$ $X_o(i) = j\text{Im}[X_{\phi_u(n)}(i)]$	odd
Time inverse	$x(N - n)$	$X_{\phi_u(n)}(N - i)$	odd
Conjugation	$x^*(n)$	$X^*_{\phi_u(n)}(i)$	-
Convolution	$x(n) * y(n)$	$X_{\phi_u(n)}(i)Y_{\phi_u(n)}(i)$	-
	$x(n)y(n)$	$X_{\phi_u(n)}(i) * Y_{\phi_u(n)}(i)$	-

* a and b are constants.

We rewrite (9.97) as

$$\mathbf{D}_{N \times 1} = \mathbf{Q}_{N \times N}\mathbf{B}_{N \times 1} \tag{9.122}$$

which gives the values of $\phi_u^{-1}(t)$ when $t = \phi_u(n)$. When $\phi_u^{-1}(t) = t$, or equivalently $\phi_u^{-1}(n) = n$, (9.122) should equal to $\phi_u(n)$,

$$\mathbf{A}_{N \times 1} = \mathbf{Q}_{N \times N}\mathbf{D}_{N \times 1}, \tag{9.123}$$

where

$$\mathbf{A}_{N \times 1} = \begin{bmatrix} \phi_u(0) & \phi_u(1) & \cdots & \phi_u(N-1) \end{bmatrix}'. \tag{9.124}$$

Considering the symmetry between $\phi_u(n)$ and $\phi_u^{-1}(n)$ and the definitions of $\mathbf{Q}_{N \times N}$ and $\mathbf{P}_{N \times N}$, we have

$$\mathbf{D}_{N \times 1} = \mathbf{P}_{N \times N}\mathbf{A}_{N \times 1} \tag{9.125}$$
$$\mathbf{B}_{N \times 1} = \mathbf{P}_{N \times N}\mathbf{D}_{N \times 1}. \tag{9.126}$$

Substituting (9.123) into (9.125), we have

$$\mathbf{D}_{N \times 1} = \mathbf{P}_{N \times N}\mathbf{Q}_{N \times N}\mathbf{D}_{N \times 1}, \tag{9.127}$$

which can be rewritten as

$$\mathbf{I}_{N \times N} = \mathbf{P}_{N \times N}\mathbf{Q}_{N \times N} = \mathbf{Q}_{N \times N}\mathbf{P}_{N \times N}. \tag{9.128}$$

Then with (9.120) and (9.128), (9.119) can be simplified to

$$\begin{aligned} \mathbf{s}'_{N \times N} &= \mathbf{Q}_{N \times N}\mathbf{E}_{N \times N}\mathbf{F}_{N \times N}\mathbf{P}_{N \times N}\mathbf{s}_{N \times 1} \\ &= \mathbf{Q}_{N \times N}\mathbf{P}_{N \times N}\mathbf{s}_{N \times 1} \\ &= \mathbf{s}_{N \times 1}. \end{aligned} \tag{9.129}$$

This proves the DHT, which means that the original signal can be recovered from the IDHT.

Since the DHT can be achieved by sampling the continuous HT, the properties of the DHT shown in Table 9.3 are similar to the properties of the HT. Because the DFT is used in the definition of the DHT (see (9.107) and (9.108)), the FFT can be used to calculate the DHT. The other processes represented by $\mathbf{P}_{N \times N}$ and $\mathbf{Q}_{N \times N}$ are interpolation operations, which means the normal spline interpolation algorithms can be used as substitutes to minimize the computation load, though some losses in accuracy are possible. This tradeoff between speed and accuracy can be decided by users based on the requirements of the applications.

9.3.4   Examples of using the DHT

Synthetic signals

Here we briefly present, as an example, the applications of the DHT for varying frequency signals. The signal is

$$x(n) = \begin{cases} e^{j\frac{\pi}{4}n}, & n \in [0, 63] \\ e^{j\frac{\pi}{2}n}, & n \in [64, 127] \\ e^{j\frac{\pi}{8}n}, & n \in [128, 255] \end{cases} \tag{9.130}$$

$$h(n) = \sum_{k=0}^{n} s(k), \tag{9.131}$$

where $s(k)$ is

$$s(k) = \begin{cases} 2, & k \in [0, 63] \\ 4, & k \in [64, 127] \\ 1, & k \in [128, 255] \end{cases}. \tag{9.132}$$

The interpolation for DHT is achieved with the cubic spline algorithm. Figure 9.7 shows the DFT and DHT of $x(n)$. In comparison, the DHT has one pulse and the DFT has three pulses, which means the bandwidth of the DHT is narrower than that of the DFT. In general, a narrow spectrum is always preferred in most signal processing applications to enhance the signal-to-noise ratio. The DHT achieves a better enhancement effect for harmonic signals. This comparison reveals the potential of the DHT for harmonic signal processing, such as voiced speech coding and speech enhancement.

Speech signals

Based on the models of the sources, speech signals are categorized into unvoiced and voiced sounds (also known as voiced speech). The unvoiced sound source can be modeled by a white noise generator. The voiced sound source can be modeled by a generator of pulses or asymmetrical triangular waves which are repeated at every fundamental period. The peak value of the source wave corresponds to the loudness of the voice. The voiced sound is one of the most common harmonic signals. The property that voiced speech repeats at every fundamental period makes the time-frequency transform a powerful tool for signal analysis. The STFT has been widely used for speech analysis with a wide window for spectral envelopes, or the overall spectral feature which reflects the resonance and antiresonance characteristics of the articulatory organs. The STFT with a narrow window is also used for fine spectral structure which reveals detailed spectral patterns for voiced sounds. A high resolution spectral structure is very important for speech processing such as coding and enhancement. The effect of enhancement highly depends on the resolution of the spectra. Low resolution makes the STFT unsuitable for the fine spectral structure analysis. On the

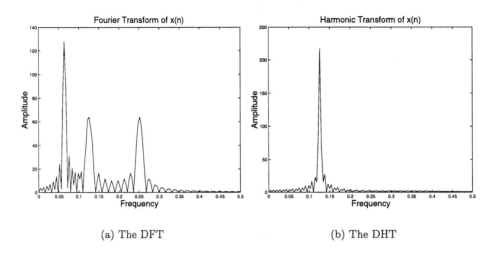

(a) The DFT   (b) The DHT

Figure 9.7   The DFT and DHT of $x(n)$.

other hand, PWD is also not suitable for voiced speech because the interference items between harmonics are overlapped on other harmonics. In this situation it is very difficult to
suppress the interference via a low pass filter with little impact on the useful components.
Voiced speech, mainly consisting of a fundamental and harmonics, can be suitably analyzed

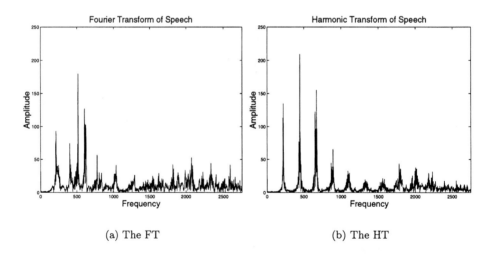

(a) The FT   (b) The HT

Figure 9.8   The FT and HT of speech part PA.

with the HT. The speech signal used for the experiment is a short sentence of a male voice
saying "Mailman, your mail". Since there is a short break in the sentence, we divide the
sentence into two parts, PA and PB, and calculate their HTs, respectively. The duration of
each part is about one second and the frequency is in hertz. The phase function of the HT,
$\phi_u(t)$, is the integral of the nominal frequency of the fundamental of the sentence, which was
extracted from the result of the sentence's STHT, to be discussed shortly. Figure 9.8 shows

the FT and the HT of PA, and Figure 9.9 shows the FT and the HT of PB. The spectrum in Figure 9.8 (b) is concentrated to several peaks which spread along the frequency axis with the same distance. It clearly reveals the fundamental and the harmonics contained in the speech signal. In comparison, the spectrum in Figure 9.8 (a) does not reveal the harmonic structure of the speech signal. Similar observations can also be made from Figure 9.9.

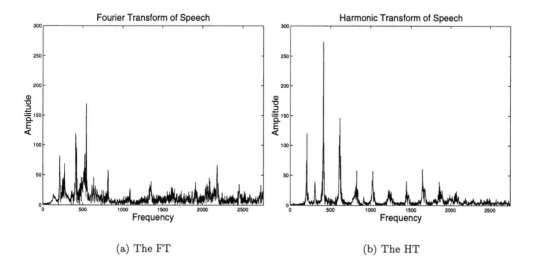

(a) The FT   (b) The HT

Figure 9.9   The FT and HT of speech part PB.

STHT

Because the STFT generally provides time-frequency representation with poor resolution for varying frequency signals, it is not an effective tool for speech signals, although certain improvements have been achieved for a better resolution [13, 16]. The HT can be used to improve the concentration of the STFT for speech analysis, for HT is able to provide concise spectra for speech signals.

Replacing the FT used in the STFT with the HT, we obtain STHT,

$$\text{STHT}_{\phi_u(t,\tau)}(t,\omega) = \int_{-\infty}^{+\infty} f(\tau)\frac{d\phi_u(t,\tau)}{d\tau}w(\tau - t)e^{-j\omega\phi_u(t,\tau)}d\tau, \qquad (9.133)$$

where $f(t)$ is the signal and $w(t)$ is the window function. When $\phi_u(t,\tau) = \tau$, (9.133) is the STFT. With a suitable unit phase function $\phi_u(t)$, the STHT can provide a better frequency resolution for harmonic signals. The best frequency resolution is reached when $\phi_u(t,\tau)$ is the phase function of the fundamental of the signal segment.

The HT can be interpreted as warping the signal in the time domain before calculating the spectrum. The warp operation is determined by $\phi_u(t,\tau)$, which may stretch or shrink the signal. Since stretching or shrinking a signal in the time domain does not preserve the energy of the signal, the STHT does not preserve the energy. Note that if a time-frequency transform satisfies the marginal conditions, it automatically satisfies the total energy requirement to reserve the energy of the signal. Since the STHT does not reserve the energy of the signal, it does not satisfy the margins.

We now discuss the implementation of the STHT when

$$\phi_u(t,\tau) = \tau + \frac{a(t)}{2}\tau^2, \tag{9.134}$$

where $a(t)$ is a function of t. Since the signal segment used in the STHT is short, (9.134) is accurate enough for almost all harmonic signals. Substituting (9.134) into the definition of the HT (9.39) gives the chirp harmonic transform (CHT),

$$\text{CHT}(\omega) = \int_{-\infty}^{+\infty}(1 + a(t)\tau)f(\tau)e^{-j\frac{a(t)}{2}\omega\tau^2}e^{-j\omega\tau}d\tau. \tag{9.135}$$

By substituting (9.134) into (9.133), we have the short-time chirp harmonic transform, (STCHT)

$$\text{STCHT}(\omega,t) = \int_{-\infty}^{+\infty}(1 + a(t)\tau)f(\tau)w(\tau - t)e^{-j\frac{a(t)}{2}\omega\tau^2}e^{-j\omega\tau}d\tau. \tag{9.136}$$

The criterion for selection of $a(t)$ is that the highest concentration should be achieved at the given time t. Exhaustive search is used to locate $a(t)$. The measure of the concentration of the STCHT is

$$M(t) = \frac{\int_{-\infty}^{+\infty}|\text{STCHT}(\omega,t)|^2 d\omega}{(\int_{-\infty}^{+\infty}|\text{STCHT}(\omega,t)|d\omega)^{\frac{3}{2}}}. \tag{9.137}$$

Because $a(t)$ is not known in advance, the method for the HT described in Section 9.3.1 cannot be used to calculate (9.136). We can get the de-chirp Fourier transform (DCFT) defined by

$$F_C(\omega,\theta) = \int_{-\infty}^{+\infty}f(\tau)e^{-j\tau^2\tan\theta}e^{-j\omega\tau}d\tau. \tag{9.138}$$

For a given θ, (9.138) is the FT of $f(\tau)e^{-j\tau^2\tan\theta}$. When $\tan\theta = \omega a(t)/2$, (9.135) and (9.138) become equivalent except for the factor $1 + a(t)\tau$. Figure 9.10 shows the DCFT of a speech segment, in which the corresponding CHTs are the values on the lines, such as OE and OF, passing through point "O". Therefore, we can obtain the CHTs for all possible values of $a(t)$ from the DCFT of the same signal. We choose the $a(t)$ whose CHT has the highest concentration. Although some CHTs obtained from the DCFT are incomplete, sufficient information is still available for comparing the concentration of CHTs to select a suitable value of $a(t)$. In summary, the CHT can be computed by the following steps.

Step 1. Estimate the range of θ, which is needed by (9.138).

Step 2. Calculate the equivalent DCFT according to (9.138).

Step 3. Obtain all possible CHTs from the DCFT.

Step 4. Select parameter $a(t)$ based on the concentration of these CHTs.

Step 5. Calculate the CHT according to (9.135).

The range of θ can be estimated by

$$|\theta| \le \frac{B_f}{T_f}, \tag{9.139}$$

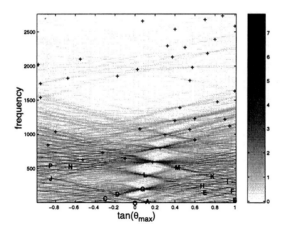

Figure 9.10   The DCFT of a speech segment.

where B_f and T_f are the bandwidth and the length of $f(t)$, respectively. The measurement of concentration for the CHT is defined by

$$M = \sqrt{\frac{L_c}{L_f}} \frac{\int_{-\infty}^{+\infty} |\text{CHT}(\omega)|^2 d\omega}{(\int_{-\infty}^{+\infty} |\text{CHT}(\omega)| d\omega)^{\frac{3}{2}}}, \tag{9.140}$$

where L_f and L_c are the lengths of the CHT calculated from (9.135) and (9.138), respectively.

Obtaining the CHT from the DCFT can be optimized to reduce the computational complexity. If the line of a CHT does not pass at least one valued peak point, the CHT can be eliminated before the concentration comparison. Each of the valued peak points, which are selected from the results of the DCFT, should satisfy the following conditions:

- Its value should be maximum among the points around it.

- Its value should not be too small, such as less than one-tenth of the maximum.

- Its frequency should fall within a certain frequency range.

In the DCFT, chirp components are represented by peak points whose values are bigger than those of their neighbors. The first condition is based on the fact that the lines of candidate CHTs should contain points representing the fundamental or harmonics. The reason for the second condition is that the fundamental or harmonics usually are strong components and the peak points of these components should be large enough to locate $a(t)$. With the second condition, we can eliminate some pseudo peak points caused by noise. For speech signals, the frequencies of the fundamental are basically within the range between 100 Hz and 200 Hz for male and from 200 Hz to 300 Hz for female. Searching within this bandwidth suffices to find a suitable $a(t)$, which is the reason for the third condition. In Figure 9.10, the valued peak points are marked with "+" or uppercase letters except point "O". The same parameters of CHTs correspond to the lines passing through origin "O" and one of the points "E", "H", "N", and "K". Figure 9.11 shows all the candidate CHTs obtained from the DCFT shown in Figure 9.10. Figure 9.12 present the M values of these candidate CHTs, in which the peaks correspond to the fundamental and harmonics.

For comparison, we use the STCHT, PWD and STFT to analyze the short sentence of a male voice. The sampling frequency is 5512.5 Hz. The window function is a Hamming

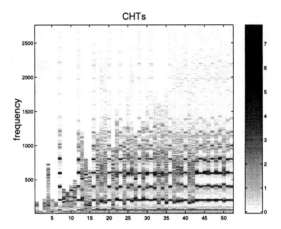

Figure 9.11   The candidate CHTs from the DCFT.

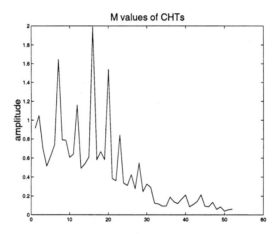

Figure 9.12   The M values of the candidate CHTs.

window with length 512. Figure 9.13 shows the STCHT, the STFT and the PWD of the voice segment. It is observed that the STCHT has the best concentration among the three time-frequency representations. The concentration of the STCHT at $t = 1.1$ is much higher than those of the other two transforms. When the frequencies of the fundamental change rapidly (i.e., between $t =1.1$ and 1.3 of the time axis, the STFT cannot provide a satisfactory representation. The PWD in Figure 9.13 (c) gives unusable information due to many cross-term interference components between harmonics. For a further comparison, Figure 9.14 gives the enlarged areas between $t = 1.1$ and 1.3 of the time axis in Figure 9.13. It is clearly shown that the STCHT provides much better representation.

The HT introduced in this section represents the signal as the sum of a fundamental and harmonics, which becomes the FT when the fundamental has a constant frequency. In comparison with the FT, the HT can provide clear representations for harmonic signals. In addition to the introduction of the continuous version, the DHT is also presented. The advantages of the HT have been demonstrated via the applications on synthetic signal processing and speech analysis. The STHT is the application of the HT on time-frequency analysis, which is the result of replacing the FT used in the STFT with the HT. The advantages of the STHT, verified by the experiments on a male voice, are an improvement in the concentration in the time-frequency analysis for harmonic signals. The procedures to find the phase function of the HT have also been outlined.

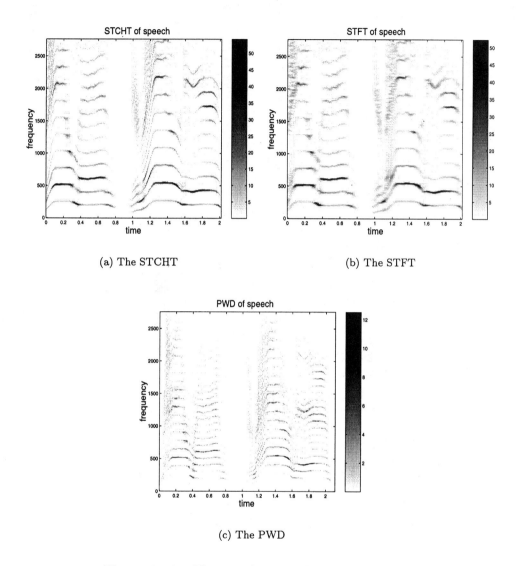

(a) The STCHT

(b) The STFT

(c) The PWD

Figure 9.13   The transforms of the speech segment.

9.4   TOMOGRAPHIC TIME-FREQUENCY TRANSFORM

In this section, we present another transform, known as the tomographic time-frequency transform (TTFT), based on the physical interpretation of the FRFT and reconstruction algorithms of computed tomography. As we know, the FT can be loosely treated as the

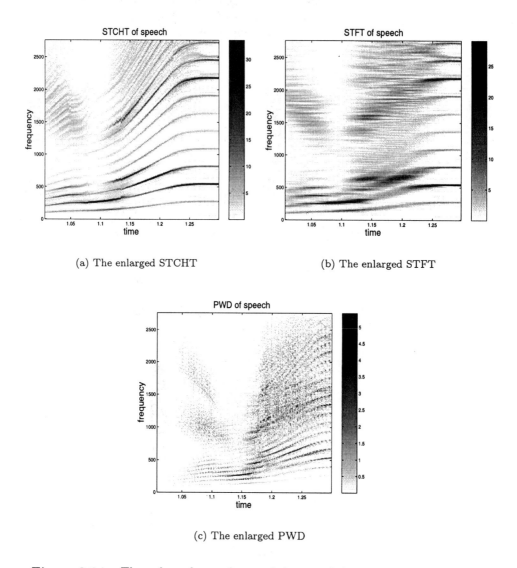

(a) The enlarged STCHT

(b) The enlarged STFT

(c) The enlarged PWD

Figure 9.14   The enlarged transforms of the speech between $t = 1.1$ and 1.3.

integral of the signal along the time axis in the time-frequency plane. Based on the same concept, the FRFT can also be loosely treated as the linear integrals of the signal at all directions in the time-frequency plane. We can define a time-frequency transform in which its linear integral for all directions is the FRFT of the signal. In other words, the relationship between the desired transform and the FRFT is similar to the one between a cross section of an object and the X-ray of this section in all directions. Then we can obtain the time-frequency representation of the signal via the reconstruction algorithms used by computed tomography.

Ideally, the resolution in the time or frequency domain should be solely determined by the signal itself. However, the window function used in many signal processing algorithms, such as in the STFT, has often undesirable effects. Two types of window effects are common. The first one is the entanglement of the signal and the window function, which means that

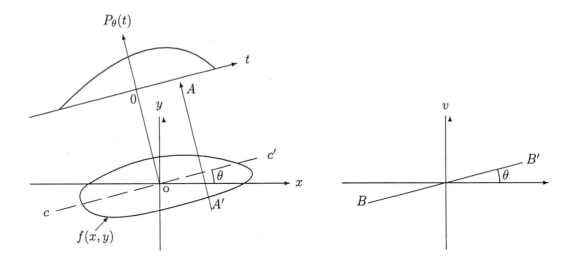

Figure 9.15 The projection and the FT along a radial line.

the outputs of the processing are a mixture of the signal and window function. The second effect is that the frequency resolution is often closely related to the width of the window function. Because the TTFT does not need any window function explicitly, these window effects can be substantially minimized. As the TTFT is based on the squared magnitudes of the FRFT, cross-term interference between components is unavoidable. However, unlike the WD, the interference of the TTFT is weak and can be suppressed effectively via an adaptive filter with little impact on the signal components, because the filtering process is applied on the partially processed data instead of the final results.

This section is organized as follows. Some basic concepts on computed tomography are introduced in Section 9.4.1. The definition of the TTFT is given in Section 9.4.2. Eliminating the interference with an adaptive filter is also discussed in this section. Several experiments are given to demonstrate the advantages of the TTFT in Section 9.4.3.

9.4.1 Computed tomography

Tomography refers to the cross-sectional image of an object from either transmitted or reflected data collected by illuminating the object from different directions. The impact of this technique in diagnostic medicine was revolutionary, since it made it possible to view the internal organs with unprecedented precision and safety. There are numerous non-medical imaging applications which lend themselves to the methods of computerized tomography. Researchers have already applied this methodology to applications such as mapping underground resources via cross borehole imaging, obtaining a cross-sectional image for non-destructive testing, determining the brightness distribution over a celestial sphere, and 3D imaging with electron microscopy.

Fundamentally, tomographic imaging deals with reconstructing an image from its projections. In the strict sense, a projection at a given angle is the integral of the image in the direction specified by that angle, as shown by $P_\theta(t)$ in Figure 9.15. However, in a loose sense, projection means the information derived from the transmitted energies when an object is illuminated from a particular angle. Although, from a purely mathematical standpoint, the solution to the problem of reconstructing a function from its projections was given by Radon as early as 1917, the current excitement in tomographic imaging originated with Hounsfield's invention of the X-ray computed tomographic scanner. His invention showed that it was possible to compute high-quality cross-sectional images with a high accuracy in spite of the fact that the projection data do not strictly satisfy the theoretical models underlying the efficiently implementable reconstruction algorithms. His invention also showed that it was possible to get an image with a high accuracy by processing a very large number of measurements with fairly complex mathematical operations.

It is perhaps fair to say that the major breakthrough in X-ray computed tomography images was mainly due to the developments in reconstruction algorithms. Hounsfield was able to use algebraic techniques to reconstruct noisy-looking images with a very good accuracy. This was followed by the application of the convolution backprojection algorithm that was first developed by Ramachandran and Lakshminarayanan [30] and later popularized by Shepp and Logan [36]. The algorithms considerably reduced the processing time for reconstruction of more accurate images.

Fourier slice theorem

For time-frequency analysis, we are interested in the mathematical basis of tomography with non-diffracting sources and parallel beam projection data. In ideal situations, projections are a set of measurements of the integrated values being straight lines through the object; these are known as line integrals. The key to tomographic imaging is the Fourier slice theorem which relates the measured projection data to the 2D FT of the cross section of the object.

The Fourier slice theorem states that the 1D FT of a parallel projection is equal to a slice of the 2D FT of the original object. It follows that, given the projection, it should be possible to estimate the object by simply performing a 2D inverse FT. In Figure 9.15 the object is represented by a 2D function $f(x, y)$ and each line integral by parameters (θ, t). The line equation can be expressed by

$$x \cos \theta + y \sin \theta = t \tag{9.141}$$

and we use this relationship to define line integral $P_\theta(t)$ as

$$P_\theta(t) = \int_{(\theta, t)} f(x, y) ds. \tag{9.142}$$

Using a delta function, this can be rewritten as

$$P_\theta(t) = \int_{-\infty}^{+\infty} \int_{-\infty}^{+\infty} f(x, y)\delta(x \cos \theta + y \sin \theta - t) dx dy. \tag{9.143}$$

Because $P_\theta(t)$ is known as the Radon transform of the function $f(x, y)$, $f(x, y)$ can be recovered from $P_\theta(t)$ by the inverse Radon transform. Let $\mathcal{R}$ and $\mathcal{R}^{-1}$ represent the Radon transform and its inverse, respectively,

$$P_\theta(t) = \mathcal{R}[f(x, y)], \quad f(x, y) = \mathcal{R}^{-1}[P_\theta(t)]. \tag{9.144}$$

Figure 9.16 gives two signals and their corresponding Radon transforms, whose dynamic range Y is compressed by

$$Y = \log(1 + P_\theta(t)).\tag{9.145}$$

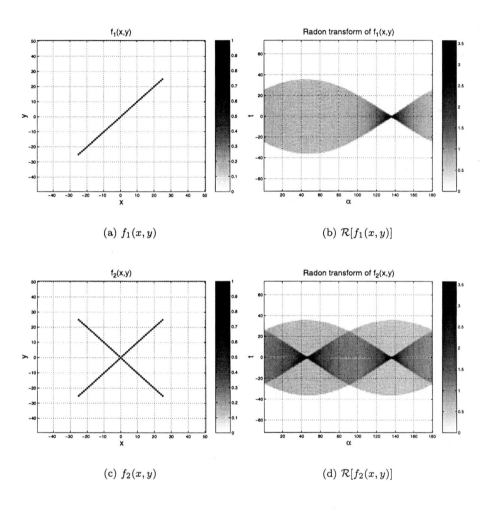

(a) $f_1(x, y)$ (b) $\mathcal{R}[f_1(x, y)]$

(c) $f_2(x, y)$ (d) $\mathcal{R}[f_2(x, y)]$

Figure 9.16 $f_1(x, y)$, $f_2(x, y)$ and their Radon transforms.

The FT of the projection $P_\theta(t)$ is

$$\begin{aligned}
S_\theta(\omega) &= \int_{-\infty}^{+\infty} P_\theta(t) e^{-j\omega t} dt \\
&= \int_{-\infty}^{+\infty} \int_{-\infty}^{+\infty} \int_{-\infty}^{+\infty} f(x, y)\delta(x\cos\theta + y\sin\theta - t)dxdy e^{-j\omega t} dt \\
&= \int_{-\infty}^{+\infty} \int_{-\infty}^{+\infty} f(x, y) \int_{-\infty}^{+\infty} \delta(x\cos\theta + y\sin\theta - t)e^{-j\omega t} dtdxdy \\
&= \int_{-\infty}^{+\infty} \int_{-\infty}^{+\infty} f(x, y) e^{-j\omega(x\cos\theta + y\sin\theta)} dxdy.
\end{aligned}\tag{9.146}$$

The right-hand side of (9.146) now represents the 2D FT of $f(x, y)$ at a spatial frequency of $(u = \omega \cos \theta, \quad v = \omega \sin \theta)$ or

$$S_\theta(\omega) = F(\omega \cos \theta, \omega \sin \theta). \tag{9.147}$$

The above result indicates that by taking the projections of an object function at angles $\theta_1, \theta_2, \ldots, \theta_k$ and calculating their FTs, we can determine the values of $F(u, v)$ on radial lines being similar to the line BB' in Figure 9.15. If an infinite number of projections are available, $F(u, v)$ would be known at all points in the uv-plane. The object function $f(x, y)$ can be recovered by using the inverse FT of $F(u, v)$,

$$f(x, y) = \frac{1}{4\pi^2} \int_{-\infty}^{+\infty} \int_{-\infty}^{+\infty} F(u, v) e^{(ux+vy)} du dv. \tag{9.148}$$

The Fourier slice theorem relates the FT of a projection to the FT of the object along a single radial. Thus given the FT of a projection at a sufficient number of angles, the projections could be assembled into a complete estimate of the 2D transform, which is then simply inverted to obtain an estimate of the object. While this provides a simple conceptual model of tomography, practical implementation requires a different approach.

Filtered backprojection algorithm

The filtered backprojection algorithm is currently being used in almost all applications of straight ray tomography. It was shown to be extremely accurate and amenable to fast implementation, which will be derived by using the Fourier slice theorem. The filtered backprojection algorithm can be given in a rather straightforward intuitive way because each projection represents a nearly independent measurement of the object. By the Fourier slice theorem, a projection gives the values of the 2D FT of the object along a single line. If the FT values of the projection are inserted into their proper place in the 2D Fourier domain of the project, a simple, albeit distorted, reconstruction can be formed by assuming the other projections to be zero in the computation of the corresponding 2D inverse FT. This shows that such a reconstruction is equivalent to the original FT of the object multiplied by a simple filter. The final reconstruction is found by adding together the 2D inverse FT of each projection. This step is widely known as backprojection since it can be perceived as smearing each filtered projection over the image plane. Figure 9.17 shows the reconstructions achieved with various numbers of projections.

The object function, $f(x, y)$, or the inverse FT can be expressed as

$$f(x, y) = \frac{1}{4\pi^2} \int_{-\infty}^{+\infty} \int_{-\infty}^{+\infty} F(u, v) e^{j(ux+vy)} du dv. \tag{9.149}$$

Changing the rectangular coordinate system (u, v) in the frequency domain into a polar coordinate system (ω, θ) with the substitutions

$$u = \omega \cos \theta, \quad v = \omega \sin \theta, \quad du dv = \omega d\omega d\theta, \tag{9.150}$$

we can write the inverse FT, or the object function $f(x, y)$ as

$$f(x, y) = \frac{1}{4\pi^2} \int_0^{2\pi} \int_0^{+\infty} F(\omega, \theta) e^{j\omega(x \cos \theta + y \sin \theta)} \omega d\omega d\theta. \tag{9.151}$$

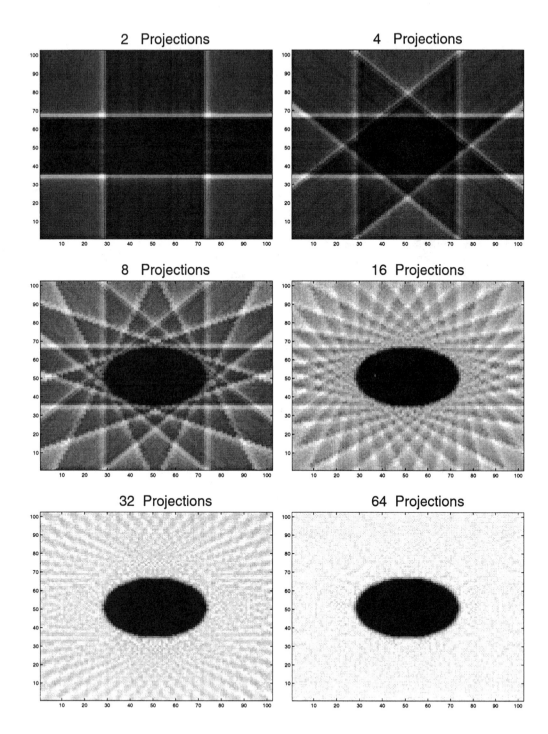

Figure 9.17   The improvement in accuracy of reconstruction with the number of projections.

This integral can be split into two terms as

$$f(x,y) = \frac{1}{4\pi^2} \int_0^\pi \int_0^{+\infty} F(\omega,\theta)e^{j\omega(x\cos\theta+y\sin\theta)}\omega d\omega d\theta$$
$$+ \frac{1}{4\pi^2} \int_0^\pi \int_0^{+\infty} F(\omega,\theta+\pi)e^{j\omega[x\cos(\theta+\pi)+y\sin(\theta+\pi)]}\omega d\omega d\theta, \quad (9.152)$$

and then using the property

$$F(\omega,\theta+\pi) = F(-\omega,\theta), \quad (9.153)$$

the above expression for $f(x,y)$ may be written as

$$f(x,y) = \frac{1}{4\pi^2} \int_0^\pi \left[\int_{-\infty}^{+\infty} F(\omega,\theta)|\omega|e^{j\omega t}d\omega \right] d\theta. \quad (9.154)$$

Here we have simplified the expression with

$$t = x\cos\theta + y\sin\theta. \quad (9.155)$$

If we substitute the FT of the projection at angle θ, $S_\theta(\omega)$, for the 2D FT $F(\omega,\theta)$, we get

$$f(x,y) = \frac{1}{4\pi^2} \int_0^\pi \left[\int_{-\infty}^{+\infty} S_\theta(\omega)|\omega|e^{j\omega t}d\omega \right] d\theta. \quad (9.156)$$

This integral in (9.156) may be expressed as

$$f(x,y) = \frac{1}{2\pi} \int_0^\pi Q_\theta(x\cos\theta + y\sin\theta)d\theta, \quad (9.157)$$

where

$$Q_\theta(t) = \frac{1}{2\pi} \int_{-\infty}^{+\infty} S_\theta(\omega)|\omega|e^{j\omega t}d\omega. \quad (9.158)$$

The estimate of $f(x,y)$, given the projection transform $S_\theta(\omega)$, has a simple form. Equation (9.158) represents a filtering operation, where the frequency response of the filter is given by $|\omega|$; therefore Q_θ is called a "filtered projection." The resulting projections for different angles θ are then collected to form the estimate of $f(x,y)$.

Equation (9.157) calls for each filtered projection, Q_θ, to be *backprojected*. This can be explained as follows. To every point (x,y) in the image plane that corresponds to a value of $t = x\cos\theta + y\sin\theta$ for a given value of θ, the filtered projection Q_θ contributes to the reconstruction of this particular value, as is illustrated in Figure 9.18. It is easily shown that for the indicated angle θ, the value of t is the same for all (x,y) on the line LM. Thus, the filtered projection, Q_θ, will make the same contribution to the reconstruction. Therefore, one could say that in the reconstruction process each filtered projection, Q_θ, is smeared back, or backprojected, over the image plane.

9.4.2   Tomographic time-frequency transform

Time-frequency representation can provide a better description of signals than classical stationary analysis methods. The STFT and WD are widely used to calculate time-frequency

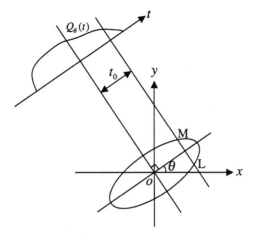

Figure 9.18   Backprojection of Q_θ to reconstruct the image.

representations [7]. For the STFT, one assumes that the time-frequency distributions of a signal within a sufficiently small interval are almost the same, so that the time-frequency distribution of the signal within that interval is treated as the spectrum of the corresponding signal segment. Such an approximation is generally not valid, particularly for signals with fast frequency variation. Another coherent problem of the STFT is the window effect because the uncertainty principle shows that no window function can achieve high resolution in both the time and frequency domains at the same time. In addition, the window function also entangles with the spectrum of the signal to be processed. The WD takes the spectrum of the signal's time-dependent auto-correlation as the time-frequency distribution of the signal. Although the WD does not have a good physical interpretation, it has many desirable characteristics in the time-frequency domain. The WD is bothered by the cross-terms between desirable signal components. Filters are used to eliminate these cross-terms, which often adversely distorts the signal components. One widely used WD with filters is the Radon–Wigner transform [31, 35, 41, 43] designed for linear FM signals. The filtering is realized by three consecutive operations on the WD of the signal: the Radon transform, filtering and the inverse Radon transform. The filtering process generally detects straight lines in the WD of the signal. The Radon–Wigner transform is widely used for linear FM signal processing and synthesis [42, 44, 45, 46].

To avoid the limitations from the STFT and WD, we attempt to define a time-frequency transform based on a simple intuition: the FT of a signal is the linear integral of the time-frequency distribution of the signal along the time axis. It is assumed that any given signal has its corresponding time-frequency distribution, which satisfies the marginal condition. From the intuition and the relation between the FT and FRFT [1, 8], we deduce that the FRFT of a signal is the Radon transform of the time-frequency distribution of the signal. Therefore, we can obtain the time-frequency distribution of the signal by the inverse Radon transform of the FRFT of the signal. Since this process is similar to the computerized tomography process, we call it the tomographic time-frequency transform (TTFT). Although both the TTFT and Radon–Wigner transform employ inverse Radon transform, the intentions behind them are totally different. The Radon–Wigner transform is a variation of the WD designed for processing linear FM signals only. This method uses the Radon transform and its inverse together with a filter to eliminate the noise and emphasize the linear FM signals. In contrast, the TTFT is based on a different concept and employs the inverse

Radon transform only to recover the time-frequency distribution from the FRFT of the signal. The TTFT is suitable for all kinds of signals because its resolutions in the time and frequency domains are determined solely by the frequency resolution of the FRFT and the computerized tomography algorithms.

In this subsection, the quadratic realization of the TTFT will be deduced. The time-frequency distribution of the signal is calculated from the energy of the FRFT (quadratic magnitude of the FRFT). As a quadratic transform, there always exist some cross-terms between the desirable signal components. An adaptive filter is used to remove those cross-terms. Because the filtering process is applied on the intermediate data (the FRFT of the signal) rather than directly on the final results, as seen in many other filtering processes, the undesirable filtering impact on the signal components can be substantially minimized.

FRFT

Although the FRFT is interpreted as rotating the signal in the time-frequency domain [1], this explanation is only based on the properties of the FRFT without further interpretation of the contributions of each multiplication factor used in the definition of (9.35). Here we discuss the impacts of every factor on the time-frequency distribution, which reveals a novel definition of the time-frequency distribution. In the following discussion, the time-frequency distribution of the signal is considered as a 3D graph in the time-frequency plane and is assumed to satisfy the marginal condition. Several properties of the time-frequency transform are listed in Table 9.4.

Table 9.4   The properties of the time-frequency transform.

Operation	No.	Signal		Time-Freq. Dist.	Coordinates*
	1	$f(t),$	$F(\omega)$	$P(t,\omega)$	(t',ω')
Shift	2	$f(t)e^{jat},$	$F(\omega - a)$	$P(t,\omega - a)$	$(t',\omega' + a)$
	3	$f(t+a),$	$F(\omega)e^{ja\omega}$	$P(t+a,\omega)$	$(t' - a,\omega')$
	4	$f(t)e^{j\frac{a}{2}t^2}$		$P(t,\omega - at)$	$(t',\omega' + at')$
	5	$F(\omega)e^{j\frac{a}{2}\omega^2}$		$P(t + a\omega,\omega)$	$(t' - a\omega',\omega')$
Scale	6	$\frac{1}{\sqrt{a}}F(\frac{\omega}{a}),$	$\sqrt{a}f(at)$	$P(at',\frac{\omega'}{a})$	$(\frac{t'}{a}, a\omega')$

a is constant.
* In each row, the value of the time-frequency transform at the coordinate is equivalent.

Table 9.5   The coordinates of PA and PB.

Signal	Plane	PA	PB
$f(t)$	ωt	$A(0,\ \omega_0)$	$B(t_0,\ \omega_0)$
	vu	$A(-\omega_0 \sin\beta,\ \omega_0 \cos\beta)$	$B(t_0 \cos\beta - \omega_0 \sin\beta,\ t_0 \sin\beta + \omega_0 \cos\beta)$
$f(v)$	vu	$C(0,\ \omega_0)$	$D(t_0,\ \omega_0)$

Suppose that $P(t,\omega)$ is the time-frequency distribution of $f(t)$. To simplify the discussion, we only consider the coordinate variations of two points, PA and PB, in the distribution. In Figure 9.19, A and B are the positions of PA and PB in the ωt-plane for $f(t)$ and C and D are the corresponding points in the vu- plane for $f(u)$, respectively. The coordinates

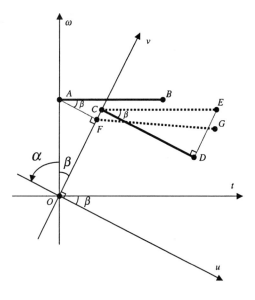

Figure 9.19   The rotation of FRFT.

of A and B in the ωt and vu planes are given in Table 9.5. By applying the multiplication factors in (9.35), the positions of PA and PB in the vu-plane move from C and D to A and B, respectively. The position variations of PA and PB corresponding to each multiplication factor are given in Table 9.6. The column *Angle* in the table is used for the angle between axes t and u, and the column *Expression* contains the mathematical expression involved in each step. We start from Step 1 in which t_0 and ω_0 are constants. The angle from axis u to axis t is $-\beta$. Step 2 is based on the property given in Table 9.4, row 2. Step 3 performs the FT and changes the signal expression is from the time domain to the frequency domain. This step adds $\frac{\pi}{2}$ to the angle and PA and PB remain at positions C and E, respectively. Step 4 is a scaling operation with the property given in row 6 of Table 9.4. Finally, Step 5 uses the property given in row 3 of Table 9.4. After this step the positions of PA and PB are moved to A and B, respectively. Since PA and PB are two arbitrary points, we can say that the time-frequency distribution of the FRFT of $f(u)$ is equal to the distribution of $f(t)$. Writing the expression that takes all steps into consideration gives

$$Y_{\frac{\pi}{2}-\beta}(v) = \frac{1}{\sqrt{2\pi \cos\beta}} e^{j\frac{v^2}{2}\tan\beta} \int_{-\infty}^{+\infty} \underbrace{f(u)}_{Step\ 1}\ e^{j\frac{u^2}{2}\tan\beta}\ e^{-jvu\frac{1}{\cos\beta}}\,du \qquad (9.159)$$

The step numbers in (9.159) correspond to the step numbers listed in Table 9.6. Let $\theta = \frac{\pi}{2} - \beta$, then the above equation becomes

$$Y_\theta(v) = \sqrt{\frac{\csc\theta}{2\pi}}\, e^{j\frac{v^2}{2}\cot\theta} \int_{-\infty}^{+\infty} f(u) e^{j\frac{u^2}{2}\cot\theta} e^{-jvu\csc\theta}\,du. \qquad (9.160)$$

Table 9.6   The paths of PA and PB in vu-plane.

Step	Angle	Expression	PA	PB
1	$-\beta$	$f(u)$	$C(0,\ \omega_0)$	$D(t_0,\ \omega_0)$
2	$-\beta$	$g(u) = f(u)e^{j\frac{u^2}{2}\tan\beta}$	$C(0,\ \omega_0)$	$E(t_0,\ \omega_0 + t_0\tan\beta)$
3	$\frac{\pi}{2} - \beta$	$G(v) = \mathcal{F}[g(u)]$	$C(0,\ \omega_0)$	$E(t_0,\ \omega_0 + t_0\tan\beta)$
4	$\frac{\pi}{2} - \beta$	$\frac{1}{\sqrt{\cos\beta}}G(\frac{v}{\cos\beta})$	$F(0,\ \omega_0\cos\beta)$	$G(\frac{t_0}{\cos\beta},$ $\omega_0\cos\beta + t_0\sin\beta)$
5	$\frac{\pi}{2} - \beta$	$\frac{1}{\sqrt{\cos\beta}}e^{j\frac{v^2}{2}\tan\beta}\,G(\frac{v}{\cos\beta})$	$A(-\omega_0\sin\beta,$ $\omega_0\cos\beta)$	$B(t_0\cos\beta - \omega_0\sin\beta,$ $\omega_0\cos\beta + t_0\sin\beta)$

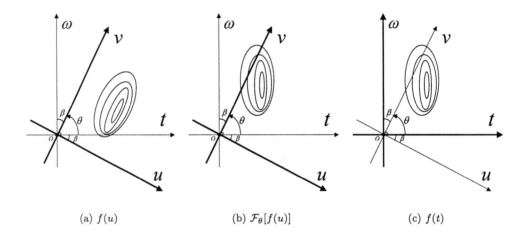

(a) $f(u)$ (b) $\mathcal{F}_\theta[f(u)]$ (c) $f(t)$

Figure 9.20   Contour plots of $f(u)$, the FRFT of $f(u)$ and $f(t)$.

The only difference between (9.35) and (9.160) is the square root factor $\sqrt{\frac{1-j\cot\theta}{2\pi}} = \sqrt{\frac{\csc\theta}{2\pi}}e^{j\frac{\theta-\frac{\pi}{2}}{2}}$. It shows that the difference between (9.35) and (9.160) is the phase constant $e^{j\frac{\alpha-\frac{\pi}{2}}{2}}$. The reason for the absence of the phase factor in (9.160) is that only magnitudes are considered in the properties of the time-frequency transform given in Table 9.4 because the time-frequency distribution is considered as energy density. In practice, the definition of the new transform can be derived from another approach.

Let us consider a signal $f(u)$, whose time-frequency distribution is illustrated in Figure 9.20 (a) which has two time-frequency planes, the vu-plane and the ωt-plane. The origins of the two planes are identical and the angle between axis u and axis t is β, and $f(t)$ is the anticlockwise rotation of $f(u)$ for angle β in the time-frequency plane. Since the angle between axis u and axis t is equal to the rotation angle of $f(t)$, the expressions of $f(t)$ and $f(u)$ are identical. The variables of $f(t)$ and $f(u)$ are in the time domain and the variables of their FRFTs such as $\mathcal{F}_\alpha[f(u)]$ are in the frequency domain.

We know that the time-frequency distribution of the FRFT, $\mathcal{F}_{\frac{\pi}{2}-\beta}[f(u)]$, in Figure 9.20 (b) is identical to the time-frequency distribution of $f(t)$ in Figure 9.20 (c). The variable of $\mathcal{F}_{\frac{\pi}{2}-\beta}[f(u)]$ is in the frequency domain and the corresponding time axis is u. Because the time-frequency distribution is assumed to satisfy the marginal conditions, the linear integration of the time-frequency distribution of $\mathcal{F}_{\frac{\pi}{2}-\beta}[f(u)]$ along axis u is the spectrum of the signal, which is $\mathcal{F}_{\frac{\pi}{2}-\beta}[f(u)]$. On the other hand, in Figure 9.20 (c), the linear integration on the time-frequency distribution of $f(t)$ along the axis u is the Radon transform of the time-frequency distribution of $f(t)$ for angle α written as $T_f(\alpha,t)|_{\alpha=\frac{\pi}{2}-\beta}$. Because the two distributions in Figure 9.20 (b) and (c) are identical, we have

$$T_f(\alpha,t)\bigg|_{\alpha=\frac{\pi}{2}-\beta} = \mathcal{F}_{\frac{\pi}{2}-\beta}[f(u)]. \tag{9.161}$$

We can get $T_f(\alpha,t)$ for all angles by varying β. The inverse Radon transform of $T_f(\alpha,t)$ is the time-frequency distribution of $f(t)$, which is also the time-frequency distribution of $f(u)$. So the time-frequency distribution of $f(u)$ can be reconstructed from its FRFT. This is the concept of the TTFT.

The expression of the TTFT can be deduced from the marginal conditions. Rewriting the marginal conditions (9.1) and (9.2) in the form of the FRFT, we have

$$\int P(t,\omega)d\omega = |x(t)|^2 = |X(0,u)|^2\bigg|_{u=t} \tag{9.162}$$

and

$$\int P(t,\omega)dt = |X(\omega)|^2 = |X(\frac{\pi}{2},u)|^2\bigg|_{u=\omega}. \tag{9.163}$$

Since FRFT rotates the signal in the time-frequency plane, we can extend the marginal conditions to an arbitrary angle α,

$$\int_{M_{\alpha,u}} P(t,\omega)ds = \mathcal{R}[P(t,\omega)] = |X(\alpha,u)|^2 = |\mathcal{F}_\alpha[x(t)]|^2, \tag{9.164}$$

where $P(t,\omega)$ is the time-frequency distribution of $x(t)$. Equation (9.164) shows that the FRFT of $x(t)$ is the Radon transform of the time-frequency distribution of $x(t)$, i.e., $P(t,\omega)$ is the inverse Radon transform of $X(\alpha,u)$. This provides the definition of the TTFT,

$$\text{TTFT}[x(t)] = P(t,\omega) = \mathcal{R}^{-1}\left\{|\mathcal{F}_\alpha[x(t)]|^2\right\} \tag{9.165}$$

which is similar to the relationship between the FRFT and WD reported in [1, 18].

9.4.3 Interference suppression

Similar to any quadratic transform, the TTFT of signals containing multicomponents also has cross-terms in the time-frequency distribution. Given a signal $x_s(t)$ containing $x(t)$ and its postponed copy $x(t-t_0)$ such as

$$x_s(t) = x(t) + x(t-t_0), \tag{9.166}$$

the magnitude of its FT is

$$\begin{aligned}|F_s(\omega)| &= |F(\omega) + e^{-j\omega t_0}F(\omega)| \\ &= |F(\omega)(1 + \cos\omega t_0 + j\sin\omega t_0)| \\ &= 2(1 + \cos\omega t_0)|F(\omega)|. \end{aligned} \tag{9.167}$$

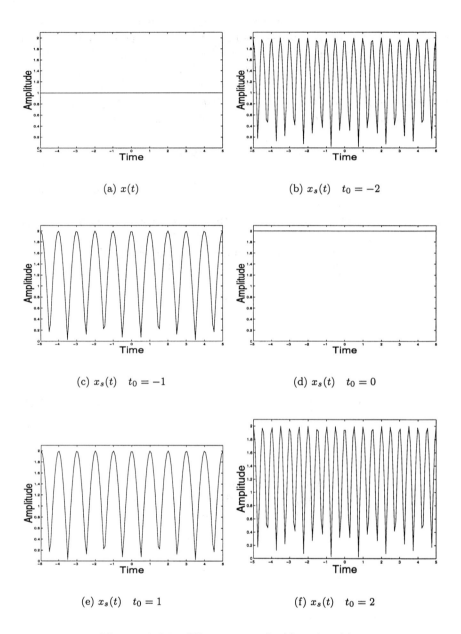

Figure 9.21   The spectra of $x(t)$ and $x_s(t)$.

The term associated with $\cos \omega t_0$ is the additional term considered to be the interference (or cross-term). For example, when $x(t) = \delta(t)$, the amplitudes of the spectra of $x(t)$ and $x_s(t)$ for different t_0 are shown in Figure 9.21. The frequency of $\cos \omega t_0$ in terms of ω is t_0, which means that if a low pass filter is employed to remove the additional term, the bandwidth of the filter should be smaller than t_0. For the TTFT, t_0 is a function of the rotation angle, which hints that adaptive filters may eliminate the interference more effectively than ordinary fixed bandwidth filters.

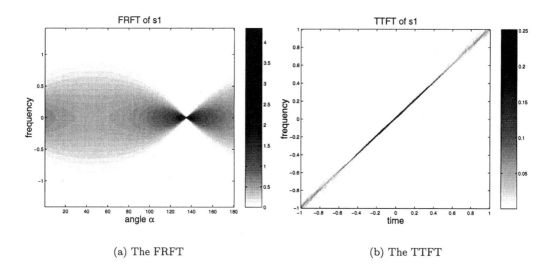

(a) The FRFT   (b) The TTFT

Figure 9.22   The distributions of $s_1(t)$.

For this case, the impulse response of the adaptive filter applied on the result of the FRFT is a Hanning window function $H(\alpha, \tau)$,

$$H(\alpha, \tau) = \begin{cases} \frac{1}{W(\alpha)}(1 + \cos\frac{2\pi\tau}{W(\alpha)}), & \tau \in [-\frac{W(\alpha)}{2}, \frac{W(\alpha)}{2}] \\ 0, & \text{others} \end{cases}, \tag{9.168}$$

where

$$W(\alpha) = \frac{A}{\max\left\{|\mathcal{F}_\alpha[x(t)]|\big|u\right\}} + B \tag{9.169}$$

and A and B are constant parameters controlling the frequency response of the filter, $\max\left\{|\mathcal{F}_\alpha[x(t)]|\big|u\right\}$ is the maximum magnitude of $\mathcal{F}_\alpha[x(t)]$ in terms of u. Applying (9.168) to (9.165) gives the filtered tomographic time-frequency transform (FTTFT),

$$\text{FTTFT}[x(t)] = \mathcal{R}^{-1}\left[\int |F_\alpha[x(t)]|^2 H(\alpha, \tau - u)d\tau\right]. \tag{9.170}$$

Examples

Let us consider two examples. The filtered back projection algorithm with Ram–Lak filter is used to calculate the inverse Radon transform. To compress the dynamic range, all the time-frequency distributions (i.e., FRFT, TTFT and FTTFT) are processed by (9.145).

The first example makes comparisons among the TTFT, STFT and WD for chirp signal analysis. The chirp signals are

$$s_1(t) = \begin{cases} \frac{1}{2}(1 + \cos\frac{2\pi t}{T})e^{j\frac{t^2}{2}}, & |t| \le \frac{T}{2} \\ 0, & \text{others} \end{cases} \tag{9.171}$$

$$s_2(t) = \begin{cases} \frac{1}{2}(1 + \cos\frac{2\pi t}{T})(e^{-j\frac{t^2}{2}} + e^{j\frac{t^2}{2}}), & |t| \le \frac{T}{2} \\ 0, & \text{others} \end{cases} \tag{9.172}$$

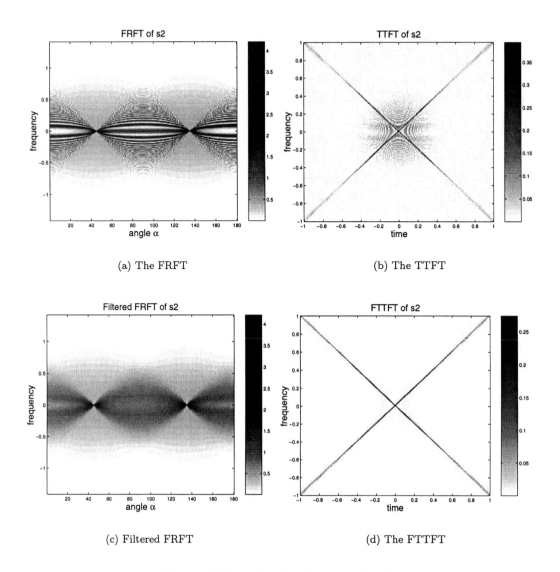

(a) The FRFT

(b) The TTFT

(c) Filtered FRFT

(d) The FTTFT

Figure 9.23   The distributions of $s_2(t)$.

in which $T = 16\sqrt{2\pi}$. The chirp signals are modulated by Hanning window functions to reduce the effects of finite duration. For simplicity the ranges of the time and the frequency axes are normalized from $[-8\sqrt{2\pi}, 8\sqrt{2\pi}]$ to $[-1, 1]$. The angle α is in degrees for all FRFTs. Figures 9.22 (a) and (b) show the FRFT and the TTFT of $s_1(t)$. The TTFT of $s_1(t)$ has very high time and frequency resolutions. The FRFT and the TTFT of $s_2(t)$ are given in Figures 9.23 (a) and (b).

The cross-terms between the two components of $s_2(t)$ are distinguishable in Figure 9.23 (b). An adaptive filter, whose characteristic $W(\alpha)$ is given in Figure 9.24, is applied on the FRFT of $s_2(t)$. The filtered FRFT is shown in Figure 9.23 (c). Applying the inverse Radon transform on the filtered FRFT of $s_2(t)$ gives the FTTFT of $s_2(t)$ in Figure 9.23 (d), in which most cross-items are eliminated with little side effect on the signal components. This ex-

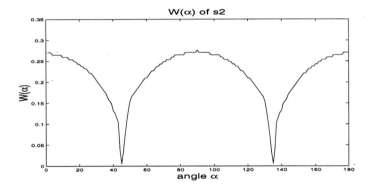

Figure 9.24   The $W(\alpha)$ of the filter.

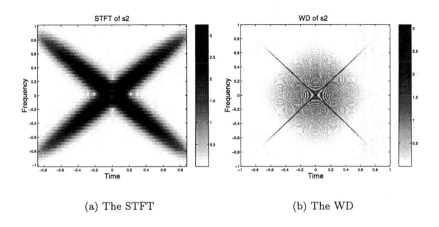

(a) The STFT   (b) The WD

Figure 9.25   The STFT and WD of $s_2(t)$.

ample illustrates the high efficiency and the low side effects of the adaptive filtering. As a reference, the STFT and the WD of $s_2(t)$ are provided in Figure 9.25.

The second example is a non-linear signal $s_{\sin}(t)$ defined by

$$s_{\sin}(t) = \begin{cases} \frac{1}{2}(1 + \cos \frac{2\pi t}{T})e^{j\frac{-T^2}{16\pi}\cos\frac{4\sqrt{2}\pi t}{T}}, & |t| \leq \frac{T}{2} \\ 0, & \text{others} \end{cases} \tag{9.173}$$

whose nominal frequency follows a sine curve. Its FRFT and TTFT are given in Figures 9.26 (a) and (b). The highest resolution is around the origin, which is related to the linear segment between -0.18 and 0.18 in time, as shown in Figure 9.26 (b). The lowest resolution occurs when $t = -0.2$ and 0.2, which corresponds to the non-linear parts of the signal. Figures 9.26 (c) and (d) show the filtered FRFT and the FTTFT. It can be seen that the filtering process does not have a perceivable impact on the resolution of the signal in the time-frequency domain. Both examples show that the resolution of the signal components in the time-frequency representation is well preserved.

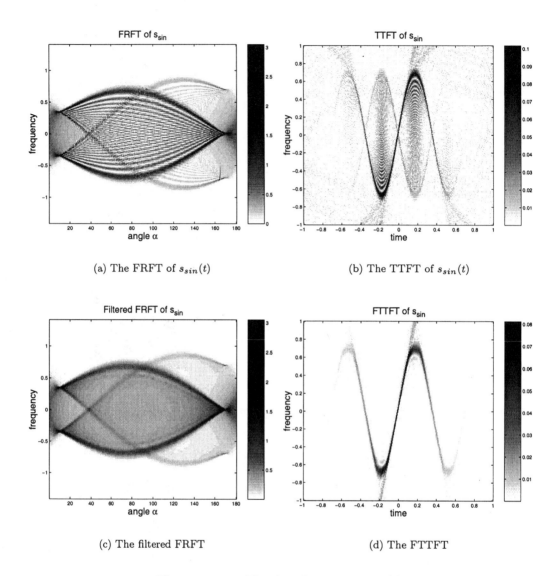

(a) The FRFT of $s_{sin}(t)$

(b) The TTFT of $s_{sin}(t)$

(c) The filtered FRFT

(d) The FTTFT

Figure 9.26   The distributions of $s_{\sin}(t)$.

9.5   CHAPTER SUMMARY

To obtain a good time-frequency distribution, we can either make improvements on the existing methods based on the characteristics of the signal or develop new transforms. The development of the STHT in Section 9.3 attempts to improve the performance of the STFT by pre-processing the signals. It is shown that for different categories of signals, different pre-processing techniques are needed. For example, the warping operation in the time domain is employed for harmonic signals. Examples show that these pre-processing operations are useful to achieve substantial improvements on resolutions in the time-frequency domain.

The TTFT presented in Section 9.4 is based on a new concept, that is, the FRFT can be treated as the linear integrals of the signal in the time-frequency domain. The time-frequency distribution of the signal can be achieved by the reconstruction algorithms of the computed tomography. The philosophy of the TTFT is different from the existing time-frequency transforms. For example, no window function is explicitly employed, thus the compromise between the time and the frequency resolutions caused by the window function is avoided. The interference suppression process imposes little impact on the signal components because the filtering operation is performed on the partially processed data instead of the final results of the transform. In theory, the resolution of the new transform should be better than that of any other time-frequency transform because the ultimate resolution is determined only by the signal itself. However, there may be some problems from the high computation load and the imperfection of the inverse Radon transform that are used to recover the time-frequency distribution.

REFERENCES

1. L. B. Almeida, The fractional Fourier transform and time-frequency representations, *IEEE Trans. Signal Process.*, vol. 42, no. 11, 3084–3091, 1994.

2. L. Auslander and R. Tolimieri, Radar ambiguity function and group theory, *SIAM J. Math. Anal*, vol. 165, no. 3, 577–601, 1985.

3. J. Bertrand and P. Bertrand, Affine time-frequency distributions, in *Time-Frequency Signal Analysis – Methods and Applications,* ed. B. Boashash, Longman-Chesire, Melbourne, Australia, 1991.

4. F. Blawatsch, Duality and classification of bilinear time-frequency signal representations, *IEEE Trans. Signal Process.*, vol. 39, no. 7, 1564–1574, 1991.

5. T. A. C. M. Claasen and W. F. G. Mecklenbrauker, The Wigner distribution – a tool for time-frequency signal analysis - Part I: Continuous-time signals, *Philips J. Res.*, vol. 35, 217–250, 1980.

6. T. A. C. M. Claasen and W. F. G. Mecklenbrauker, The Wigner distribution - a tool for time-frequency signal analysis - Part III: Relations with other time-frequency signal transformations, *Philips J. Res.*, vol. 35, no. 6, 372–389, 1980.

7. L. Cohen, Time-frequency distributions - a review, *Proc. IEEE*, vol. 77, no. 7, 941–981, 1989.

8. M. F. Erden, M. A. Kutay and H. M. Ozaktas, Repeated filtering in consecutive fractional Fourier domains and its application to signal restoration, *IEEE Trans. Signal Process.*, vol. 47, no. 5, 1458–1462, 1999.

9. P. Flandrin, Time-frequency processing of bat sonar signals, animal sonar systems symposium, *Helsinger (DK)*, Sept. 10-19. 1986, also in *Animal Sonar - Processes and Performance,* P. E. Nachtigall and P.W.B. Moore eds., 797–802, Plenum Press, New York, 1988.

10. O. D. Grace, Instantaneous power spectra, *J. Acoust Soc. Ant.,* vol. 69, 191–198, 1981.

11. F. Hlawatsch and P. Flandrin, The interference structure of the Wigner distribution and related time-frequency signal representatins, in *The Wigner Distribution – Theory and Applications in Signal Processing,* W. Mecklenbrauker, ed., Elsevier Science Publishers, North Holland, 1992.

12. H. Inuzuka, T. Ishiguroand and S. Mizuno, Elimination of cross-components in the Wigner distribution of the exponentially swept data by varying the sampling rate, *Conference Record – IEEE Instrumentation & Measurement Technology*, 2, 717–720, 1994.

13. D. L. Jones and T. W. Parks, A high resolution data-adaptive time-frequency representation, *IEEE Trans. Acoust., Speech. Signal. Process.*, vol. 38, no. 12, 2127–2135, 1990.

14. D. L. Jones and T. W. Parks, A resolution comparison of several time-frequency representations, *IEEE Trans. Signal Process.*, vol. 40, no. 2, 413–420, 1992.

15. S. Kadamte and G. F. Boudreaux-Bartels, A comparison of the existence of 'cross terms' in the Wigner distribution and the squared magnitude of the wavelet transform and the short-time Fourier transform, *IEEE Trans. Signal Process.*, vol. 40, no. 10, 2498–2517, 1992.

16. A. S. Kayhan, A. El-Jaroudi and L. F. Chaparro, Data-adaptive evolutionary spectral estimation, *IEEE Trans. Signal Process.*, vol. 43, no. 1, 204–213, 1995.

17. J. R. Kiauder, A. C. Price, S. Darlinglon and W. J. Albersheim, The theory and design of chirp radars, *Bell System Tech. J.*, vol. 39, 745–808, 1960.

18. A. W. Lohmann and B. H. Soffer, Relationships between the Radon–Wigner and fractional Fourier transforms, *J. Opt. Soc. Amer. A.*, vol. 11 no. 6, 1798–1801, 1994.

19. S. V. Marid and E. L. Titlebaum, Frequency hop multiple access codes based upon the theory of cubic congruences, *IEEE Trans. Aerospace, Elect. Syst.*, vol. 26, no. 6, 1035–1039, 1990.

20. W. F. G. Mecklenbráuker, A tutorial on non-parametric bilinear time-frequency signal representations, *Les Houches, Session XLV, 1985, Signal Processing,* eds, J. L. Lacoume, T. S. Durrani and R. Stora, 277–336, 1987.

21. G. Mourgues, M. R. Feix, J. C. Andrieux and P. Bertrand, Not necessary but sufficient conditions for the positivity of generalized Wigner functions, *J. Math. Phys.*, vol. 26, 2554–2555, 1985.

22. S. N. Nawab and T. F. Quatieri, *Advanced Topics in Signal Processing,* J. S. Lim and A. N. Oppenheim. eds., Prentice-Hall, Englewood Cliffs, NJ, 1988.

23. H. M. Ozaktas and D. Mendlovic, The fractional Fourier transform as a tool for analyzing beam propagation and spherical mirror resonators, *Opt. Lett.*, vol. 19, 1678–1680, 1994.

24. A. Papoulis, *Signal Analysis,* McGraw-Hill Book Co., New York. 1977.

25. P. Pellat-Finet and G. Bonnet, Fractional-order Fourier transform and Fourier optics, *Opt. Commun.*, vol. 111, 141–154, 1994.

26. S. Qian and D. Chen, *Joint time-frequency analysis: methods and applications,* Prentice Hall PTR, Upper Saddle River, NJ, 1996.

27. S. Qian and J. M. Morris, Wigner distribution decomposition and cross-terms deleted representation, *Signal Processing*, vol. 27, no. 2, 125–144, 1992.

28. L. Qiu, Elimination of the cross-term for two-component signals in the Wigner distribution, *International J. of Electronics*, vol. 78. 1091–1099, 1995.

29. L. R. Rabiner and R. W. Schafer, Digital Processing of Speech Signals, Prentice-Hall, Inc., Englewood Cliffs, NJ, 1978.

30. G. N. Ramachandran and A. V. Lakshminarayanan, Three dimensional reconstructions from radiographs and electron micrographs: Application of convolution instead of Fourier transforms, *Proc. Nat. Acad. Sci.*, vol. 68, 2236–2240, 1971.

31. I. Raveh and D. Mendlovic, New properties of the Radon transform of the cross Wigner ambiguity distribution function, *IEEE Trans. Signal Process.*, vol. 47, no. 7, 2077–2080, 1999.

32. M. G. Raymer, M. Bech and D. F. Mcalister, Complex wave-field reconstruction using phase-space tomography, *Phys. Rev. Lett.*, vol. 72, 1137–1140, 1994.

33. A. W. Rihaczek, Design of zigzag FM zignals, *IEEE Trans. AER EL*, vol. AES-4, 1968.

34. O. Rioul and P. Flandrin, Time-scale energy distributions: a general class extending wavelet transforms, *IEEE Trans. Signal Process.*, vol. 40, no. 7, 1746–1757, 1992.

35. B. Ristic and B. Boashash, Kernel design for time-frequency signal analysis using the Radon transform, IEEE Trans. Signal Process., vol.41, no. 5, 1996–2008, 1993.

36. L. A. Shepp and B. F. Logan, The Fourier reconstruction of a head section, *IEEE Trans. Nucl. Sci.,* vol. NS-21, 21–43, 1974.

37. D. T. Smithey, M. Beck and M. G. Raymer,

Measurement of the Wiger distribution and the density matrix of a light mode using optical homodyne tomography application to squeezed states and the vacuum, *Phys. Rev. Lett.*, vol. 70, 1244–1247, 1993.

38. L. Stankovic, Method for improved distribution concentration in the time-frequency analysis of multicomponent signals using the L-Wigner distribution, *IEEE Trans. Signal Process*, vol. 43, no. 5, 1262–1268, 1995.

39. H. H. Szu and J. A. Blodgett, Wigner distribution and ambiguity function, in *Optics in Four Dimensions*, ed. L. M. Narducci, American Institute of Physics, New York, pp. 355–81, 1981.

40. H. L. VanTrees, *Detection, Estimation and Modulation Theory, Part III*, J. Wiley & Sons Publ., New York, 1971.

41. X. G. Xia, Y. Owechko, B. H. Soffer and R. M. Matic, On generalized-marginal time-frequency distributions, *IEEE Trans. Signal Process.*, vol.44, no. 11, 2882–2886, 1996.

42. A. L. Warrick and P. A. Delaney, Detection of linear features using a localized Radon transform with a wavelet filter, *ICASSP-97*, vol. 4, 2769–2772, 1997.

43. J. C. Wood and D. T. Barry, Tomographic time-frequency analysis and its application toward time-varying filtering and adaptive kernel design for multicomponent linear-FM signals, *IEEE Trans. Signal Process.*, vol. 42, no. 8, 2094–2104, 1994.

44. J. C. Wood and D. T. Barry, Linear signal synthesis using the Radon–Wigner transform, *IEEE Trans. Signal Process.*, vol. 42, no. 8, 2105–2111, 1994.

45. J. C. Wood and D. T. Barry, Radon transformation of time-frequency distributions for analysis of multicomponent signals, *IEEE Trans. Signal Process.*, vol. 42, no. 11, 3166–3177, 1994.

46. J. C. Wood and D. T. Barry, Time-frequency analysis of skeletal muscle and cardiac vibrations, *Proceedings of the IEEE*, vol. 84, no. 9, 1281–1294, 1996.

47. F. Zhang, G. Bi and Y. Chen, Tomography time-frequency transform, *IEEE Trans. Signal Process.* vol. 50, no. 6, 1289–1297, 2002.

Index

Applied and Numerical Harmonic Analysis

Forthcoming Titles

E. Prestini: *The Evolution of Applied Harmonic Analysis* (ISBN 0-8176-4125-4)

J.J. Benedetto and A. Zayed: *Sampling, Wavelets, and Tomography* (0-8176-4304-4)

J.A. Hogan and J.D. Lakey: *Time-Frequency and Time-Scale Methods* (ISBN 0-8176-4276-5)

W. Freeden and V. Michel: *Multiscale Potential Theory* (ISBN 0-8176-4105-X)

C. Cabrelli and U. Molter: *Harmonic Analysis and Fractal Geometry* (ISBN 0-8176-4118-1)